SALT MARSHES

Salt marshes are highly dynamic and important ecosystems that dampen impacts of coastal storms and are an integral part of tidal wetland systems, which sequester half of all global marine carbon. They are now being threatened due to sea-level rise, decreased sediment influx, and human encroachment. This book provides a comprehensive review of the latest salt marsh science, investigating their functions and how they are responding to stresses through formation of salt pannes and pools, headward erosion of tidal creeks, marsh-edge erosion, ice-fracturing, and ice-rafted sedimentation. Written by experts in marsh ecology, coastal geomorphology, wetland biology, estuarine hydrodynamics, and coastal sedimentation, it provides a multidisciplinary summary of recent advancements in our knowledge of salt marshes. The future of wetlands and potential deterioration of salt marshes is also considered, providing a go-to reference for graduate students and researchers studying these coastal systems, as well as marsh managers and restoration scientists.

DUNCAN M. FITZGERALD is a professor in the Department of Earth and Environment, Boston University. He is a sedimentologist and coastal geomorphologist, whose work focuses on estuaries and tidal inlet and barrier island dynamics and evolution. During the past 15 years, he has been working on salt marshes and the impact of sea-level rise on these systems. He is a fellow of the Geological Society of America and sits on the board of the Coastal Education and Research Foundation.

ZOE J. HUGHES is a coastal oceanographer and geomorphologist and is an assistant research professor at Boston University, where she has worked since 2004. She began her career looking at tidal inlets and sandy barriers but has since expanded to other coastal systems, including estuarine and marsh systems along the Gulf of Mexico and Eastern Seaboard of the United States. Through modeling and field data collection, she researches the interaction of hydrodynamics and sediment transport along coastlines, especially shorelines that incorporate channelized systems such as salt marshes.

SALT MARSHES

Function, Dynamics, and Stresses

Edited by

DUNCAN M. FITZGERALD

Boston University

ZOE J. HUGHES

Boston University

CAMBRIDGE
UNIVERSITY PRESS

University Printing House, Cambridge CB2 8BS, United Kingdom

One Liberty Plaza, 20th Floor, New York, NY 10006, USA

477 Williamstown Road, Port Melbourne, VIC 3207, Australia

314–321, 3rd Floor, Plot 3, Splendor Forum, Jasola District Centre, New Delhi – 110025, India

79 Anson Road, #06–04/06, Singapore 079906

Cambridge University Press is part of the University of Cambridge.

It furthers the University's mission by disseminating knowledge in the pursuit of
education, learning, and research at the highest international levels of excellence.

www.cambridge.org
Information on this title: www.cambridge.org/9781107186286
DOI: 10.1017/9781316888933

© Cambridge University Press 2021

First published 2021

Printed in the United Kingdom by TJ Books Limited, Padstow Cornwall

A catalogue record for this publication is available from the British Library.

Library of Congress Cataloging-in-Publication Data
Names: FitzGerald, Duncan M., editor.
Title: Salt marshes : function, dynamics, and stresses / edited by Duncan M. FitzGerald, Boston University,
Zoe J. Hughes, Boston University.
Description: Cambridge, United Kingdom ; New York, NY : Cambridge University Press, 2021. | Includes
bibliographical references and index.
Identifiers: LCCN 2020016007 (print) | LCCN 2020016008 (ebook) | ISBN 9781107186286 (hardback) |
ISBN 9781316888933 (epub)
Subjects: LCSH: Salt marshes. | Salt marsh ecology.
Classification: LCC GB621 .S33 2022 (print) | LCC GB621 (ebook) | DDC 577.69–dc23
LC record available at https://lccn.loc.gov/2020016007
LC ebook record available at https://lccn.loc.gov/2020016008

ISBN 978-1-107-18628-6 Hardback

Contents

List of Contributors   *page* viii
Acknowledgments   xii

1   State of Salt Marshes   1
     DUNCAN M. FITZGERALD AND ZOE J. HUGHES

Part I Marsh Function   7

2   Salt Marsh Distribution, Vegetation, and Evolution   9
     DANIEL F. BELKNAP AND JOSEPH T. KELLEY

3   Salt Marsh Formation   31
     ANTONIO B. RODRIGUEZ AND BRENT A. MCKEE

4   Salt Marsh Hydrodynamics   53
     ANDREA D'ALPAOS, ALVISE FINOTELLO,
     GUILLAUME C. H. GOODWIN, AND SIMON M. MUDD

5   Community Ecology of Salt Marshes   82
     STEVEN C. PENNINGS AND QIANG HE

6   The Role of Marshes in Coastal Nutrient Dynamics and Loss   113
     ANNE E. GIBLIN, ROBINSON W. FULWEILER,
     AND CHARLES S. HOPKINSON

Part II Marsh Dynamics   155

7   Marsh Equilibrium Theory: Implications for Responses to Rising
    Sea Level   157
    JAMES T. MORRIS, DONALD R. CAHOON, JOHN C. CALLAWAY,
    CHRISTOPHER CRAFT, SCOTT C. NEUBAUER, AND
    NATHANIEL B. WESTON

8   Salt Marsh Ecogeomorphic Processes and Dynamics   178
    CAROL A. WILSON, GERARDO M. E. PERILLO, AND ZOE J. HUGHES

9   Salt Marsh Sediments as Recorders of Holocene Relative Sea-Level
    Change   225
    W. ROLAND GEHRELS AND ANDREW C. KEMP

10 Storm Processes and Salt Marsh Dynamics   257
    KATHERINE A. CASTAGNO, JEFFREY P. DONNELLY,
    AND JONATHAN D. WOODRUFF

11 Understanding Marsh Dynamics: Modeling Approaches   278
    SERGIO FAGHERAZZI, WILLIAM KEARNEY, GIULIO MARIOTTI,
    NICOLETTA LEONARDI, AND WILLIAM NARDIN

12 Understanding Marsh Dynamics: Laboratory Approaches   300
    CHARLIE E. L. THOMPSON, SARAH FARRON, JAMES TEMPEST,
    IRIS MÖLLER, MARTIN SOLAN, AND JASMIN GODBOLD

Part III Marsh Response to Stress   335

13 Climatic Impacts on Salt Marsh Vegetation   337
    KATRINA L. POPPE AND JOHN M. RYBCZYK

14 Impacts of Exotic and Native Species Invading Tidal Marshes   367
    DAVID M. BURDICK, GREGG E. MOORE, AND KATHARYN E. BOYER

15 Marsh Edge Erosion   388
    MICHELE BENDONI, IOANNIS Y. GEORGIOU, AND ALYSSA B. NOVAK

16   Upland Migration of North American Salt Marshes   423
     DANTE D. TORIO AND GAIL L. CHMURA

17   Restoration of Tidal Marshes   443
     JOHN DAY, DAVID M. BURDICK, CARLES IBÁÑEZ,
     WILLIAM J. MITSCH, TRACY ELSEY-QUIRK, AND SOFIA RIVAES

18   Impacts of Climate Change and Sea Level Rise   476
     ZOE J. HUGHES, DUNCAN M. FITZGERALD, AND CAROL A. WILSON

Index   482

Colour plates can be found between pages 276 and 277

Contributors

Daniel F. Belknap
School of Earth and Climate Sciences, University of Maine, Orono

Michele Bendoni
Deptartment of Civil and Environmental Engineering, University of Florence, Florence, Italy

Katharyn E. Boyer
Estuary & Ocean Science Center, San Francisco State University, California

David M. Burdick
University of New Hampshire, Department of Biological Sciences, Jackson Estuarine Laboratory, Durham

Donald R. Cahoon
US Geological Survey, Patuxent Wildlife Research Center, Beltsville, Maryland

John C. Callaway
Environmental Science, University of San Francisco, California

Katherine A. Castagno
Department of Geology & Geophysics, Woods Hole Oceanographic Institution, Massachusetts

Gail L. Chmura
Centre for Climate and Global Change Research, McGill University, Montreal, Quebec, Canada

Christopher Craft
School of Public and Environmental Affairs, Indiana University, Bloomington

Andrea D'Alpaos
Dipartimento Di Geoscienze, University of Padua, Italy

John Day
Department of Oceanography and Coastal Sciences, Louisiana State University, Baton Rouge

Jeffrey P. Donnelly
Department of Geology and Geophysics, Woods Hole Oceanographic Institution, Massachusetts

Tracy Elsey-Quirk
Department of Oceanography and Coastal Sciences, Louisiana State University, Baton Rouge

Sergio Fagherazzi
Department of Earth and Environment, Boston University, Massachusetts

Sarah Farron
Department of Earth and Environment, Boston University, Massachusetts

Alvise Finotello
Department of Geosciences, University of Padua, Italy

Duncan M. FitzGerald
Department of Earth and Environment, Boston University, Massachusetts

Robinson W. Fulweiler
Department of Earth and Environment, Boston University, Massachusetts

Ioannis Y. Georgiou
University of New Orleans, Louisiana

W. Roland Gehrels
Environment Department, University of York, UK

Anne E. Giblin
The Ecosystems Center, Marine Biological Laboratory, Woods Hole, Massachusetts

Jasmin Godbold
Ocean and Earth Science, University of Southampton, UK

Guillaume C. II. Goodwin
School of Geosciences, University of Edinburgh, UK

Qiang He
Nicholas School of the Environment, Beaufort, New Carolina

Charles S. Hopkinson
Department of Marine Sciences, University of Georgia, Athens

Zoe J. Hughes
Department of Earth and Environment, Boston University, Massachusetts

Carles Ibáñez
IRTA Aquatic Ecosystems, St. Carles de la Ràpita, Catalonia, Spain

William Kearney
Department of Earth and Environment, Boston University, Massachusetts

Joseph T. Kelley
School of Earth and Climate Sciences, University of Maine, Orono

Andrew C. Kemp
Department of Earth and Ocean Sciences, Tufts University, Massachusetts

Nicoletta Leonardi
Department of Geography and Planning, University of Liverpool, Liverpool, UK

Giulio Mariotti
College of the Coast and Environment, Louisiana State University

Brent A. McKee
Institute of Marine Sciences, University of North Carolina at Chapel Hill

William J. Mitsch
Everglades Wetland Research Park, Florida Gulf Coast University, Naples

Iris Möller
Department of Geography, University of Cambridge, UK

Gregg E. Moore
Department of Biological Sciences, University of New Hampshire, Durham

James T. Morris
Baruch Institute, University of South Carolina, Columbia

Simon M. Mudd
School of Geosciences, University of Edinburgh, UK

William Nardin
Department of Earth and Environment, Boston University, Massachusetts

Scott C. Neubauer
Department of Biology, Virginia Commonwealth University, Richmond

Alyssa B. Novak
Department of Earth and Environment, Boston University, Massachusetts

Steven C. Pennings
Department of Biology and Biochemistry, University of Houston, Texas

Gerardo M. E. Perillo
Instituto Argentino De Oceanografia, Bahía Blanca, Argentina

Katrina L. Poppe
Huxley College of the Environment, Western Washington University, Bellingham, Washington

Sofia Rivaes
Riet Vell Natural Reserve, SEO/BirdLife Technical Office in the Ebro Delta, Amposta, Spain

Antonio B. Rodriguez
Institute of Marine Sciences, University of North Carolina at Chapel Hill

John M. Rybczyk
Huxley College of the Environment, Western Washington University, Bellingham, Washington

Martin Solan
Ocean and Earth Science, University of Southampton, UK

James Tempest
University of Cambridge, UK

Charlie E. L. Thompson
Ocean and Earth Science, University of Southampton, UK

Dante D. Torio
Centre for Climate and Global Change Research, McGill University, Montreal, Quebec, Canada

Nathaniel B. Weston
Department of Geography and the Environment, Villanova University, Pennsylvania

Carol A. Wilson
Department of Geology and Geophysics, Louisiana State University, Baton Rouge

Jonathan D. Woodruff
Department of Geosciences, University of Massachusetts, Amherst

Acknowledgments

In addition to thanking the editors, copyeditors, and staff at Cambridge University Press, we very gratefully acknowledge those who kindly gave their time to review chapters for us: Shimon Anisfeld, Debra Ayres, Don Cahoon, John Callaway, John Day, Simon Englehart, Sergio Fagherazzi, Rusty Feginn, Jon French, Ioannis Georgiou, Glenn Guntenspergen, Andrew Kemp, Paul Kemp, Nicole Khan, Giulio Mariotti, Scott Neubauer, Daria Nikitina, Alyssa Novak, Aleja Ortiz, Steven Pennings, Denise Reed, Anthony Rodriguez, Kristin Saltsonstall, Lisa Schile, Amanda Spivak, Lorie Staver, and Charlotte Thompson. This book has been made stronger by your rigorous peer reviews – thank you.

We acknowledge funding through grants from the Dept. of the Interior (DOI/NWF 32324-F2), the National Science Foundation (EAR1832177, EAR1904470, EAR1800810), the Massachusetts Department of Energy and Environmental Affairs (award RFR ENV 18 POL 04 to the Town of Newbury and RFR ENV 19 MVP 02 to the Town of Essex), and the National Park Service (P09AC00232).

1

State of Salt Marshes

DUNCAN M. FITZGERALD AND ZOE J. HUGHES

Salt marshes are expected to undergo substantial change or, potentially, disappear in the next couple of centuries as a result of rising sea level. Increasingly, scientists are asking the question: how long can they survive? This book draws on global expertise to look at how salt marshes evolved, how they function, and how they are responding to the stresses caused by social and environmental change. These environments occur throughout the world: behind barrier islands, bordering estuaries, and dominating lower delta plains (Fig. 1.1) in warm to cool latitudes ($\geq 30°$ latitude). Up until now, previous loss and degradation of coastal marshes has been related to a variety of human actions including dredging and filling, reduction in sediment supplies, and hydrocarbon withdrawal, as well as other causes. However, in the future the greatest impact to marshes will be a consequence of climate change, especially sea-level rise (SLR). Most of the present marshes formed under very different sedimentation and SLR regimes compared to those that occur today. During their formation and throughout their evolution, the rate of SLR was relatively slow and steady, between 0.2 and 1.6 mm/year (Table 1.1). The sustainability of marshes is now threatened by an acceleration in SLR to rates many times greater than those under which they initiated and have evolved. For example, the Romney marsh, which is located north of Boston, Massachusetts, contains a 2-m-thick peat that began forming 3.1 ka BP when sea level was rising at about 0.8 mm/year, a rate that slowed to 0.52 mm/year around 1 ka BP (Donnelly 2006). The rate of SLR in Boston Harbor is now 2.85 mm/year (NOAA 2019), which far exceeds the rate occurring when the Romney marsh built to a supratidal elevation. Eventually, SLR, along with marsh-edge erosion, will outpace the ability of most marshes to accrete vertically (Crosby et al. 2016) and/or compensate for marsh loss by expanding into uplands (Kirwan et al. 2016, Farron 2018).

Over the short term, some researchers believe that biogeomorphic feedbacks will improve marsh survival as increased mineral sedimentation on the marsh platform will occur due to longer periods of tidal flooding, and resulting from increased biomass (Morris et al. 2002; Mudd et al. 2009). This will be further enhanced as plant productivity responds to warmer temperatures (Kirwan et al. 2009) and higher carbon dioxide concentrations (Langley et al. 2009; Ratliff et al. 2015). Although this will offer some relief to the problem, increased sedimentation rates will actually depend on the availability of suspended sediment, which is likely to be diminishing due to progressively lower volumes of riverine sediment reaching the coastal ocean (Syvitski et al. 2005; Weston 2014). Some

1

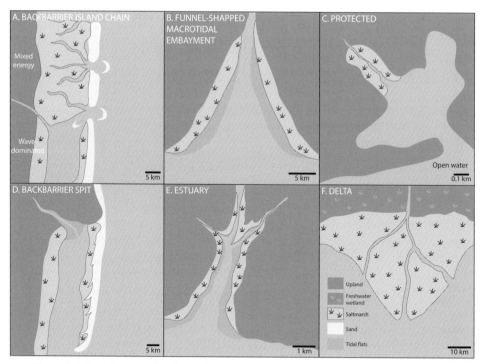

Figure 1.1 Broad types of salt marsh environments, including: (A) *Backbarrier Island Chains* (e.g., East and Gulf Coasts of USA; Algarve, Portugal; Frisian Islands, Germany); (B) *Funnel-shaped Macro-tidal Embayment* (e.g., The Wash, England; Mouth of Elbe River, Germany; Mont St. Michael Bay, France; Nushagak Bay, AK); (C) *Protected* (e.g., Sunborn Cove, Gouldsboro, ME; Etang de Toulvern, Bretagne, France); (D) *Backbarrier Spit* (e.g., Long Beach, WA; Cape Romain, SC; Hashirikotan barrier spit, Japan); (E) *Estuary* (e.g., Delaware and Chesapeake Bays; Rivers Esk and Eden, UK; Columbia River, USA; Lérez Estuary, Spain); (F) *Deltaic* (e.g., Mississippi River delta, LA; Yukon River delta, AK). (A black and white version of this figure will appear in some formats. For the color version, please refer to the plate section.)

investigators (Hopkinson et al. 2018) have suggested that wave-induced marsh edge retreat may, in fact, offer a benefit in that the eroded sediment can be transported to the marsh surface. Marsh loss caused by edge erosion is partly offset by upland marsh migration during SLR. Finally, over the past decade, it has been recognized that marshes are not static, vertically accreting platforms, resulting from a simple balance of inorganic sedimentation and belowground biomass production or decomposition. Rather, recent research has demonstrated that marshes are highly dynamic ecosystems that respond to storms and numerous interconnected hydrological, biological, sedimentological, and geochemical processes. These responses include the formation and expansion of salt pannes and pools, headward erosion of tidal creeks, landscape-scale feedbacks to fauna-induced devegetation or bioturbation, marsh edge calving and erosion, storm-related deposition, ice fracturing, and ice-rafted sedimentation. In totality, climate change is affecting all of these marsh processes, but in a differential manner related to their geologic and climatic setting.

Table 1.1 *Examples of rates of sea-level rise during salt marsh formation (from FitzGerald and Hughes 2019)*

Location	Curve Database	Rates (mm/yr)	Reference
Chezzetcook, Nova Scotia	Basal peats & forams	1.7 from 1000 to 1800 AD 1.6 from 1800 to 1900 AD	Gehrels et al. 2005
Phippsburg, ME	Basal peats & forams	0.5–1.4 from 5.7 to 3.0 ka BP 0.2–0.8 from 3.0 ka BP to present	Gehrels et al. 1996
Wells, ME	Basal peats & forams	0.7–2.2 from 5.7 to 3.5 ka BP 0.0–0.6 from 3.5 ka BP to present	Gehrels et al. 1996
Northern MA	Basal peats	0.80 +/− 0.25 from 3.3 to 1.0 ka BP 0.52 +/− 0.62 from 1.0 to 0.15-0.5 ka BP	Donnelly 2006
Hudson River, NY	N/A	1.2 +/− 0.2 during Late Holocene	Pardi et al. 1984
Delaware	Basal peats	1.2 +/− 0.2 during Late Holocene	Belknap and Kraft 1977; Nikitina et al. 2000
Eastern Shore, VA	Basal peats	0.9 +/− 0.3 during Late Holocene	Engelhart 2009
Central to Southern, NC	Basal peats	0.82 +/− 0.02 since 3903–3389 cal a BP	Horton et al. 2009
West Brittany, France	Basal peats	0.90 +/− 0.12 since 6.3 ka BP	Stéphan et al. 2015
Northern Scotland	Basal peats & other	Stable with +/− 40.0 cm of change during past 2 ka	Gehrels et al. 2006a; Barlow et al. 2014
Ho Bugt, Denmark	Basal peats & other	1.0 during the past 4 ka	Gehrels et al. 2006b
Pounawea, New Zealand	Peats & forams	0.3 +/− 0.3 from 1500 to 1900 AD	Gehrels et al. 2005
East Coast, South Korea	N/A	0.74 from 5.4 ka BP to Present	Lee et al. 2008

Fortunately, dramatic advances in computing power have allowed increasingly complex numerical simulations to project the future evolution and sustainability of salt marshes. Marsh modeling has become an essential tool for predicting how marsh systems will respond to greater frequencies and durations of tidal inundation and in quantifying tipping points, when marshes will ultimately begin to disintegrate. To complement this, physical models of marshes, including flume studies, are shedding light on the mechanics of marsh erosion, particularly in combination with bioturbation and geochemical processes (Möller et al. 2014, Farron 2018, Reef et al. 2018).

In other research, the utilization of radioisotopic dating (Pb-210 and Cs-137), Surface Elevation Tables (SET), and the study of tide-indicator microfossils in combination with statistical analyses (transfer functions; Gehrels 2000) have produced more accurate rates of marsh accretion and have expanded our understanding of marsh growth in response to SLR, globally. We are also increasingly able to identify the sources and patterns of suspended sediment delivery to marshes and to define how marsh platforms will segment and measure rates of edge erosion, with most of these findings involving a combination of field and modeling techniques.

Based on publication rates, the number of marsh studies, have increased tenfold during the past couple of decades, as the importance of marshes as contributors of detritus and nutrients to the coastal ocean, unique coastal habitats, nursery grounds of shellfish and finfish, ameliorators of storm surges and wave erosion, assimilators of upland pollutants, and of their economic value and beauty have been increasingly recognized by scientists and the coastal population. The significant advancements in our understanding of salt marsh processes and how this knowledge base is being used to study the dramatic stresses that marshes will face during the coming century are the impetuses for this book. From the viewpoint of wetland experts from around the world, this volume explores and summarizes many facets of marsh research and provides the current state of knowledge of salt marsh science.

References

Barlow, N. L. M., Long, A. J., Saher, M. H., Gehrels, W. R., Garnett, M. H., and Scaife, R. G. 2014. Salt-marsh reconstructions of relative sea-level change in the North Atlantic during the last 2000 years. *Quaternary Science Reviews.* 99:1–16.

Belknap, D. F., and Kraft, J. C. 1977. Holocene relative sea-level changes and coastal stratigraphic units on the northwest flank of the Baltimore Canyon trough geosyncline. *Journal of Sedimentary Research.* 47:610–629.

Crosby, S. C., Sax, D. F., Palmer, M. E., Booth, H. S., Deegan, L. A., Bertness, M. D., and Leslie, H. M. 2016. Salt marsh persistence is threatened by predicted sea-level rise. *Estuarine and Coastal Shelf Science*, 181:93–99.

Donnelly, J. P. 2006. A revised late Holocene sea-level record for Northern Massachusetts, USA. *Journal of Coastal Research*, 22:1051–1061.

Engelhart, S. E., Horton, B. P., Douglas, B. C., Peltier, W. R., and Törnqvist, T. E. 2009. Spatial variability of late Holocene and 20th century sea-level rise along the Atlantic coast of the United States. *Geology*, 37:1115–1118.

Farron, S. 2018. *Morphodynamic responses of salt marshes to sea-level rise: upland expansion, drainage evolution, and biological feedbacks.* PhD thesis, Boston Univ., Boston, MA.

FitzGerald, D. M. and Hughes, Z. 2019. Marsh processes and their response to climate change and sea-level rise. *Annual Review of Earth and Planetary Sciences*, 47:481–517.

Gehrels, W. R. 1996. Integrated high-precision analyses of Holocene relative sea-level changes: Lessons from the coast of Maine. *Geological Society of America, Bulletin*, 108:1073–1088.

Gehrels, W. R. 2000. Using foraminiferal transfer functions to produce high-resolution sea-level records from saltmarsh deposits, Maine, USA. *The Holocene* 10:367–376.

Gehrels, W. R., Kirby, J. R., Prokoph, A., Newnham, R. M., Achterberg, E. P., Evans, H., Black, S., and Scott, D. B. 2005. Onset of recent rapid sea-level rise in the western Atlantic Ocean. *Quaternary Science Reviews.* 24:2083–2100.

Gehrels, W. R., Marshall, W. A., Gehrels M. J., Larsen, G., Kirby, J. R., Eiriksson, J., Heinemeier, J., and Shimmield, T. 2006a. Rapid sea-level rise in the North Atlantic Ocean since the first half of the 19th century. *Holocene*, 16:948–964.

Gehrels, W. R. Szkornik, K., Bartholdy, J., Kirby, J. R., Bradley, S. L., Marshall, W. A., Heinemeier, J., and Pedersen, J. B. T. 2006b. Late Holocene sea-level changes and isostasy in western Denmark. *Quaternary Research*, 66:288–302

Gehrels, W. R. Hayward, B. W., Newnham, R. M., and Southall, K. E. 2008. A 20th century sea-level acceleration in New Zealand. *Geophysical Research Letters* 35: L02717.

Hopkinson, C. S., Morris, J. T., Fagherazzi, S., Wollheim, W. M., and Raymond, P. A. 2018. Lateral marsh edge erosion as a source of sediments for vertical marsh accretion. *Journal of Geophysical Research, Biogeosciences*, 123:2444–2465.

Horton, B. P., Peltier, W. R., Culver, S. J., Drummond, R., Engelhart, S. E., Kemp, A. C, Mallinson D, et al. 2009. Holocene sea-level changes along the North Carolina Coastline and their implications for glacial isostatic adjustment models. *Quaternary Science Reviews*, 28:1725–1736.

Kirwan, M. L., and Temmerman, S. 2009. Coastal marsh response to historical and future sea-level acceleration. *Quaternary Science Reviews*, 28:1801–1808.

Kirwan, M. L., Walters, D., Reay, W., Carr, J. 2016. Sea level driven marsh expansion in a coupled model of marsh erosion and migration. *Geophysical Research Letters*, 43: 4366–4373.

Langley, J. A., McKee, K. L., Cahoon, D. R., Cherry, J. A., and Megonigal, J. P. 2009. Elevated CO_2 stimulates marsh elevation gain, counterbalancing sea-level rise. *Proceedings of the National Academy of Sciences*, 106:6182–6186.

Lee, Y. G.; Choi, J. M., and Oertel, G. F. 2008. Postglacial sea-level change of the Korean southern sea shelf. *Journal of Coastal Research*, 24:118–132.

Möller, I., Kudella, M., Franziska, R., Spencer, T., Paul, M., Van Wesenbeeck B. K., Wolters G., et al. 2014. Wave attenuation over coastal salt marshes under storm surge conditions. *Nature Geoscience*, 7:727–731.

Morris, J. T, Sundareshwar, P. V., Nietch, C. T., Kjerfve, B., and Cahoon, D. R. 2002. Responses of coastal wetlands to rising sea level. *Journal of Ecology*, 83:2869–2877.

Mudd, S. M., Howell, S. M., Morris, J. T. 2009. Impact of dynamic feedbacks between sedimentation, sea-level rise, and biomass production on near-surface marsh stratigraphy and carbon accumulation. *Estuarine, Coastal and Shelf Science*, 82:377–389.

Nikitina, D. L, Pizzuto, J. E., Schwimmer, R. A., Ramsey, K. W. 2000. An updated Holocene sea-level curve for the Delaware coast. *Marine Geology*, 171:7–20.

NOAA. 2019. Tide levels: https://tidesandcurrents.noaa.gov/waterlevels.html?id=8443970&units=metric&bdate=20171226&edate=20180104&timezone=GMT&datum=MLLW&interval=6&action=

Pardi, R. R., Tomecek, L., and Newman W. S. 1984. Queens College radiocarbon measurements IV. *Radiocarbon*, 26:412–430.

Ratliff, K. M., Braswell, A. E., and Marani, M. 2015. Spatial response of coastal marshes to increased atmospheric CO_2. *Proceedings of the National Academy of Sciences*, 112:15580–15584.

Reef, R., Schuerch, M., Christie, E. K., Möller, I., and Spencer, T. 2018. The effect of vegetation height and biomass on the sediment budget of a European saltmarsh. *Estuarine, Coastal and Shelf Science*, 202:125–133.

Stéphan, P., Goslin, J., Pailler Y., Manceau, R., Suanez, S., Van Vliet-Lanoë, B., Hénaff, A., and Delacourt C. 2015. Holocene salt-marsh sedimentary infilling and relative sea-level changes in West Brittany (France) using foraminifera-based transfer functions. *Boreas*, 44:153–177.

Syvitski, J. P. M., Vörösmarty, C. J., Kettner, A. J., and Green, P. 2005. Impact of humans on the flux of terrestrial sediment to the global coastal ocean. *Science*, 308:376–380.

Weston, N. 2014. Declining sediments and rising seas: an unfortunate convergence for tidal wetlands. *Journal of Estuaries and Coasts*, 37:1–23.

Part I

Marsh Function

2

Salt Marsh Distribution, Vegetation, and Evolution

DANIEL F. BELKNAP AND JOSEPH T. KELLEY

2.1 Introduction

Salt marshes are common globally in low-lying coastal environments. Their geological settings and ecosystems vary widely by latitude and climatic settings (Chapman, 1960). Allen (2000) provides a comprehensive sketch of European salt marshes, while Rogers and Woodroffe (2014) give a recent summary of the subject. Woodwell et al. (1973) suggest that there are more than 38 million hectares (380,000 km^2) of salt marshes worldwide, but specific delineation of distributions is incomplete, particularly in Asia, Africa, and South America. That area is greater than the total area of coastal American states from New Jersey to South Carolina. This chapter concentrates on the east coast of North America as containing examples of well-studied environments, with a few additional examples.

Salt marshes are intertidal wetlands dominated by herbaceous plants. They are sometimes referred to as tidal marshes, which is a more encompassing term considering estuarine and lagoonal settings that extend to brackish conditions (Belknap, 2003; Bertness, 2007). These tidal marshes are environments of autochthonous growth within favorable estuarine, lagoonal, or deltaic settings where they are sheltered from high-energy wave and tidal erosion. They often contain varying amounts of allochthonous inorganic and inorganic input as well. They may grow upon a substrate of rock or till in paraglacial regions (e.g., Daly et al. 2007); on glaciomarine sediments in Atlantic Canada and New England (Bloom, 1964; Scott and Greenberg, 1983; Belknap et al. 1987; van de Plassche, 1991; Gehrels et al. 2004); on generally sandy substrates along the Long Island outwash plains (Rampino and Sanders, 1980; Clark, 1986) and Atlantic Coastal Plain (e.g., Kraft et al. 1979; Frey and Basan, 1985); or on shallow soils over limestone on the west-central Florida coast (Hine et al. 1988).

Salt marshes host halophytic (salt-tolerant) grasses, rushes, sedges, and forbs that are able to survive frequent saltwater immersion, which excludes upland vegetation. Mature marshes are dominated by living plants overlying fibrous peats, the partially preserved remnants (subfossils) of former marsh plants, which sometimes accumulate to several meters thick. Marshes usually stand higher than flats, from mid-tide to highest spring tides, and are often best developed as a broad flat expanse within a few decimeters of mean high water (MHW). Varying amounts of inorganic sediment comprise the matrix of the internal sediments, but fibrous in situ roots, rhizomes, and detrital organics make up their distinctive fabric and supply strength sufficient to accommodate new plant growth. The term "peat" is a useful description of this fibrous fabric, but should be considered a field term, as the strict definition of >60% dry weight organic carbon (Neuendorf et al. 2005) is rarely met other than in freshwater bog peats. Marsh sediments have been used as tools for stratigraphic reconstruction of coastal evolution (Kraft, 1971) and bases for interpretation of Holocene sea-level change (Redfield and Rubin, 1962; Belknap and Kraft, 1977; van de Plaasche, 1986; Gehrels et al. 1996) using radiocarbon dating of peats with an indicative meaning (van de Plassche, 1986) that allows reconstruction of former sea levels.

Salt marshes have great importance for their ecosystems services (e.g., Tiner, 2009). They are important habitats for many animals. They are particularly important for the well-being of migratory birds. They serve as a nursery for juvenile fish. They exhibit very high primary productivity (Roberts, 1979; Snow, 1980) and export abundant detrital organic matter to the base of the coastal food web (e.g., Orson et al. 1998). Marshes sequester large masses of organic carbon that would otherwise add to atmosphere CO_2 content, estimated at 210 g/m^2/yr, and store 430 Tg in the upper 50 cm of marsh soils globally (Chmura, 2003). This perhaps overly precise estimate should be tempered with an understanding of deeper-seated decay and release of CO_2. The morphology of broad marshes provides a barrier to storms by mitigating wave and storm-surge impacts on the coastline (Moller et al. 2014). At present, increasing human coastal development and effects of sea-level rise are decreasing the areas of salt marshes that would otherwise be in dynamic equilibrium with rising sea level (Silliman et al. 2009; Kirwan et al. 2016).

2.2 Distribution

Salt marsh biogeography is controlled by vegetation associations, exhibiting strong latitudinal control, but aridity can also be a limitation (Silliman et al. 2005). Adam (1990) and Rogers and Woodroffe (2014) list six main types, as follows:

1. Tropical marshes may be in competition with mangroves, or be limited by salt flats at upper intertidal zones.
2. Dry coast marshes are predominantly halophytic because of saline soil conditions.
3. Arctic marshes have low diversity because of the stringent climate conditions, such as freezing, snow and ice cover, and seasonally low light levels.
4. Boreal marshes are intermediate between 3 and 5, having somewhat greater diversity than arctic marshes.
5. Temperate marshes have physiological and environmental limitations less severe than 1–3. They have the greatest diversity of species. Adam (1990) identifies several longitudinal subgroups, but notes definite similarities shared by temperate marshes in both the northern and southern hemispheres.
6. West Atlantic marshes are temperate types with notable similarities over a wide span of tidal ranges and climate subzones. They are dominated by *Spartina* species, particularly the ubiquitous *Spartina alterniflora* in the low marsh.

The east coast of North America contains wide stretches of tidal marshes in embayment, estuarine, and back-barrier settings, from Newfoundland to Florida, some of which are discussed in Section 2 4 on **Morphodynamics**. The Gulf of Mexico coast is similarly well endowed, from Florida to northern Mexico, particularly in the Mississippi Delta, which contains about 40% of marshlands in the conterminous USA.

2.3 Methods of Analysis

Tidal marshes are autogenic systems (created by their flora and fauna), and so are studied for biology and ecology, as well as the geological focus of this chapter. In fact, a multi-disciplinary approach is the most fruitful way to understand modern processes and stratigraphic evolution. Geomorphology is a starting point, relating tidal marshes to environments, such as estuarine, deltaic, back-barrier, and open-coast settings (Allen 2000), as well as to slope, tidal range, wave energy, and sediment supply conditions. Mapping of subenvironments is traditionally accomplished through air photo analysis coupled with field mapping (e.g., Jacobson, 1988; Kelley et al. 1992). Lidar is increasingly used to digitally quantify the three-dimensional aspects of marsh surfaces to decimeter precision and to characterize subenvironments (Morris et al. 2005; Wang et al. 2009; Hladik and Alber, 2012).

Geological studies often involve coring, usually through employment of devices that are portable across the marsh surface, such as pipes, gouge augers, or vibracores. Almost every coastal geology program has produced numerous theses and dissertations that use coring as the basis for marsh stratigraphy and/or development of sea-level curves. Cores obtain sediment samples and (sub)fossils of peat, wood, foraminifera, and shells. Major sediment changes are notable from facies representing: (1) sandy flats and creek thalwegs, (2) muddy tidal flat and creek margins, (3) marsh peats, and (4) basal substrates and upland margins. More subtle changes between low marsh, high marsh, higher high marsh, brackish or transitional marsh, and paleo-salt-pools require more detailed analysis of plant remains and other paleoenvironmental indicators.

Dating of marsh accumulation horizons and rates of buildup are usually measured either through radiometric dating, by comparison of marker horizons, or the measurement of surficial changes. Radiocarbon dating of peats has been a standard tool in marsh studies since the work of Redfield and Rubin (1962), with ongoing improvements to the radiocarbon dating technique, sampling, and stratigraphic analysis. Analysis of ^{210}Pb and ^{137}Cs in salt marshes provides a window to the past ca. 150–200 years and post-1963 horizon respectively (DeLaune et al. 1989; Chmura et al. 2001). Other isotopes may supplement these in the future (Boyd and Summerfield, 2017). In the short term (years to decades), establishment of marker horizons such as aluminum "glitter" (Harrison and Bloom, 1977), ground feldspar (Cahoon and Turner, 1989), or brick dust (Wood et al. 1989) provide a visible layer that can be recovered in short surficial cores. Goodman et al. (2007) were able to recover brick dust horizons in Maine high marshes established 17 years earlier. They found accumulation rates of ca. 2–3 mm/yr, similar to the decadal rates of sea-level rise recorded at nearby tide gauges. The limitation of marker horizons is that they document surficial accumulation rate, which may be different from the overall change in marsh elevation relative to a fixed level. Conversely, the establishment of a deeply anchored post supporting a sedimentation–erosion table (SET) provides precise measurements of seasonal and longer changes to the elevation of the marsh surface, which includes accumulation and near-surface subsidence (autocompaction) (Boumans and Day, 1993; Cahoon et al. 1995, 1996). Natural markers include pollen, which can be compared to regional records for a relative or remotely constrained dating technique (e.g., Chmura et al. 2001). Any of these techniques must be linked to tide gauges or other measurements for a comparison with sea-level change.

One major goal associated with salt marsh studies is the development of local relative sea-level curves, which require radiocarbon datable materials of known provenance (also known as *indicative meaning*: van de Plassche, 1986). Early studies dated bulk peats of ca. 100 g mass (Redfield and Rubin, 1962; Bloom, 1964; Belknap and Kraft, 1977) with inherent limitations of precision and accuracy. However, it was clear that avoidance of compaction displacement was important (Bloom, 1964), so Belknap and Kraft (1977) emphasized the importance of using basal peats along transects on top of incompactible substrate. Improvements in radiocarbon analysis by accelerator mass spectrometry (AMS) allowed dating of samples of grams or less mass, to the level of individual plant fragments, improving the accuracy of time horizon placement within an environment (e.g., Gehrels, 1994; Gehrels et al. 1996). More detailed local analysis and regional comparisons have been made possible through the combination of salt-marsh foraminifer transfer functions (statistical models of indicative meaning) with AMS dating of plant fragments (Gehrels, 2000; Gehrels et al. 2005; Engelhart et al. 2011; Engelhardt and Horton, 2012).

2.4 Morphodynamics

2.4.1 Geomorphology

The development and evolution of tidal marshes consist of a dynamic interplay between processes of aggradation and degradation. These processes vary laterally on the margins and the surface of marshes, and vary depending on their morphology. Kelley et al. (1988) noted the variety of types that occur in bedrock-framed New England marshes, but the geomorphic types can be generalized to much of North America.

Fringing marshes are meters to tens of meters in shore-normal width, extending as patches or more extended stretches along upland shores. They are generally steep, and have rather sharp vegetation zones (e.g., St. George River, Thomaston, Maine; Belknap et al. 2004). Bluff-toe fringing marshes may be young and ephemeral, related to cycles of bluff failures, colonization, and continued erosion (Kelley and Hay, 1986).

Broad marshes extend tens of meters to kilometers in shore-normal width, and may occupy large stretches of interrelated wetlands. They generally require low energy protected environments, such as back-barrier systems or estuaries. *Back-barrier broad marshes* are a common feature of the Atlantic coast from Florida to New England. They can range from those that fill most of the available area in tide-dominated mixed energy

(Hayes, 1979) barrier systems such as Georgia and South Carolina, to bands around open lagoons in wave-dominated mixed energy systems, such as North Carolina, Delaware, New Jersey, and Long Island barrier systems. Most barriers have a fringing marsh on the lagoon side of the barrier that is more or less influenced by storm washover fans (Hayes and Kana, 1976; Leatherman, 1979) or former tidal deltas (Kraft, 1971; Kraft et al. 1979). *Estuarine broad marshes* may be protected by narrow and thin sandy barriers, such as in southern Delaware Bay (Kraft, 1971, Kraft et al. 1979), or may be exposed to the limited wave conditions of large estuaries, such as Chesapeake Bay and northern Delaware Bay (Nikitina et al. 2014). Broad marshes are generally low in slope with extensive portions graded to near mean high water (±30 cm or less).

Fluvial-minor marshes (Kelley et al. 1988) occur in smaller estuarine and tidal-creek settings such as in Maine (Gehrels, 1984), Connecticut (Orson et al. 1987; Gehrels and van de Plassche, 1991), and the margins of many estuarine and back-barrier systems of the Atlantic coast. They are elongated marginal to the channel and the uplands, and can vary from the steeper fringing type to the flatter broad marsh. They may fill the drowned valley system nearly completely, and have an internal stratigraphy complicated by migration of the main channel (e.g., Morse River and Sprague River marshes, south-central Maine coast, Gehrels et al. 1996).

2.4.2 Processes of Aggradation

Aggradation processes in salt marshes include both organic and inorganic accumulation (Fig. 2.1). Salt marshes accumulate primarily at a rate determined by sea-level rise and creation of accommodation space, that is, as space for saturation and resulting lower rates of decay is created. There are varying opinions on marsh growth or degradation, and regional variations. Vertical accretion potential of undisturbed marshes is generally considered to be capable of keeping up with sea-level rise (Orson et al. 1985; Stevenson et al. 1986; Reed, 1990, 1995, 2002, Cahoon and Reed, 1995; Cahoon et al. 1996; Roman et al. 1997; Kirwin et al. 2016), but human disturbances may be driving recent losses (e.g., Delaune et al. 1983, 1989; Cahoon and Turner, 1989; Kennish, 2001; Silliman et al. 2009). Mudd (2011) suggests that a rapid influx of inorganic sediment (floods, storms, human disturbance) may result in a pulse of marsh growth. Kirwan et al. (2011) suggest that European colonization of North America released sediment in runoff that resulted in rapid expansion of salt marshes, with converse implications for modern rates of sediment supply. Wood et al. (1989) and Goodman et al. (2007) found modern accumulation in Maine high marshes of ca. 2–3 mm/yr, similar to the decadal rates of sea-level rise recorded at nearby tide gauges. This is consistent with a compilation of global values of 2.5 ± 1.4 mm/yr (FitzGerald et al. 2008). Because of increased inorganic sediment input, low marsh values average 5.8 ± 2.8 mm/yr, and fluvial-margin marshes are 7.3 ± 3.2 mm/yr (FitzGerald et al. 2008).

Vegetation growth is the most visible and often dominant process of marsh buildup, and provides the framework of interwoven roots and rhizomes that make the resulting peat stronger (Bertness, 1992; FitzGerald et al. 2008). Organic accumulation includes both

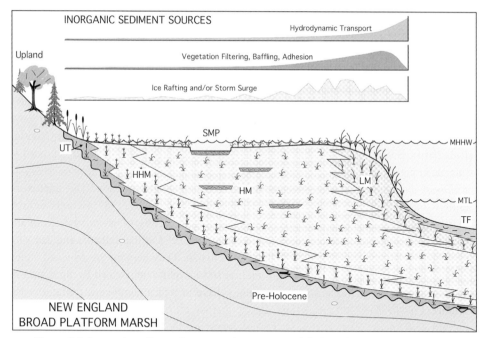

Figure 2.1 Inorganic sediment inputs, environments, and facies of a New England broad salt marsh. UT: Upland Transition; HHM: Higher High Marsh; HM: High Marsh; LM: Low Marsh; TF: Tidal Flat; SMP: Salt Marsh Pool; MTL: Mean Tide Level; MHHW: Mean Higher High Water.

autochthonous (in situ) growth of aboveground and belowground plant components, but allochthonous (transported) organic accumulation must also be accounted in a sediment budget. This can include nearby plant stems, leaves, flowers, and seeds, shed onto the marsh surface particularly after the growth season. However, distant sources may also play a significant role, such as beach grass, eelgrass, seaweed, and upland organic material. Fibrous peat forms and preserves roots and rhizomes as well as the allochthonous detritus that is preserved beyond a single growth season.

Inorganic sediment is brought to the marsh through both rapid and long-term processes (Fig. 2.1). Twice-daily tides flood the marsh with seawater containing suspended fine sediments from fluvial, estuarine, and lagoonal waters, as well as resuspended tidal-flat sediments. This hydrodynamic transport is most effective on the outer margins of the marsh, as well as near tidal creeks. Tidal currents predominate in channels. Flow is progressively channelized during late ebb and early flood stages, while there is sluggish sheet flow at high water over the broad marsh. The turbid water flooding the marsh is slowed in the boundary layer at the marsh surface, which is thickened by the roughness of the vegetation baffling. Leonard and Luther (1995) demonstrated that flow velocity in the marsh canopy was inversely proportional to plant stem density and distance from tidal creeks, with resulting effects on suspended matter transport. Davidson-Arnott et al. (2002) found that the highest suspended concentrations occurred at spring tidal range, and measured deposition was highest at middle portion of their macrotidal marsh, decreasing into the interior. Vegetation can directly filter the water, as it seeps into the unsaturated surface. Sediment is also trapped on the aboveground plant components as an adhering film, which is quite visible after a spring tide. Additional sediment may be trapped by pelletization by invertebrates (mollusks, decapods, worms). Analysis of peats demonstrates a clear relationship between loss-on-ignition (a proxy for proportion of organic content) and distance from the margin, with the greatest inorganic component on the outer marsh (Ward et al. 2008), as schematized in Figure 2.1. Broad marshes receive inorganic sediment from tidal creeks, which may have slightly coarser-grained levees, and fining toward the interiors. There is little documentation of upland slope runoff as a source of inorganic sediment to marshes in most settings. Wood et al. (1989) found sediment accumulation rates up to 80 mm/yr in low marsh-tidal flat colonization zone, 5 mm/yr in fringing marshes, and 2 mm/yr or less in broad marshes in Maine. Thus, broad marsh accumulates in equilibrium with sea-level rise and local compaction of marsh, while low marsh can colonize and accumulate much more rapidly

Storm surges may bring coarser material onto the marshes at irregular intervals, sometimes leaving identifiable layers millimeters to centimeters thick even in broad marshes Stumpf (1983). On the Mississippi Delta, Turner et al. (2006) found that sediment introduction from two large hurricanes were much more important than annual Mississippi River sediment contributions and dwarfed sediment input from human-built river diversions. Donnelly et al. (2001) have used these layers as proxies for impact of intense hurricanes in southern New England, but northeast storms are also potential sources. Finally, in northern New England (and perhaps farther south in periods of colder climate) ice rafting is a significant source of a variety of grain sizes to the marsh. Wood et al. (1989) documented ice rafting as constituting 20% or more of the inorganic fraction of some Maine marshes, while Argow et al. (2011) documented ice rafting as the source of approximately 5% of the annual volume of sediment needed by marshes in southern Maine to keep pace with rising sea level.

Highly fibrous New England marsh peats often contain less than 10% organics by dry weight, although they are 90% water and organics by volume (Wood et al. 1989). New England marshes are firm and fibrous, able to support human steps and even cattle grazing. Besides their more fibrous nature, this may be in part due to pre-compaction by ice loading (Argow and FitzGerald, 2006). Georgia and Florida marshes tend to have muddy or sandy substrates with less dense vegetation than New England marshes, while mid-Atlantic marshes are intermediate in composition.

2.4.3 Processes of Degradation

Degradation of marshes can be both vertical and lateral (Letzsch and Frey, 1980; Gedan and Silliman, 2009). Wave action may cause physical erosion at the margins of marshes; this may be exacerbated by boat wakes and other human activity. Kraft et al. (1976) and Swisher (1982) documented the historic shrinkage of marsh islands in Rehoboth Bay lagoon, Delaware, due to wave erosion. Kearney and Stevenson (1991) made similar measurements in Chesapeake Bay. Many similar changes are documented in back-barrier settings (e.g., Sepanik and McBride, 2015) or deltaic locations (Day et al. 2000; Barras et al. 2004). FitzGerald et al. (2006) relate erosion rates to degree of exposure to wave energy, largely dependent on tidal inlet size and width of the lagoon, though sediment deprivation is a central problem in some deltaic wetlands (Day et al. 2000; Syvitski and Saito, 2007).

Maps and air photos show distinctive meanders as well as dendritic patterns in many marshes; however, there is little evidence of distinct channel migration from air photos. Longer-term studies of maps show indications of meandering and change (Jacobson, 1988), but lack of accuracy in historic maps has not allowed firm establishment of such change. Dendritic patterns of creeks within broad marshes suggest an ebb-dominant drainage in intermittent creek flow (French and Stoddart, 1992). Allen (2000) suggests that tidal creeks are in short term (decades to centuries) dynamic equilibrium, neither infilling with sediment nor increasing in size through erosion. Stratigraphic evidence for shifting tidal creeks is a better indicator in the longer term (Gehrels et al. 1996). Despite lack of clear evidence for meander, blocks constantly slump from margins of creeks, particularly in New England type marshes. This may be due to high tidal range and ice effects. However, rapid growth and sediment accumulation in low-marsh environments build up and heal the slumps in a few years, without leading to progressive cutbank lateral displacement. Along much of the Atlantic coast creek margin degradation is enhanced by burrowing by fiddler crabs (Frey and Howard, 1969; Katz, 1980) and muskrats (Meredith et al. 1985; Yelverton and Hackney, 1986). A recent explosive increase in invasive European green crabs (*Carcinas maenas*) in Maine has led to localized lateral burrowing and degradation of marsh edges as well as denuding the low-marsh surface (Belknap and Wilson, 2014, 2015). Another crab, *Sesarma reticulatum* may be to blame for recent rapid denudation of Cape Cod marshes (National Park Service, 2017).

Vertical subsidence is one component of degradation. In undisturbed marshes there is a relative balance reached between accretion (Harrison and Bloom, 1977) and autocompaction (Kaye and Barghoorn, 1964). Autocompaction is the compression of water and gas-filled organic structures by gradually increasing overburden. This causes dewatering and increasing density. Kaye and Barghoorn (1964) used the cross-sectional shape of logs to infer compaction to 40% of original vertical measure, albeit in mainly freshwater coastal peat. Bloom (1964) found compaction to as much as 13% of original in sedge peats. Belknap and Kraft (1977) and Belknap et al. (1994) used radiocarbon-dated peats to demonstrate compaction displacement of isochronous paleo-marsh surfaces that increased with thickness, toward the axis of incised-valley-fill sections. Organic materials also degrade, especially through microbial decomposition of labile organic compounds. However, cellulose is relatively stable in the saturated anoxic and lowered pH conditions of most salt marshes. A recently identified mechanism for loss of belowground structure and strength of living biomass is elevated nutrients (Darby and Turner, 2008; Turner et al. 2009). Marshes may be compacted by overburden, such as under dredge spoil piles (Jacobson, 1988), by human and livestock trampling, and by ice. Ice loading and trampling may rebound elastically to some degree (Argow and FitzGerald, 2006) because of the strength of the fibrous peat. Even the design of experiments on the marsh surface must be planned with walkways to avoid inadvertent compaction that can disturb measurements (Davidson-Arnott et al. 2002).

Twentieth century losses in marsh area in the Mississippi Delta are related to dredging and spoils disposal, sediment exclusion by artificial levees, rapid subsidence, and other factors (Delaune et al. 1994; Day et al. 2007), but the threshold level for the rate of subsidence plus sea-level rise is not well known (Parkinson et al. 2017). There is concern that recent increases in the rate of sea-level rise could cause Atlantic coast marshes to drown in place, and the increase in the area of salt-marsh pools is one mechanism cited (Kearney et al. 1988, 2002; Hartig et al. 2002). Other studies suggest that salt marsh pools are part of a normal, cyclical dynamic process of the surface of New England marshes (Wilson et al. 2009, 2010) and may not necessarily indicate drowning.

2.5 Vegetation and Ecosystems

Tidal marshes are dominated by halophytes that tolerate saltwater inundation and saline soil pore waters. These elevated salinities exclude most upland competitor plants. The halophytes are primarily perennial herbaceous plants; those with no permanent woody stem undergo an annual dieback of aboveground parts, and regenerate and propagate to a large degree vegetatively from rhizomes. In salt marshes the belowground parts contribute biomass, and are preserved in the peat after death because of slowed decomposition. The aboveground stems, leaves, and flowers may become part of the peat section through nearby accumulation, or may be exported to the marine environment.

Grasses (Gramineae) such as *Spartina alterniflora* and *Spartina patens* are the most predominant vegetation of low marsh and of the high marsh in cooler latitudes. *Phragmites australis* is a 2+ m tall grass predominant in transitional to upland fringes in much of the northern mid-Atlantic. Rushes (Juncaceae) are commonly found in high and higher high marsh settings. Sedges (Cyperaceae) are usually found in brackish to upland fringe settings. Forbs are a diverse group of herbaceous plants other than grasses, rushes, and sedges, such as Compositae *Solidago sempirvirens* (seaside goldenrod) and *Potentilla* spp. Woody plants grow in marshes as low woody-stemmed perennials. The flowering shrubs *Rosa rugosa*, *Iva fructescens,* and *Baccharus halimifolia* are found in the higher-high marsh to transitional environments of tidal marshes.

Patterns of zonation in salt marshes (Fig. 2.2) are controlled by plant ecology (Niering and Warren, 1980; Bertness and Ellison, 1987). Halophytes respond most clearly to frequency of inundation, and may produce a sharp lateral zonation on steep fringing marshes in high tidal ranges (Belknap et al. 2004). Broad marshes in lower tidal ranges tend to exhibit a mosaic pattern, controlled by distance from tidal creeks, marsh edge, and potentially by depth to substrate.

Low salt marsh has restricted diversity, often formed by monocultures of *Spartina alterniflora* (salt marsh cordgrass) in much of North America. *S. alterniflora* in the low marsh is generally the tall form; a short variety is sometimes found in high marsh,

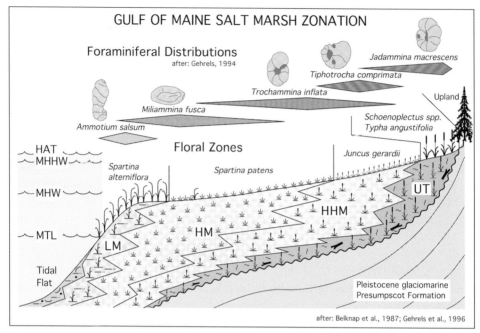

Figure 2.2 Plant and foraminiferal zonation of northern Gulf of Maine salt marshes, after Gehrels (1994) and Gehrels et al. (1996). HAT: Highest Astronomical Tide; MHHW: Mean Higher High Water; MHW: Mean High Water; MTL: Mean Tide Level; LM: Low Marsh; HM: High Marsh; HHM: Higher High Marsh: UT: Upland Transition (sometimes Basal Peat, e.g., Belknap and Kraft, 1977).

particularly near pools. Low salt marsh occurs between Mean Sea Level (MSL) and MHW (Fig. 2.2). It is flooded virtually every tidal cycle.

High salt marsh is more diverse, generally from MHW to Mean Higher High Water (MHHW), flooded every fortnight during spring tides. This diversity varies by latitude and climate zone. In New England the high marsh is dominated by *Spartina patens* grass, in the mid-Atlantic by *S. patens* and *Distichlis spicata*, and in Georgia and Florida by the rush *Juncus roemarianus*. On the California coast, it is dominated by *Salicornia pacifica*, while *Pucinellia maritima* is common in the British Isles.

Higher high marsh is flooded a few times per month, occurring above MHW to Spring High Water. The ecology also varies by latitude. In Maine the dominant plant is generally *Juncus gerardii*, while the Mid-Atlantic is less distinct, with the grass *Distichlis spicata* and shrubs *I. fructescens* and *B. halimifolia* important components. *Solidago* spp. and other forbs are associated plants in this zone, but rarely form a major part of the suite of species.

Upland Fringe is a brackish to freshwater system, flooded only during storms or highest astronomical tides (HAT), several times per year. Sedges such as *Schoenoplectus* spp. (syn. *Scirpus*, SDNHM, 2006) may form a distinct upper zone on fringing marshes, or transition to cattails (*Typha angustifolia*) that predominate on the landward side of the Upland Fringe in Maine. In the mid-Atlantic coast *P. australis* is a dominant Upland Fringe plant in a dense monoculture, with thick rhizomes that stifle other competitors. *P. australis* is an aggressive colonizer of disturbed lands, and in the last few decades it has expanded greatly in mid-Atlantic marshes, on dredge spoil islands, and is being carried to marshes in northern New England, particularly near major roads (perhaps as seeds hitchhiking on vehicles).

Marshes may also grade from primarily saline systems, as described above, to more brackish and freshwater systems across a transect or along an estuary. Many species of sedges such as *Schoenoplectus* (syn. *Scirpus*) and *Carex* characterize these transitional brackish marshes.

Salt pools are distinct lacunae in the high marsh, nearly constantly flooded, while pannes are intermittently flooded low, sparsely vegetated areas in high marsh, with variable salinities that vary from fresh to hypersaline. Pools are a critical ecosystem for larval and juvenile fish (e.g., Dionne et al. 1999). The origins of pools are controversial, but usually ascribed to abandoned tidal creeks, ice plucking, or "rotten spots" caused by wracks of dead grasses deposited by storms. They may contain submerged aquatic vegetation, such as *Zostera marina* or *Ruppia maritima*, which has a distinctive drupe that can be preserved in stratigraphic cores (Wilson et al. 2009; 2010). Pools and pannes may contain cyanobacterial mats, and may be fringed by *Salicornia* spp. (*S. virginica* in the mid-Atlantic, and *S. europaea* in New England) and other forbs.

Agglutinated foraminifera are often well preserved, and have become critical tools in identification of marsh paleoenvironments. Their tests are composed of silt grains cemented together by organic material, rather than composed of calcium carbonate as in marine foraminifera, and thus can be preserved in the acidic Holocene peats. Scott and Medioli (1978) first studied salt marsh foraminifera with a goal of identifying former sea levels in Atlantic Canada. Gehrels (1994, 2000) and Gehrels et al. (1996) used Scott and Medioli's techniques to extend such research into Maine. The use of salt marsh agglutinated foraminifera is now a widely used technique around the world (e.g., Horton et al. 1999). Foraminifera are more closely confined by frequency of flooding than the plants, and may provide a finer discrimination of vertical zonation. The current modern approach uses statistical analysis of foraminiferal abundances to create transfer function models of indicative meaning in the sampled levels (Horton et al. 1999; Gehrels, 2000), with applications to reconstructing Holocene sea-level curves (e.g., Engelhart and Horton, 2011, 2012). Figure 2.2 shows simplified salt marsh foraminiferal zones in the northern Gulf of Maine, named for the predominant taxa (after Gehrels, 1994). *Ammotium salsum* predominates in the low marsh. *Miliammina fusca* is a major component of the assemblages of the lower part of the high marsh, and is common in the low marsh. *Trochammina inflata* has a wide range in the high and higher high marshes, with peak abundance near the transition between the two. *Tiphotrocha comprimata* is characteristic of the higher high marsh. *Jadammina macrescens* ranges from the higher high marsh to the upland transition, and becomes essentially a monoculture at HAT.

2.6 Environmental Evolution and Stratigraphy

Marshes exist at intersections of coastal and upland environments, but also must be considered in terms of stratigraphic and evolutionary successions. Marshes may border and transition to a variety of other environments (e.g., flats to marsh, marsh to upland, salt marsh over freshwater marsh and bog). Stratigraphy of marshes can be determined by changes in sediment types, but also by detailed analysis of plant and other remains. Subfossil plant remains (roots, rhizomes, seeds) are commonly preserved well enough to identify species (Kraft et al. 1979; Wilson et al. 2010). Foraminifera, insect parts, diatoms, and pollen are also preserved and identifiable. Calcareous shells are not commonly preserved, because of the acidic conditions.

There are three primary historic models of stratigraphic evolution of marshes. Mudge (1858) envisioned marshes building up in an area of subsidence, and secondarily as transgressing onto the landward margin. Shaler (1885) interpreted his observations as marsh colonizing and prograding out over tidal flats. Redfield (1965, 1972) produced a more complete model, based on extensive coring, of a combination of progradation onto tidal flats as well as upbuilding and transgression in different parts of the system (Fig. 2.3). Redfield and Rubin (1962) used this understanding to produce an early detailed sea-level curve for southeastern Massachusetts, using radiocarbon dating of fossil peats in the transgressive sequence that was a model for many studies all over the world.

With increasing capability for deep coring and offshore sampling it became clear that the Mudge–Shaler–Redfield models applied only to the shallow late Holocene record. The deeper and offshore record was truncated by erosion and decoupled from earlier marsh

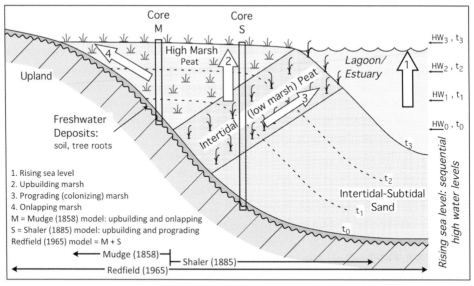

Figure 2.3 Simplistic historical models of salt marsh development and stratigraphy. The Redfield (1965, 1972) model combines both the Mudge (1858) model of upbuilding and upland onlap and well as the Shaler (1885) model of salt marsh upbuilding as well as colonizing and prograding out over tidal flats. They can be accommodated in the same system with sea-level rise and sufficient autochthonous growth coupled with input of inorganic material. Note that the transgression of intertidal deposits over freshwater deposits (e.g., Core S) is difficult to accept in a modern concept.

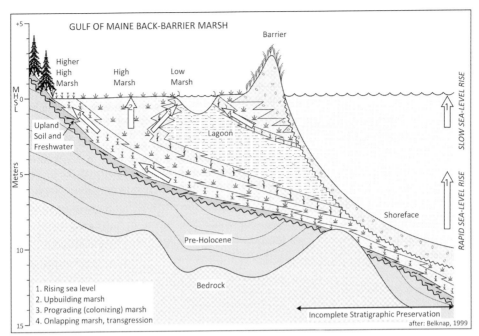

Figure 2.4 Model of transgressive back-barrier marsh development in the Gulf of Maine, after Belknap (1999). Note that the Redfield (1965) and previous models (see Fig. 2.3) explain the upper and more landward portions of the stratigraphy (generally from arrow 3 landward), but older deeper, and especially offshore sections show a "root" of dominantly transgressive stratigraphy, and sometimes discontinuous pockets of preserved peat (e.g., Belknap and Kraft, 1985; Belknap et al. 1994). The continuously transgressive sections may be explained by more rapid rates of mid-Holocene sea-level rise.

facies during sea-level rise. In fact, the rapid rates of mid-Holocene sea-level rise and transgression preclude thick, completely preserved sections of barrier and back-barrier facies, including marshes (Fig. 2.4) (e.g., Shepard, 1960; Curray, 1964; Swift, 1968, 1975; Belknap and Kraft, 1981, 1985; Belknap et al. 2002, 2005) Only in the isostatically complicated relative sea-level changes of the Gulf of Maine and Atlantic Canada are there "slowstands" and other changes in the rate of sea-level rise that may enhance the opportunity for greater preservation (e.g., Stea et al. 2001; Kelley et al. 2010, 2013). Evolution and preservation within incised valley estuarine facies involves rates of sea-level rise, but also localized pre-Holocene drainage morphologies and production of localized tidal ravinement unconformities (Belknap et al. 1994).

The surface of New England marshes in particular are marked by numerous salt pools and pannes. There have been various opinions on the significance of these features. One school of thought is that they are features of an evolutionary development from youth to old age, and represent a degradation and potential drowning of marshes, an inability to keep up with sea-level rise, or other degrading factors (e.g., Frey and Basan, 1985; Mudd, 2011). However recent studies by Wilson et al. (2009, 2010) demonstrate development, growth, drainage, and recolonization of pools, suggesting that the surface of the marsh may be more dynamic than recognized in the older literature. This does not conflict with the general stratigraphic evolution of the Redfield (1965, 1972) model, but rather emphasizes the dynamic nature of theses systems, in adjusting to rising sea level and landward shifting transgression, while maintaining the marsh environment over the long term.

Coupled to this stratigraphic interpretation is the concept of preservation potential. What part of a marsh section will be preserved during transgression? There are numerous examples of peats and tree stumps on barrier beachfaces (e.g., Hussey, 1959; Kraft, 1971; Kraft et al. 1979) that demonstrate partial preservation of the marsh facies after transgression by the overriding barrier. In addition, recovery of preserved sections in deep estuarine and offshore cores (e.g., Rampino and Sanders, 1980; Belknap and Kraft, 1985; Belknap et al. 1994) shows that shoreface and tidal channel erosion may not completely erase the record of marshes in ongoing transgression. Piecing together these more isolated samples of deeper datable peats with the semi-continuous record in intact marshes allows for more complete local relative sea-level curves (e.g., Belknap and Kraft, 1977; Belknap et al. 1987; Kelley et al. 2010).

2.7 Conclusions

Tidal salt marshes are crucial components of coastal ecosystems. They have ecosystem functions that are irreplaceable for fish, migratory birds, mammals, and invertebrates. They mitigate storms, filter turbid and polluted water, and amass large volumes of organic carbon that might otherwise be carried to the atmosphere as CO_2 gas. These highly productive systems are naturally valuable, but also important to humans as natural landscapes, wildlife refuges, recreation, fishing, and hunting. In the past they were used for pasture and cash crops of hay in diked meadows, especially in the northeastern Gulf of Maine and the Bay of Fundy (Smith and Bridges, 1982; Smith et al. 1989). At present, marshes are under increasing stress because of human alteration and encroachment, and may be at risk from climate change and rising sea level. Salt marshes have long been a fundamental source of geological and paleoenvironmental data that can document relative sea-level changes, resiliency in the face of change, and relationships to other coastal systems. Continued study with existing and improved tools is crucial for a wide variety of disciplines (Orson et al. 1998), and for coastal communities and society in general.

References

Adam, P. 1990. *Saltmarsh Ecology*. Cambridge University Press, Cambridge, UK.

Allen, J. R. L. 2000. Morphodynamics of Holocene salt marshes: a review sketch from the Atlantic and Southern North Sea coasts of Europe. *Quaternary Science Reviews*, 19: 1155–1231.

Aman, J., and Grimes, K. W. 2016. Measuring impacts on invasive European Green Crabs on Maine Salt Marshes: a novel approach: Report to the Maine Outdoor Heritage Fund.

Argow, B. A., and FitzGerald, D. M. 2006. Winter processes on northern salt marshes: evaluating the impact of in-situ peat compaction due to ice loading, Wells, ME. *Estuarine, Coastal and Shelf Science*, 69: 360–369.

Argow, B. A., Hughes, Z. J., and FitzGerald, D. M. 2011. Ice raft formation, sediment load, and theoretical potential for ice-rafted sediment influx on northern coastal wetlands. *Continental Shelf Research*, 31: 1294–1395.

Barras, J., Beville, S., Britsch, D., Hartley, S., Hawes, S., Johnston, J., Reed, D., Roy, K., Sapkota, S., and Suhaayda, J. 2004. Historical and projected coastal Louisiana land changes: 1978–2050: U.S. Geological Survey Open-File Report OFR 03-334.

Belknap, D. F. 1999. Sea-level rise and Gulf of Maine salt marshes. *Gulf of Maine NEWS, Regional Association for Research on the Gulf of Maine*, Spring, 1999: 1, 8–10.

Belknap, D. F. 2003. Salt marshes. In: G. Middleton, ed., *Encyclopedia of Sediments and Sedimentary Rocks*. Kluwer Academic Publishers, Dordrecht, pp. 586–588.

Belknap, D. F., Andersen, B. G., Anderson, R. S., Anderson, W. A., Borns, H. W., Jr., Jacobson, G., Jr., et al. 1987. Late Quaternary sea-level changes in Maine. In: D. Nummedal, O. H. Pilkey, Jr. and J. D. Howard, eds., *Sea-Level Fluctuation and Coastal Evolution*, Society of Economic Paleontologists and Mineralogists Special Publication, No. 41, pp. 71–85.

Belknap, D. F., Gontz, A. M., and Kelley, J. T. 2005. Paleodeltas and preservation potential on a paraglacial coast – evolution of eastern Penobscot Bay, Maine. Chapter 16. In: D. M. FitzGerald and J. Knight, eds., *High Resolution Morphodynamics and Sedimentary Evolution of Estuaries*. Springer, Dordrecht, pp. 335–360.

Belknap, D. F., Kelley, J. T., FitzGerald, D. M., and Buynevich, I. 2004. Quaternary Sea-level Changes and Coastal Evolution in Eastern and Central Coastal Maine, Field Trip Guidebook, International Geological Correlation Program #495, Quaternary Land-Ocean Interactions: Driving Mechanisms and Coastal Responses, Conference and Field Trip, Bar Harbor, ME, October 14–17, 2004, Dept. Earth Sciences, UniMaine, Orono.

Belknap, D. F., Kelley, J. T., and Gontz, A. M. 2002. Evolution of the glaciated shelf and coastline of the northern Gulf of Maine, USA. *Journal of Coastal Research Special Issue*, 36: 37–55.

Belknap, D. F., and Kraft, J. C. 1977. Holocene relative sea-level changes and coastal stratigraphic units on the northwest flank of the Baltimore Canyon Trough geosyncline. *Journal of Sedimentary Petrology*, 47: 610–629.

Belknap, D. F., and Kraft, J. C. 1981. Preservation potential of transgressive coastal lithosomes on the U.S. Atlantic Shelf. *Marine Geology*, 42: 429–442.

Belknap, D. F., and Kraft, J. C. 1985. Influence of antecedent geology on stratigraphic preservation potential and evolution of Delaware's barrier systems. *Marine Geology*, 63: 235–262.

Belknap, D. F., Kraft, J. C., and Dunn, R. K. 1994. Transgressive valley-fill lithosomes: Delaware and Maine: In: Boyd, R., Zaitlin, B. A. and Dalrymple, R., eds., *Incised Valley Fill Systems*, SEPM Special Pub. **51**: 303–320.

Belknap, D. F., and Wilson, K. R. 2014. Invasive green crab impacts on salt marshes in Maine – sudden increase in erosion potential. *Geological Society of America Abstracts with Programs*, 46, no. 1, Abstract 55-9: 104.

Belknap, D. F., and Wilson, K. R. 2015. Effects of invasive Green Crabs on salt marshes in Maine. *Geological Society of America Abstracts with Programs*, 47, no. 1, Abstract 65-8: 127–128.

Bertness, M. D. 1992. The ecology of a New England salt marsh. *American Scientist*, 80: 260–268.

Bertness, M. D. 2007. *Atlantic Shorelines: Natural History and Ecology*. Princeton University Press.

Bertness, M. D., and Ellison, A. M. 1987. Determinants of pattern in a New England salt marsh plant community. *Ecological Monographs*, 57: 129–147.

Bloom, A. L. 1964. Peat accumulation and compaction in a Connecticut coastal marsh. *Journal of Sedimentary Petrology*, 34: 599–603.

Boumans, R. M., and Day, J. W., Jr. 1993. High precision measurement of surface elevation in shallow coastal areas using a sediment-erosion table. *Estuaries*, 16: 375–380.

Boyd, B., and Sommerfield, C. K. 2017. Detection of fallout [241]Am in U.S. Atlantic salt marsh soils. *Estuarine, Coastal and Shelf Science*, 196: 373–378.

Cahoon, D. R., Lynch, J. C., and Powell, A. N. 1996. Marsh vertical accretion in a Southern California estuary U.S.A. *Estuarine, Coastal and Shelf Science*, 43: 19–32.

Cahoon, D. R., and Reed, D. J. 1995. Relationships among marsh surface topography, hydroperiod, and soil accretion in a deteriorating Louisiana salt marsh. *Journal of Coastal Research*, 11: 357–369.

Cahoon, D. R., Reed, D. J., and Day, J. W. Jr.. 1995. Estimating shallow subsidence in microtidal salt marshes of the southeastern United States: Kaye and Barghoorn revisited. *Marine Geology*, 128: 1–9.

Cahoon, D. R., and Turner, R. E. 1989. Accretion and canal impacts in a rapidly subsiding wetland II. Feldspar marker horizon technique. *Estuaries*, 12: 260–268.

Chapman, V. J. 1960. *Salt Marshes and Salt Deserts of the World*, Interscience Publishes, Inc, New York.

Chmura, G. L., Anisfled, S. C., Cahoon, D. R., and Lynch, J. C. 2003. Global carbon sequestration in tidal, saline wetland soils. *Global Biogeochemical Cycles*, 17: 1111–1133.

Chmura, G. L., Helmer, L. L., Beecher, C. B., and Sunderland, E. M. 2001. Historical rates of salt marsh accretion on the outer Bay of Fundy. *Canadian Journal of Earth Sciences*, 38: 1081–1092.

Clark, J. S. 1986. Late-Holocene vegetation and coastal processes at a Long Island tidal marsh. *Journal of Ecology*, 74: 561–578.

Curray, J. R. 1964. Transgressions and regressions. In: Miller, R. L., ed., *Papers in Marine Geology*, MacMillan, New York, pp. 175–203.

Daly, J. F., Belknap, D. F., Kelley, J. T., and Bell, T. 2007. Late Holocene sea-level change around Newfoundland. *Canadian Journal of Earth Sciences*, 44: 1453–1465.

Darby, F. A., and R. E. Turner. 2008. Effects of eutrophication to salt marsh roots, rhizomes, and soils. *Marine Ecology Progress Series*, 363: 63–70.

Davidson-Arnott, R. G. D., van Proosdij, D. V. Ollerhead, J., and Schostak, L. 2002. Hydrodynamics and sedimentation in salt marshes: examples from a macrotidal marsh, Bay of Fundy. *Geomorphology*, 48: 209–231.

Day, J. D., Britsch, L. D., Hawes, S., Shaffer, G. P., Reed, D. J., and Cahoon, D. 2000. Pattern and process of land loss in the Mississippi Delta: a spatial and temporal analysis of wetland habitat change. *Estuaries*, 23: 425–438.

Day, J. D., Boesch, D. F., Clairain, E. J., Kemp, G. P., Laska, S. B., Mitsch, W. J., Orth, K., et al. 2007. Restoration of the Mississippi Delta: Lessons from Hurricanes Katrina and Rita. *Science*, 315: 1679–1684.

DeLaune, R. D., Baumann, R. H., and Gosselink, J. G. 1983. Relationships among vertical accretion, coastal submergence and erosion in a Louisiana Gulf Coast marsh. *Journal of Sedimentary Petrology*, 53: 147–157.

DeLaune, R. D., Nyman, J. A., and Patrick, Jr., W. H. 1994. Peat collapse, ponding and wetland loss in a rapidly submerging coastal marsh. *Journal of Coastal Research*, 10: 1021–1030.

DeLaune, R. D., Whitcomb, J. H., Patrick, W. H. Jr., Pardue, J. H., and Pezeshki, S. R. 1989. Accretion and canal impacts in a rapidly subsiding wetland I. [137]Cs and ^{210}Pb techniques. *Estuaries*, 12: 247–259.

Dionne, M., Short, F. T., and Burdick, D. M. 1999. Fish utilization of restored, created, and reference salt-marsh habitat in the Gulf of Maine. *American Fisheries Society Symposium*, 22: 84–404.

Donnelly, J. P., Bryant, S. S., Butler, J., Dowling, J., Fan, L., Hausmann, N., Newby, P., Shuman, B., Stern, J., and Webb, T. III. 2001. 700 yr sedimentary record of intense hurricane landfalls in southern New England. *Geological Society of America Bulletin*, 113: 714–727.

Engelhart, S. E., and Horton, B. P. 2012. Holocene sea level database for the Atlantic coast of the United States. *Quaternary Science Reviews*, 54: 12–25.

Engelhart, S. E., Horton, B. P., and Kemp, A. C. 2011. Holocene sea levels along the United States' Atlantic coast. *Oceanography*, 24: 70–79.

FitzGerald, D. M., Buynevich, I., and Argow, B. 2006. Model of tidal inlet and barrier island dynamics in a regime of accelerated sea-level rise. *Journal of Coastal Research, Special Issue*, 39: 789–795.

FitzGerald, D. M., Fenster, M. S., Argow, B. A., and Buynevich, I. V. 2008. Coastal impacts due to sea-level rise. *Annual Review of Earth and Planetary Sciences*, 36, 601–647.

French, J. R., and Stoddart, D. R. 1992, Hydrodynamics of salt marsh creek systems: implications for marsh morphological development and material exchange. *Earth Surface Processes and Landforms*, 17: 235–252.

Frey, R. W., and Basan, P. B 1985. Coastal salt marshes. In: Davis, R. A., Jr., ed., *Coastal Sedimentary Environments*, Springer-Verlag, New York, pp. 225–301.

Frey, R. W., and Howard, J. D. 1969. A profile of biogenic sedimentary structures in a Holocene barrier island-salt marsh complex, Georgia. *Transactions of the Gulf Coast Association Geological Society*, 19: 427–444.

Gedan, K. B., and Silliman, B. R. 2009. Patterns of salt marsh loss within coastal regions of North America. In: Silliman, B., Grosholz, E., and Bertness, M.D., eds., *Human Impacts on Salt Marshes: A Global Perspective*, University of California Press, Los Angeles, CA, pp. 253–265.

Gehrels, W. R. 1994. Determining relative sea-level change from salt-marsh foraminifera and plant zones on the coast of Maine, USA. *Journal of Coastal Research*, 10: 990–1009.

Gehrels, W. R. 2000. Using foraminiferal transfer functions to produce high-resolution sea-level records from salt-marsh deposits, Maine, USA. *The Holocene*, 10: 367–376.

Gehrels, W. R., Belknap, D. F., and Kelley, J. T., 1996. Integrated high-precision analyses of Holocene relative sea-level changes: lessons from the coast of Maine. *Geological Society of America Bulletin*, 108: 1073–1088.

Gehrels, W. R., Kirby, J. R., Prokoph, A., Newnham, R. W., Achterberg, E. P., Evans, H., Black, S., and Scott, D. B. 2005. Onset of rapid sea-level rise in the western Atlantic Ocean. *Quaternary Science Reviews*, 24: 2083–2100.

Gehrels, W. R., Milne, G. A., Kirby, J. R., Patterson, R. T., and Belknap, D. F. 2004. Late Holocene sea-level changes and isostatic crustal movements in Atlantic Canada. *Quaternary International, Special Issue – International Geological Correlation Program, Project 437 "Late Quaternary Highstands," Barbados*, 120: 79–89.

Gehrels, W. R., and van de Plassche, O. 1991. Origin of the paleovalley system underlying Hammock River Marsh, Clinton, Connecticut. *Journal of Coastal Research, Special Issue,* 11: 73–83.

Goodman, J. E., Wood, M. E., and Gehrels, W. R. 2007. A 17-yr record of sediment accumulation in the salt marshes of Maine (USA). *Marine Geology*, 242: 109–121.

Harrison, E. Z., and Bloom, A. L. 1977. Sedimentation rates on tidal salt marshes in Connecticut. *Journal of Sedimentary Petrology*, 47: 1484–1490.

Hartig, E. K., Gornitz, V., Kolker, A., Mushacke, F., and Fallon, D. 2002. Anthropogenic and climate-change impacts on salt marshes of Jamaica Bay, New York City. *Wetlands*, 22: 71–89.

Hayes, M. O., and Kana, T. W. 1976. Terrigenous clastic depositional environments, Technical Report No. *11-CRD Coastal Research Division, Department of Geology*, University of South Carolina, Columbia.

Hine, A. C., Belknap, D. F., Hutton, J. G., Osking, E. B., and Evans, M. W. 1988. Recent geologic history and modern sedimentary processes along an incipient, low-energy, epicontinental-sea coastline: northwest Florida. *Journal of Sedimentary Petrology*, 58: 567–579.

Hladik, C., and Alber, M. 2012. Accuracy assessment and correction of a LIDAR-derived salt marsh digital elevation model. *Remote Sensing of Environment*, 121: 224–235.

Horton, B. P., Edwards, R. J., and Lloyd, J. M. 1999. Foraminiferal-based transfer function: implications for sea-level studies. *Journal of Foraminiferal Research*, 29, 117–129.

Hussey, A. M., II. 1959. Age of intertidal tree stumps at Wells Beach and Kennebunk Beach, Maine. *Journal of Sedimentary Petrology*, 29: 464–465.

Jacobson, H. A. 1988. Historical development of the saltmarsh at Wells, Maine. *Earth Surface Processes and Landforms*, 13: 475–486.

Katz, L. C. 1980. Effects of burrowing by the fiddler crab *Uca pugnax* (Smith). *Estuarine and Coastal Marine Science*, 11: 233–237.

Kaye, C. A., and Barghoorn, E. S. 1964. Late Quaternary sea-level change and crustal rise at Boston, Massachusetts, with notes on the autocompaction of peat. *Geological Society of America Bulletin*, 75: 63–68.

Kearney, M. S., Grace, R. E., and Stevenson, J. C. 1988. Marsh loss in Nanticoke Estuary, Chesapeake Bay. *Geographical Review*, 78: 205–220.

Kearney, M. S., Rogers, A. S., Townshend, J. R. G., Rizzo, E., Stutzer, D., Stevenson, J. C., and Sundborg, K., 2002. Landsat imagery shows decline of coastal marshes in Chesapeake and Delaware Bays. *Eos*, 83: 173, 177–178.

Kearney, M. S., and Stevenson, J. C. 1991. Island land loss and marsh vertical accretion rate evidence for historical sea-level changes in Chesapeake Bay. *Journal of Coastal Research*, 7: 403–415.

Kelley, J. T., Almquist-Jacobson, H., Jacobson, G. H., Jr., Gehrels, W. R., and Schneider, Z. 1992. The geologic and vegetative development of tidal marshes at Wells, Maine, USA. Research Report to the Wells National Estuarine Research Reserve and the National Oceanic and Atmospheric Administration.

Kelley, J. T., Belknap, D. F., and Claesson, S. 2010. Drowned coastal deposits with associated archaeological remains from a sea-level "slowstand," Northwestern Gulf of Maine, USA. *Geology*, 38: 695–698

Kelley, J. T., Belknap, D. F., Kelley, A. R., and Claesson, S. H. 2013. A model for drowned terrestrial habitats with associated archeological remains in the northwestern Gulf of Maine, USA. *Marine Geology*, 338: 1–16.

Kelley, J. T., Belknap, D. F., Jacobson, G. L., Jr., and Jacobson, H. A. 1988. The morphology and origin of salt marshes along the glaciated coastline of Maine, USA. *Journal of Coastal Research*, 4: 649–665.

Kelley, J. T., and Hay, B. W. B. 1986. Bunganuc Bluffs, Day 3, Stop 6. In: Kelley, J. T. and Kelley, A. R., eds. *Coastal Processes and Quaternary Stratigraphy Northern and Central Coastal Maine, Society of Economic Paleontologists and Mineralogists Eastern Section Field Trip Guidebook*, pp. 66–74.

Kennish, M. J. 2001. Salt marsh systems in the U.S.: a review of anthropogenic impacts. *Journal of Coastal Research*, 17: 731–748.

Kirwan, M. L., Murray, A. B., Donnelly, J. P., and Corbett, D. R. 2011. Rapid wetland expansion during European settlement and its implication for marsh survival under modern sediment delivery rates. *Geology*, 39: 507–510.

Kirwan, M. L., Temmerman, S., Skeehan, E. E., Guntenspergen, G. R., and Fagherazzi, S. 2016. Overestimation of marsh vulnerability to sea level rise. *Nature Climate Change*, 6: 253–260.

Kraft, J. C. 1971. Sedimentary facies patterns and geologic history of a Holocene marine transgression. *Geological Society of America Bulletin*, 82: 2131–2158.

Kraft, J. C., Allen, E. A., Belknap, D. F., John, C. J., and Maurmeyer, E. M. 1976. Delaware's Changing Shorelines. Technical Report #1, Delaware Coastal Zone Management Program, Dover.

Kraft, J. C., Allen, E. A., Belknap, D. F., John, C. J. and Maurmeyer, E. M. 1979. Processes and morphologic evolution of an estuarine and coastal barrier system, In: Leatherman, S. P., ed., *Barrier Islands*, Academic Press, New York, pp. 149–183.

Leatherman, S. P. 1979. Migration of Assateague Island, Maryland, by inlet and overwash processes. *Geology*, 7: 104–107.

Leonard, L. A., and Luther, M. E. 1995. Flow hydrodynamics in tidal marsh canopies. *Limnology and Oceanography*, 40: 1474–1484.

Letzsch, S. W., and Frey, R. W. 1980. Deposition and erosion in a Holocene salt marsh, Sapelo Island, Georgia. *Journal of Sedimentary Petrology*, 50: 529–542.

Meredith, W. H., Saveikis, D. E., and Stachecki, C. J. 1985. Guidelines for "Open Marsh Water Management" in Delaware's salt marshes – objectives, system designs, and installation. *Wetlands*, 5: 119–133.

Moller, I., Kudella, M., Rupprecht, F., Spencer, T., Paul, M., van Wesenbeeck, B., Wolters, G., et al. Wave attenuation over coastal salt marshes under storm surge conditions. 2014. *Nature Geoscience*, 7: 727–731.

Morris, J. T., Porter, D., Neet, M., Noble, P. A., Schmidt, L., Lapine, L. A., and Jensen, J. R. 2005. Integrating LIDAR elevation data, multispectral imagery and neural network modeling for marsh characterization. *International Journal of Remote Sensing*, 26: 5221–5234.

Mudd, S. M. 2011. The life and death of salt marshes in response to anthropogenic disturbance of sediment supply. *Geology*, 39: 511–512.

Mudge, B. F. 1858. The salt marsh formations of Lynn. *Proceedings of Essex Institute*, 2: 117–119.

National Park Service – Cape Cod National Seashore, 2017. Crab-driven vegetation losses: www.nps.gov/caco/learn/nature/crab-driven-vegetation-losses.htm

Neuendorf, K. K. E, Mehl, J. P. Jr., and Jackson, J. A. 2005. *Glossary of Geology* 5th Edn., American Geological Institute, Alexandria, VA.

Niering, W. A., and Warren, R. S. 1980. Vegetation patterns and processes in New England salt marshes. *Bioscience*, 30: 301–307.

Nikitina, D. L., Kemp, A. C., Horton, B. P., Vane, C. H., van de Plassche, O., and Engelhardt, S. E. 2014. Storm erosion during the past 2000 years along the north shore of Delaware Bay, USA. *Geomorphology*, 208: 160–172.

Orson, R., Panageotou, W., Leatherman, S. P. 1985. Response of tidal salt marshes of the U.S. Atlantic and Gulf coasts to rising sea levels. *Journal of Coastal Research*, 1: 29–7.

Orson, R. A., Warren, R. S., and Niering, W. A. 1987. Development of a tidal marsh in a New England river valley. *Estuaries*, 10: 20–27.

Orson, R. A., Warren, R. S., Niering W. A., and Van Patten, P., eds. 1998. Research in New England Marsh-Estuarine Ecosystems, Directions and Priorities into the Next

Millennium: Summary of a Sea Grant Workshop, May 15–17, 1997, 61 pp., Connecticut College, New London, CT. Connecticut Sea Grant College Program, Groton, CT: 5-11.

Parkinson, R. W., Craft, C., DeLaune, R. D., Donoghue, J. F., Kearney, M., Meeder, J. F., Morris, J., and Turner, R. E. 2017. Marsh vulnerability to sea-level rise. *Nature Climate Change*, 7: 756.

Rampino, M. R., and Sanders, J. E. 1980. Holocene transgression in south-central Long Island, New York. *Journal of Sedimentary Petrology*, 50: 1063–1080.

Redfield, A. C. 1965. Ontogeny of a salt marsh estuary. *Science*, 147: 50–55.

Redfield, A. C. 1972. Development of a New England salt marsh. *Ecological Monographs*, 42: 201–237.

Redfield, A. C., and Rubin. M. 1962. The age of salt marsh peat and its relation to recent changes in sea level at Barnstable, Massachusetts. *Proceedings of the National Academy of Sciences*, 48: 1728–1735.

Reed, D. J. 1989. Patterns of sediment deposition in subsiding coastal salt marshes, Terrebone Bay, Louisiana: the role of winter storms. *Estuaries*, 12: 222–227.

Reed, D. J. 1990. The impact of sea-level rise on coastal salt marshes. *Progress in Physical Geography*, 14: 465–481.

Reed, D. J. 1995. The response of coastal marshes to sea-level rise: survival or submergence? *Earth Surface Processes and Landforms*, 20: 39–48.

Reed, D. J. 2002. Sea-level rise and coastal marsh sustainability: geological and ecological factors in the Mississippi delta. *Geomorphology*, 48: 233–243.

Roberts, M. F. 1979. *The Tidemarsh Guide*, E. P. Dutton, New York.

Rogers, K., and Woodroffe, C. D. 2014. Tidal flats and salt marshes. In: G. Masselink, and R. Gehrels, eds., *Global Environments and Global Change*, John Wiley and Sons, Ltd., Chichester, UK, pp. 227–250.

Roman, C. T., Peck, J. A., Allen, J. R., King, J. W., and Appleby, P. G. 1997. Accretion of a New England (U.S.A.) salt marsh in response to inlet migration, storms and sea-level rise. *Estuarine, Coastal and Shelf Science*, 45: 717–727.

SDNHM (San Diego Natural History Museum). 2006. *www.sdplantatlas.org/NameChanges.aspx*, Genus *Scirpus* is now *Schoenoplectus*.

Schwimmer, R. A. 2001. Rates and processes of marsh shoreline erosion in Rehoboth Bay, Delaware, USA. *Journal of Coastal Research*, 17: 672–683.

Scott, D. B., and Greenberg, D. A. 1983. Relative sea-level rise and tidal development in the Fundy tidal system. *Canadian Journal of Earth Sciences*, 20: 1554–1564.

Scott, D. B., and Medioli, F. S. 1978. Vertical zonations of marsh foraminifera as accurate indicators of former sea levels. *Nature*, 272: 528–531.

Sepanik, J. M., and McBride, R. A. 2015. Increasing rate of salt-marsh loss in a barrier-island system: Parramore and Cedar Islands, Virginia, from 1957 to 2012, Section 1.6: pp. 392–401 of: McBride, R. A. Fenster, M. S., Seminack, C. T., Richardson, T. M., Sepanik, J. M., Hanley, J. T., Bundick, J. A. and Tedder, E., Holocene barrier-island geology and morphodynamics of the Maryland and Virginia open-ocean coasts: Fenwick, Assateague, Chincoteague, Wallops, Cedar and Parramore Islands, in Brezinski, D. K., Halka, J. P., and Ortt, R. A., Jr., eds., Tripping from the Fall Line: Field Excursions for the GSA Annual Meeting, Baltimore, 2015: Geological Society of America Field Guide 40, Boulder, CO: 309–424.

Shaler, N. S. 1885. Preliminary report on sea-coast swamps of the Eastern United States: U.S. Geological Survey 6th Annual Report, 1885: pp. 353–398.

Shepard, F. P., 1960, Gulf coast barriers. In: F. P. Shepard, F. B. Phleger, and T. H. von Andel, eds., *Recent Sediments, Northwest Gulf of Mexico*, American Association of Petroleum Geologists, Tulsa, Oklahoma, pp. 56–81.

Silliman, B. R., Grosholz, E. D., and Bertness, M. D., (eds.). 2009. *Human Impacts on Salt Marshes: a global perspective*. University of California Press, Berkeley, CA.

Silliman, B. R., Van der Kopple, J., Bertness, M. D., Stanton, I. E., and Mendelssohn, I. A. 2005. Drought, snails, and large-scale dieoff of southern U.S. salt marshes: *Ecology*, 310: 1803–1806.

Smith, D. C., and Bridges, A. E., 1982. Salt marsh dikes (dykes) as a factor in eastern Maine agriculture. *Maine Historical Society Quarterly*, 21: 219–226.

Smith, D. C., Konrad, V., Koularis, H., Borns, H. W., Jr., and Hawes, E. 1989. Salt marshes as a factor in the agriculture of northeastern North America. *Agricultural History*, 63: 270–294.

Snow, J. O. 1980. *Secrets of a Salt Marsh*. Guy Gannett Pub. Co, Portland, ME.

Stea, R. R., Fader, G. B. J., Scott, D. B., and Wu, P. 2001. Glaciation and relative sea-level change in Maritime Canada. In: T. K. Weddle, and M. J. Retelle, eds., *Deglacial History and Relative Sea-Level Changes Northern New England and Adjacent Canada*, Geological Society of America Special Paper 351: 35–49.

Stevenson, J. C., Ward, L. G., and Kearney, M. S. 1986. Vertical accretion in marshes with varying rates of sea level rise. In: D. A. Wolfe, ed., *Estuarine Variability*, Academic Press, New York, pp. 241–259.

Stumpf, R. P. 1983. The process of sedimentation on the surface of a salt marsh. *Estuarine, Coastal and Shelf Science*, 17: 495–508.

Swift, D. J. P. 1968. Coastal erosion and transgressive stratigraphy. *Journal of Geology*, 77: 444–456.

Swift, D. J. P. 1975. Barrier island genesis: evidence from the central Atlantic shelf, eastern U.S.A. *Sedimentary Geology*, 14: 1–43.

Swisher, M. L. 1982. *The rates and causes of shore erosion around a coastal lagoon, Rehoboth Bay, Delaware*: M.S. thesis, Dept. Geology, University of Delaware, Newark.

Syvitski, J. P. M., and Saito, Y. 2007. Morphodynamics of deltas under the influence of humans. *Global and Planetary Change*, 57: 261–282.

Tiner, R. W. 2009. *Field Guide to Tidal Wetland Plants of the Northeastern United States and Neighboring Canada*. University of Massachusetts Press, Amherst, MA.

Turner, R. E., Baustain, J. J., Swenson, E. M., and Spicer, J. S. 2006. Wetland sedimentation from Hurricanes Katrina and Rita. *Science*, 314: 449–452.

Turner, R. E., Howes, B. L., Teal, J. M., Milan, C. S., Swenson, E. M., and Goehringer Toner, D. 2009. Salt marshes and eutrophication: an unsustainable outcome. *Limnology and Oceanography*, 54: 1634–1642.

van de Plasscche, O. 1986. *Sea-level Research: a Manual for the Collection and Evaluation of Data*. Geo Books, Norwich, England.

van de Plassche, O. 1991. Late Holocene sea-level fluctuations on the shore of Connecticut inferred from transgressive and regressive overlap boundaries in salt-marsh deposits: Origin of the paleovalley system underlying Hammock River Marsh, Clinton, Connecticut. *Journal of Coastal Research, Special Issue* 11: 159–179.

Wang, C., Meneti, M., Stoll, M.-P., Feola, A., Belluco, E., and Marani, M. 2009. Separation of ground and low vegetations signatures in LiDAR measurements of salt-marsh environments. *IEEE Transactions on Geoscience and Remote Sensing*, 47: 2014–2023.

Ward, L. G., Zaprowski, B. J., Trainer, K. D., and Davis, P. T. 2008. Stratigraphy, pollen history and geochronology of tidal marshes in a Gulf of Maine estuarine system: climatic and relative sea level impacts. *Marine Geology*, 256: 1–17.

Wilson, K. R., Kelley, J. T., Croitoru, A., Dionne, M., Belknap, D. F., and Steneck, R. S. 2009. Stratigraphic and ecophysical characterizations of salt pools: dynamic features of the Webhannet Estuary salt marsh, Wells, Maine, USA. *Estuaries and Coasts*, 32: 855–870.

Wilson, K. R., Kelley, J. T., Tanner, B. R., and Belknap, D. F. 2010. Probing the origins and stratigraphic signature of salt pools from north-temperate marshes in Maine, U.S.A. *Journal of Coastal Research*, 26: 1007–1026.

Wood, M. E., Kelley, J. T., and Belknap, D. F. 1989. Pattern of sediment accumulation in the tidal marshes of Maine. *Estuaries*, 12: 237–246.

Woodwell, G. M., Rich, P. H., and Hall, C. A. S. 1973. Carbon in estuaries. In: G. M. Woodwell, and E. Pecan, eds., Carbon and the Biosphere, U.S. Atomic Energy Commission, Springfield, VA, USA, pp. 221–240.

Yelverton, G. F., and Hackney, C. T. 1986. Flux of dissolved organic carbon and pore water through the substrate of a *Spartina alterniflora* marsh in North Carolina. *Estuarine, Coastal, and Shelf Science*, 22: 255–267.

3

Salt Marsh Formation

ANTONIO B. RODRIGUEZ AND BRENT A. MCKEE

3.1 Introduction

Historical records show a massive decline in salt marsh area (Pendleton et al. 2012), > 50% in many locations, such as sites in Australia (Saintilan and Williams, 2000; Rogers et al. 2006), the British Isles (Baily and Pearson, 2007), and New England, USA (Bertness et al. 2002). These losses are mainly fueled by an underappreciation of the large contributions of salt marsh to maintaining healthy and productive estuaries. Prior to the middle twentieth century, the value of salt marsh primarily depended on its potential for reclamation. Davis (1910) proclaimed that "...[salt marshes] are conspicuous, being generally unutilized for any purpose except for making a small amount of inferior hay, hence they are practically desert places, except where land values are sufficiently high to make it worth while to raise the surface above high tide level for building purposes, or to dike out the tides." We now view salt marsh as a valuable estuarine habitat because it provides coastal protection from waves (Shepard et al. 2011), erosion control (Neumeier and Ciavola, 2004), water purification (Sousa et al. 2008), fish and bird habitat (Peterson and Turner, 1994; Van Eerden et al. 2005), carbon sequestration (Mcleod et al. 2011), and tourism/recreation (Barbier et al. 2011; Altieri et al. 2012). Salt marsh is also a coastal depositional environment that can accrete vertically over millennial time scales at rates equal to, or greater than, sea-level rise (Gehrels et al. 1996; Ouyang and Lee, 2014). The relatively high accretion rates and resistance of salt marshes to erosion (Mudd et al. 2010) make them valuable sites for preserving records of sea level (van de Plassche et al. 1998; Engelhart et al. 2011; Kemp et al. 2017), storms (Donnelly et al. 2001; Boldt et al. 2010; de Groot et al. 2011), and tsunamis (Morton et al. 2007; Komatsubara et al. 2008) in their sediments. Salt marsh loss and associated services have been pervasive globally, mainly due to the direct (grazing, ditching, pollution, etc.) and indirect (climate change) effects of human activities, resulting in the recent emphasis on restoration, conservation, and management (Lotze et al. 2006; Airoldi and Beck, 2007; Gedan et al. 2009). Although recent focus has been on better understanding of those mechanisms and processes that are related to salt marsh degradation, reviewing salt marsh formation and the different modes of salt marsh expansion will aid efforts aimed at preserving and increasing salt marsh habitat area and extracting climate and tectonic information from their sedimentary records.

3.2 Conditions Conducive to Salt Marsh Formation and Expansion

The successful colonization of substrate by salt marsh species and subsequent growth depends on interacting abiotic and biotic factors (Redfield, 1965; Engels et al. 2011). Salt marsh formation, persistence, and expansion requires relatively low wave and current energy and the intertidal-flat pioneer zone and salt marsh must maintain an intertidal elevation with sedimentation $\geq$ erosion (Dijkema, 1997). Successful establishment of salt marsh vegetation on tidal flats increases with increasing tidal-flat area, which also dampens wave power (Bouma et al. 2016). The seasonal variability of the tidal-flat-bed elevation due to varying rates of sedimentation and erosion also must be low to allow roots to take hold and limit burial of seedlings (Bouma et al. 2016). On established salt marshes, the submergence frequency and duration, wave- and tidal-current energy, and salinity decreases with increasing elevation and distance from the shoreline. These physiochemical gradients create distinct salt marsh zones from the shoreline to the upland boundary (Odum, 1988; Engels and Jensen, 2009).

Salt marsh biota contain both marine and terrestrial characteristics making them stress tolerant of salinities between 12 and 35 ppt (Odum, 1988; Flowers and Colmer, 2008). Although growth and survival are greatest in freshwater (Adams, 1963; Phleger, 1971), salt marsh species are excluded from tidal freshwater areas due to interspecific competition but thrive in higher salinities where physical stress is too great for freshwater marsh plants to survive (Snow and Vince, 1984, Crain et al. 2004; Engels and Jensen, 2009; Engles et al. 2011). Salt marsh vegetation traps sediments by reducing turbulence, which enhances particle settling and lessens erosion (Stumpf, 1983; Neumeier and Ciavola; 2004; Neumeier and Amos, 2006; Mudd et al. 2010). This reduced physical disturbance causes the vegetation to grow better (Bruno, 2000; Van der Wal et al. 2008) and increases the efficiency of the salt marsh sediment trap. Salt marsh formation is thus controlled by both geomorphological and ecological feedbacks (Fagherazzi et al. 2012; Kirwan and Megonigal, 2013).

Broadly, salt marsh formation and evolution strongly depend on relative changes in sediment accommodation and sediment accumulation. Accommodation is defined by Jervey (1988) as the amount of space that is available for sediments to accumulate in. Accommodation is always below base level (the level of erosion) and at the coast is closely tied to sea level. Near the shoreline, in the shallow water parts of estuaries, base level is also influenced by tidal currents and locally generated wind-wave power. At yearly and shorter time scales, storms are the principle driver of changes in salt marsh sediment accommodation because they erode shorelines and shallow substrates (Leonardi et al. 2015). Nearshore bathymetry also controls wave power near estuarine shorelines, with conditions being more conducive to deposition (low wave power) as the fetch and depth of adjacent tidal flats decreases (Mariotti and Fagerazzi, 2013).

Relative sea level (RSL), defined as changes in the volume of water in the ocean plus local tectonic changes (e.g., uplift and subsidence), is the principle driver of salt marsh sediment accommodation at decadal and longer time scales. Rising RSL creates sediment accommodation along the upland boundary of the salt marsh promoting landward salt marsh expansion. In addition, rising RSL increases sediment accommodation across the extant salt marsh, and depending on the rate of rise, tidal range, and suspended sediment concentration, could either enhance vertical accretion or, if the rate of RSL rise is too high, it could result in salt marsh loss through excessive soil waterlogging, adversely impacting soil chemistry (increase in salinity, decrease in oxygen), plant growth, and maintenance of an intertidal elevation (Reed, 1995; Morris et al. 2002; Kirwan et al. 2010). Conversion of salt marsh to subtidal flat is less likely in areas with a high tidal range and high suspended sediment supply (Kirwan et al. 2010). Accumulating marsh sediment is mainly composed of organic material from plant growth and mineral matter that settles onto the marsh from the water column and through ice rafting in middle to high latitudes. Sediment availability in the estuary is important for marsh formation, and increasing suspended sediment concentrations promotes subtidal flat accretion and its transformation to a marsh platform (Marani et al. 2010; Gunnell et al. 2013; Bouma et al. 2016)

3.3 Salt Marsh Classification

Salt marshes are associated with a variety of different estuarine landscapes and are commonly placed into groups based on geomorphology (Kelley et al. 1988; Allen, 2000), continental geography (Adam, 1990), and/or vegetation assemblage (Chapman, 1960). While these groupings are useful for characterizing the spatial distribution and associated genera of salt marsh, they are limited in their application to understanding shifts in salt marsh area and location under those constantly varying conditions associated with global change, such as warming, sea-level rise, and coastal development. Here, we simplify salt marsh classification, basing different types on their mode of formation and morphogenetic relationships between salt marsh and adjacent environments. We classify salt marsh into three general end-member groups, including: (1) Fringing, (2) Patch and (3) Deltaic (Fig. 3.1). Differences among salt marsh classes are mainly defined by their associated ecotones, ontogeny, and the processes that build and/or convert substrate to intertidal elevations. Fringing salt marsh initially forms on upland substrate and is located between the mainland and the open-water parts of estuaries (Fig. 3.1A). Patch salt marsh initially forms on intertidal flats of flood-tidal deltas or wash-over deposits in wave-dominated estuaries and on tidal sand ridges in tide-dominated estuaries (Fig. 3.1B). Patch salt marsh can also form on intertidal oyster reefs, but that succession of environments is relatively understudied. Deltaic salt marsh is found at the heads of estuaries and along shoreline protuberances where rivers discharge directly into basins. Deltaic salt marsh is commonly associated with the lower delta plain, including interdistributary bays and distributary-channel mouth bars (Fig. 3.1C). The physiochemical conditions that promote growth of salt marsh vegetation can be widespread, making these classifications not inclusive of all settings where salt marsh vegetation grows, such as ditches, inter-ridge swales, and coastal ponds.

3.4 Fringing Salt Marsh

Fringing salt marsh forms adjacent to the upland, typically along the protected shorelines of drowned river valleys and tidal creeks (Fig. 3.1A). This type of salt marsh initially forms as

A. Fringing Marsh

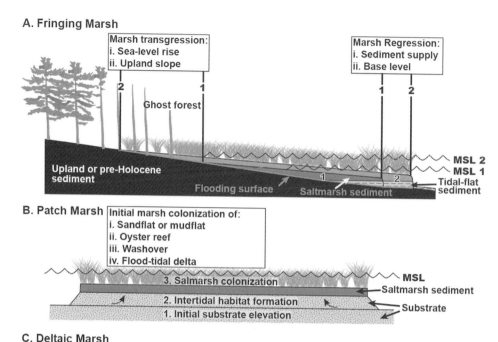

B. Patch Marsh

C. Deltaic Marsh

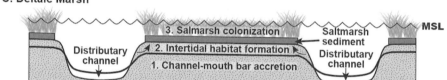

Figure 3.1 Salt marsh is grouped into three general classifications. Fringing marsh (A) forms with salt marsh transgression of the upland, which is modulated by the rate of RSL rise and upland slope. Fringing marsh also forms with salt marsh regression at the shoreline, which is modulated by base level and sediment supply. Patch marsh (B) is disconnected from the upland shoreline of the estuary and forms when salt marsh vegetation colonizes intertidal substrate such as tidal flat, oyster reef, washover fan, or flood-tidal delta bar. Deltaic marsh (C) forms at river mouths where sedimentation at the delta front and delta plain forms intertidal habitat suitable for salt marsh colonization.

RSL rise inundates upland areas. The notion that a salt marsh expands landward with RSL rise is not new (e.g., Davis, 1910 and Redfield, 1965), and in addition to geological data sets that show salt marsh landward expansion over millennia (Gardner and Porter, 2001; Tornqvist et al. 2004; Kemp et al. 2017), salt marsh transgression over the past 100 years is apparent from historic maps, aerial photographs, and satellite imagery (Williams et al. 1999; Feagin et al. 2010; Raabe and Stumpf, 2016). The stratigraphy of the landward portions of all fringing marshes shows a deepening-upward succession of environments with upland soil or old strata (Pleistocene or older) observed directly below salt marsh peat (Davis, 1910; Johnson, 1919). The contact between fringing salt marsh and underlying strata near the upland boundary is always a flooding surface (Fig. 3.1A). As sea-level rises the upland inundates, salinizes, and becomes tidal, making conditions conducive to salt marsh colonization. The rate of salt marsh transgression increases as the rate of RSL rise increases and/or as the upland gradient decreases (Belknap and Kraft, 1985; Davis and Clifton, 1987; Kraft et al. 1992; Theuerkauf et al. 2015; Kirwan et al. 2016). In some locations, however, the upland slope is steep (Kraft et al. 1992; Theuerkauf and Rodriguez, 2017) and fronted by a narrow fringing marsh that provides little wave reduction (Möller et al. 2014). In that morphological setting, storm waves could erode the upland and lower its slope. This would make any predictive model of salt marsh formation that relies only on geometrical relationships between RSL rise and the slope of the land being inundated problematic.

Fringing salt marsh can also expand basinward when sedimentation rates exceed the rate sediment accommodation is created (Fig. 3.1A). In most estuaries, wind-driven waves control the depth of erosion by establishing a local wave-base level, which ultimately controls the level beneath which sediments can accumulate (Shideler, 1984; Nichols, 1989; Simms and Rodriguez, 2015). As a result, in estuarine central basins, sediment accumulation generally keeps pace with RSL rise and average water depth remains constant over millennia (Nichols 1989).

The bathymetry adjacent to fringing marsh shorelines is more dynamic than in the central estuarine basin because the preservation of strata is more variable at the shoreline. For example, changing shoreline morphology can produce a local sheltering of wind waves (e.g., spit accretion) and a change in prevailing wind direction can decrease wave-base near a shoreline or transform it from one extreme to another (Gunnell et al. 2013). Under conditions of high sediment input, any decrease in base level can produce a rapid and areally extensive decrease in water depth. This shallowing is different than what is observed near a river mouth, where flow divergence results in sedimentation, formation of clinoforms, and a gradational increase in water depth (Simms at al., 2018). Rather, at the fringing salt marsh shoreline adjacent flats rapidly accrete vertically to an equilibrium elevation where the new intertidal flat is colonized by salt marsh (Fagherazzi et al. 2006). Basinward of this expanded salt marsh there can be a sharp bathymetric transition between the new shoreline position and the adjacent subtidal flat, the depth of which is in equilibrium with wave base (Fagherazzi et al. 2006). Slumping of steep upland terrane along estuarine shorelines can also supply nearshore areas with sediment forming new intertidal flats that rapidly become colonized with salt marsh vegetation (Kelley et al. 1988).

The basinward expansion of fringing marsh occurs episodically, as conditions on the adjacent tidal flat cycle between being conducive for deposition or favorable for resuspension. In the British Isles, cyclic basinward expansion forms a terraced fringing marsh morphology (Allen, 2000; Fig. 3.2). Each terrace represents an episode of marsh colonization of an accreting wave-cut platform, vertical marsh accretion, and finally a transition to an erosional regime with cliff formation and landward migration of the shoreline. As conditions basinward of the cliffed shoreline change to depositional, the cycle begins again as new salt marsh colonizes the accreting platform. This new marsh is at a lower elevation than the landward adjacent older marsh because it is less mature, and even though the new marsh platform is accreting at a greater rate than the landward

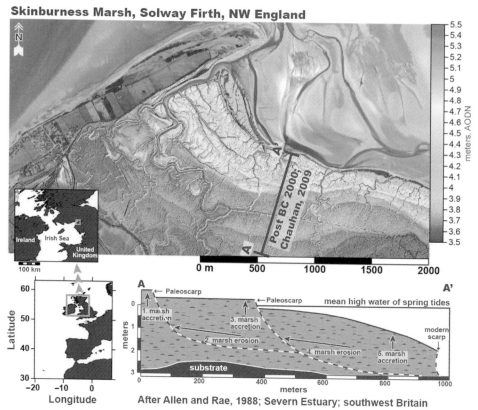

Figure 3.2 Digital elevation model of the Skinburness (Cumbria, UK) fringing salt marsh shows a series of terraces 0.5 m–0.1 m in relief. The delineated area accreted discontinuously over millennial time scales (Singh Chauhan, 2009). Cross section A–A′ is from the margin of the Severn Estuary, another terraced marsh in the UK, and illustrates that the terraces are related to periods of marsh accretion and erosion (after Allen and Rae, 1987). AODN = Above Ordnance Datum Newlyn. Data used to create the digital elevation model was downloaded from data.gov.uk. (A black and white version of this figure will appear in some formats. For the color version, please refer to the plate section.)

higher-elevation marsh terrace, its elevation has not caught up to the older marsh yet (Pethick, 1981; Allen, 2000). In addition to intrinsic factors promoting variable sedimentation rates across the marsh and adjacent tidal flat (Singh Chauhan, 2009; Allen and Haslett; 2014), older marsh terraces are also at a higher elevation due to RSL fall during the Holocene, upon which the marsh accretion–degradation cycles are superimposed (Shennan and Horton, 2002; Allen, 2000; Allen and Haslett; 2014).

In areas of Holocene RSL rise and rapid marsh accretion the same accretion–degradation cycles are recognized during basinward marsh expansion, but are not associated with a terraced marsh surface expression. Schwimmer and Pizzuto (2000) identified two accretion–degradation cycles in Rehobooth Bay, Delaware, USA under conditions of RSL rise (Fig. 3.3). The more recent episode of fringing marsh basinward expansion

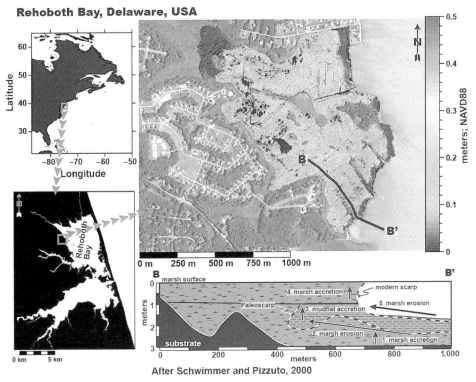

Rehoboth Bay, Delaware, USA

After Schwimmer and Pizzuto, 2000

Figure 3.3 Digital elevation model of fringing salt marsh around Rehoboth Bay, DE, shows higher elevation at the shoreline and along the margins of channels. Cross section B–B′ shows discontinuous bayward accretion of the salt marsh shoreline and a buried regional erosional scarp with no surface expression (i.e., a terrace). At around 700 yr BP the marsh shoreline transitioned from erosion to accretion and the paleoscarp formed (Schwimmer and Pizzuto, 2000). NAVD88 = North American Vertical Datum of 1988.

completely buried the paleoscarp that formed during the initial period of erosion (Fig. 3.3). The relief of the preserved paleoscarp is similar to the modern scarp and the paleoscarp is buried in ~1 m of salt marsh sediment. Salt marsh surface topography is relatively flat and shows no evidence of the episodic salt marsh regression preserved in the stratigraphic record, as the current salt marsh elevation achieved equilibrium with sea level.

3.5 Patch Salt Marsh

Patch salt marsh forms in the center of estuaries or in backbarrier environments and patch marshes initially grow on tidal flats not uplands (Fig. 3.1B). The formation of patch marshes is commonly event driven. Storms can cause sandflat migration and vertical accretion to intertidal elevations. Salt marsh will subsequently colonize that intertidal sandflat. Storms can also cause barrier island overwash, washover deposition, and subsequent salt marsh colonization of the new intertidal substrate. Patch marshes are thinner and have existed for a shorter period of time than fringing marshes, which initially formed and expanded landward with the estuary and can reach >5 m depth (Allen and Posamentier, 1993; Belknap et al. 1994; Gehrels et al. 1996). Patch marshes currently in equilibrium with sea level generally show a shallowing-upward succession of depositional environments with mud- or sandflat, oyster reef, flood-tidal delta, or washover fan sediments at the base (Fig. 3.1B).

As the bottom of the estuary accretes above mean sea level (MSL), away from the margin, patch salt marsh formation can occur. In many locations, salt marsh will colonize intertidal flat and oysters will subsequently settle along the margin of the salt marsh and form reefs (e.g., Ridge et al. 2017). Patch salt marsh substrate, however, is not only specific to sand- and mudflats. In warmer latitudes, vertically building oyster reefs can also be colonized directly by salt marsh vegetation (Fig. 3.4). Patch oyster reefs can accrete rapidly (10 cm/yr; Rodriguez et al. 2014) to their maximum intertidal elevation, which is close to MSL (Ridge et al. 2017) and the minimum elevation required for salt marsh accretion (Morris et al. 2002). When the elevation of an oyster reef is above MSL, salt marsh can out compete the oysters. As the salt marsh vegetation begins to colonize, an increase in plant-stem density will reduce flow, increase sediment trapping, and provide positive feedback conducive to salt marsh accretion and detrimental to the health of the oysters that are already stressed from limited inundation associated with their position on top of the reef. In the Shallotte River Estuary, North Carolina, USA, we sampled 48 cm of salt marsh sediment above an intertidal oyster reef in core SRE-MARSH-1, taken from the center of a patch salt marsh with a fringe of oyster reef surrounding its margin (Fig. 3.4). No articulated oysters were preserved at or directly below the contact (45–55-cm depth), indicating productivity on top of the reef was low when marsh colonization initiated. The flanks of the oyster reef remained productive after the top of the reef transformed into salt marsh (Fig. 3.4).

Patch salt marsh also commonly forms in backbarrier environments, where formation of intertidal substrate is directly linked to those changes in barrier-island environments that take place during landward migration (Johnson, 1919; Kraft, 1971; Godfrey and Godfrey, 1974; Theuerkauf and Rodriguez, 2017). RSL rise principally forces barrier islands to migrate landward and during migration, storm overwash creates new intertidal sandflats through washover, and/or flood-tidal delta deposition. After the storm, the barrier island will regain elevation through dune accretion, which will prevent subsequent overwash, and tidal inlets will close or migrate down drift. These post-storm morphological changes to the barrier reduce the tidal energy and stabilizes the backbarrier intertidal sandflats associated with the washover or the flood-tidal delta interdistributary-channel bars, creating conditions conducive to the growth of salt marsh vegetation.

Cheeseman Inlet, located on Bogue Banks, North Carolina, USA, is an example of how changing barrier-island morphology forms new patch salt marsh (Fig. 3.5). Cheeseman Inlet formed after a storm breached the barrier island and connected Bogue Sound with the Atlantic Ocean for 10–15 years (Fisher, 1962) before it closed in CE 1806 (Fig. 3.5A). After Cheeseman Inlet closed, tidal energy in Bogue Sound decreased, and salt marsh colonized the relic flood-tidal delta (Fig. 3.5B). Cores CI-17-2 and CI-17-1, from two individual patch salt marshes, sampled 33 and 16 cm of salt marsh sediment above

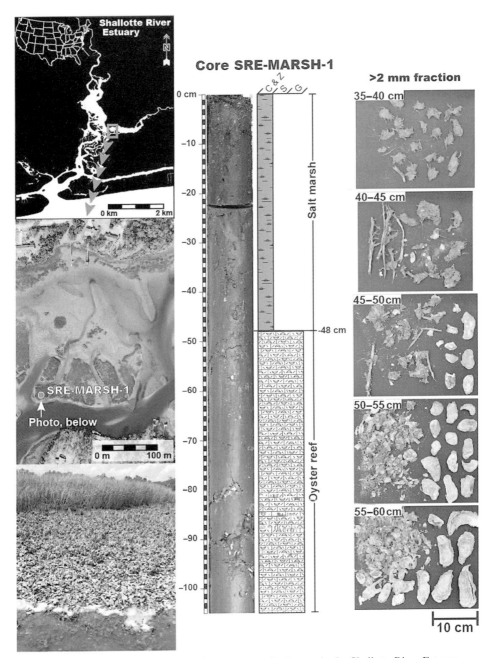

Figure 3.4 Patch salt marsh formed on oyster-reef substrate in the Shallotte River Estuary, NC. Core SRE-MARSH-1, collected from the center of the patch salt marsh, was sampled continuously into 5-cm long sections (10-cm diameter). Those samples were washed through a 2-mm sieve and photographed. The core sampled 48-cm of salt marsh above a >57-cm thick oyster reef. From the top of the core, interval 55–60 cm was the first to sample large articulated oysters (>10-cm long) and no salt marsh vegetation, interpreted to represent the end of a productive oyster reef.

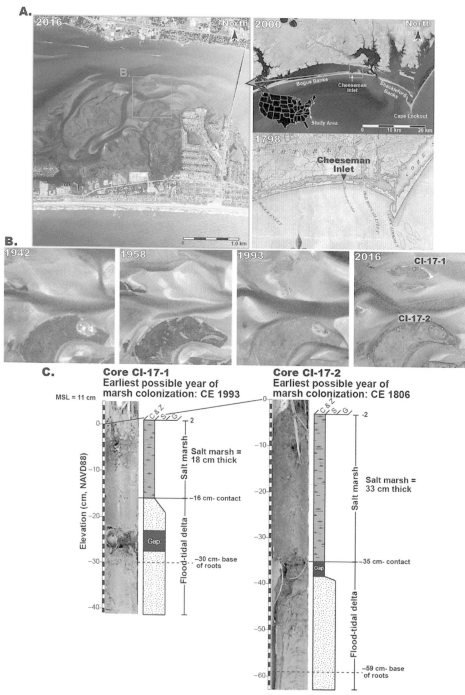

Figure 3.5 (A) Cheeseman Inlet formed around CE 1790 at Bogue Banks, NC. (B) Patch salt marsh colonized the flood-tidal delta soon after the inlet closed in CE1806 and distal parts of the flood tidal delta continue to form new patch salt marsh as physiochemical conditions become conducive for the growth of salt marsh vegetation. (C) Cores CI-17-1 and CI-17-2 were collected from younger and older patch salt marshes, respectively, and accreted vertically at the rates similar to RSL rise. CE = Common Era.

flood-tidal delta sand, respectively (Fig. 3.5C). Salt marsh likely formed immediately after conditions were conducive for vegetative growth. Core CI-17-2 was collected from a part of the flood-tidal delta colonized by salt marsh shortly after Cheeseman Inlet closed, and the thickness of the salt marsh sampled is within the 43 $\pm$12-cm range of RSL rise from CE 1806 to 2005 (Kemp et al. 2017). From CE 1942 to 1993, a sandflat along the northern part of the relic flood-tidal delta accreted to an intertidal elevation and was subsequently colonized with salt marsh vegetation sometime between CE 1958 and 1993, based on historical aerial photos (Fig. 3.5). Core CI-17-1 was collected from this younger patch salt marsh and the thickness of salt marsh sampled is close to the 11 $\pm$5 or 22 $\pm$5 cm of RSL rise (linear MSL) measured from CE 1993-2017 and CE 1958-2017, respectively, at the NOAA 8656483 tide gauge, located 7.3 km from the patch salt marsh. Based on historical maps and aerial photos that document the time of sandflat formation and marsh thicknesses that match the increase in sea level that occurred since sandflat formation, it is likely that both patch salt marshes accreted at average rates of RSL rise. In addition, the timing of initial salt marsh colonization, extrapolated from sea-level records and maps, suggests that natural patch salt marsh formation can occur rapidly in less than a decade.

Similar to fringing salt marsh, backbarrier patch salt marsh can transgress the upland margin with RSL rise, expanding towards the center of the barrier island. This expansion towards the barrier, however, is in a seaward direction and will be short-lived because the planar contact between salt marsh and the overlying barrier-island migrates landward with RSL rise and storms. Ultimately, this will decrease the area of backbarrier patch salt marsh by burying it in sand through aeolian and overwash processes (Theuerkauf and Rodriguez, 2017). The formation of backbarrier intertidal substrate for colonization by salt marsh vegetation will have a net landward movement as barrier-island transgression continues, and the lagoon narrows and shallows. Eventually, the lagoon will decrease in size, fill with sediment, and transform into an extensive fringing salt marsh confined to the area between the barrier and the upland when the mainland fringing salt marsh and backbarrier patch salt marshes merge.

3.6 Deltaic Salt Marsh

Deltaic salt marsh exists at the heads of estuaries (bayhead deltas) and along river-dominated open-ocean shorelines (wave-, tide-, and river-dominated deltas; Galloway, 1975). Deltaic salt marsh formation is controlled by the deposition of a river's load at the margin of a basin, the associated accretion of intertidal flats, and subsequent changes in delta morphology and hydrology that make physiochemical conditions suitable for salt marsh vegetation to colonize (Fig. 3.1C). Salt marsh can be widespread in the delta-plain, delta-front, and interdistributary bay environments. The delta plain extends from the first distributary channel to the shoreline and the delta front encompasses the shoreline, distributary mouth bars, and adjacent basinward-sloping bed (Bhattacharya, 2006). Commonly, the upper delta plain is composed of freshwater wetland and the lower delta plain contains salt marsh, with the demarcation being strongly controlled by discharge, slope, and tidal range. The delta plain extends seaward as distributary-mouth bars accrete, become intertidal, and force the distributary channel to bifurcate and/or deflect in the direction of longshore transport. Delta lobes form and accrete by the coalescing of distributary-mouth bars and the formation of numerous orders of distributary channels (Olariu and Bhattacharya, 2006). Salt marsh forms in low-energy settings on distributary-mouth bars, crevasse-splays, and/or along the margins of interdistributary bays and back-bar lagoons.

Salt marsh formation at river dominated deltas, like the Mississippi Delta, is linked with delta-lobe building, abandonment, and deterioration (Penland et al. 1988; Roberts, 1997). Reed (2002) presented a conceptual model that incorporated ecological processes into the "delta cycle" (Roberts, 1997). As a delta lobe forms, salinity of the coalescing distributary bars is low and freshwater marshes form rapidly (Cahoon et al. 2011; Oliver and Edmonds, 2017). After the delta lobe is fully developed, the river avulses, sedimentation < RSL rise, the area becomes a brackish interdistributary bay, and intertidal freshwater wetlands convert into salt marsh. Most salt marsh in the modern Mississippi Delta is located within interdistributary bays (Howes et al. 2010). The interdistributary bay continues to expand and salt marsh area declines with RSL rise. Eventually, accommodation becomes high enough that the river shifts back into the area, sedimentation > RSL rise, and the autogenic delta cycle begins again (Penland et al. 1988; Roberts, 1997). Human activities, like fluid withdrawal, damming, and levee construction disrupt this cycle and offset the balance between marsh formation and destruction, leading to massive wetland loss and an increase in the vulnerability of coastal populations to flooding and inundation (Kennish, 2001; Blum and Roberts, 2009; Syvitski et al. 2009).

Bayhead deltas are located at the heads of estuaries, confined to incised valleys (Dalrymple et al. 1992). Deltaic salt marsh formation occurs on the lower delta plain and distributary mouth bars, which are protected in the estuary from erosive open-ocean waves and currents (Saintilan and Hashimoto, 1999; White et al. 2002). Deltaic salt marsh is less extensive at bayhead deltas than open-ocean deltas because river loads and the space available for intertidal channel-mouth bars is lower. Similar to the delta cycle of Penland et al. (1988), bayhead deltas and associated salt marsh also experience cycles of rapid transgression followed by stability and growth (Thomas and Anderson, 1994; Rodriguez et al. 2010; Anderson et al. 2016; Simms et al. 2018). These autogenic cycles are due to RSL rise forcing the bayhead delta to migrate landward through the irregular dendritic drainage network of the lower watershed. Tributary junctions are associated with an abrupt increase in river gradient, causing a decrease in the sediment accommodation/sediment supply ratio and bayhead delta growth and stability (Simms and Rodriguez, 2014).

Human activities, like impoundments, can decrease wetland accretion in bayhead deltas (White et al. 2002; McKee et al. 2006; Canuel et al. 2009; Jalowska et al. 2015; Jalowska et al. 2017); however, land-use changes can have the opposite effect. Clearing of land for farming during initial human settlement resulted in increased runoff, high river sediment load, and delta plain accretion from legacy sedimentation (James, 2013; Watson and Byrne, 2013; Jalowska et al. 2015). Bayhead deltas can respond rapidly to changes in land use. Sediment from landscape erosion at a silviculture operation in the Newport River water-shed, North Carolina, USA, took <3 years to reach the bayhead delta. The addition of this new sediment source to the river load resulted in delta-plain salt marsh formation to

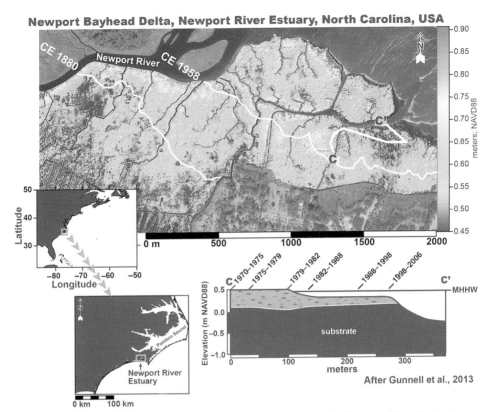

Figure 3.6 Digital elevation model of deltaic salt marsh at the Newport River Bayhead Delta, NC showing bayward accretion. Cross section C–C′ showing variations in deltaic salt marsh thickness and ages based on a time series of aerial photographs and cores (see Gunnell et al. 2013 for additional information) NAVD88 = North American Vertical Datum of 1988. (A black and white version of this figure will appear in some formats. For the color version, please refer to the plate section.)

abruptly increase from 10,000 m^2 yr^{-1} to 15,000 m^2 yr^{-1} (Mattheus et al. 2009; Fig. 3.6). Similar to the other types of salt marsh, the emergence of vegetation occurred rapidly, but was not punctuated by erosional events, like fringing salt marsh commonly is (Figs. 3.2, 3.3, and 3.6). Salt marsh formation at the Newport had an along-shore trajectory, promoted by marsh promontories that shielded adjacent mudflats from erosive forces in the already sheltered estuary (Fagherazzi, 2013; Gunnell et al. 2013; Fig. 3.6).

3.7 Salt Marsh Restoration and Rehabilitation

Salt marsh formation via restoration and rehabilitation is commonly carried out to counter losses from human impacts such as salt marsh reclamation, burial, excavation, pollution, development, and shoreline armoring, in addition to alteration of hydrology, sedimentation, and subsidence (Kennish, 2001; Adam, 2002). After restoration, the full development of salt marsh ecosystem function (vegetation, macroinvertebrate populations, fish and bird use) can take 20 years (Craft, 2000; Warren et al. 2002), while soil development and carbon accumulation can take more than a century before it is on par with a natural marsh (Craft et al. 2002). Salt marsh rehabilitation commonly takes the form of changing the physio-chemical conditions around the margin of an estuary or delta, making them more conducive to salt marsh growth. Nearshore breakwater sills and groins protect salt marsh from erosion, decrease wave power, and increase sedimentation providing new intertidal substrate for salt marsh expansion (Adam 2002; Currin et al. 2008). Removal or modification of embankments, dikes, or walls at salt marsh sites returns tidal flow, sediment, and salt marsh vegetation to reclaimed areas (Bakker et al. 2002; Williams and Orr, 2002). In many deltaic salt marshes, thin-layer deposition of dredged material (Ford et al. 1999; Graham and Mendelssohn, 2013) and/or reconnection of the river to the delta plain (Day et al. 2007) conserves intertidal elevations and provides additional intertidal substrate for salt marsh colonization through increased sedimentation. Salt marsh restoration takes the form of planting salt marsh vegetation at sites where the requirements for salt marsh establishment are met (Broome, 1988). Restoration projects span a wide range of scales from <0.1 ha to >6,000 ha for the South Bay Salt Pond Restoration Project, San Francisco, California, USA (in progress). Salt marsh formation can also occur by introducing new species to an area, such as *Spartina alterniflora* that was introduced to China in 1979. *Spartina* was planted in the Yangtze Delta during the 1990s to increase reclamation by accelerating the accretion of tidal flats and offshore sands (Chung et al. 2004; Zhang et al. 2004; Xiao et al. 2010). This strategy worked well due to the high sediment load of the river, and a small restoration project in the Yangtze Delta expanded rapidly from 108 m^2 in 1988 to 12.74 km^2 by 2001 (Zhang et al. 2004).

3.8 Conclusions

Specific physiochemical conditions are required for salt marsh formation, the most important being brackish water, intertidal elevation, and low hydrodynamic energy. Most salt marsh falls into one of three different formation classifications, including fringing, patch, and deltaic. Fringing salt marsh expands landward with sea-level rise and bayward as tidal flats accrete and large expanses are colonized with salt marsh vegetation. Many fringing marshes are expanding landward through transgression and eroding along the shoreline, with the net growth or decay being determined by those relative rates. Patch salt marsh is initially disconnected from the upland and forms on intertidal substrates including flood-tidal deltas, washover fans, tidal flats, and oyster reefs. The conditions that promote patch salt marsh formation and degradation are tightly linked with the evolution of adjacent environments (e.g., barrier island rollover and oyster reef accretion) and changing hydrodynamics (e.g., a migrating tidal inlet) and storminess. Deltaic salt marsh forms on channel-mouth bars and within the lower delta plain including interdistributary bays. Deltaic salt marsh formation and degradation is coupled with autogenic delta processes (avulsion), sea-level rise, and human modifications to rivers (impoundments, levees, land-use change).

The conditions required for salt marsh formation and health often run counter to the needs and desires of humans. This has driven the massive historical decline in salt marsh area, which is projected to accelerate with climate change. Salt marsh restoration and rehabilitation is commonly employed to return some of the lost ecosystem services, but it is unlikely to keep pace with declining salt marsh area. Eventually, accelerating sea-level rise will adversely impact most salt marshes, the most resilient being in areas where sedimentation is high, such as flood tidal deltas, prograding river deltas, or along the margins of estuaries with high suspended-sediment concentrations. Landward migration will extend the life of some fringing salt marshes but increasing development along estuarine shorelines and steep upland gradients will be limiting factors.

References

Adam, P. 1990. *Saltmarsh Ecology*. Cambridge University Press, Cambridge; New York.

Adam, P. 2002. Saltmarshes in a time of change. *Environmental Conservation*, 29: 39–61.

Adams, D. A. 1963. Factors influencing vascular plant zonation in North Carolina Salt Marshes. *Ecology*, 44: 445–456.

Airoldi, L., and Beck, M. W. 2007. Loss, status and trends for coastal marine habitats of Europe. *Oceanography and Marine Biology: An Annual Review*, 45: 345–405.

Allen, G. P., and Posamentier, H. W. 1993. Sequence stratigraphy and facies model of an incised valley fill; the Gironde Estuary, France. *Journal of Sedimentary Research*, 63: 378–391.

Allen, J., and Rae, J. 1987. Late Flandrian shoreline oscillations in the Severn Estuary: a geomorphological and stratigraphical reconnaissance. *Philosophical Transactions of the Royal Society of London B: Biological Sciences*, 315: 185–230.

Allen, J. R. L. 2000. Morphodynamics of Holocene salt marshes: a review sketch from the Atlantic and Southern North Sea coasts of Europe. *Quaternary Science Reviews*, 19: 1155–1231.

Allen, J. R. L., and Haslett, S. K. 2012. Salt-marsh evolution at Northwick and Aust warths, Severn Estuary, UK: a case of constrained autocyclicity. *Atlantic Geology*, 50: 1–17.

Altieri, A. H., Bertness, M. D., Coverdale, T. C., Herrmann, N. C., and Angelini, C. 2012. A trophic cascade triggers collapse of a salt-marsh ecosystem with intensive recreational fishing. *Ecology*, 93: 1402–1410.

Amos, C. L., Feeney, T., Sutherland, T. F., and Luternauer, J. L. 1997. The stability of fine-grained sediments from the Fraser River Delta. *Estuarine, Coastal and Shelf Science*, 45: 507–524.

Anderson, J. B., Wallace, D. J., Simms, A. R., Rodriguez, A. B., Weight, R. W. R., and Taha, Z. P. 2016. Recycling sediments between source and sink during a eustatic cycle: Systems of late Quaternary northwestern Gulf of Mexico Basin. *Earth-Science Reviews*, 153: 111–138.

Bahattacharya, J. P. 2006. Deltas. In: *Facies Models Revisited*. Eds H. W. Posamentier and R. G. Walker., Society for Sedimentary Geology, Tulsa, pp. 237–292.

Baily, B., and Pearson, A. W. 2007. Change detection mapping and analysis of salt marsh areas of Central Southern England from Hurst Castle Spit to Pagham Harbour. *Journal of Coastal Research*, 23: 1549–1564.

Bakker, J., Esselink, P., Dijkema, K., Van Duin, W., and De Jong, D. 2002. Restoration of salt marshes in the Netherlands. *Hydrobiologia*, 478: 29–51.

Barbier, E. B., Hacker, S. D., Kennedy, C., Koch, E. W., Stier, A. C., and Silliman, B. R. 2011. The value of estuarine and coastal ecosystem services. *Ecological Monographs*, 81: 169–193.

Belknap, D. F., and Kraft, J. C. 1985. Influence of antecedent geology on stratigraphic preservation potential and evolution of Delaware's barrier systems. *Marine Geology*, 63: 235–262.

Belknap, D. F., Kraft, J. C., and Dunn, R. K. 1994. Transgressive valley-fill lithosomes: Delaware and Maine. In: *Incised-Valley Systems: Origin and Sedimentary Sequences*. Eds R. W. Dalrymple, R. Boyd and B. A. Zaitlin., *SEPM, Special Publication 51*, SEPM, Tulsa, pp. 303–320.

Bertness, M. D., Ewanchuk, P. J., and Silliman, B. R. 2002. Anthropogenic modification of New England salt marsh landscapes. *Proceedings of the National Academy of Sciences of the USA*, 99: 1395–1398.

Blum, M. D., and Roberts, H. H. 2009. Drowning of the Mississippi Delta due to insufficient sediment supply and global sea-level rise. *Nature Geoscience*, 2: 488–491.

Boldt, K. V., Lane, P., Woodruff, J. D., and Donnelly, J. P. 2010. Calibrating a sedimentary record of overwash from Southeastern New England using modeled historic hurricane surges. *Marine Geology*, 275: 127–139.

Bouma, T. J., van Belzen, J., Balke, T., van Dalen, J., Klaassen, P., Hartog, A. M., Callaghan, D. P., et al. 2016. Short-term mudflat dynamics drive long-term cyclic salt marsh dynamics. *Limnology and Oceanography*, 61: 2261–2275.

Broome, S. W., Seneca, E. D., and Woodhouse, W. W. 1988. Tidal salt marsh restoration. *Aquatic Botany*, 32: 1–22.

Bruno, J. F. 2000. Facilitation of cobble beach plant communities through habitat modification by *Spartina alterniflora*. *Ecology*, 81: 1179–1192.

Cahoon, D. R., White, D. A., and Lynch, J. C. 2011. Sediment infilling and wetland formation dynamics in an active crevasse splay of the Mississippi River delta. *Geomorphology*, 131: 57–68.

Canuel, E. A., Lerberg, E. J., Dickhut, R. M., Kuehl, S. A., Bianchi, T. S., and Wakeham, S. G. 2009. Changes in sediment and organic carbon accumulation in a highly-disturbed ecosystem: the Sacramento-San Joaquin River Delta California, USA. *Marine Pollution Bulletin*, 59: 154–63.

Chapman, V. J. 1960. *Salt Marshes and Salt Deserts of the World*. L. Hill, London.

Chung, C. H., Zhuo, R. Z., and Xu, G. W. 2004. Creation of Spartina plantations for reclaiming Dongtai, China, tidal flats and offshore sands. *Ecological Engineering*, 23: 135–150.

Craft, C. 2000. Co-development of wetland soils and benthic invertebrate communities following salt marsh creation. *Wetlands Ecology and Management*, 8: 197–207.

Craft, C., Broome, S., and Campbell, C. 2002. Fifteen years of vegetation and soil development after brackish-water marsh creation. *Restoration Ecology*, 10: 248–258.

Crain, C. M., Silliman, B. R., Bertness, S. L., and Bertness, M. D. 2004. Physical and biotic drivers of plant distribution across estuarine salinity gradients. *Ecology*, 85: 2539–2549.

Currin, C. A., Delano, P. C., and Valdes-Weaver, L. M. 2008. Utilization of a citizen monitoring protocol to assess the structure and function of natural and stabilized fringing salt marshes in North Carolina. *Wetlands Ecology Management*, 16: 97–118.

Dalrymple, R. W., Zaitlin, B. A., and Boyd, R. 1992. Estuarine facies models: conceptual basis and stratigraphic implications. *Journal of Sedimentary Petrology*, 62: 1130–1146.

Davis, C. A. 1910. Salt marsh formation near Boston and its geological significance. *Economic Geology*, 5: 623–639.

Davis, R. A., and Clifton, H. E. 1987. Sea-level change and the preservation potential of wave-dominated and tide-dominated coastal sequences. In: *Sea-level Fluctuation and Coastal Evolution.* Eds D. Nummedal, O. H. Pilkey Jr., and J. D. Howard., Special Publications of SEPM 41, Tulsa, pp. 167–178.

Day, J. W., Boesch, D. F., Clairain, E. J., Kemp, G. P., Laska, S. B., Mitsch, W. J., Orth, K., et al. 2007. Restoration of the Mississippi delta: lessons from Hurricanes Katrina and Rita. *Science*, 315: 1679–1684.

de Groot, A. V., Veeneklaas, R. M., and Bakker, J. P. 2011. Sand in the salt marsh: Contribution of high-energy conditions to salt-marsh accretion. *Marine Geology*, 282: 240–254.

Dijkema, K. S. 1997. Impact prognosis for salt marshes from subsidence by gas extraction in the Wadden Sea. *Journal of Coastal Research*, 13: 1294–1304.

Donnelly, J. P., Roll, S., Wengren, M., Butler, J., Lederer, R., and Webb, I. I. I. T. 2001. Sedimentary evidence of intense hurricane strikes from New Jersey. *Geology*, 29: 615–618.

Engelhart, S. E., Horton, B. P., and Kemp, A. C. 2011. Holocene sea level changes along the United States' Atlantic Coast. *Oceanography*, 24: 70–79.

Engels, J. G., and Jensen, K. 2010. Role of biotic interactions and physical factors in determining the distribution of marsh species along an estuarine salinity gradient. *Oikos*, 119: 679–685.

Engels, J. G., Rink, F., and Jensen, K. 2011. Stress tolerance and biotic interactions determine plant zonation patterns in estuarine marshes during seedling emergence and early establishment. *Journal of Ecology*, 99: 277–287.

Fagherazzi, S. 2013. The ephemeral life of a salt marsh. *Geology*, 41: 943–944.

Fagherazzi, S., Carniello, L., D'Alpaos, L., and Defina, A. 2006. Critical bifurcation of shallow microtidal landforms in tidal flats and salt marshes. *Proceedings of the National Academy of Sciences of the USA*, 103: 8337–8341.

Fagherazzi, S., Kirwan, M. L., Mudd, S. M., Guntenspergen, G. R., Temmerman, S., D'Alpaos, A., van de Koppel, et al. 2012. Numerical models of salt marsh evolution: Ecological, geomorphic, and climatic factors. *Reviews of Geophysics*, 50: RG1002.

Feagin, R. A., Martinez, M. L., Mendoza-Gonzalez, G., and Costanza, R. 2010. Salt marsh zonal migration and ecosystem service change in response to global sea level rise: a case study from an urban region. *Ecology and Society*, 15(4): 14.

Fisher, J. J. 1962. *Geomorphic Expression of Former Inlets along the Outer Banks of North Carolina*, University of North Carolina at Chapel Hill.

Flowers, T. J., and Colmer, T. D. 2008. Salinity tolerance in halophytes. *New Phytologist*, 179: 945–963.

Ford, M. A., Cahoon, D. R., and Lynch, J. C. 1999. Restoring marsh elevation in a rapidly subsiding salt marsh by thin-layer deposition of dredged material. *Ecological Engineering*, 12: 189–205.

Galloway, W. E. 1975. Process framework for describing the morphologic and stratigraphic evolution of deltaic depositional systems. In: *Deltas Models for Exploration*, Ed M. L. Broussard., Houston Geological Society, Houston, pp. 87–98.

Gardner, L. R., and Porter, D. E. 2001. Stratigraphy and geologic history of a southeastern salt marsh basin, North Inlet, South Carolina, USA. *Wetlands Ecology and Management*, 9: 371–385.

Gedan, K. B., Silliman, B. R., and Bertness, M. D. 2009. Centuries of human-driven change in salt marsh ecosystems. *Annual Review of Marine Science*, 1: 117–141.

Gehrels, R. W., Belknap, D. F., and Kelley, J. T. 1996. Integrated high-precision analyses of Holocene relative sea-level changes: lessons from the coast of Maine. *GSA Bulletin*, 108: 1073–1088.

Godfrey, P. J., and Godfrey, M. M. 1974. The role of overwash and inlet dynamics in the formation of salt marshes on North Carolina barrier islands. In: *Ecology of Halophytes.* Eds R. J. Reimold and W. H. Queen., Academic Press, Inc., New York, pp. 407–427.

Graham, S. A., and Mendelssohn, I. A. 2013. Functional assessment of differential sediment slurry applications in a deteriorating brackish marsh. *Ecological Engineering,* 51: 264–274.

Gunnell, J. R., Rodriguez, A. B., and McKee, B. A. 2013. How a marsh is built from the bottom up. *Geology,* 41: 859–862.

Jalowska, A. M., McKee, B. A., Laceby, J. P., and Rodriguez, A. B. 2017. Tracing the sources, fate, and recycling of fine sediments across a river-delta interface. *Catena,* 154: 95–106.

Jalowska, A. M., Rodriguez, A. B., and McKee, B. A. 2015. Responses of the Roanoke Bayhead Delta to variations in sea level rise and sediment supply during the Holocene and Anthropocene. *Anthropocene,* 9: 41–55.

James, L. A. 2013. Legacy sediment: definitions and processes of episodically produced anthropogenic sediment. *Anthropocene,* 2: 16–26.

Jervey, M. T. 1988. Quantitative geological modeling of siliciclastic rock sequences and their seismic expression. In: *Sea-Level Changes: An Integrated Approach.* Eds C. K. Wilgus, B. S. Hastings, C. A. Ross, H. W. Posamentier, J. C. Van Wagoner, and C. G. S. C. Kendall. *Special Publication 42,* SEPM, Tulsa, pp. 47–69.

Johnson, D. W. 1919. *Shore Processes and Shoreline Development.* John Wiley & Sons, Incorporated, Boston.

Kelley, J. T., Belknap, D. F., Jacobson, G. L., and Heather, A. J. 1988. The morphology and origin of salt marshes along the glaciated coastline of Maine, USA. *Journal of Coastal Research,* 4: 649–666.

Kemp, A. C., Horton, B. P., Corbett, D. R., Culver, S. J., Edwards, R. J., and van de Plassche, O. 2017. The relative utility of foraminifera and diatoms for reconstructing late Holocene sea-level change in North Carolina, USA. *Quaternary Research,* 71: 9–21.

Kennish, M. J. 2001. Coastal salt marsh systems in the U.S.: A review of anthropogenic impacts. *Journal of Coastal Research,* 17: 731–748.

Kirwan, M. L., Guntenspergen, G. R., D'Alpaos, A., Morris, J. T., Mudd, S. M., and Temmerman, S. 2010. Limits on the adaptability of coastal marshes to rising sea level. *Geophysical Research Letters,* 37: L23401.

Kirwan, M. L., and Megonigal, J. P. 2013. Tidal wetland stability in the face of human impacts and sea-level rise. *Nature,* 504: 53.

Kirwan, M. L., Walters, D. C., Reay, W. G., and Carr, J. A. 2016. Sea level driven marsh expansion in a coupled model of marsh erosion and migration. *Geophysical Research Letters,* 43: 4366–4373.

Komatsubara, J., Fujiwara, O., Takada, K., Sawai, Y., Aung, T. T., and Kamataki, T. 2008. Historical tsunamis and storms recorded in a coastal lowland, Shizuoka Prefecture, along the Pacific Coast of Japan. *Sedimentology,* 55: 1703–1716.

Kraft, J. C. 1971. Sedimentary facies patterns and geologic history of a Holocene marine transgression. *Geological Society of America Bulletin,* 82: 2131–2158.

Kraft, J. C., Yi, H. L., and Khalequzzaman, M. 1992. Geologic and human factors in the decline of the tidal salt marsh lithosome: the Delaware estuary and Atlantic coastal zone. *Sedimentary Geology,* 80: 233–246.

Leonardi, N., and Fagherazzi, S. 2015. Local variability in erosional resistance affects large scale morphodynamic response of salt marshes to wind waves and extreme events. *Geophysical Research Letters,* 42: 5872–5879.

Lotze, H. K., Lenihan, H. S., Bourque, B. J., Bradbury, R. H., Cooke, R. G., Kay, M. C., Kidwell, S. M., et al. 2006. Depletion, degradation, and recovery potential of estuaries and coastal seas. *Science*, 312: 1806–1809.

Marani, M., D'Alpaos, A., Lanzoni, S., Carniello, L., and Rinaldo, A. 2010. The importance of being coupled: stable states and catastrophic shifts in tidal biomorphodynamics. *Journal of Geophysical Research: Earth Surface*, 115: F04004, doi:10.1029/2009JF001600.

Mariotti, G., and Fagherazzi, S. 2013. Critical width of tidal flats triggers marsh collapse in the absence of sea-level rise. *Proceedings of the National Academy of Sciences of the USA*, 110: 5353–5356.

Mattheus, C. R., Rodriguez, A. B., and McKee, B. A. 2009. Direct connectivity between upstream and downstream promotes rapid response of lower coastal-plain rivers to land-use change. *Geophysical Research Letters*, 36: L20401, doi:10.1029/2009GL039995.

McKee, L. J., Ganju, N. K., and Schoellhamer, D. H. 2006. Estimates of suspended sediment entering San Francisco Bay from the Sacramento and San Joaquin Delta, San Francisco Bay, California. *Journal of Hydrology*, 323: 335–352.

McLeod, E., Chmura, G. L., Bouillon, S., Salm, R., Björk, M., Duarte, C. M., Lovelock, C. E., et al. 2011. A blueprint for blue carbon: toward an improved understanding of the role of vegetated coastal habitats in sequestering CO_2. *Frontiers in Ecology and the Environment*, 9: 552–560.

Möller, I., Kudella, M., Rupprecht, F., Spencer, T., Paul, M., van Wesenbeeck, B. K., Wolters, G., et al. 2014. Wave attenuation over coastal salt marshes under storm surge conditions. *Nature Geoscience*, 7: 727–731.

Morales, J. A. 1997. Evolution and facies architecture of the mesotidal Guadiana River delta S.W. Spain-Portugal. *Marine Geology*, 138: 127–148.

Morris, J. T., Sundareshwar, P. V., Nietch, C. T., Kjerfve, B., and Cahoon, D. R. 2002. Responses of coastal wetlands to rising sea level. *Ecology*, 83: 2869–2877.

Morton, R. A., Gelfenbaum, G., and Jaffe, B. E. 2007. Physical criteria for distinguishing sandy tsunami and storm deposits using modern examples. *Sedimentary Geology*, 200: 184–207.

Mudd, S. M., D'Alpaos, A., and Morris, J. T. 2010. How does vegetation affect sedimentation on tidal marshes? Investigating particle capture and hydrodynamic controls on biologically mediated sedimentation. *Journal of Geophysical Research*, 115: F03029, doi:10.1029/2009JF001566.

Neumeier, U., and Amos, C. L. 2006. The influence of vegetation on turbulence and flow velocities in European salt-marshes. *Sedimentology*, 53: 259–277.

Neumeier, U., and Ciavola, P. 2004. Flow resistance and associated sedimentary processes in a *Spartina maritima* salt-marsh. *Journal of Coastal Research*, 20: 435–447.

Nichols, M. M. 1989. Sediment accumulation rates and relative sea-level rise in lagoons. *Marine Geology*, 88: 201–219.

Odum, W. E. 1988. Comparative ecology of tidal freshwater and salt marshes. *Annual Review of Ecology and Systematics*, 19: 147–176.

Olariu, C., and Bhattacharya, J. P. 2006. Terminal distributary channels and delta front architecture of river-dominated delta systems. *Journal of Sedimentary Research*, 76: 212–233.

Olliver, E. A., and Edmonds, D. A. 2017. Defining the ecogeomorphic succession of land building for freshwater, intertidal wetlands in Wax Lake Delta, Louisiana. *Estuarine, Coastal and Shelf Science*, 196: 45–57.

Ouyang, X., and Lee, S. Y. 2014. Updated estimates of carbon accumulation rates in coastal marsh sediments. *Biogeosciences*, 11: 5057–5071.

Pendleton, L., Donato, D. C., Murray, B. C., Crooks, S., Jenkins, W. A., Sifleet, S., Craft, C., et al. 2012. Estimating global "blue carbon" emissions from conversion and degradation of vegetated coastal ecosystems. *PLoS One*, 7: e43542.

Penland, S., Boyd, R., and Suter, J. R. 1988. Transgressive depositional systems of the Mississippi Delta plain; a model for barrier shoreline and shelf sand development. *Journal of Sedimentary Research*, 58: 932–949.

Peterson, G. W., and Turner, R. E. 1994. The value of salt marsh edge vs. interior as a habitat for fish and decapod crustaceans in a Louisiana tidal marsh. *Estuaries* 17: 235–262.

Pethick, J. S. 1981. Long-term accretion rates on tidal salt marshes. *Journal of Sedimentary Research*, 51: 571–577.

Phleger, C. F. 1971. Effect of salinity on growth of a salt marsh grass. *Ecology*, 52: 908–911.

Raabe, E. A., and Stumpf, R. P. 2015. Expansion of tidal marsh in response to sea-level rise: Gulf Coast of Florida, USA. *Estuaries and Coasts*, 39: 145–157.

Redfield, A. C. 1965. Ontogeny of a salt marsh estuary. *Science*, 147: 50–55.

Reed, D. J. 2002. Sea-level rise and coastal marsh sustainability: geological and ecological factors in the Mississippi delta plain. *Geomorphology*, 48: 233–243.

Ridge, J. T., Rodriguez, A. B., and Fodrie, F. J. 2017. Salt marsh and fringing oyster reef transgression in a shallow temperate estuary: implications for restoration, conservation and blue carbon. *Estuaries and Coasts*, 40: 1013–1027.

Roberts, H. H. 1997. Dynamic changes of the Holocene Mississippi River Delta Plain: the delta cycle. *Journal of Coastal Research*, 13: 605–627.

Rodriguez, A. B., Anderson, J. B., Banfield, L. B., Taviani, M., Abdulah, K., and Snow, J. N. 2000. Identification of a −15m middle Wisconsin shoreline on the Texas inner continental shelf. *Palaeogeography, Palaeoclimatology, Palaeoecology*, 158: 25–43.

Rodriguez, A. B., Fodrie, F. J., Ridge, J. T., Lindquist, N. L., Theuerkauf, E. J., Coleman, S. E., et al. 2014. Oyster reefs can outpace sea-level rise. *Nature Climate Change*, 4: 493–497.

Rodriguez, A. B., Simms, A. R., and Anderson, J. B. 2010. Bay-head deltas across the northern Gulf of Mexico back step in response to the 8.2 ka cooling event. *Quaternary Science Reviews*, 29: 3983–3993.

Rogers, K., Wilton, K. M., and Saintilan, N. 2006. Vegetation change and surface elevation dynamics in estuarine wetlands of southeast Australia. *Estuarine, Coastal and Shelf Science*, 66: 559–569.

Saintilan, N., and Hashimoto, T. R. 1999. Mangrove-saltmarsh dynamics on a bay-head delta in the Hawkesbury River estuary, New South Wales, Australia. *Hydrobiologia*, 413: 95–102.

Saintilan, N., and Williams, R. 2010. Short Note: The decline of saltmarsh in southeast Australia: Results of recent surveys. *Wetlands Australia Journal*, 18: 49–54.

Schwimmer, R. A., and Pizzuto, J. E. 2000. A model for the evolution of marsh shorelines. *Journal of Sedimentary Research*, 70: 1026–1035.

Shennan, I., and Horton, B. 2002. Holocene land- and sea-level changes in Great Britain. *Journal of Quaternary Science*, 17: 511–526.

Shepard, C. C., Crain, C. M., and Beck, M. W. 2011. The protective role of coastal marshes: A systematic review and meta-analysis. *PLoS ONE*, 6: e27374.

Shideler, G. L. 1984. Suspended sediment responses in a wind-dominated estuary of the Texas Gulf Coast. *Journal of Sedimentary Petrology*, 54: 731–745.

Simms, A. R., and Rodriguez, A. B. 2014. Where do coastlines stabilize following rapid retreat? *Geophysical Research Letters*, 41: 1698–1703.

Simms, A. R., and Rodriguez, A. B. 2015. The Influence of valley morphology on the rate of Bayhead Delta Progradation. *Journal of Sedimentary Research*, 85: 38–44.

Simms, A. R., Rodriguez, A. B., and Anderson, J. B. 2018. Bayhead deltas and shorelines: Insights from modern and ancient examples. *Sedimentary Geology*, 374: 17–35.

Singh Chauhan, P. P. 2009. Autocyclic erosion in tidal marshes. *Geomorphology*, 110: 45–57.

Snow, A. A., and Vince, S. W. 1984. Plant Zonation in an Alaskan Salt Marsh: II. An experimental study of the role of edaphic conditions. *Journal of Ecology*, 72: 669–684.

Sousa, A. I., Lillebø, A. I., Caçador, I., and Pardal, M. A. 2008. Contribution of *Spartina maritima* to the reduction of eutrophication in estuarine systems. *Environmental Pollution*, 156: 628–635.

Stumpf, R. P. 1983. The process of sedimentation on the surface of a salt marsh. *Estuarine, Coastal and Shelf Science*, 17: 495–508.

Syvitski, J. P. M., Kettner, A. J., Overeem, I., Hutton, E. W. H., Hannon, M. T., Brakenridge, G. R., Day, J., et al. 2009. Sinking deltas due to human activities. *Nature Geoscience*, 2: 681–686.

Ta, T. K. O., Nguyen, V. L., Tateishi, M., Kobayashi, I., Saito, Y., and Nakamura, T. 2002. Sediment facies and Late Holocene progradation of the Mekong River Delta in Bentre Province, southern Vietnam: an example of evolution from a tide-dominated to a tide- and wave-dominated delta. *Sedimentary Geology*, 152: 313–325.

Theuerkauf, E. J., and Rodriguez, A. B. 2017. Placing barrier-island transgression in a blue-carbon context. *Earth's Future*, 5: 789–810.

Theuerkauf, E. J., Stephens, J. D., Ridge, J. T., Fodrie, F. J., and Rodriguez, A. B. 2015. Carbon export from fringing saltmarsh shoreline erosion overwhelms carbon storage across a critical width threshold. *Estuarine, Coastal and Shelf Science*, 164: 367–378.

Thomas, M. A., and Anderson, J. B. 1994. Sea-level controls on the facies architecture of the Trinity/Sabine incised-valley system, Texas continental shelf. In: *Incised-Valley Systems: Origin and Sedimentary Sequences*. Eds R. W. Dalrymple, R. Boyd, and B. A. Zaitlin., *SEPM, Special Publication 51*, SEPM, Tulsa, pp. 63–82.

Törnqvist, T. E., Gonzalez, J. L., Newsom, L., van der Borg, K., de Jong, A. F. M., and Kurnik, C. W. 2004. Deciphering Holocene sea-level history on the U.S. Gulf Coast: a high-resolution record from the Mississippi Delta. *Geological Society of America Bulletin*, 116: 1026–1039.

van de Plassche, O., van der Borg, K., and de Jong, A. F. M 1998. Sea level-climate correlation during the past 1400 yr. *Geology*, 26: 319–322.

Van der Wal, D., Wielemaker-Van den Dool, A., and Herman, P. M. J. 2008. Spatial patterns, rates and mechanisms of saltmarsh cycles Westerschelde, the Netherlands. *Estuarine, Coastal and Shelf Science*, 76: 357–368.

Van Eerden, M. R., Drent, R. H., Stahl, J., and Bakker, J. P. 2005. Connecting seas: western Palaearctic continental flyway for water birds in the perspective of changing land use and climate. *Global Change Biology*, 11: 894–908.

Warren, R. S., Fell, P. E., Rozsa, R., Brawley, A. H., Orsted, A. C., Olson, E. T., Swamy, V., and Niering, W. A. 2002. Salt marsh restoration in Connecticut: 20 years of science and management. *Restoration Ecology*, 10: 497–513.

Watson, E. B., and Byrne, R. 2013. Late Holocene marsh expansion in Southern San Francisco Bay, California. *Estuaries and Coasts*, 36: 643–653.

White, W. A., Morton, R. A., and Holmes, C. W. 2002. A comparison of factors controlling sedimentation rates and wetland loss in fluvial-deltaic sytems, Texas Gulf coast. *Geomorphology*, 44: 47–66.

Williams, K., Ewel, K. C., Stumpf, R. P., Putz, F. E., and Workman, T. W. 1999. Sea-level rise and coastal forest retreat on the west coast of Florida, USA. *Ecology*, 80: 2045–2063.

Williams, P. B., and Orr, M. K. 2002. Physical evolution of restored breached levee salt marshes in the San Francisco Bay Estuary. *Restoration Ecology*, 10: 527–542.

Xiao, D., Zhang, L., and Zhu, Z. 2010. The range expansion patterns of *Spartina alterniflora* on salt marshes in the Yangtze Estuary, China. *Estuarine, Coastal and Shelf Science*, 88: 99–104.

Yang, S. L., Li, H., Ysebaert, T., Bouma, T. J., Zhang, W. X., Wang, Y. Y., Li, P., et al. 2008. Spatial and temporal variations in sediment grain size in tidal wetlands, Yangtze Delta: on the role of physical and biotic controls. *Estuarine, Coastal and Shelf Science*, 77: 657–671.

Zhang, R. S., Shen, Y. M., Lu, L. Y., Yan, S. G., Wang, Y. H., Li, J. L., and Zhang, Z. L. 2004. Formation of *Spartina alterniflora* salt marshes on the coast of Jiangsu Province, China. *Ecological Engineering*, 23: 95–105.

4

Salt Marsh Hydrodynamics

ANDREA D'ALPAOS, ALVISE FINOTELLO, GUILLAUME C. H. GOODWIN,
AND SIMON M. MUDD

4.1 Introduction

Salt marshes occupy the intertidal zone and support rich ecosystems of salt-tolerant plants and other biota (Costanza et al. 1997; Mitsch and Gosselink, 2000). These ecosystems contain channel networks that dissect marsh platforms, just as terrestrial river networks dissect hillslopes. In contrast to upland landscapes, marsh platforms are very low relief, are inundated by tides, and the channels that dissect them experience bidirectional flows (D'Alpaos et al. 2005; Hughes, 2012; Coco et al. 2013). These conditions are also present in intertidal mudflats, yet marsh platforms sit at different elevations (Fagherazzi et al. 2006), have different characteristics of their channel networks (Rinaldo et al. 1999a, 1999b; Kleinhans et al. 2009), and different hydrodynamics (Fagherazzi et al. 2012). The fundamental difference is the presence of marsh vegetation which has a profound effect on flow within marsh canopies (Nepf, 2012).

The transition between mudflat and marsh is determined by the success of pioneer species, which are resilient to tidal flows once established. Pioneer plants require calm conditions to establish themselves on intertidal flats (Balke et al. 2011; Hu et al. 2015b). As a result, salt marshes are only found along low-energy coasts, where swell has been damped by reefs, sandbars, dunes, or barrier islands. Such environments have specific regional hydrodynamic conditions dominated by short period wind waves and tidal currents. Salt marsh ecosystems and their biogeomorphological features further add to these distinctive hydrodynamics. In this chapter, we explore how the unique attributes of salt marsh ecosystems are reflected in their hydrodynamic behavior.

The majority of marsh platforms lie between the mean water level and mean high tide (e.g., Allen, 2000; Friedrichs and Perry, 2001). The elevation of marsh platforms within this frame depends on sediment supply, rate of relative sea-level rise (driven by eustatism and local subsidence), tidal range, and marsh growth stage (Fig. 4.1). Channels that dissect marsh platforms lie at lower elevations, and mudflats lie at even lower elevations. Mudflats tend to be situated just above mean low tide (Allen, 2000; Defina et al. 2007; Marani et al. 2010). Although, the distinct elevations of marsh platforms, channels, and mudflats drive distinct hydrodynamic conditions, all of these settings face tidal forcing. The mudflat environment, which is frequently inundated, often experiences waves. These waves impinge on marsh edges and play a critical role on the expansion and/or contraction of the marsh platform (Marani et al. 2011; Bendoni et al. 2014; Leonardi et al. 2016b).

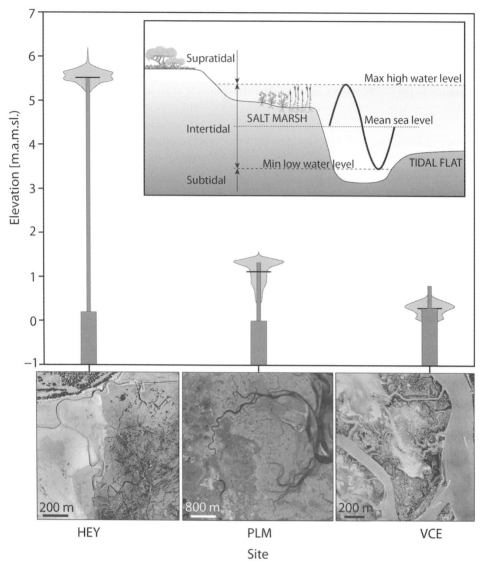

Figure 4.1 Elevation distribution of the marsh platforms (bulbous top) compared to Mean High Water (MWH, thin horizontal line) and Mean Sea Level (MSL, medium thick vertical line). Elevations are presented in metres above above mean sea level (m.a.m.s.l.) relative to the local datum. The sites included are are HEY: Heysham, UK; PLM: Plum Island, MA, USA; VCE: Venice (San Felice Island), Italy. Heysham and Plum Island marshes contain areas which are only rarely flooded, being above MHW. Marshes in Venice, on the other hand, are very frequently submerged, with parts below MSL being the process of disappearing. (Images for HEY and PLM: Google Earth). Inset: zonation of the tidal region (modified from Marani et al., 2010)

4.2 Salt Marsh Platforms and Channel Networks: an Intertwined System

Salt-marsh platforms are commonly dissected by networks of tidal channels (Fig. 4.2) (Fagherazzi et al. 1999; Rinaldo et al. 1999a, 1999b; D'Alpaos et al. 2005; Hughes, 2012) whose morphological complexity varies from simple linear channels to far more complex dendritic networks and meandering channels (Pye and French, 1993; Hughes, 2012) (Fig. 4.3). These channel networks exert a primary control on the ecomorphodynamic evolution of salt marsh platforms, facilitating the exchange of water, sediments, and nutrients with the adjacent tidal flats and the open sea (D'Alpaos et al. 2005; Coco et al. 2013).

Tidal channel networks can either form from previously unchannelized salt marsh platforms, or can be inherited from incipient channels dissecting tidal flats (Redfield, 1972; Marani et al. 2002; Stefanon et al. 2010). A common mechanism in the latter case is when the lateral edges of incipient channels are progressively colonized by halophytic vegetation (Fig. 4.4). When channels are inherited from mudflats, they tend to maintain their position in the tidal frame through aggradation and may also prograde horizontally in the estuarine landscapes (Pestrong, 1972; Temmerman et al. 2007; Schwarz et al. 2014; Belliard et al. 2015).

Conversely, channel development on marsh platforms is driven by concentration of flow, particularly during the ebb phase of tidal flows. Flow concentration is promoted by the presence of either small, random perturbations in the elevation of the marsh platform, or local differences in vegetation density. Flow concentration produces a local excess in bed shear stress that results in the progressive erosion of tidal channel tips. Positive feedbacks between flow concentration and channel incision result in headward erosion of tidal channels, allowing them to grow landward. As channels grow, their cross-sectional area changes to accommodate the local tidal prism (French and Stoddart, 1992; Beeftink and Rozema, 1993; Fagherazzi and Furbish, 2001; D'Alpaos et al. 2006; Temmerman et al. 2007; Hughes et al. 2009) (Fig. 4.5). This causes branching and meandering, resulting in an increase in the complexity of the channel network.

Both initial network incision and adaptation to the tidal prism are generally agreed to occur over short timescales if compared to the overall evolution of the marsh platform (Gabet, 1998; D'Alpaos et al. 2007; Stefanon et al. 2010). Once the channel network has formed, its overall structure only undergoes minor, slow modifications as long as the relative sea level remains stable (Stefanon et al. 2010). This slow evolution, however, can sometimes be interrupted by channel piracy or meander cutoff (Hughes, 2012; Brivio et al. 2016; D'Alpaos et al. 2017; Ghinassi et al. 2017; Finotello et al. 2018). Meander cutoff events drastically change both the drainage area and tidal prism flowing through a given channel cross section, thus leading to a modification of the overall network structure over annual to decadal timescales (D'Alpaos et al. 2005; Hughes et al. 2009; Ghinassi et al. 2018).

Channel evolution, both in time and space, depends on a number of factors, related to both the physical and biological features of the landscapes they cut through (e.g., marsh geometry, bank erodibility, bioturbation, presence and type of vegetation), as well as on

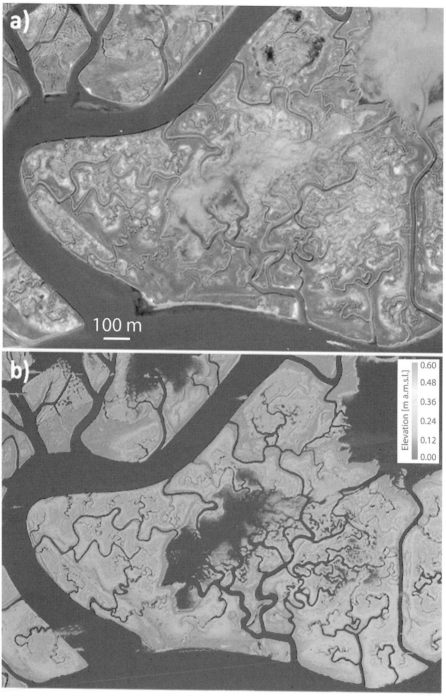

Figure 4.2 (a) Aerial photo and (b) color-coded elevation map derived from Lidar data (higher elevations in red, lower elevations in blue) of the San Felice salt marsh in the Venice lagoon, Italy (at approximately 45.79599°N, 1051253°E) showing tidal channels cutting through the marsh platform together with deposition patterns. (A black and white version of this figure will appear in some formats. For the color version, please refer to the plate section.)

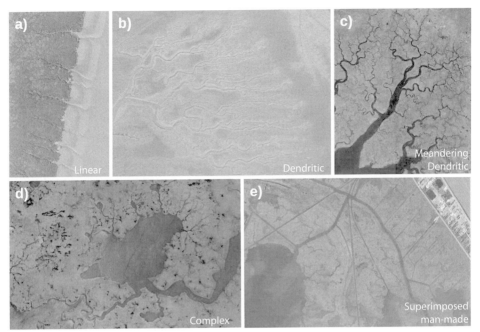

Figure 4.3 Examples of channel network types, according to the classification by Pye & French (1993). (a) Linear channels developing into linear dendritic channels in the Parrett Estuary, UK; (b) Dendritic network on the Wirral, UK; (c) meandering dendritic network east of the Suwannee River delta, FL; (d) Complex network in Biloxi marsh, LA. Such networks develop when tidal pools are connected to tidal creeks through creek incision or pool expansion; (e) Superimposed man-made network of linear channels near Port Sulphur, LA. Such features are generated when economic activities, such as gas pipelines, necessitate access. (Image source: ©GoogleEarth)

local hydrodynamics (e.g., water and solid discharges) (Garofalo, 1980; Gabet, 1998; Fagherazzi et al. 2004; Finotello et al. 2018). These interactions, along with the strong influence of bidirectional flow, suggest that tidal channel networks feature different scaling metrics compared to their fluvial cousins (Fagherazzi et al. 1999; Rinaldo et al. 1999a, 1999b; Marani et al. 2003). This is due to the complex interplay of the various processes involved in the network development and evolution, most of them acting at different overlapping spatial and temporal scales (Marani et al. 2002, 2003; Fagherazzi and Sun, 2004; Coco et al. 2013).

4.3 Hydrodynamics: Tidal Flows

Tidal flow over intertidal platforms can be described by the depth-averaged two-dimensional shallow water equations of momentum and continuity (e.g., Dronkers, 1964; Dronkers, 2016):

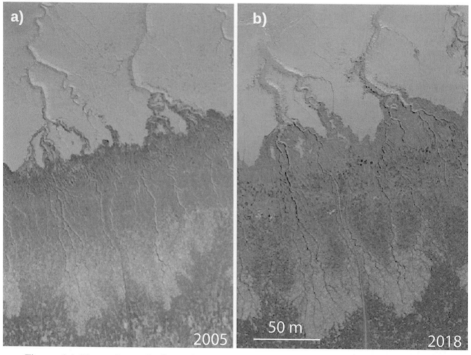

Figure 4.4 Vegetation colonizes rill channels in the mega-tidal Severn estuary, near Avonmouth, UK. On the high marsh, lighter vegetation patches expand as the drainage network develops landward. (a) Ortho-image taken on 06/07/2005; (b) Ortho-image taken on 05/03/2018. Image © Google Earth

$$\frac{\partial u}{\partial t} + \left(u \frac{\partial u}{\partial x} + v \frac{\partial u}{\partial y} \right) = -g \frac{\partial H}{\partial x} - g \frac{u \,|\boldsymbol{u}|}{\chi^2 D} \tag{4.1.a}$$

$$\frac{\partial v}{\partial t} + \left(u \frac{\partial v}{\partial x} + v \frac{\partial v}{\partial y} \right) = -g \frac{\partial H}{\partial y} - g \frac{v \,|\boldsymbol{v}|}{\chi^2 D} \tag{4.1.b}$$

$$\frac{\partial H}{\partial t} + \frac{\partial}{\partial x}(D\, u) + \frac{\partial}{\partial y}(D\, v) = 0 \tag{4.1.c}$$

where x and y are the horizontal coordinates in an orthogonal $\{x,y,z\}$ coordinate system where the vertical direction z is positive upward; u and v are the depth-averaged flow velocity components along the x and y directions, respectively; H represents the elevation of the water free surface, g is the gravitational acceleration, and D denotes the local flow depth given by $D = H - z_b$, where z_b is the local elevation of the intertidal platform.

Modeling salt marsh hydrodynamics and its temporal evolution in response to morphological changes, e.g., for long-term morphodynamic simulations, requires the formulation of simplified modeling frameworks retaining, however, the most relevant features of the governing processes. Toward this goal, Rinaldo et al. (1999a, 1999b) proposed a simplified

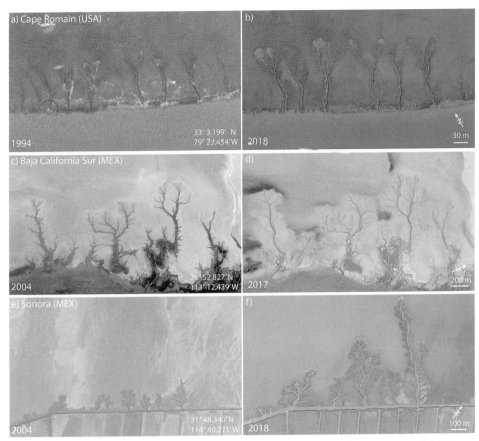

Figure 4.5 Examples of headward growth of tidal channel networks on vegetated (a,b), partially vegetated (c,d), and unvegetated intertidal platforms. (a,b) Salt marshes in Cape Romain, South Carolina (USA) – see also Hughes et al. (2009); (c,d) marsh platform in San Ignacio Lagoon, Baja California Sur (MEX); (e,f) tidal channels developing along abandoned shrimp farms, Colorado River Delta, Sonora (MEX). Image © Google Earth

description of the key hydrodynamic properties of the flow over an intertidal platform. The simplified model is based on the assumption that balance holds in the momentum equations between water surface slope and friction:

$$\frac{\partial h_1}{\partial x} = -\frac{\lambda}{D} u \quad ; \qquad \frac{\partial h_1}{\partial y} = -\frac{\lambda}{D} v \qquad (4.2)$$

where $h_1(x, t)$ is the local deviation of the water surface from its instantaneous spatially averaged value, $h_0(t)$, referenced to mean sea level (MSL), i.e., $H(x, t) = h_0(t) + h_1(x, t)$ (see Fig. 4.6a); D is the local flow depth over the marsh surface that can be written as $D(x, t) = h_0(t) + h_1(x,t) - z_b$, with z_b the local bottom elevation of unchanneled areas, referenced to the MSL; $\lambda = 8/(3\pi)U_0/\chi^2$ is a bottom friction coefficient which depends on Chézy's friction parameter, χ, and on a characteristic value of the maximum tidal current, U_0, assumed

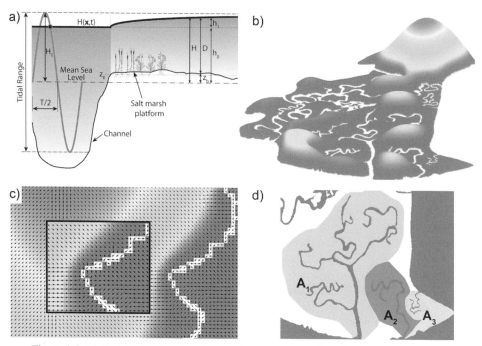

Figure 4.6 (a) Sketch of water elevation and bottom topography of a typical intertidal area according to the simplified hydrodynamic model developed by Rinaldo et al. (1999a). (b) Vertically exaggerated representation of the water surface elevation field $h_1(x, t)$ for a given instant during the ebb phase of the tide, derived by solving the Poisson boundary value problem (equation 4.3). (c) Flow directions over the marsh platform obtained, at any location, by considering the steepest descent direction of water surface topography. (d) Watershed delineation obtained of the basis of flow directions computed by using the Poisson model. Painted areas represent the watersheds related to the inlets of the three channel networks, and the portion of the domain in white is drained by the surrounding channels. Watershed divides are the loci of points characterized by local maxima of water surface elevations, in analogy with the definition of their fluvial counterparts as clearly emerges from panels (b) and (d).

to be constant over the considered marsh platform. Further assuming that the spatial variations of the instantaneous water surface elevations over the marsh, $h_1(\boldsymbol{x}, t)$, are significantly smaller than the instantaneous average water depths ($h_1 \ll D$); and that fluctuations of marsh platform elevations $z_b(\boldsymbol{x})$, around mean marsh elevation, z_0, are significantly smaller than the instantaneous average water depth ($z_b(\boldsymbol{x}) - z_0 \ll D$), the field of free surface elevations can be computed by solving the Poisson equation (see Rinaldo et al. 1999a; Marani et al. 2003:

$$\nabla^2 h_1 = \frac{\lambda}{(h_0 - z_0)^2} \frac{\partial h_0}{\partial t} \tag{4.3}$$

Assuming tidal propagation to be much faster within the channel network than over the shallow, friction-dominated marsh areas, i.e., considering local water surface elevations within the channel network as spatially independent, $H(x,t) = h_0(t)$, or a flat water level,

$h_l = 0$, within the network, allows one to determine the field of free surface elevations over the unchanneled marsh platform, at any instant t of the tidal cycle, by solving the Poisson boundary value problem (4.3) (Fig. 4.6b).

Differences in free surface elevation between the channel and the adjoining intertidal platform drive the flow field therein. A no-flux condition needs to be prescribed at the boundaries of the intertidal platform, aiming at representing either a physical boundary or a watershed divide (e.g., Rinaldo et al. 1999b; Marani et al. 2003; D'Alpaos et al. 2005). On the basis of the resulting water surface topography, flow directions can be obtained at any location on the intertidal areas by determining the steepest descent direction (Fig. 4.6c), and watersheds related to any channel cross section may be thus identified (Fig. 4.6d). The simplified Poisson model we have described applies, in principle, to relatively short tidal basins, that is, when the length of the basin is much smaller than the tidal wavelength (Lanzoni and Seminara, 1998). Nevertheless, as thoroughly discussed by Marani et al. (2003) by comparison with observations and complete hydrodynamic simulations, the Poisson model leads to quite robust estimates of drainage directions and watersheds, and, through the use of the continuity equation (Rinaldo et al. 1999b), of the landscape-forming discharges, even when the hypothesis of a short tidal basin is not strictly met.

In addition, we recall that the influence of friction on the momentum balance is taken into account in equations (4.1.a) and (4.1.b) by means of a Chézy friction law, being χ the Chézy coefficient for hydraulic roughness. Although the original model (Rinaldo et al. 1999a) was developed by considering a spatially constant friction coefficient, (Van Oyen et al. 2012, 2014) relaxed such an assumption and extended the Poisson model by accounting for the impact of spatial variations in vegetation cover and species, by modifying the spatial distribution of the friction coefficient. The latter can be further adjusted by considering the effect that the physical properties of vegetation species (e.g., stem density and height) colonizing the marsh platform have on χ, (e.g., Mudd et al. 2010) thus allowing one to reproduce more directly the impact of vegetation on the modeled flow field. Further information on the effect of vegetation on flow over marsh surfaces may be found in the following paragraphs. It is also worth recalling that Di Silvio et al. (2010) suggested to apply the simplified shallow water equations adopted in the Poisson model to a whole tidal basin (including the channels) by relaxing the assumption of a spatially uniform water depth, and that subsequently Mariotti (2018) applied a similar modeling framework by considering a spatially variable water depth to a mesotidal salt marsh system. Interestingly, Mariotti (2018)'s model predicted reasonably well flow velocities within the channels despite the assumption of negligible inertial effects was not strictly met.

While more complex, fully fledged finite element models could be employed to investigate the evolution of vegetated intertidal platforms (e.g., Carniello et al. 2011; Belliard et al. 2015), the use of a simplified approach such as the one presented here typically reduces computational effort and allows one to reproduce, over long temporal scales, the main ecogeomorphic characteristics of vegetated tidal marshes, and the feedbacks between morphology and hydrodynamics, through sediment transport, mediated by the presence of vegetation over intertidal platforms (D'Alpaos et al. 2007b; Van Oyen et al. 2012; D'Alpaos and Marani, 2016).

4.4 Relevant Indicators of Hydrodynamics and Morphology

Marsh hydrodynamics are closely linked to marsh morphology. The interplay between hydrodynamics and both channel and platform structure, all in the context of flows, settling, and erosion modulated by vegetation is a key characteristic of salt marsh landscapes. These interactions account for both the ecogeomorphic patterns (Da Lio et al. 2013; D'Alpaos & Marani, 2016) that develop over the unchanneled marsh portions and the plano-altimetric configuration of the channels that cut through the marsh platform. The tidal prism provides a relevant example of this strong link between hydrodynamics and morphology. The tidal prism, P, is defined as the volume of water that flows through a given cross section during half of a tidal cycle, i.e., during flood or ebb. The tidal prism can be computed as follows:

$$P = \int_0^{T/2} Q(t)dt \qquad (4.4)$$

where T is the tidal period and $Q(t)$ is the discharge flowing through the considered cross section. When tidal propagation effects can be neglected, the tidal prism can be computed as:

$$P = \int_A \left(h_{max}(x) - \max\{z(x), h_{min}(x)\} \right) dx \qquad (4.5)$$

where A is the drainage area of the considered tidal basin, $h_{max}(x)$ and $h_{min}(x)$ are the maximum and the minimum tidal levels reached during spring tides, respectively, and $max\{z(x), h_{min}(x)\}$ represents the maximum between local topographic elevation, $z(x)$, and $h_{min}(x)$. The tidal prism depends on the area of the marsh platform drained by the considered cross section, on marsh topography, and on the maximum elevation locally attained by the tide. It also depends on the geometry of the channels cutting through the marsh, since the volume of water contained in the network is also included in the computation (Fig. 4.7).

It has long been recognized that a power law relation holds between the tidal prism, P, and the channel cross-sectional area, Ω, for a large number of tidal systems in dynamic equilibrium (O'Brien, 1969; Jarrett, 1976; Marchi, 1990). More recently, Friedrichs (1995), Rinaldo et al. (1999b), Lanzoni and Seminara (2002), van der Wegen et al. (2008), and D'Alpaos et al. (2010) explored, in several tidal systems, the relationship between Ω (computed as channel cross-sectional area at mean sea level) and spring (i.e., maximum

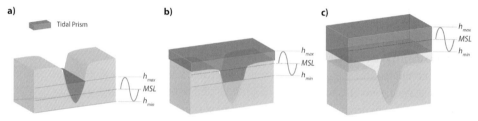

Figure 4.7 Cartoon depicting the tidal prism for different channels depending on the relative elevation of the channels and the intertidal elevations.

astronomical) peak discharge, Q_{max}. Those authors highlighted that an almost linear relationship between Ω and Q_{max} (which is directly related to the tidal prism) also exists for sheltered channel cross sections (D'Alpaos et al. 2009). This finding is in accordance with empirical observations (e.g., Myrick and Leopold, 1963; Nichols et al. 1991) suggesting a proportionality $\Omega \propto Q_{max}^q$ with the scaling coefficient q in the range 0.85–1.20. Such a relationship was theoretically explained by Friedrichs (1995) by relating the equilibrium cross-sectional geometry to the so-called stability shear stress, i.e., the total bottom shear stress just necessary to maintain a null along-channel gradient in net sediment transport. It is worth recalling that D'Alpaos et al. (2010) used field evidence and numerical model results to test the broad applicability of tidal prism cross-sectional area relations to sheltered cross sections within the Venice Lagoon. They found that values of the exponent α of the relation $\Omega = k\, P^{\alpha}$ meet the value $\alpha = 6/7$, empirically observed by O'Brien (1969) and theoretically derived by Marchi (1990). Therefore, D'Alpaos et al. (2010) suggested to term the tidal prism cross-sectional area relation the "O'Brien–Jarrett–Marchi" (OBJM) law (D'Alpaos et al. 2009).

An empirical relationship has also been shown to exist between channel cross-sectional area, Ω, and its drainage area, A, i.e., $\Omega \propto A$. This is essentially a generalization of the OBJM law in which one can consider the cross-sectional area, Ω, to be related to the drainage area, A, computed, e.g., according to the procedure outlined by Rinaldo et al. (1999a), instead of being related to either the tidal prism, P, or the maximum peak discharge, Q_{max}.

Another indicator that couples hydrodynamics and morphology is represented by the drainage density of tidal networks. Analyses carried out by Marani et al. (2003) emphasized that the classic morphological description of fluvial drainage networks (Horton, 1945; Strahler, 1957) fails to distinctively characterize the morphometric features of a given tidal network and the relationship with the salt marsh it wanders through. Echoing previous studies in fluvial geomorphology (Tucker et al. 2001), Marani et al. (2003) proposed a new approach to address the drainage density that relies on the statistics of the unchanneled flow lengths L (i.e., flow-path lengths connecting any unchanneled site over the marsh to the nearest channel, see Fig. 4.6c). Flow-path lengths are determined on the basis of drainage directions defined by the steepest descent of water surface topography. The latter can be computed, to a first-order of approximation, by neglecting both temporal changes in the average water level (h_0) and gradients of mean flow depth across the marsh plain, thus assuming the right-hand side term in Eq. (4.3) to be constant and considering the drainage directions as time invariant ($\nabla h_1(\mathbf{x}, t) = \nabla h_1(\mathbf{x})$, see Rinaldo et al. 1999a). While the classic Hortonian drainage density (total channelized length divided by the watershed area) is a measure of the degree of channelization of a given watershed and, therefore, it represents a poorly distinctive measure of how the catchment is dissected by the channel network, the drainage density determined as the inverse of the mean flow distance from any unchanneled site to the nearest tidal channel, indicates how efficiently the network feeds (drains) its watershed during flood (ebb) (Tucker et al. 2001; Marani et al. 2003).

Interestingly, by running laboratory experiments to unravel tidal network response to changes in the forcing rates of sea level rise, Stefanon et al. (2012) found that a linear relationship holds between the drainage density, *sensu* Marani et al. (2003), and the

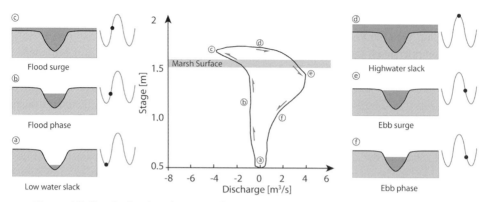

Figure 4.8 Sketch showing the stage–discharge relationship typically observed in a salt marsh channel (modified from Fagherazzi et al. 2013)

landscape-forming tidal prism. Although Stefanon et al.'s (2012) experiments were carried out over unvegetated platforms, the results are relevant to salt marsh landscapes because the linear relationship suggests that changes in tidal prism rapidly influence network efficiency in draining the platform and the related transport of water, sediments, and nutrients.

The functional relationship between salt marsh morphology and its hydrodynamic response in terms of flow velocities (and discharges) is still poorly characterized if compared to fluvial systems (Fagherazzi et al. 2008). Early studies on salt marsh channels highlighted the presence of a hysteresis in the stage-discharge (or stage-velocity) relationship, with different values of discharge attained, for the same stage value, during the ebb and the flood phases of the tidal cycle (Myrick and Leopold, 1963; Bayliss-Smith et al. 1979; French and Stoddart, 1992; Pethick, 1992).

The stage-discharge (stage-velocity) relationship in salt-marsh creek systems is characterized by the presence of a discharge (velocity) peak, also known as "surge," during both the ebb and the flood phases (Fig. 4.8). The presence of this peak is determined by the storage capacity of the salt marsh area (Boon, 1975), which can be expressed by means of the basin hypsometric curve $S(h)$, representing the surface S inundated at a given tidal stage h. According to Boon's (1975) theory, the discharge flowing through the mouth of a channel can be expressed as:

$$Q = S(h)\frac{dh}{dt} \tag{4.6}$$

with asymmetries in flow velocities between the ebb and the flood phases determined by overtides generated in shallow water creek system (Pethick, 1980). The position of discharge (velocity) surges also depends on branching character of the channel network (Pethick, 1980).

Different flow velocities between the channels and the marsh platform, due to different flow resistance, strongly influence the distribution of travel times, leading to a characteristic stage–discharge relationship with the flood and ebb discharge peaks located slightly above and below the marsh surface elevation, respectively (Fagherazzi et al. 2008). Tidal asymmetry, together with different velocity maxima observed during the ebb and flood phases, are persistent characteristics of channel flow in salt marsh creeks and exert a fundamental control on the net sediment transport therein (Boon, 1975). Such a control has a strong impact on both sedimentary structure (Tambroni et al. 2017) and the long-term morphodynamics (Boon, 1975) of salt marsh systems.

In wave-dominated environments typical of the North American Atlantic coast, wave power has been shown to exert an influence on the shape of the marsh edge, with high wave powers leading to more jagged outlines than with low wave powers (Leonardi and Fagherazzi, 2014). In particular, Leonardi et al. (2016a) demonstrated that distinct fractal dimensions and outline irregularities characterize salt marsh edges retreating at different rates, thus ultimately reflecting the incoming wave power density (Leonardi et al. 2016b).

Finally, on the basis of the water surface elevation field computed, e.g., by solving the Poisson boundary value problem (3), one can determine the distribution of bottom shear stresses due to tidal currents at every point $\mathbf{x}$ on the marsh platform as follows:

$$\tau = -\gamma \cdot D \cdot \nabla h_1 \qquad (4.7)$$

where $\tau(\mathbf{x}; t)$ is the local value of the bottom shear stress, γ is the specific weight of water, and D is the local water depth. Fig. 4.9 portrays the results of the analysis of the spatial distribution of the local bottom shear stress for a case study in the Venice Lagoon (see Figs. 4.2 and 4.6). We observe that marsh portions close to the tips of the channel network and to pronounced channel bends display higher values of the shear stress. The probability

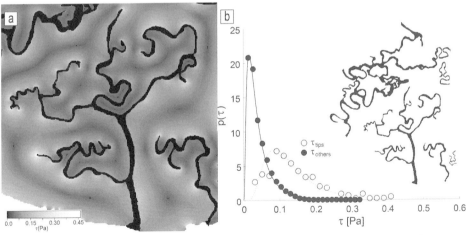

Figure 4.9 (a) Spatial distribution of the bottom shear stresses over salt marsh platforms adjacent to the network dissecting the southern part of the San Felice salt marsh in the Venice Lagoon. The higher values of the shear stress are observed to occur at the tips of the channel network and in correspondence of pronounced channel bends. (b) Probability density functions of the bottom shear stress, $p(\tau)$, both at the tips (τ_{tips}) and in the remaining part of the sites adjacent to the tidal network (τ_{others}). The mean bottom shear stress value acting at the tips of the network for the investigated zone is 0.12Pa, much larger than the shear stress acting on other points.

density functions of bottom shear stresses at the channel network heads, τ_{tips}, and at all other adjacent sites with the exception of the tips, τ_{others}, confirms previous observations (D'Alpaos et al. 2005; and Fig. 4.9b), furthermore supporting the speculation that head-ward erosion and tributary addition (possibly originating at sites where the stress increases along bends) are the main processes driving early channel development (Pethick, 1969; Steel and Pye, 1997; Allen, 2000; D'Alpaos et al. 2007a; Hughes et al. 2009). Erosional activities can thus be primarily expected in those parts of the basin where the local value of the bottom shear stress $\tau(\mathbf{x})$ is larger than a threshold value for erosion, τ_{CR} (Rinaldo et al. 1993, 1995; Rigon et al. 1994). This suggests that channel headward growth, driven by the spatial distribution of local shear stress, is the chief land-forming agent for network formation on marsh platforms (e.g., Hughes et al. 2009). When embedded in tidal network evolution models (Fagherazzi and Sun, 2004; D'Alpaos et al. 2005) this assumption was found to produce reasonable structures of tidal drainage densities and associated features within tidal landscapes (D'Alpaos et al. 2005, 2007b).

4.5 Mutual Interactions between Hydrodynamics and Vegetation

We now move onto the marsh platform, where vegetation plays a key role in determining how water floods the platform during the rising tide, and how this water ebbs on the falling tide. The presence of this vegetation means that the hydrodynamics of the platform differs substantially from tidal channels and mudflats, and one could argue that beyond the tidal forcing, vegetation is the dominant driver of hydrodynamics on the platform. Marsh vegetation exhibits a wide variety of both stem and leaf morphologies, from the broadly cylindrical *Juncus roemerianus*, the cordgrass *Spartina alterniflora*, and the more complex morphologies of vegetation like *Salicornia europaea* and various *Limonium* species. These all play a role in the flow of water over the marsh surface through both drag, which slows flow within the vegetation canopy, and also via turbu-lence (Leonard and Luther, 1995), which not only plays an important role in dispersing seeds, sediment, nutrients, and pollutants (Nepf, 1999), but also dissipates energy (Chen et al. 2016). The drag and turbulent energy dissipation on marsh platforms lead to reduced flow velocities and, in addition, can substantially alter the spatial distribution of velocities while the platform is flooded.

4.5.1 Drag, Turbulence, and Marsh Vegetation

The drag forces exerted by marsh plants slow flow considerably compared to the velocities in marsh creeks; in some macrotidal marshes peak flow velocities can sometimes exceed 0.4m/s but typically peak flow velocities are on the order of 0.1 m/s (Leonard and Luther, 1995; Yang, 1998; Shi et al. 2000; Temmerman et al. 2005; Leonard and Croft, 2006; Neumeier and Amos, 2006; Torres and Styles, 2007). These low velocities mean that, unlike in tidal creeks, inertia is insignificant (Rinaldo et al. 1999a) and water flowing over the marsh surface can be adequately described using the balance between gravitational forces (driven by astronomic and wind-driven tides) and drag forces (Nepf, 1999):

$$(1 - a\,d)C_b U^2 + \frac{1}{2} C_d a d \left(\frac{h}{d}\right) U^2 = g h \frac{\partial h}{\partial x} \tag{4.8}$$

where g (dimensions length per time, dimensions hence force denoted as [M]ass, [L]ength, and [T]ime) is gravitational acceleration, C_b [dimensionless] is the coefficient of bed drag, C_d [dimensionless] is the bulk drag coefficient of the vegetation, h [L] is the elevation of the water surface, a [L^{-1}] is the projected area of plant stems per unit area, d [L] is the stem diameter, and U [L T^{-1}] is the flow velocity. In analogy with the Poisson hydrodynamic model (equation (4.3); Rinaldo et al. 1999a) described above, this formulation assumes that a balance holds between water surface slope and friction.

Tanino and Nepf (2008) performed laboratory experiments using cylinders and found that under such conditions the drag coefficient can be described by:

$$C_d = 2\left(\frac{\alpha_0}{Re_c} + \alpha_1\right) \tag{4.9}$$

where Re_c is the Reynolds number based on cylinder diameter ($Re_c = U\,d/\nu$ where ν [L^2 T^{-1}] is the kinematic viscosity of water), α_0 and α_1 are empirical coefficients. The first empirical coefficient, α_0, was determined to be a constant, whereas the second, α_1, was a function of the properties of the solid fraction of the cylinders (Tanino and Nepf, 2008).

The drag equations suggested by Tanino and Nepf (2009) are most appropriate for application to vegetation types that can be approximated as cylinders, e.g., *J. roemerianus*. Both terms in equation (4.9) are modulated by the properties of marsh vegetation. The properties that play a role are the projected area per unit volume of stems and leaves and the stem diameter.

As mentioned previously, a number of vegetation types do not lend themselves to approximation as cylinders (e.g., *Salicornia virginica, Limonium serotinum*). However, equation (4.9), and other similar equations (e.g., Baptist et al. 2007), have been applied to ecosystems with upright grasses such as *S. alterniflora* (e.g., Mudd et al. 2010) and *Spartina anglica* (e.g., Temmerman et al. 2007); consistency between field measurements and models based on equation (4.9) suggests that use of that equation is reasonable for such vegetation types.

The relative influence of velocity on the drag force exerted by plant leaves and stems is a function of the flexibility of the vegetation. Emergent vegetation, such as that which lives on marsh platforms, tends to have rounder stems than submerged vegetation, such as seagrass, and the rounder cross sections lead to increased stiffness (Nepf, 2012). Nepf (2012) suggested that the drag force can be described as proportional to velocity to the $2 + E$ power, where E is a function of the rigidity of the vegetation: $E \sim 0$ for rigid vegetation whereas $E \sim -2$ for flexible vegetation.

Given the importance of plant characteristics in determining drag forces on the marsh platform as well as turbulent energy with the flows, Mudd et al. (2004, 2010) attempted to quantify how marsh biomass could be used as a predictor for the plant characteristics most important to drag: projected plant area per unit volume and stem diameter. In marshes in South Carolina, where detailed biomass and plant characteristic measurements have been carried out for over several decades as part of a Long Term Ecological Research (LTER) site (Morris et al. 2013), Mudd et al. (2004, 2010) found that projected plant area per unit volume and stem diameter could be approximated by a power law function of biomass.

In addition to stem diameter, stem density is important in determining the characteristics of turbulence on the marsh surface. Turbulent kinetic energy was first measured in marshes in situ by Leonard and Luther (1995), and subsequent laboratory (Nepf, 1999; Folkard, 2011) and field studies (Neumeier and Amos, 2006) have demonstrated that aboveground plant biomass such as leaves and stems help set the length scale for turbulent eddies. This length scale is the smaller of either the stem diameter or the average spacing between stems (Tanino and Nepf, 2008), in all but the densest plant canopies stem diameter is likely to drive the length scale of turbulent eddies. Although the bed of the marsh platform does generate turbulence, the energy from bed-derived turbulence is dwarfed by that derived from plant–flow interactions (López and García, 2001).

The structure of the plant canopy may also play a role in estimating drag forces: Silinski et al. (2015) studied different canopy structures (young *Scirpus maritimus*, adult *S. maritimus,* and adult *S. tabernaemontani*) and found that at shallow depths and for long-period waves, drag forces depart from predictions, particularly for plants with more complex canopies.

In an emergent plant canopy, according to Nepf (1999) the turbulent kinetic energy generated per unit mass of water is:

$$k = \alpha_k^2 U^2 (C_d a \, d_c)^{\frac{2}{3}}$$
(4.10)

where α_k is an empirical coefficient reported to be 0.9 derived from laboratory experiments, and $a{\cdot}d_c$ is the plant population density. This equation applies to flows where form drag dominates, which is the case in rigid canopies. In flexible vegetation, skin drag plays an increasing role and equation (4.10) will overestimate the turbulent kinetic energy in the canopy (Nikora and Nikora, 2007).

Furthermore, vegetation may affect the direction of flow: Chen et al. (2016) adapted the work of Hu and Chen (2005) to show that wave refraction induces flow rotation when waves propagate from a mudflat to a salt marsh, according to Snell's law:

$$\frac{\sin(\theta_1)}{\sin(\theta_2)} = \frac{C_1}{C_2} = \frac{n_1}{n_2}$$
(4.11)

where θ is the flow direction, $C = (g*h)^{0.5}$ is the phase celerity and n is the refractive index for the mudflat (n_1) and the salt marsh (n_2). For a marsh of dense *S. alterniflora* with a stem height of 1.5–2.0 m, the refractive index of the marsh is 1.2 compared to 1.0 for the tidal flat, causing wave refraction, assuming the elevation of the marsh and mudflat are the same.

4.5.2 Patches, Patterns, and Hydrodynamics

In some marsh platforms, vegetation does not form a continuous cover but instead occurs as discrete patches, occasionally forming patterned ground (Rietkerk and van de Koppel, 2008). This is most common where mudflats are being converted to marsh platforms via

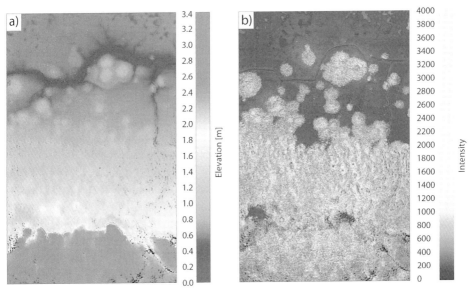

Figure 4.10 Morphological progression of a salt marsh platform in the Solway Firth, UK. In the top section, discontinuous vegetation patches have established on the intertidal mudflat and are anchoring the position of tidal creeks. In the early stages of marsh formation, vegetated patches have little influence on the mudflat topography. In the middle section, the low marsh has formed a continuous platform, although residual pools and channels are still visible. In the lower section, the high marsh sits approximately 1 m above the high marsh and is delineated by a subvertical scarp. At the foot of the scarp, erosion pools are visible. (a) Elevation map obtained by Terrestrial Laser Scanner (TLS) on 01/06/2017; (b) Intensity map obtained by TLS on 01/06/2017. High intensity values correspond to vegetated areas. (A black and white version of this figure will appear in some formats. For the color version, please refer to the plate section.)

colonization of pioneer vegetation (Fig. 4.10). The topography of the patches at a given time, which may resemble a dome, an elevated plateau or follow the topography of the tidal flat, is a good proxy of the ongoing regional morphodynamic processes and represents a suitable indicator for the possible future expansion of vegetation (Balke et al. 2012). Because the formation and coalescence of vegetated patches is important in the conversion of mudflats to marshes, a number of authors have explored the feedbacks between flow around and through these patches and the influence of that flow on the potential for further colonization and plant recruitment (Temmerman et al. 2007).

The most obvious consequence of patches is to divert flow. The diversion of flow causes acceleration around the patch (Vandenbruwaene et al. 2011). The spatially heterogeneous flow field that results can affect plant productivity; van Wesenbeeck et al. (2008) found, using both experiments and field measurements, that *S. anglica* was more productive within vegetation patches than in interstitial gullies.

Because vegetation patches slow flow in a spatially heterogeneous manner, they create shear that leads to turbulence. White and Nepf (2007) performed experiments on arrays of rigid cylinders to investigate the generation of turbulence amongst patches:

they found a two-layer structure of shear where near the array of cylinders there is a rapidly changing shear layer, the thickness of which is set by bed friction and the water depth. This layers sets the scale of shedding vortices, which can sweep fluid into the patches as they pass (White and Nepf, 2007), creating a system by which sediment, nutrients, and pollutants can be exchanged between the patch and the unobstructed flow around the patch.

One of the most challenging aspects of predicting flow across vegetated marsh platforms is correctly quantifying flow parameters such as drag coefficients. Equation (4.8) can be used to back calculate drag coefficients, using water surface slopes, but in general the water surface slope on marshes is very low (e.g., Leopold et al. 1993) and it may be difficult to get accurate measurements. In consequence, water surface slope and drag coefficient are not typically reported in studies of marsh hydrodynamics. Flow velocities through marsh canopies and between patches are regularly collected, however, and the velocity differences can be used to calculate drag coefficients. White and Nepf (2008) found that given the difference between flow velocities the drag coefficient, C_d, can be estimated with:

$$C_d = c_f U_c^2 \left(h \, a \, U_v^2 \right)^{-1}$$

(4.12)

where c_f [dimensionless] is the bed friction coefficient (this can be estimated from bed particle size), h [L] is the flow depth in the channel, and U_c and U_v [L T^{-1}] are flow velocities in the area between the patches and in the vegetated patch, respectively.

Patches influence the turbulent intensity and as flow converges again behind patches this can lead to sedimentation (Fig. 4.11). Experimental studies suggest a peak in turbulent energy directly behind patches and one further downstream that scales with the drag coefficient, the diameter of the patch, and the projected area per unit volume of the patch (Chen et al. 2012). The region between these peaks is typically where sediment is deposited (Chen et al. 2012).

Conversion of mudflats to vegetated marshes occurs frequently, and this conversion results in sedimentation patterns that are not seen in fully vegetated marsh surfaces. In many cases, pioneer species can colonize mudflats when seedlings stay dry for a particular period, often called the "window of opportunity" (Balke et al. 2011; Hu et al. 2015b). These tussocks can then reduce flow velocities and encourage sedimentation (Fig. 4.11). Initial tussocks can spread and make way for continuous meadows (e.g., Callaway and Josselyn, 1992).

Before continuous meadows form, the marsh surface undergoes an evolution that relates to changing hydrodynamics brought about by patches. Flow resistance within patches diverts flow between them, which can scour channels (e.g., Temmerman et al. 2005). In addition to eroding sediment from the channels (Temmerman et al. 2007), flow concentration can also inhibit settling or germination of seeds, and can create a hostile environment for seedlings (Bouma et al. 2009; Hu et al. 2015b). A stabilizing feedback is that as the marsh becomes more channelized, it becomes better drained thus reducing flow concentration and, therefore, allowing establishment of marsh vegetation in previously hostile channels or areas of flow concentration (Temmerman et al. 2005). This feedback

Figure 4.11 Drag tails downstream of *Suaeda maritima* individuals in the Mont-Saint-Michel bay, France, showing deposition behind the plants. Flow direction is from top left to bottom right.

sets the scale of channel spacing and is sufficiently strong to mean that densely spaced channels on mudflats may evolve into a marsh surface with a much lower drainage density (e.g., Schwarz et al. 2014).

4.5.3 Storm Surge, Wave Attenuation, and Marsh Vegetation

Salt marshes play an important role in defending coastlines against storm surges (large-scale rise in water levels associated with high winds pushing seawater toward the coast and with low pressure weather systems) and waves (e.g., Wamsley et al. 2010; Shepard et al. 2011). In fact, authors have proposed that marsh conservation can play a large role in defending coastal and estuarine cities against future sea-level rise due to their ability to damp storm-driven increases in water level (Temmerman et al. 2013). Coastlines vulnerable to damaging storms, for example along the Atlantic and Gulf coasts of the USA, rely on protection provided by marshes to reduce damage along heavily populated coastlines (e.g., Gedan et al. 2011).

The effects of vegetative damping can be substantial. For example, during Hurricane Rita in the Gulf Coast of the USA, vegetated marshes damped storm surge by 1 m for every

4 km of lateral distance (Wamsley et al. 2010). However, this damping is not a panacea as it depends on the characteristic vegetation height and spatial density, as well as on both the geometry of the wetland and the intensity of storm-winds (Wamsley et al. 2010; Temmerman et al. 2013; Hu et al. 2015a).

Numerical modeling has predicted storm surge attenuation as well; a study mimicking conditions along the Scheldt Estuary in the Netherlands and in Belgium found that over a 20-km long estuary, marsh vegetation could attenuate a 2.5-m high storm surge at a rate of 0.26 m/km, and that storm surges could still be significantly attenuated even if up to 50% of existing marshes would die off (Temmerman et al. 2012).

Significant wave height can also be reduced if waves propagate over the marsh surface, as does the total energy of the waves (Möller et al. 1999). This effect strongly depends on water depth and vegetation type (Rupprecht et al. 2017), with tall canopy vegetation providing better attenuation until a critical orbital velocity is reached (e.g., 0.42 m/s for *Elymus athericus*), above which stems of tall vegetation are at risk of breaking and thus limiting their ability to attenuate wave energy.

4.5.4 Vegetation and Flow in Tidal Channels

Tidal channels tend to be devoid of vegetation but this does not mean that flows within the channels are not influenced by plants. Both the geometry and flow characteristics within tidal channels are modulated by plants on the marsh platform. As we have seen in earlier sections, the tidal prism drives flow through the tidal channels. However, the timing of this water entering and leaving the tidal watershed is heavily influenced by vegetation. Up to 50% of the tidal prism flows on the marsh surface, with the remainder of flow moving through the tidal channels, and this is particularly evident in macrotidal settings (e.g., Temmerman et al. 2005). As we have seen, drag forces due to plants can substantially slow the flow on the marsh platform, thus leading to flow concentration within the channels, while water depth is smaller than canopy height (Temmerman et al. 2005). At larger depths, sheet flow develops, decreasing the relative importance of channel flow (Vandenbruwaene et al. 2013). There are some feedbacks between this slow flow on the platform and the nature of the tidal prism because the slowing of flow encourages sediment trapping (see Section 4.5.5). The greater the trapping of sediment, the higher the elevation of the marsh platform, which can reduce the tidal prism. This in turn leads to less flow through tidal channels, which will make them shallower (D'Alpaos et al. 2006). The density of tidal channels is also influenced by vegetation (and by the tidal prism as we have highlighted in previous sections). In the Scheldt estuary, marshes that have developed from mudflats with initially low drainage density are characterized by a much greater density of channels once the vegetated platform is established (Vandenbruwaene et al. 2013). Furthermore, the presence of vegetation has been shown to increase the network efficiency (Kearney and Fagherazzi, 2016), where efficiency is defined as the ratio between the Hortonian length ℓ_H (i.e., the inverse of Hortonian drainage density, Horton, 1945) and the mean unchanneled path length ℓ_m (Marani et al. 2003). However, the presence of branching and meandering tidal networks over unvegetated tidal mudflats and the high variability of the mean unchanneled lengths not only from one salt marsh to another in the same tidal system (e.g., Marani et al. 2003) but also within the same salt marsh and within distances of a few hundred meters calls for more detailed investigations on the effects of vegetation and sediment cohesion on network form and function.

4.5.5 Wave Dynamics and Sediment Remobilization

Waves propagating in shallow water generate shear stress on the bed, often calculated using the formulas of Soulsby (1997) and Soulsby and Clarke (2005) (Callaghan et al. 2010; Townend, 2010; Carniello et al. 2012). If the bottom shear stress τ exceeds a critical value τ_c, sediment will be resuspended and made available for transport. τ_c is strongly influenced by grain size and composition (Ahmad et al. 2011), compaction (Salehi and Strom, 2012), vegetation cover, and the presence of biofilm (Amos et al. 2004; Valentine et al. 2014), and may vary seasonally but also with depth (Amos et al. 2010). Tidal flats are strongly affected by wave-induced scouring, and the eroded sediment provides material for sedimentation, usually on the salt marsh platform. On marsh platforms, where vegetation plays a stronger role in sediment biostabilization (Julian and Torres, 2006), scour is much less common. This, in addition to turbulence generation and sediment capture by canopies, contributes to the high average deposition rates on marsh platforms. In time, this process reduces the inundation depth on the marsh platform, further accentuating the role of friction on salt marsh flows. In drowning marshes, it has been argued that waterlogging promotes the formation of pools (Hartig et al. 2002), thus allowing scour to occur on the platform.

Wave action is, however, not limited to the seabed: by breaking against the marsh scarp, waves cause blocks to topple onto the tidal flat (Bendoni et al. 2014, 2016). Fallen blocks are then eroded and contribute to the regeneration of the tidal flat (Francalanci et al. 2013). The impact of waves on the marsh scarp strongly depends on water depth (Tonelli et al. 2010). Indeed, Marani et al. (2011) showed, on the basis of a theoretical framework based on dimensional analyses tested through field observations and wind–wave modeling, that the rate of marsh edge erosion is linearly related to the incident wave power density. Because the incident wave power density is a quadratic function of wave height, that in turns depends on wind velocity, water depths, and fetch, an increase in one of the above recalled driving factors leads to nonlinear increases in wave power and therefore in marsh lateral retreat (Marani et al. 2011; Leonardi et al. 2016a, 2016b).

4.6 Conclusions

Salt marshes exhibit unique hydrodynamics due to a combination of factors not found elsewhere. These environments feature bidirectional flows, wetting, and drying on a daily timescale, slack water occurring approximately at high and low tide, and maximum ebb (flood) velocities occurring at mid-tide on the falling (rising) tide. Such a peculiar hydrodynamic behavior leads to a hysteresis in the stage–velocity relationship within in salt-marsh channels, where, for the same stage value, flow velocity displays different magnitudes during the ebb and flood phases of the tidal cycle. Because salt-marsh plat-forms and tidal channels cutting through them form an intertwined system, the overall marsh hydrodynamics are strongly influenced by salt marsh elevation and by the ecogeo-morphic patterns that characterize these systems. Indeed, marsh surface vegetation, and the related ecogeomorphic zonation patterns, lead to both drag and turbulence, and vegetative drag interacts with bidirectional flow to create velocity asymmetries during flood and ebb tides. In addition, these densely vegetated surfaces are frequently dissected by a dense network of channels that lacks the scale invariance characteristics typically observed in fluvial patterns. Tidal networks are shaped by numerous physical and biological processes acting at overlapping spatial and temporal scales, which hinder these patterns to develop scale-invariant features. Several features of the tidal channels, in fact, display strong variations in space, unheard of in fluvial morphology, with the spatial scale of persistence of geomorphic features being often limited to a few tens of meters. This further increases the peculiarity of salt-marsh hydrodynamics, and further challenges scientists to unravel the role of the various processes responsible for the development of the distinctive and beautiful ecogeomorphic patterns displayed by salt-marsh systems, whose ecological and socioeconomic importance can hardly be overestimated.

References

Ahmad, M. F., Dong, P., Mamat, M., Nik, W. B. W., and Mohd, M. H. 2011. The critical shear stresses for sand and mud mixture. *Applied Mathematical Sciences*, **5**: 53–71.

Allen, J. R. L. 2000. Morphodynamics of Holocene salt marshes: A review sketch from the Atlantic and Southern North Sea coasts of Europe. *Quaternary Science Reviews*, **19**: 1155–1231.

Amos, C. L., Bergamasco, A., Umgiesser, G., Cappucci, S., Cloutier, D., Denat, L., Flind, M., Bonardi, M., and Cristante, S. 2004. The stability of tidal flats in Venice Lagoon–the results of in-situ measurements using two benthic, annular flumes. *Journal of Marine Systems*, **51**: 211–241.

Amos, C. L., Umgiesser, G., Ferrarin, C., Thompson, C. E. L. C. , Whitehouse, R. J. S., Sutherland, T. F., and Bergamasco A. 2010. The erosion rates of cohesive sediments in Venice lagoon, Italy. *Continental Shelf Research*, **30**: 859–870.

Balke, T., Bouma, T. J., Horstman, E. M., Webb, E. L., Erftemeijer, P. L. A., and Herman, P. M. J. 2011. Windows of opportunity: thresholds to mangrove seedling establishment on tidal flats. *Marine Ecology Progress Series*, **440**: 1–9.

Balke, T., Klaassen, P. C., Garbutt, A., Van der Wal, D., Herman, P. M. J., and Bouma, T. J. 2012. Conditional outcome of ecosystem engineering: A case study on tussocks of the salt marsh pioneer *Spartina anglica*. *Geomorphology*, **153–154**: 232–238.

Baptist, M. J., Babovic, V., Rodríguez-Uthurburu, J., Keijzer, M., Uittenbogaard, R. E., Mynett, A., and, Verwey A. 2007. On inducing equations for vegetation resistance. *Journal of Hydraulic Research*, **45**: 435–450.

Bayliss-Smith, T. P., Healey, R., Lailey, R., Spencer, T., and, Stoddart, D. R. 1979. Tidal flows in salt marsh creeks. *Estuarine and Coastal Marine Science*, **9**: 235–255.

Beeftink, W. G., and Rozema, J. 1993. The nature and functioning of salt marshes. In: *Pollution of the North Sea*, Salomons, W, Bayne, B. L., Duursma, E. K., and Forstner, U. (eds). Springer, Berlin, Heidelberg, pp. 59–87.

Belliard, J.-P., Toffolon, M., Carniello, L., and D'Alpaos, A. 2015. An ecogeomorphic model of tidal channel initiation and elaboration in progressive marsh accretional contexts. *Journal of Geophysical Research: Earth Surface*, **120**: 1040–1064.

Bendoni, M., Francalanci, S., Cappietti, L., and Solari, L. 2014. On salt marshes retreat: Experiments and modeling toppling failures induced by wind waves. *Journal of Geophysical Research: Earth Surface*, **119**: 603–620.

Bendoni, M., Mel, R., Lanzoni, S., Francalanci, S., and Oumeraci, H. 2016. Insights into lateral marsh retreat mechanism through localized field measurements. *Water Resources Research*, **52**: 1446–1464.

Boon, J. D. I. 1975. Tidal discharge asymmetry in a salt marsh drainage system. *Limnology and Oceanography*, **20**: 71–80.

Bouma, T. J., Friedrichs, M., Van Wesenbeeck, B. K., Temmerman, S., Graf, G., and Herman, P. M. J. 2009. Density-dependent linkage of scale-dependent feedbacks: a flume study on the intertidal macrophyte *Spartina anglica*. *Oikos*, **118**: 260–268.

Brivio, L., Ghinassi, M., D'Alpaos, A., Finotello, A., Fontana, A., Roner, M., and Howes, N. 2016. Aggradation and lateral migration shaping geometry of a tidal point bar: An example from salt marshes of the Northern Venice Lagoon (Italy). *Sedimentary Geology*, **343**: 141–155.

Callaghan, D. P., Bouma, T. J., Klaassen, P., van der Wal, D., Stive, M. J. F., and Herman, P. M. J. 2010. Hydrodynamic forcing on salt-marsh development: Distinguishing the relative importance of waves and tidal flows. *Estuarine, Coastal and Shelf Science*, **89**: 73–88.

Callaway, J. C., and Josselyn, M. N. 1992. The introduction and spread of smooth cordgrass *Spartina alterniflora* in South San Francisco Bay. *Estuaries*, **15**: 218–226.

Carniello, L., D'Alpaos, A., and Defina, A. 2011. Modeling wind waves and tidal flows in shallow micro-tidal basins. *Estuarine, Coastal and Shelf Science*, **92**: 263–276.

Carniello, L., Defina, A., and D'Alpaos, L. 2012. Modeling sand-mud transport induced by tidal currents and wind waves in shallow microtidal basins: Application to the Venice Lagoon (Italy). *Estuarine, Coastal and Shelf Science*, **102–103**: 105–115.

Chen, Y., Li, Y., Cai, T., Thompson, C., and Li, Y. 2016. A comparison of biohydrodynamic interaction within mangrove and saltmarsh boundaries. *Earth Surface Processes and Landforms*, **41**: 1967–1979.

Chen, Z., Ortiz, A., Zong, L., and Nepf, H. 2012. The wake structure behind a porous obstruction and its implications for deposition near a finite patch of emergent vegetation. *Water Resources Research*, **48**: 1–12.

Coco, G., Zhou Z., van Maanen, B., Olabarrieta, M., Tinoco, R., and Townend, I. H. 2013. Morphodynamics of tidal networks: Advances and challenges. *Marine Geology*, **346**: 1–16.

Costanza, R., d'Arge, R., de Groot, R., Farber, S., Grasso, M., Hannon, B., Limburg, K., et al. 1997. The value of the world's ecosystem services and natural capital. *Nature*, **387**: 253–260.

D'Alpaos, A., Ghinassi, M., Finotello, A., Brivio, L., Bellucci, L. G. L. G., and Marani, M. 2017. Tidal meander migration and dynamics: A case study from the Venice Lagoon. *Marine and Petroleum Geology*, **87**: 80–90.

D'Alpaos, A., Lanzoni, S., Marani, M., Bonometto, A., Cecconi, G., and Rinaldo, A. 2007a. Spontaneous tidal network formation within a constructed salt marsh: Observations and morphodynamic modelling. *Geomorphology*, **91**: 186–197. DOI: 10.1016/j.geomorph.2007.04.013

D'Alpaos, A., Lanzoni, S., Marani, M., Fagherazzi, S., and Rinaldo, A. 2005. Tidal network ontogeny: Channel initiation and early development. *Journal of Geophysical Research: Earth Surface*, **110**: 1–14.

D'Alpaos, A., Lanzoni, S., Marani, M., and Rinaldo, A. 2007b. Landscape evolution in tidal embayments: Modeling the interplay of erosion, sedimentation, and vegetation dynamics. *Journal of Geophysical Research: Earth Surface*, **112**: 1–17.

D'Alpaos A., Lanzoni S., Marani M., and Rinaldo A. 2009. On the O'Brien–Jarrett–Marchi law. *Rendiconti Lincei*, **20**: 225–236.

D'Alpaos, A., Lanzoni, S., Marani, M., and Rinaldo, A. 2010. On the tidal prism-channel area relations. *Journal of Geophysical Research: Earth Surface*, **115**: 1–13.

D'Alpaos, A., Lanzoni, S., Mudd, S. M., and Fagherazzi, S. 2006. Modeling the influence of hydroperiod and vegetation on the cross-sectional formation of tidal channels. *Estuarine, Coastal and Shelf Science*, **69**: 311–324.

D'Alpaos, A., and Marani, M. 2016. Reading the signatures of biologic-geomorphic feedbacks in salt-marsh landscapes. *Advances in Water Resources*, **93**: 265–275.

Defina, A., Carniello, L., Fagherazzi, S., and D'Alpaos, L. 2007. Self-organization of shallow basins in tidal flats and salt marshes. *Journal of Geophysical Research: Earth Surface*, **112**: 1–11.

Di Silvio, G., Dall'Angelo, C., Bonaldo, D., Fasolato, G., Dall'Angelo, C., Bonaldo, D., and Fasolato, G. 2010. Long term model of planimetric and bathymetric evolution of a tidal lagoon. *Continental Shelf Research*, **30**: 894–903.

Dronkers, J. 2016. *Dynamic of Coastal System*. 2nd edn. World Scientific, Singapore.

Dronkers, J. J. 1964. *Tidal Computations in Rivers and Coastal Waters*. North Holland, Amsterdam.

Fagherazzi, S., Bortoluzzi, A., Dietrich, W. E., Adami, A., Lanzoni, S., Marani, M., and Rinaldo, A. 1999. Tidal networks 1. Automatic network extraction and preliminary scaling features from digital terrain maps. *Water Resources Research*, **35**: 3891–3904.

Fagherazzi, S., Carniello, L., D'Alpaos, L., and Defina, A. 2006. Critical bifurcation of shallow microtidal landforms in tidal flats and salt marshes. *Proceedings of the National Academy of Sciences of the United States of America*, **103**: 8337–8341.

Fagherazzi, S., and Furbish, D. J. 2001. On the shape and widening of salt marsh creeks. *Journal of Geophysical Research*, **106**: 991.

Fagherazzi, S., Gabet, E. J., and, Furbish, D. J. 2004. The effect of bidirectional flow on tidal channel planforms. *Earth Surface Processes and Landforms*, **29**: 295–309.

Fagherazzi, S., Hannion, M., and D'Odorico, P. 2008. Geomorphic structure of tidal hydrodynamics in salt marsh creeks. *Water Resources Research*, **44**: 1–12.

Fagherazzi, S., Kirwan, M. L., Mudd, S. M., Guntenspergen, G. R., Temmerman, S., D'Alpaos, A., van de Koppel, J. et al. 2012. Numerical models of salt marsh evolution: Ecological, geomorphic, and climatic factors. *Reviews of Geophysics*, **50**: 1–28.

Fagherazzi, S., and Sun, T. 2004. A stochastic model for the formation of channel networks in tidal marshes. *Geophysical Research Letters*, **31**: 1–4.

Fagherazzi, S., Wiberg, P. L., Temmerman, S., Struyf, E., Zhao, Y., and Raymond, P. A. 2013. Fluxes of water, sediments, and biogeochemical compounds in salt marshes. *Ecological Processes*, **2**: 1–16.

Finotello, A., Lanzoni, S., Ghinassi, M., Marani, M., Rinaldo, A., and D'Alpaos, A., 2018. Field migration rates of tidal meanders recapitulate fluvial morphodynamics. *Proceedings of the National Academy of Sciences of the United States of America*, **115**: 1463–1468.

Folkard, A. M. 2011. Flow regimes in gaps within stands of flexible vegetation: Laboratory flume simulations. *Environmental Fluid Mechanics*, **11**: 289–306.

Francalanci, S., Bendoni, M., Rinaldi, M., and Solari, L. 2013. Ecomorphodynamic evolution of salt marshes: Experimental observations of bank retreat processes. *Geomorphology*, **195**: 53–65.

French, J. R., and Stoddart, D. R. 1992. Hydrodynamics of salt marsh creek systems: Implications for marsh morphological development and material exchange. *Earth Surface Processes and Landforms*, **17**: 235–252.

Friedrichs, C. T. 1995. Stability shear stress and equilibrium cross-sectional of sheltered tidal channels. *Journal of Coastal Research*, **11**: 1062–1074.

Friedrichs, C. T., and Perry, J. E. 2001. Tidal salt marsh morphodynamics: A synthesis. *Journal of Coastal Research*, **SI**: 7–37.

Gabet, E. J. 1998. Lateral migration and bank erosion in a saltmarsh tidal channel in San Francisco Bay, California. *Estuaries* **21**: 745–753.

Garofalo, D. 1980. The influence of wetland vegetation on tidal stream channel migration and morphology. *Estuaries*, **3**: 258–270.

Gedan, K. B., Kirwan, M. L., Wolanski, E., Barbier, E. B., and Silliman, B. R. 2011. The present and future role of coastal wetland vegetation in protecting shorelines: answering recent challenges to the paradigm. *Climatic Change*, **106**: 7–29.

Ghinassi, M., D'alpaos, A., Gasparotto, A., Carniello, L., Brivio, L., Finotello, A., Roner, M. et al. 2018. Morphodynamic evolution and stratal architecture of translating tidal point bars: Inferences from the northern Venice Lagoon (Italy). *Sedimentology*, **65**: 1354–1377.

Hartig, E. K., Gornitz, V., Kolker, A., Mushacke, F., and Fallon, D. 2002. Anthropogenic and climate-change impacts on salt marshes of Jamaica Bay, New York City. *Wetlands*, **22**: 71–89.

Horton, R. E. 1945. Erosional development of streams and their drainage basins; Hydrophysical approach to quantitative morphology. *Geological Society of America Bulletin*, **56**: 151–180.

Hu, K., Chen, Q., and Wang, H. 2015a. A numerical study of vegetation impact on reducing storm surge by wetlands in a semi-enclosed estuary. *Coastal Engineering*, **95**: 66–76.

Hu, X., and Chen, C. T. 2005. Refraction of water waves by periodic cylinder arrays. *Physical Review Letters*, **95**: 1–4.

Hu, Z., van Belzen, J., van der Wal, D., Balke, T., Wang, Z. B., Stive, M. and Bouma, T. J. 2015b. Windows of opportunity for salt marsh vegetation establishment on bare tidal flats: The importance of temporal and spatial variability in hydrodynamic forcing. *Journal of Geophysical Research: Biogeosciences*, **120**: 1450–1469.

Hughes, Z. J. 2012. Tidal channels on tidal flats and marshes. In: *Principles of Tidal Sedimentology*, Davis, R. A. and Dalrymple, R. W. (eds). Springer, Dordrecht, pp. 269–300.

Hughes, Z. J., FitzGerald, D. M., Wilson, C. A., Pennings, S. C., Więski, K., and Mahadevan, A. 2009. Rapid headward erosion of marsh creeks in response to relative sea level rise. *Geophysical Research Letters*, **36**: 1–5.

Jarrett, J. T. 1976. Tidal prism-inlet area relationships. *Joural of Waterways and Harbors*, **95**: 43–52.

Julian, J. P., and Torres, R. 2006. Hydraulic erosion of cohesive riverbanks. *Geomorphology*, **76**: 193–206.

Kearney, W. S., and Fagherazzi, S. 2016. Salt marsh vegetation promotes efficient tidal channel networks. *Nature Communications*, **7**: 1–7.

Kleinhans, M. G., Schuurman, F., Bakx, W., and Markies, H. 2009. Meandering channel dynamics in highly cohesive sediment on an intertidal mud flat in the Westerschelde estuary, the Netherlands. *Geomorphology*, **105**: 261–276.

Lanzoni, S., and Seminara, G. 1998. On tide propagation in convergent estuaries. *Journal of Geophysical Research: Oceans*, **103**: 30793–30812.

Lanzoni, S., and Seminara, G. 2002. Long-term evolution and morphodynamic equilibrium of tidal channels. *Journal of Geophysical Research*, **107**: 1–13.

Leonard, L. A., and Croft, A. L. 2006. The effect of standing biomass on flow velocity and turbulence in *Spartina alterniflora* canopies. *Estuarine, Coastal and Shelf Science*, **69**: 325–336.

Leonard, L. A., and Luther, M. E. 1995. Flow hydrodynamics in tidal marsh canopies. *Limnology and Oceanography*, **40**: 1474–1484.

Leonardi, N., Defne, Z., Ganju, N. K., and Fagherazzi, S. 2016a. Salt marsh erosion rates and boundary features in a shallow Bay. *Journal of Geophysical Research: Earth Surface*, **121**: 1861–1875.

Leonardi, N., and Fagherazzi, S. 2014. How waves shape salt marshes. *Geology*, **42**: 887–890.

Leonardi, N., Ganju, N. K., and Fagherazzi, S. 2016b. A linear relationship between wave power and erosion determines salt-marsh resilience to violent storms and hurricanes. *Proceedings of the National Academy of Sciences*, **113**: 64–68.

Leopold, L. B., Collins, J. N., and Collins, L. M. 1993. Hydrology of some tidal channels in estuarine marshland near San Francisco. *Catena*, **20**: 469–493.

Da Lio, C., D'Alpaos, A., and Marani, M. 2013. The secret gardener: vegetation and the emergence of biogeomorphic patterns in tidal environments. *Philosophical Transactions. Series A, Mathematical, Physical, and Engineering Sciences*, **371**: 20120367.

López, F., and García, M. H. 2001. Mean flow and turbulence structure of open-channel flow through non-emergent vegetation. *Journal of Hydraulic Engineering*, **127**: 392–402.

Marani, M., Belluco, E., D'Alpaos, A., Defina, A., Lanzoni, S., and Rinaldo, A. 2003. On the drainage density of tidal networks. *Water Resources Research*, **39**: 1–11.

Marani, M., D'Alpaos, A., Lanzoni, S., Carniello, L., and Rinaldo, A. 2010. The importance of being coupled: Stable states and catastrophic shifts in tidal biomorphodynamics. *Journal of Geophysical Research: Earth Surface*, **115**: 1–15.

Marani M., D'Alpaos, A., Lanzoni, S., and Santalucia, M. 2011. Understanding and predicting wave erosion of marsh edges. *Geophysical Research Letters*, **38**: 1–5.

Marani, M., Lanzoni, S., Zandolin, D., Seminara, G., and Rinaldo, A. 2002. Tidal meanders. *Water Resources Research*, **38**: 7–14.

Marchi, E. 1990. Sulla stabilità delle bocche lagunari a marea. *Rendiconti Lincei*, **1**: 137–150.

Mariotti, G. 2018. Marsh channel morphological response to sea level rise and sediment supply. *Estuarine, Coastal and Shelf Science*, **209**: 89–101.

Mitsch, W. J., and Gosselink, J. G. 2000. The value of wetlands: importance of scale and landscape setting. *Ecological Economics*, **35**: 25–33.

Möller, I., Spencer, T., French, J. R., Leggett, D. J., and Dixon, M. 1999. Wave transformation over saltmarshes: A field and numerical modelling study from North Norfolk, England. *Estuarine, Coastal and Shelf Science*, **49**: 411–426.

Morris, J. T., Sundberg, K., and Hopkinson, C. S. 2013. Salt marsh primary production and its responses to relative sea level and nutrients in estuaries at Plum Island, Massachusetts, and North Inlet, South Carolina, USA. *Oceanography*, **26**: 78–84.

Mudd, S. M., D'Alpaos, A., and Morris, J. T. 2010. How does vegetation affect sedimentation on tidal marshes? Investigating particle capture and hydrodynamic controls on biologically mediated sedimentation. *Journal of Geophysical Research: Earth Surface*, **115**: 1–14.

Mudd, S. M., Fagherazzi, S., Morris, J. T., and Furbish, D. J. 2004. Flow, sedimentation, and biomass production on a vegetated salt marsh in South Carolina: Toward a predictive model of marsh morphologic and ecologic evolution. In: *The*

Ecogeomorphology of Tidal Marshes, Coastal and Estuarine Studies n. 59, Fagherazzi, S., Marani, M., and Blum, L. K. (eds). American Geophysical Union, Washington, D.C., pp. 165–188.

Myrick, R. M., and Leopold, L. B. 1963. Hydraulic geometry of a small tidal estuary. *United States Geological Survey Professional Paper* **422**: 1–18.

Nepf, H. M. 1999. Drag, turbulence, and diffusion in flow through emergent vegetation. *Water Resources Research*, **35**: 479–489.

Nepf, H. M. 2012. Hydrodynamics of vegetated channels. *Journal of Hydraulic Research*, **50**: 262–279.

Neumeier, U., and Amos, C. L. 2006. The influence of vegetation on turbulence and flow velocities in European salt-marshes. *Sedimentology*, **53**: 259–277.

Nichols, M. M., Johnson, G. H., and Peebles, P. C. 1991. Modern sediments and facies model for a microtidal coastal plain estuary, the James Estuary, Virginia. *Journal of Sedimentary Petrology*, **61**: 883–899.

Nikora, N., and Nikora, V. 2007. A viscous drag concept for flow resistance in vegetated channels. *Proceedings of the 32nd IAHR Congress*, Venice.

O'Brien, M. P. 1969. Equilibrium flow areas of inlets on sandy coasts. *Journal of Waterways and Harbors*, **95**: 43–52.

Van Oyen, T., Carniello, L., D'Alpaos, A., Temmerman, S., Troch, P., and Lanzoni, S. 2014. An approximate solution to the flow field on vegetated intertidal platforms: Applicability and limitations. *Journal of Geophysical Research F: Earth Surface*, **119**: 1682–1703.

Van Oyen, T., Lanzoni, S., D'Alpaos, A., Temmerman, S., Troch, P., and Carniello, L. 2012. A simplified model for frictionally dominated tidal flows. *Geophysical Research Letters*, **39**: 1–6.

Pestrong, R. 1972. Tidal-flat sedimentation at cooley landing, Southwest San Francisco bay. *Sedimentary Geology*, **8**: 251–288.

Pethick, J. 1992. Saltmarsh geomorphology. In: *Saltmarshes: Morphodynamics, Conservation and Engineering Significance*, Allen, J. R. L., and Pye, K. (eds). Cambridge University Press, Cambridge, pp. 41–62.

Pethick, J. S. 1969. Drainage in salt marshes. In: *The Coastline of England and Wales*. 3rd edn. J. R. Steers (ed.). Cambridge University Press: Cambridge, pp. 752–730.

Pethick, J. S. 1980. Velocity surges and asymmetry in tidal channels. *Estuarine and Coastal Marine Science*, **11**: 331–345.

Pye, K., and French, P. 1993. Erosion & Accretion Processes on British Salt Marshes. Cambridge Environmental Research Consultants.

Redfield, A. C. 1972. Development of a New England salt marsh. *Ecological Monographs*, **42**: 201–237.

Rietkerk, M., and van de Koppel, J. 2008. Regular pattern formation in real ecosystems. *Trends in Ecology and Evolution*, **23**: 169–175.

Rigon, R., Rinaldo, A., and Rodriguez-Iturbe, I. 1994. On landscape self-organization. *Journal of Geophysical Research: Solid Earth*, **99**: 11971–11993.

Rinaldo, A., Dietrich, W. E., Rigon, R., Vogel, G. K., and Rodriguez-Iturbe, I. 1995. Geomorphological signatures of varying climate. *Nature*, **374**: 632–635.

Rinaldo, A., Fagherazzi, S., Lanzoni, S., Marani, M., and Dietrich, W. E. 1999a. Tidal networks 2. Watershed delineation and comparative network morphology. *Water Resources Research*, **35**: 3905–3917.

Rinaldo, A., Fagherazzi, S., Lanzoni, S., Marani, M., and Dietrich, W. E. 1999b. Tidal networks 3. Landscape-forming discharges and studies in empirical geomorphic relationships. *Water Resources Research*, **35**: 3919–3929.

Rinaldo, A., Rodriguez-Iturbe, I., Rigon, R., Ijjasz-Vasquez, E., and Bras, R. L. 1993. Self-organized fractal river networks. *Physical Review Letters*, **70**: 822–825.

Rupprecht, F., Möller, I., Paul, M., Kudella, M., Spencer, T., van Wesenbeeck, B. K., Wolters, G., et al. 2017. Vegetation-wave interactions in salt marshes under storm surge conditions. *Ecological Engineering*, **100**: 301–315.

Salehi, M., and Strom, K. 2012. Measurement of critical shear stress for mud mixtures in the San Jacinto estuary under different wave and current combinations. *Continental Shelf Research*, **47**: 78–92.

Schwarz, C., Ye, Q., Wal, D., Zhang, L., Bouma, T., Ysebaert, T., and Herman, P. 2014. Impacts of salt marsh plants on tidal channels initiation and inheritance. *Journal of Geophysical Research: Earth Surface*, **119**: 385–400.

Shepard, C. C., Crain, C. M., and Beck, M. W. 2011. The protective role of coastal marshes: A systematic review and meta-analysis. *PLOS ONE* **6**: e27374.

Shi, Z., Hamilton, L. J., and Wolanski, E. 2000. Near-bed currents and suspended sediment transport in saltmarsh canopies. *Journal of Coastal Research*, **16**: 909–914.

Silinski A., Heuner M., Schoelynck J., Puijalon S., Schröder U., Fuchs E., Troch P., et al. 2015. Effects of wind waves versus ship waves on tidal marsh plants: A flume study on different life stages of *Scirpus maritimus*. *PLOS ONE*, **10**: 1–16.

Soulsby, R. L. 1997. *Dynamics of Marine Sands*. Thomas Telford Publications, London.

Soulsby, R. L., and Clarke, S. 2005. Bed shear-stresses under combined waves and currents on smooth and rough beds. *Hydraulics Research Report*, **1905**: TR 137.

Steel, T. J., and Pye, K. 1997. The development of salt marsh creek networks: Evidence from the UK. *Canadian Coastal Conference*, pp. 1–16.

Stefanon, L., Carniello, L., D'Alpaos, A., Lanzoni, S., D'Alpaos, A., and Lanzoni, S. 2010. Experimental analysis of tidal network growth and development. *Continental Shelf Research* **30**: 950–962.

Stefanon, L., Carniello, L., D'Alpaos, A., and Rinaldo, A. 2012. Signatures of sea level changes on tidal geomorphology: Experiments on network incision and retreat. *Geophysical Research Letters*, **39**: 1–6.

Strahler, A. N. 1957. Quantitative analysis of watershed geomorphology. *Eos, Transactions American Geophysical Union*, **38**: 913–920.

Tambroni, N., Luchi, R., and Seminara, G. 2017. Can tide dominance be inferred from the point bar pattern of tidal meandering channels? *Journal of Geophysical Research : Earth Surface*, **122**: 1–21.

Tanino, Y., and Nepf, H. M. 2008. Lateral dispersion in random cylinder arrays at high Reynolds number. *Journal of Fluid Mechanics*, **600**: 339–371.

Tanino, Y., and Nepf, H. M. 2009. Laboratory investigation of lateral dispersion within dense arrays of randomly distributed cylinders at transitional Reynolds number. *Physics of Fluids*, **21**: 1–13.

Temmerman, S., Bouma, T. J., Govers, G., Wang, Z. B., De Vries, M. B., Herman, P. M. J., De Vries, M. B., and Herman, P. M. J. 2005. Impact of vegetation on flow routing and sedimentation patterns: Three-dimensional modeling for a tidal marsh. *Journal of Geophysical Research: Earth Surface*, **110**: 1–18.

Temmerman, S., Bouma, T. J., Van de Koppel, J., Van der Wal, D., De Vries, M. B., and Herman, P. M. J. 2007. Vegetation causes channel erosion in a tidal landscape. *Geology*, **35**: 631–634.

Temmerman, S., Meire, P., Bouma, T. J., Herman, P. M. J., Ysebaert, T., and De Vriend, H. J. 2013. Ecosystem-based coastal defence in the face of global change. *Nature*, **504**: 79–83.

Temmerman, S., De Vries, M. B., and Bouma, T. J. 2012. Coastal marsh die-off and reduced attenuation of coastal floods: A model analysis. *Global and Planetary Change*, **92–93**: 267–274.

Tonelli, M., Fagherazzi, S., and Petti, M. 2010. Modeling wave impact on salt marsh boundaries. *Journal of Geophysical Research: Oceans*, **115**: 1–17.

Torres, R., and Styles, R. 2007. Effects of topographic structure on salt marsh currents. *Journal of Geophysical Research: Earth Surface*, **112**: F02023.

Townend, I. H. 2010. An exploration of equilibrium in Venice Lagoon using an idealised form model. *Continental Shelf Research*, **30**: 984–999.

Tucker, G. E., Catani, F., Rinaldo, A., and Bras, R. L. 2001. Statistical analysis of drainage density from digital terrain data. *Geomorphology*, **36**: 187–202.

Valentine, K., Mariotti, G., and Fagherazzi, S. 2014. Repeated erosion of cohesive sediments with biofilms. *Advances in Geosciences*, **39**: 9–14.

Vandenbruwaene, W., Bouma, T. J., Meire, P., and Temmerman, S. 2013. Bio-geomorphic effects on tidal channel evolution: Impact of vegetation establishment and tidal prism change. *Earth Surface Processes and Landforms*, **38**: 122–132.

Vandenbruwaene, W., Temmerman, S., Bouma, T. J., Klaassen, P. C., de Vries, M. B., Callaghan, D. P., van Steeg, P. et al. 2011. Flow interaction with dynamic vegetation patches: Implications for biogeomorphic evolution of a tidal landscape. *Journal of Geophysical Research: Earth Surface*, **116**: 1–13.

Wamsley, T. V., Cialone, M. A., Smith, J. M., Atkinson, J. H., and Rosati, J. D. 2010. The potential of wetlands in reducing storm surge. *Ocean Engineering*, **37**: 59–68.

van der Wegen, M., Wang, Z. B., Savenije, H. H. G., and Roelvink, J. A. 2008. Long-term morphodynamic evoluation and energy dissipation in a coastal plain, tidal embayment. *Journal of Geophysical Research: Earth Surface*, **113**: 1–22.

van Wesenbeeck, B. K., van De K. oppel, J., Herman, P. M. J., and Bouma, T. J. 2008. Does scale-dependent feedback explain spatial complexity in salt-marsh ecosystems? *Oikos*, **117**: 152–159.

White, B. L., and Nepf, H. M. 2007. Shear instability and coherent structures in shallow flow adjacent to a porous layer. *Journal of Fluid Mechanics*, **593**: 1–32.

White, B. L., and Nepf, H. M. 2008. A vortex-based model of velocity and shear stress in a partially vegetated shallow channel. *Water Resources Research*, **44**: 1–15.

Yang, S. L. 1998. The role of scirpus marsh in attenuation of hydrodynamics and retention of fine sediment in the Yangtze estuary. *Estuarine, Coastal and Shelf Science*, **47**: 227–233.

5

Community Ecology of Salt Marshes

STEVEN C. PENNINGS AND QIANG HE

Salt marshes have been useful study systems for community ecologists. They are amenable to experimental manipulation, and the simplicity and strong abiotic gradients of salt marshes lead to clear patterns and experimental outcomes. Many early ecologists believed that salt marsh ecosystems were primarily controlled by bottom-up factors (i.e., that nutrients, salinity, and other abiotic factors were the primary factors regulating productivity, and that productivity in turn regulated ecosystem trophic structure). More recently, many ecologists have argued that consumers have an important role in structuring salt marsh ecosystems through "top-down" processes. A simple conceptual approach, which we take here, is to think of salt marsh communities as being structured by bottom-up, top-down, and non-trophic processes.

5.1 Bottom-up Processes

5.1.1 Physiological Controls

Salt marshes are periodically flooded with seawater. The limited physiological tolerance of most organisms to salinity and cycles of flooding and drying creates a strong "filter" that excludes most of the world's organisms from salt marsh habitats. Like tropical mangrove forests, salt marshes are species-poor in plants, fungi, mammals, insects, and other "terrestrial arthropods" compared to productive terrestrial habitats, species-poor in algae and most marine invertebrates compared to rocky intertidal habitats, and species-poor in amphibians and turtles compared to freshwater habitats. The few species that can survive within the salt marsh, however, often flourish, achieving high densities and biomass.

Flooding and salinity vary across a marsh, and gradients in these stressors define the potential distribution – the fundamental niche – of each salt marsh species. Salt marsh plant species vary in their tolerance to salinity and flooding (Rozema et al. 1985), making more stressful parts of the marsh unsuitable for some species, and limiting the productivity of others. Similarly, animals of terrestrial origin may not be able to continuously inhabit regularly flooded portions of the marsh, and animals of marine origin may not be able to continuously inhabit rarely flooded portions of the marsh.

Although ecologists often refer to "flooding" and "salinity" gradients, the proximate physiological stressors that limit the abundance of species within a salt marsh are more complicated (Mendelssohn and Morris 2000). Flooding of soils leads to anoxia, which sets the stage for a chain of redox reactions that create toxic compounds such as hydrogen sulfide (Chapter 6). Conversely, dry soils may be stressful for organisms adapted to wet conditions, and drying of salt marsh soils that are usually wet can create acidic soils and mobilize toxic metals (McKee et al. 2004). Salt not only creates osmotic problems for marsh organisms, but also can be directly toxic or can limit uptake of nitrogen (Mendelssohn and Morris 2000).

Abiotic factors can be thought of as "bottom-up controls" on marsh structure and function. Historically, bottom-up controls were a primary focus of salt marsh ecology. A rich body of work examined how flooding (Wiegert et al. 1983), salinity (Linthurst 1980), nitrogen availability (Valiela et al. 1978), and sulfides (Bradley and Dunn 1989) affected plant production (Mendelssohn 1979). From these studies and their successors grew a deep understanding of how abiotic conditions mediate productivity and species richness of salt marshes (Mendelssohn and Morris 2000).

More recent work has begun to compare these physiological tolerances across a broader range of species. For example, comparisons of 6–12 wetland grass species in the genus *Spartina* showed that species found in the low intertidal were better able to transport oxygen internally, remove dissolved oxygen from the environment, respire at lower levels of limited oxygen, and tolerate sulfide (Maricle et al. 2006, Maricle and Lee 2007). Conversely, species characteristic of high marsh habitats were more tolerant of drought or salinity stress (Maricle et al. 2007). Other studies have begun to determine thresholds of physiological tolerances to better characterize the niches of different species (He et al. 2017a).

Despite these advances, the limits of our understanding of bottom-up control on salt marsh production have been revealed by unexpected episodes of large-scale marsh dieback in southeastern and Gulf Coast salt marshes of the USA (McKee et al. 2004). Although these events appeared to be triggered by drought, and hence to have a bottom-up component, the proximate and ultimate causes of the dieback events remain elusive (Alber et al. 2008).

Another knowledge gap about physiological controls on marsh function lies at the interfaces between ecology, hydrology, and biogeochemistry. We have only a basic understanding, for example, of how tidal flooding and groundwater flows interact with evapotranspiration to create marsh areas with different salinity regimes that support different plant species (Wilson et al. 2015). Moreover, plants and abundant marsh invertebrates affect marsh abiotic conditions in ways that are only beginning to be understood, creating important feedbacks between biota and bottom-up factors (Section 5.3.3).

5.1.2 Variation in Primary Production

Ecologists have been long fascinated with the prodigious levels of primary production in salt marshes (Teal 1962). Although bottom-up factors play a key role in determining the productivity of salt marsh plants, they do not act alone. Ecosystem engineers such as fiddler crabs or mussels can enhance the productivity of marsh plants (Section 5.3.3). Disturbance (Section 5.3.5) and herbivory (Section 5.2.1) remove biomass, which may stimulate or reduce primary production. These factors interact to create differences in primary production across elevation, along estuaries and across latitude.

Primary production varies across marsh elevation, in large part due to variation in bottom-up conditions. In Atlantic and Gulf Coast marshes of the USA, the dominant plant, *Spartina alterniflora*, is tallest at creekbanks and decreases in height and productivity at higher marsh elevations due to salinity, sulfide, and nutrient stress (Mendelssohn and Morris 2000, Morris et al. 2002). Standing biomass and productivity then increase again toward the terrestrial border of the marsh where abiotic conditions are least stressful and plants such as *Iva frutescens* and *Phragmites australis* dominate (Bertness et al. 1992b, Windham and Lathrop 1999). Similar gradients in abiotic conditions and disturbance lead to elevational patterns in standing biomass and productivity in marshes worldwide, although details of the elevational pattern may vary. Primary production also varies along the estuarine gradient as a function of salinity and sulfide stress, with productivity greatest in low-salinity tidal marshes (Odum 1988, Więski et al. 2010). Finally, primary production also varies across latitude, largely as a function of growing season length and temperature (Turner 1976, Kirwan et al. 2009). Latitudinal variation in productivity is likely also modulated by ice disturbance at high latitudes (Hardwick-Witman 1985) and herbivory, especially at low latitudes (Pennings et al. 2009). Variation among sites in tide range leads to further geographic variation in biomass (Steever et al. 1976, Liu et al. 2016). The emerging theme in each of these cases is spatial variation – different marshes are likely to vary considerably in productivity, and one should be cautious in extrapolating estimates of productivity from one location to another.

Historically, interest in primary production in salt marshes was stimulated by the desire to understand their high productivity (Cebrian 1999), and by an inability to account for the fate of all the primary production, leading to the hypothesis that salt marshes subsidized secondary production in adjacent habitats (Teal 1962). These two research programs led to a number of estimates of marsh primary production (Hopkinson et al. 1978, Gallagher et al. 1980, Schubauer and Hopkinson 1984, Dai and Wiegert 1996a, b) and of carbon export from salt marshes (de Bettencourt et al. 1994, Cebrian 1999). Along the way, researchers realized that benthic microalgae growing among marsh plants or on mudflats just below the plants are also highly productive, and, because they lack the difficult-to-digest structural material of higher plants, may be as or more important than higher plants in supporting parts of the marsh food web (Haines and Montague 1979, Currin et al. 1995, Page 1997).

Some of the more exciting recent work on marsh primary production has focused on how productivity varies in space and time. Morris and colleagues have turned to niche theory to better explain variation in plant productivity across elevation within a marsh (Morris et al. 2002, Morris et al. 2013, Voss et al. 2013). Long-term data sets have revealed that plant biomass in salt marshes can vary several-fold among years as a function of sea level, river discharge, and temperature (Morris et al. 1990, Więski and Pennings 2014b, O'Donnell and Schalles 2016). At other sites, however, variation in standing biomass among years remains to be explained (Teal and Howes 1996).

Considerable knowledge gaps remain. We lack an understanding of how multiple abiotic factors, ecosystem engineering by organisms, disturbance, and herbivory interact to mediate plant productivity. Mathematical models of salt marsh plant productivity are still in their infancy (Zheng et al. 2016). There is a need for models that integrate all the factors affecting primary production, and that can be used to predict how marshes will respond to future global changes.

5.1.3 Secondary Production and Food Webs

Salt marshes are nurseries and feeding grounds for fish and crustacean species. Salt marshes support secondary production due to their high primary production (Boesch and Turner 1984) and because they provide a refuge from predation (Section 5.2.2).

For a long time, secondary production in marshes was thought to be primarily based on detritus of dead plants–a "brown food web." Some marsh animals feed on decaying plant material (Stiven and Kuenzler 1979, Langdon and Newell 1990). The biomass and energy obtained by these animals can be transferred via a series of predator–prey relationships (e.g., predation by fish and crustaceans that enter and leave the marsh with the tides) from marshes to estuaries and even to the open sea, a process called a "trophic relay" (Minello and Zimmerman 1992, Kneib 2000). Salt marshes may also enhance offshore food webs by directly exporting organic matter, although this function of salt marshes is highly variable, depending on hydrology and geomorphological settings among other factors (Tobias and Neubauer 2009).

Recent studies have revealed that secondary production and food webs in marshes are based on more than plant detritus. Living vascular plants and algae in marsh sediments are also important contributors to secondary production, though their relative contributions may differ by consumer species (Kwak and Zedler 1997). Some animals such as grasshoppers and geese feed on living plants (He and Silliman 2016). Many other marsh animals feed on benthic microalgae (Haines 1976). Consumers of marsh plants and algae are eaten in turn by primary and top predators (e.g., spiders, blue crabs, sea otters, and shorebirds), forming a "green food web."

Despite an expanding body of recent literature, we still lack an understanding of the complete structure of salt marsh food webs and their complexity across space and time. While marsh ecologists recognize the importance of both green and brown food webs, the relative importance of the two has been rarely assessed (Schrama et al. 2012). Spatiotemporal patterns in secondary production and their drivers have also been little studied. Although it has been a challenge to study mobile consumers in the thick vegetation and turbid water characteristic of marsh habitats, technological developments are providing marsh ecologists with promising tools for overcoming these challenges (Pautzke et al. 2010).

5.1.4 Biodiversity-Ecosystem Functioning

One of the major advances in ecology over the past three decades has been experimental confirmation that species richness affects ecosystem function (Naeem et al. 1994, Tilman et al. 1996, Naeem 2002, Worm et al. 2006). Although salt marshes have relatively low species richness, pursuing a better understanding of biodiversity-ecosystem functioning (BEF) relationships in salt marshes is important, for two reasons. First, the largest gains in ecosystem function in BEF relationships typically occur at low levels of species richness typical of those found in salt marshes (Tilman et al. 1996). Second, even if species richness is low, genetic diversity within dominant species can also affect ecosystem functioning (Zerebecki et al. 2017).

BEF relationships can be observed at any trophic level. Predator diversity can weaken predation pressure if predators harm each other (Finke and Denno 2004), or can strengthen it if different predators are active at different times (Griffin and Silliman 2011). Plant diversity usually has positive effects on ecosystem function. Callaway et al. (2003) planted plots with 1, 3, and 6 species of plants in a California salt marsh. Productivity and nitrogen accumulation increased with species richness, largely due to the selection effect, meaning that the best-performing species were more likely to be present in more diverse mixtures (Callaway et al. 2003). Because the best-performing species were also competitively dominant and most abundant in the field, marsh function would be sustained if only the top-performing three or even one species were present (Sullivan et al. 2007). As the authors point out, however, if more functions are measured, it is likely that more species will be found to play important roles. Thus, in a Georgia salt marsh, snails, crabs, and fungi affected different ecosystem functions. Although individual functions tended not to be affected by species richness, multi-functionality increased with species richness (Hensel and Silliman 2013).

Salt marsh plant communities are often dominated by single species, and these species can be genetically diverse at small spatial scales (Richards et al. 2004, Travis and Hester 2005, Hughes and Lotterhos 2014). Genotypes vary in their morphological and growth traits (Seliskar et al. 2002, Travis and Grace 2010). As a result, different genotypes interact in different ways with competitors and consumers (Zerebecki et al. 2017), and mixtures of different genotypes out-perform genetically uniform stands (Wang et al. 2012). This has clear implications for restoration projects: seed additions or plantings should include a range of plant genotypes.

5.2 Top-down Processes

5.2.1 Herbivory

Salt marsh plants – both angiosperms and microalgae – are eaten by a variety of herbivores of both marine and terrestrial origin (Pfeiffer and Wiegert 1981, Pennings and Bertness 2001) (see Fig. 5.1). Common herbivores in salt marshes include crabs (Crichton 1960), snails (Silliman and Zieman 2001), insects (Denno 1980, Stiling and Strong 1983), geese (Zacheis et al. 2001), rodents (Lynch et al. 1947, Gedan et al. 2009), hares (Van Der Wal et al. 2000), and horses (Levin et al. 2002). In Europe, sheep and cattle are grazed in marshes (Ranwell 1961, Jensen 1985). Herbivory affects plant production, species composition, and succession (He and Silliman 2016). Its role is context dependent, varying with elevation and latitude. In general, studies of herbivory in salt marshes have been conducted

Figure 5.1 A: The orthopteran *Orchelimum fidicinium* eating *S. alterniflora* in a Georgia salt marsh. B: Horses, *Equus ferus*, grazing on *Spartina alterniflora* in a North Carolina salt marsh. C: Sheep *Ovis aries* grazing in a French salt marsh. D: The parasitic plant *Cuscuta* sp. parasitizing host plants in a salt marsh in Colombia. E: The decapod *Helice tientsinensis* feeding on *Suaeda salsa* in a Chinese salt marsh. F: The crane, *Grus japonensis*, searching for prey in a Chinese salt marsh. Photo credits: A, B, E, F: Qiang He; C, D: Steven Pennings. (A black and white version of this figure will appear in some formats. For the color version, please refer to the plate section.)

by community ecologists, rather than ecosystem ecologists. As a result, our growing understanding of herbivory in salt marshes has yet to fully inform studies of primary production, energy flow, and other marsh functions.

Experimental studies have shown that natural levels of grazing can markedly reduce plant biomass and reproduction in salt marshes (Rowcliffe et al. 1998, He and Silliman 2016). In extreme cases, geese, crabs, and snails can transform vegetated salt marshes into unvegetated mudflats (Srivastava and Jefferies 1996, Silliman et al. 2005, Holdredge et al. 2008, Altieri et al. 2012, He et al. 2017b). Because most estimates of primary production in salt marshes do not account for production lost to herbivores, they probably underestimate primary production in areas where herbivores are abundant (Silliman and Bortolus 2003). In addition, herbivores commonly suppress sexual reproduction of salt marsh plants (Ellison 1991, Bertness and Shumway 1992, Rand 2004, He and Silliman 2016, Li and Pennings 2017). Grazing can also suppress benthic microalgal biomass, although results are variable (Pace et al. 1979, Darley et al. 1981, Armitage and Fong 2004).

When herbivores feed selectively, they alter plant composition by suppressing favored food plants (Zacheis et al. 2001) and changing the strength and nature of plant–plant interactions (He and Silliman 2016). For example, grazing by snow geese in Canada reduced plant species richness (Bazely and Jefferies 1986), and grazing by crabs in China affected plant zonation patterns (He et al. 2015). Selective feeding can also affect succession and plant invasions (Gedan et al. 2009, Li et al. 2014b). For example, in the Netherlands, herbivory by geese and hares retarded succession in the low marsh, but did not affect species composition in the high marsh (Van Der Wal et al. 2000, Kuijper and Bakker 2005). Conversely, spread of an invasive *Spartina* hybrid in California was facilitated because geese preferred to feed on native *Spartina* (Grosholz 2010). Herbivory-induced changes in salt marsh vegetation can affect multiple other marsh species by changing abiotic conditions and habitat structure (Levin et al. 2002).

Because abiotic conditions in the marsh affect herbivores both directly and indirectly (through effects on plant quality), and because predator abundance varies across the marsh, the importance of herbivory varies across abiotic gradients within the marsh (Hacker and Bertness 1995, Denno et al. 2005, Alberti et al. 2007a, Marczak et al. 2013, Li and Pennings 2016).

The effect of some herbivores also varies with latitude (He and Silliman 2016). Small ectothermic herbivores damaged plants more strongly at low versus high latitudes, despite better plant defenses at low latitudes (Pennings and Silliman 2005, Pennings et al. 2009, Wieski and Pennings 2014a). Latitudinal variation in the top-down impact of larger-bodied, endothermic herbivores, however, was not detected (He and Silliman 2016).

5.2.2 Predation

One reason that salt marshes are nursery grounds for many species is that the thick vegetation limits access by predators. Nevertheless, the abundance of prey in the nursery attracts predators that use marshes as feeding grounds. As a result, predation affects the behavior, abundance, and distribution of a range of salt marsh animals, both marine and terrestrial (Denno et al. 2003, Cresswell et al. 2010). Predation affects salt marsh animals both through direct consumption and by changing prey behavior (Kimbro 2012).

Predation in salt marshes varies spatially and temporally, and is mediated by a variety of abiotic and biotic factors that include tidal cycles, weather and climate, prey abundance, body size, and habitat structure. There are many examples that illustrate this spatial and temporal variation. Many large fish and crustaceans can only utilize the marsh habitat and prey on benthic invertebrates when the marsh is flooded by tides (Minello et al. 2003), and their predation intensity often decreases with increasing elevation (Lewis and Eby 2002). Redshanks (a type of shorebird) forage in riskier locations at low temperatures but safer locations when it is raining (Hilton et al. 1999). With increasing prey abundance, the proportion of planthoppers killed by wolf spiders first increased and then decreased, while mortality of three other sap-feeding insects steadily decreased, indicating different functional responses of predators (Denno et al. 2003). Blue crabs were much more effective at feeding on marsh periwinkles when their size ratios (crab width: snail length) were greater than six (Schindler et al. 1994). Finally, the presence of habitat structure, such as plants, pools, and creeks, also affects where and to what extent predators can access prey (Lewis and Eby 2002, Finke and Denno 2006).

Predation not only affects the abundance and distribution of prey, but also has cascading impacts on lower trophic levels in salt marshes. Such "trophic cascades" are common in salt marshes (He and Silliman 2016). In a New Jersey salt marsh, for example, the presence of arthropod predators enhanced productivity of the grass *S. alterniflora* by suppressing the population of planthopper herbivores. Moreover, higher predator diversity promoted antagonistic interactions among predators, weakened their collective, top-down control on herbivore populations, and dampened their positive effects on *S. alterniflora* productivity (Finke and Denno 2004). In southeastern US salt marshes, a trophic cascade generated by blue crab predation on periwinkle snails positively affected *S. alterniflora* productivity (Silliman and Bertness 2002). In New England salt marshes, predator removal led to increases in the abundance of a grazing crab that denuded salt marsh vegetation on creek edges (Bertness et al. 2014). Omnivores can also generate trophic cascades if they feed preferentially on herbivores rather than plants (Ho and Pennings 2008).

Our understanding of predation and its impacts in salt marshes is still insufficient in many areas. As noted above (Section 5.1.3), we lack an understanding of the complete structure of salt marsh food webs and their complexity across space and time. There is also a need for a greater understanding of the impacts of trophic cascades, apex predators, and predator diversity in salt marshes (He and Silliman 2016). The predators that have been investigated in previous studies are often small, primary predators, such as spiders, crabs or fish, that are easy to work with. Salt marsh habitats, however, still retain top predators, such as alligators, large-bodied shorebirds, sharks, and sea otters (Nifong and Silliman 2013), and there is great potential for taking advantage of this to understand the ecosystem roles of large predators.

5.2.3 Disease and Parasitism

Disease and parasitism likely play important roles in the ecology of salt marshes, but our understanding of how parasites affect host populations and communities remains rudimentary. The importance of many parasites has probably been overlooked due to their small size and cryptic lifestyle. It is likely, however, that parasites of both plants and animals are abundant and ecologically important in salt marshes.

Salt marsh plants are hosts of a variety of parasites, such as bacteria, fungi, and parasitic plants (Raybould et al. 1998). Parasitic plants often preferentially attack dominant plant species, reducing competition, altering plant zonation patterns, and increasing plant diversity and heterogeneity (Pennings and Callaway 1996, Callaway and Pennings 1998, Grewell 2008b, a). The impact of bacterial and fungal pathogens on ecological processes is not well studied, although decomposer fungi can suppress plant biomass (Silliman and Newell 2003), and pathogenic fungi associated with introduced *S. alterniflora* have been implicated in the decline of native plants in China (Li et al. 2014a). A few studies have looked at mycorrhizae in salt marshes (Daleo et al. 2008), but in general plant–soil feedbacks are also not well studied.

Salt marsh insects are attacked by a variety of parasitoids, and the top-down effect of these parasitoids can affect herbivore populations, herbivory rates, and plant performance in a trophic cascade (Moon and Stiling 2004, Stiling and Moon 2005). "Marine" animals in salt marshes and estuaries also have a wide variety of parasites (Lafferty et al. 2006, Blakeslee et al. 2012, Byers et al. 2014, Malek and Byers 2017). These parasites can be abundant (Pung et al. 2002), with a biomass exceeding that of top predators (Kuris et al. 2008), and their importance can be increased by anthropogenic eutrophication (Johnson and Heard 2017). Although the impact of parasites on host invertebrate populations is not well known, trematode parasites castrate salt marsh snails (Lafferty 1993), increase mortality of blue crabs several-fold (Shields and Squyars 2000), and alter behavior of salt marsh fish in ways that increase vulnerability to predation by the parasite's final host (Lafferty and Morris 1996).

5.2.4 Interaction of Top-down and Bottom-up Factors

Top-down processes in salt marshes are often regulated by bottom-up factors, such as nutrients, salinity, and flooding. Bottom-up factors may amplify, overwhelm, weaken, or have no effects on the strength of top-down control (Elschot et al. 2017).

Nutrients often amplify the impact of herbivory in salt marshes. In New England, nutrients increased insect herbivory, and this eliminated the direct, positive effects of nutrients on plant biomass (Bertness et al. 2008). The amplifying effect of nutrients on top-down control was found to be general across multiple herbivore species in both natural and livestock-grazed salt marshes (He and Silliman 2015, He and Silliman 2016). The strength of nutrient effects on herbivory, however, varied from site to site. In particular, nitrogen addition more strongly affected herbivory at higher latitudes (He and Silliman 2015). This was because nitrogen more strongly increased plant foliar nitrogen at higher latitudes and higher plant nitrogen concentrations generally stimulate herbivory. In contrast to the general result that nutrients amplify top-down control by herbivores, nutrients weakened the top-down impact of wolf-spider predation on planthoppers in a salt marsh, likely because planthoppers could better escape predation when occupying high-quality plants (Denno et al. 2003).

Salinity and flooding also affect top-down control in salt marshes. Increased salinities exacerbated top-down control of *S. alterniflora* by periwinkle snails in southeastern US salt marshes, likely because salinity stress weakened the resistance of *S. alterniflora* to grazing (Silliman et al. 2005). This effect of salinity on top-down control, however, was not detected in an Argentinian salt marsh (Daleo et al. 2015). Marine consumers (e.g., marsh crabs and blue crabs) often more strongly suppress lower trophic levels in marsh areas that are flooded more frequently (Alberti et al. 2010b, He et al. 2015, He and Silliman 2016). In contrast, terrestrial consumers, such as raccoons and rabbits, should more strongly suppress lower trophic levels in less frequently flooded marsh areas.

How climate affects top-down control in salt marshes has been gaining attention recently. Although we know of no studies that have experimentally examined the effect of warming on top-down control in salt marshes, increasing temperature along latitudinal gradients was found in a meta-analysis to enhance the negative effect of ectothermic herbivores on plant biomass, but to have no effect on that of endothermic herbivores (He and Silliman 2016). The impact of crab grazing in salt marshes often decreases in drier soils (Alberti et al. 2007b, He and Silliman 2016). Grazing impacts, however, may intensify during seasonal or multiyear droughts and act in concert with drought stress to weaken plants (Silliman et al. 2005, He et al. 2017b). Further studies are crucial to improving our current understanding of the impact of changing climate on herbivory, predation, and trophic cascades in marshes.

We are only beginning to understand how bottom-up and top-down factors interact to affect salt marsh structure and function (He and Silliman 2016). Salt marsh ecologists still know little about the relative impacts of abiotic factors such as temperature and salinity on species at different trophic levels (Moon et al. 2000, Moon and Stiling 2002, 2004), and how bottom-up and top-down factors interact in different ways to affect ecosystem functions and services of salt marshes. Integrating both top-down and bottom-up processes into salt marsh ecology also requires that we understand their relative importance and interactions across multiple temporal and spatial scales.

5.3 Non-trophic Processes

5.3.1 Behavior

Conditions in salt marshes vary over the course of a day, over a tidal cycle, and seasonally. Animals in salt marshes adjust their behavior on all these timescales so as to minimize abiotic stress, competition, and predation risk. Better understanding these behaviors will shed new insight into ecological interactions, but is slowed by the difficulties in observing behavior in dense vegetation and turbid water.

Behavior of many organisms varies over the course of the day. For example, the crab *Sesarma reticulatum* is most active just after dusk, whereas its relative *Armases* (=*Sesarma*) *cinereum* is active throughout more of the night (Seiple 1981). Other organisms move on and off the marsh as the tide rises, such that a suite of "terrestrial" organisms like birds and raccoons is periodically replaced on the marsh surface with a suite of "marine" organisms like swimming crabs, shrimp, and fish. Less mobile organisms like snails and insects may migrate up the stems of marsh plants as the tide rises in order to escape submergence and marine predators, or simply tolerate inundation (Foster and Treherne 1976, Hovel et al. 2001). Because the diel and tidal cycles have different periods, organisms that respond to both will be most active when the two cycles align (Seiple 1981). Organisms may also migrate across the marsh seasonally. Spiders and insects may overwinter in the high marsh, which is relatively stable year-round, and then migrate to the low marsh during summer to exploit higher-quality but ephemeral resources (Döbel et al. 1990, Denno et al. 1996). Similarly, migratory vertebrates like birds and large fish may be part of the food web only at certain times of the year (Pautzke et al. 2010).

Organisms also respond behaviorally to each other, avoiding competitors and predators. Crowding by conspecifics or congenerics stimulates emigration in *Prokelisia* planthoppers, with effects differing among species depending on competitive ability (Denno and Roderick 1992). Snails avoid areas occupied by dominant competitors (Lee and Silliman 2006) and avoid crab predators by climbing higher on marsh plants (Davidson et al. 2015). Small fish and shrimp remain in wet areas on the marsh platform at low tide so as to avoid predation from larger fish that move from the marsh platform into small creeks at low tide (Kneib 1987).

5.3.2 Competition

Because salt marsh communities are typified by relatively few species that occur at fairly high densities, they have been useful model systems for studying inter- and intra-specific competition (Pennings and Bertness 2001). The simplest conceptual model of competition that has been applied to salt marshes posits a tradeoff between competitive ability and stress tolerance. This model has proved very useful in explaining the distributions of salt marsh plants across the intertidal and along the estuary; how well it can explain the distribution patterns of marsh animals is unknown.

At any single site, salt marsh plants are typically arranged in distinct bands, or "zones," from the low to the high intertidal (Chapman 1974). In some cases, such as in the Northeast USA, this is explained by a tradeoff between competitive ability and stress tolerance. Species that are strong competitors occupy the high intertidal zone, which is not very stressful, and species that are poor competitors, but better at tolerating flooding with seawater, occupy lower intertidal zones (Bertness 1991b, a, Bertness et al. 1992b, Pennings and Moore 2001). A similar tradeoff between competitive ability and stress tolerance can explain the turnover in plant species from fresh to brackish to salt marshes along the salinity gradient of the estuary. Strong competitors occupy tidal fresh marshes, and exclude poor competitors from these habitats. The inferior competitors are more tolerant of saline conditions, and occupy brackish and salt marshes (Crain et al. 2004, Engels and Jensen 2010, Guo and Pennings 2012). Because these tradeoffs are a function of abiotic conditions, changing conditions – for example by increasing nutrient availability – can change competitive hierarchies and zonation patterns (Levine et al. 1998, Pennings et al. 2002, Schoolmaster and Stagg 2018). Although this simple model of a tradeoff between competition and stress tolerance along a single stress gradient has been very useful, a number of studies have accumulated where it only partially explains plant distribution patterns (Pennings and Callaway 1992, Fariña et al. 2009, Guo and Pennings 2012, Fariña et al. 2017). In these cases, it is likely that abiotic gradients are more complicated, and a more sophisticated model is needed that considers multiple abiotic gradients that simultaneously affect plant performance (Keddy 1990).

Animals in salt marshes are also affected by competition, but this has been less studied (Pennings and Bertness 2001). The snail *Melampus bidentatus* is common at high latitudes on the Atlantic Coast of the USA but rare at low latitudes due to competition from the snail *Littoraria irrorata* (Lee and Silliman 2006). An introduced snail in California marshes outcompetes a native snail by more efficiently using diatoms as food (Byers 2000). Fiddler crab species compete in the laboratory (Teal 1958) but it is less clear if competition determines their distributions in the field. Stem-boring insects living inside marsh grasses kill potential competitors that they encounter within the same stem, but it is unclear whether this has a strong influence on distribution patterns (Stiling and Strong 1983, 1984).

Competition can also trade off with dispersal ability to explain changes in distributions of both plant and animal species over time. Marsh patches that are disturbed may initially be occupied by plants that are poor competitors but good at dispersal (Bertness et al. 1992a). These "fugitive species" are eventually outcompeted by competitive dominants as the latter reinvade (Section 5.3.5). Similarly, ephemeral but high-quality marsh habitats are colonized by herbivorous insects that are good at dispersal, whereas permanent but lower-quality habitats are occupied by better competitors (Denno et al. 2000).

5.3.3 Facilitation and Mutualism

Organisms in salt marshes can positively affect one another (Hughes et al. 2014). *S. alterniflora* production at low elevations in Atlantic Coast salt marshes, for example, is stimulated by ribbed mussels (*Geukensia demissa*) (Fig. 5.2), because mussels bind sediments, prevent erosion, and increase soil nutrients (Bertness 1984). At upper elevations *S. alterniflora* production is also stimulated by ribbed mussels, because mussel mounts improve drainage and reduce soil salinity (Angelini et al. 2016). Mussels also facilitate invertebrate biomass and diversity (Crotty et al. 2018). Similarly, burrowing by fiddler crabs reduces soil waterlogging, alters nutrient cycling, and enhances *S. alterniflora* growth (Montague 1982, Bertness 1985). Plants can also affect each other positively. For example, in a New England salt marsh, the presence of *Juncus gerardii* ameliorated soil salinities and increased growth of *I. frutescens* (Bertness and Hacker 1994). Besides amelioration of abiotic stresses, amelioration of consumer pressure can also be an important but overlooked mechanism of facilitations in salt marshes (He and Cui 2015).

Figure 5.2 A: Mussels, *Geukensia demissa*, facilitating the grass *S. alterniflora* in a US salt marsh. B: Zonation between the grass *S. alterniflora* on the left and the rush *Juncus roemerianus* on the right, in a Georgia salt marsh. C: The shrub *Tamarix chinensis* (center) facilitates the forb *Suaeda glauca* (arrow) beneath its canopy in a Chinese salt marsh. The salt-tolerant forb *S. salsa* occurs outside the shrub canopy. D: The arc-shaped head of an eroding creek in a Georgia salt marsh is both cleared of vegetation and excavated by the crab *Sesarma reticulatum*. E: Slumping of the creekbank in a Georgia salt marsh. F: Wrack disturbance along the creekbank of a Georgia salt marsh. Photo credits: A, B, C: Qiang He; D, E, F: Steven Pennings. (A black and white version of this figure will appear in some formats. For the color version, please refer to the plate section.)

Facilitative interactions vary with species and along environmental gradients. Species that are competitive dominants and those that are more tolerant of abiotic stresses are likely to be beneficiaries and benefactors, respectively (He et al. 2012). For example, in a California marsh, *Arthrocnemum subterminale* facilitated winter annuals that were sensitive to salt stress, but competed with annuals that were salt tolerant (Callaway 1994). Plant phylogenetic relatedness can also affect the net outcome of plant interactions (Zhang et al. 2017), though studies from salt marshes are few. At local scales, facilitation often increases and competition decreases with increasing abiotic and biotic stress, as predicted by the stress-gradient hypothesis (Bertness and Callaway 1994, He et al. 2013). At broader scales, plant interactions in New England were more facilitative in warmer, southern sites than in cooler, northern sites, likely because salinity stress was higher and the ameliorating effect of neighbors on salinity stress was stronger at warmer sites (Bertness and Ewanchuk 2002). This latitudinal pattern in facilitation, however, did not extend to geographic scales, because low-latitude marshes lacked species sensitive to salinity. So although low-latitude marshes were saltier, they were structured by competition rather than facilitation (Pennings et al. 2003). Other outstanding questions in understanding facilitation include how the relative importance of competition and facilitation varies across the full spectrum of stress or environmental gradients, and to what extent competition and facilitation affect the realized niches of salt marsh species (He and Bertness 2014).

Facilitative interactions have important consequences for salt marsh ecosystems, affecting not only species distributions (He et al. 2012) and diversity (Hacker and Bertness 1999), but also salt marsh function (Bilkovic et al. 2017) and resilience to environmental change (Angelini et al. 2016). Understanding facilitation is also of importance to marsh management practices (Chapter 16). For instance, salt marsh restoration may be improved by encouraging facilitative interactions among plants (Callaway et al. 2003, Silliman et al. 2015) or between invertebrates and plants (Bilkovic et al. 2017).

5.3.4 Ecosystem Engineering

Ecosystem engineers are organisms that actively or passively create or modify habitats. In non-trophic, facilitative interactions, benefactors are actually one type of ecosystem engineer. As described above (Section 3.3), salt marsh plants can buffer waves, trap sediments, oxygenate soils, reduce solar radiation, reduce soil water evaporation, reduce heat stress, and ameliorate salt buildup at surface soils. Indeed, sediment trapping by ecosystem engineering plants is a primary mechanism driving the formation of marshes (Chapter 3). When ecosystem engineers are introduced outside their native range, they can have large effects on natural systems. Examples include the introduction of *S. alterniflora* to Europe and China, and the introduction of novel genotypes of *Phragmites australis* to the USA (Chapter 13). Similarly, some marsh animals are also ecosystem engineers. For example, mussels can bind sediments, prevent erosion, or increase soil nutrients, and burrowing crabs can improve soil drainage, ameliorate soil anoxic stress, and affect soil biogeochemistry, such as decomposition and carbon sequestration (Section 5.3.3). Interestingly, by oxygenating soils, burrowing crabs have been shown to promote mycorrhizal associations on salt marsh plant roots, thereby activating a mycorrhizal mutualism (Daleo et al. 2007). Additionally, ecosystem engineers can also create novel biotic structures that are utilized by other species as shelters from predation or competition (Section 5.3.3). In a Chinese salt marsh, for example, the presence of shrubs creates a shelter microhabitat for crabs to avoid predation by shorebirds (He and Cui 2015). In all these cases, ecosystem engineers enhance habitat quality and positively affect intra- or interspecific organisms.

Although the positive effects of ecosystem engineers have attracted much attention, the negative effects of ecosystem engineers should not be neglected. For example, burrowing crabs promote marsh erosion in Argentina (Escapa et al. 2007) and New England (Coverdale et al. 2013). In other cases, engineering of tidal creeks by burrowing crabs may enhance salt marsh primary production by promoting drainage efficiency (Vu et al. 2017), but this depends on how common the crabs are and where they are located. Similarly, goose grubbing can promote erosion in coastal marshes in British Columbia (Kirwan et al. 2008), and bioturbation and herbivory by infaunal polychaetes may drive creek erosion in salt marshes in southeast England (Morris et al. 2004, Paramor and Hughes 2004). Ecosystem engineers can also negatively affect plants. For instance, by burying seeds, burrowing crabs can suppress the recruitment of pioneer plants in bare patches in a salt marsh in Argentina (Alberti et al. 2010a).

A knowledge gap in our understanding of ecosystem engineers is how their effects vary with context. A burrowing crab in Argentina, for example, benefits plants by engineering in anoxic stressful habitats, but harms them by herbivory in benign habitats (Daleo and Iribarne 2009). This finding supports the hypothesis that ecosystem engineers ameliorate physical stress in extreme physical environments, although it is contrary to the hypothesis that in physically benign environments where competitor and consumer pressure is typically high, ecosystem engineers provide competitor- or predator-free space (Crain and Bertness 2006). These hypotheses, however, need to be further tested empirically.

5.3.5 *Disturbance*

Disturbances are temporary events that alter community or ecosystem structure, often by creating or destroying habitat. Disturbances in salt marshes include deposition of sediments by storms, deposition of wrack by tides, erosion by ice, creekbank calving, and fire.

Storms mobilize sediments that can be deposited into salt marshes in layers several centimeters thick (Rejmanek et al. 1988, Nyman et al. 1995, Donnelly et al. 2001a, 2001b). This is likely to smother marsh animals and vegetation. Deposition of sediment may also create entirely new habitat suitable for salt marsh vegetation by raising the elevation of intertidal mudflats (Osgood et al. 1995).

Wrack, which consists of dead plant material, algae, and seagrass leaves, floats into the marsh at high tide and is trapped by vegetation (Ranwell 1961, Beeftink 1977, Reidenbaugh and Banta 1980, Hackney and Bishop 1981). Thick wrack mats can crush taller vegetation, and will kill plants if they remain in place long enough (Bertness and Ellison 1987, Valiela and Rietsma 1995, Li and Pennings 2016). At the same time, wrack is beneficial to marsh organisms: it shades the soil and limits evaporative concentration of salts, provides food and habitat for marsh invertebrates, and provides nutrients to the soil as it decomposes (Pennings and Richards 1998, Hanley et al. 2017).

At high latitudes, water on the marsh surface or edge may freeze in the winter (Redfield 1972). When the ice breaks into blocks that are carried away by the tides, these blocks may carry chunks of marsh sediment with them. This process erodes the surface and edge of the marsh, and can move blocks of marsh soil either down into the bottom of creeks or up onto the top of the marsh platform, in either case moving organisms to intertidal elevations to which they are not adapted (Richard 1978, Hardwick-Witman 1985). Scars on the marsh surface created by the removal of marsh sediments become waterlogged, which inhibits vegetation recovery (Ewanchuk and Bertness 2003).

Marsh edges – either facing open water or along creekbanks – often have a steep slope. Because the top few decimeters of the marsh surface are reinforced by a network of live and dead plant roots and rhizomes, these edges often collapse in large blocks that slide down the face of the creekbank, a process known as calving or mass wasting (Basan and Frey 1977, Frey and Basan 1978). This process is most noticeable where wave action tends to undercut the marsh edge by eroding soft sediments below the rooting zone (van de Koppel et al. 2005). The result is that chunks of marsh are transported to lower elevations that are not suitable for associated organisms. This process may be accelerated by eutrophication (Deegan et al. 2012).

Finally, many marshes are disturbed by fire (Turner 1987, Taylor et al. 1994, Bortolus and Iribarne 1999). Grasses and rushes do not drop dead leaves, and so may accumulate standing dead biomass that will burn if it becomes dry enough. The effects vary depending on whether the fire is intense enough to kill roots and rhizomes (Baldwin and Mendelssohn 1998).

Whatever their nature, disturbances create or free up resources for organisms. The sequence of recovery of disturbed areas through succession is influenced both by colonization of new organisms (Ellison 1987) and by the ability of existing organisms to survive and regrow following the disturbance (Bertness and Ellison 1987, Li and Pennings 2017). The pathway and speed of succession is influenced by physical stress (Bertness et al. 1992a, Shumway and Bertness 1994), competition (Brewer et al. 1998), facilitation (Bertness and Shumway 1993), and consumers (Angelini and Silliman 2012, He et al. 2017b).

Disturbance in salt marshes has been primarily been studied either by geologists or ecologists, and there is considerable scope for better integrating their approaches. In addition, most of the ecological work on disturbance has focused on community ecology, and there is considerable scope for understanding how disturbance affects nutrient cycling and primary production across the marsh landscape. Finally, we need a better understanding of the importance of physical disturbances in affecting marsh communities and ecosystem functions at broad temporal and spatial scales.

5.4 Scaling-up

5.4.1 Scaling-up to the Landscape

Historically, salt marsh ecologists have focused their work at single sites, typically chosen for their convenience. Within those sites, most of the focus was on variation in marsh structure and function across elevation. Other studies focused on variation along the estuarine salinity gradient (Odum 1988). Even early comparative studies, however, found that different marshes – even of a single "salinity type" and within a single geographic area – looked and worked differently (Ganong 1903, Penfound and Hathaway 1938, Purer 1942). With the advent of remote sensing and additional comparative studies of marshes, it is now clear that there is tremendous variation among marshes in structure and function (Schalles et al. 2013). This variation stems from differences in fetch, elevation, tide range, freshwater input, sediment input, and other factors. As a result, edge erosion (Chapter 11), primary production (Schalles et al. 2013), the dominant plant and animal species (Schalles et al. 2013), interactions among plants species (Bertness and Ewanchuk 2002), physical disturbance (Brewer and Bertness 1996, Li and Pennings 2016), and recruitment (B. R. Silliman, unpublished data) can all vary severalfold among nearby marshes.

This creates two related problems for studies of marsh ecology. First, it is dangerous to assume that any site is "typical" without a good understanding of variation among sites. As a result, how general the results of any particular study are is not always clear. Second, variability among sites makes it problematic to scale up from studies at single sites to the entire landscape. The solution to both problems is to deliberately stratify observational and experimental work across the landscape so as to document the variation that exists at the landscape scale.

5.4.2 Geographic Patterns

Salt marshes occur worldwide on low-energy coastlines, except in the tropics where they are replaced by mangroves (Chapman 1974). As discussed above, species richness is low due to strong abiotic "filters" that exclude most species from salt marsh habitats. Many salt marsh species, however, have large geographic distributions.

Bottom-up factors interact with competition to determine plant composition at the largest geographic scales. As precipitation declines and porewater salinities increase, graminoid-dominated marshes give way to succulent-dominated marshes and then to unvegetated salt flats, and as winter temperatures rise above the lethal threshold for mangroves, marshes are replaced by mangroves (Osland et al. 2013, Cavanaugh et al. 2014, Feher et al. 2017, Gabler et al. 2017). Within a vegetation type, climate also affects bottom-up processes. For example, salt marshes vary across latitude in salinity stress (Bertness and Pennings 2000), soil organic content (Craft 2007), productivity (Turner 1976, Kirwan et al. 2009), disturbance (Pennings and Bertness 2001), and plant composition (Pennings et al. 2003, Rachlin et al. 2012). As discussed above, variation in climate across latitude also correlates with plant–herbivore interactions (Pennings and Silliman 2005, He and Silliman 2016) and with plant–plant interactions (Bertness and Ewanchuk 2002).

Because many marsh species are widely distributed, their distributions encompass strong abiotic gradients. Geographic variation in abiotic conditions, species interactions and dispersal has led to geographic variation in population structure, hybridization, and cryptic speciation within widely distributed species (Ainouche et al. 2003, 2009, Blum et al. 2007, Diaz-Ferguson et al. 2010, Bernik et al. 2016).

5.5 Human Alterations of Salt Marsh Ecology in the Anthropocene

Ecological processes that structure salt marshes are not constant over time. For centuries, salt marshes have been used by humans for grazing livestock and for salt production. But in the Anthropocene, salt marshes, such as those in coastal China (He et al. 2014), are facing pressure from human activities at unprecedented scales and intensities. Human activities can change both bottom-up and top-down processes in salt marshes.

A variety of human activities affect marshes through bottom-up processes. At individual sites, filling and ditching, shoreline hardening, and increased inputs of nutrients and pollutants change marsh hydrology, edaphic conditions, and water quality. At larger scales, marsh bottom-up processes are altered by climate warming, drought, sea-level rise, and rising CO_2 concentrations (Chapter 17).

Human activities also affect marsh top-down and non-trophic processes. In New England salt marshes, for instance, overharvesting of large-bodied predators has caused outbreaks of smaller herbivorous crabs that denude and erode marsh creekbanks (Altieri et al. 2012). Human introductions of exotic plants, animals, and diseases also impact native marsh communities by creating novel interactions with introduced species (Chapter 13).

Understanding human alterations of marsh ecology is challenging, because it involves investigating how bottom-up, top-down, and non-trophic processes interact with multiple human activities. Drought or eutrophication, for instance, can impact marsh primary production not only by changing marsh physical conditions, but also by simultaneously altering the top-down impacts of grazers (Section 2.4) or competition between native and exotic species (Section 3.2). Despite these challenges, salt marsh ecology must evolve from a pure study of nature to one that fully integrates human dimensions, so that it can better inform conservation and management activities in an increasingly human-dominated biosphere.

Acknowledgements

We thank J. Byers, J. Goeke, K. Lafferty, D. Strong, and Y. Zhang for comments on the manuscript. This material is based upon work supported by the National Science Foundation through the Georgia Coastal Ecosystems Long-Term Ecological Research program under Grant No. OCE-1237140. This is contribution number xxx from the University of Georgia Marine Institute.

References

Ainouche, M. L., Baumel, A., Salmon, A., and Yannic, G. 2003. Hybridization, polyploidy and speciation in *Spartina* (Poaceae). *New Phytologist*, **161**: 165–172.

Ainouche, M. L., Fortune, P. M., Salmon, A., Parisod, C., Grandbastien, M.-A., Fukunaga, K., Ricou, M., and Misset, M.-T. 2009. Hybridization, polyploidy and invasion: lessons from *Spartina* (Poaceae). *Biological Invasions*, **11**: 1159–1173.

Alber, M., Swenson, E. M., Adamowicz, S. C., and Mendelssohn, I. A. 2008. Salt marsh dieback: an overview of recent events in the US. *Estuarine, Coastal and Shelf Science*, **80**: 1–11.

Alberti, J., Escapa, M., Daleo, P., Casariego, A. and Iribarne, O. 2010a. Crab bioturbation and herbivory reduce pre- and post-germination success of Sarcocornia perennis in bare patches of SW Atlantic salt marshes. *Marine Ecology Progress Series*, **400**: 55–61.

Alberti, J., Escapa, M., Daleo, P., Iribarne O., Silliman, B., and Bertness, M. 2007a. Local and geographic variation in grazing intensity by herbivorous crabs in SW Atlantic salt marshes. *Marine Ecology Progress Series*, **349**: 235–243.

Alberti, J., Méndez Casariego, A., Daleo, P., Fanjul, E., Silliman, B. R., Bertness, M. D., and Iribarne, O. 2010b. Abiotic stress mediates top-down and bottom-up control in a Southwestern Atlantic salt marsh. *Oecologia*, **163**: 181–191.

Alberti, J., Montemayor, D., Alvarez, F., Casariego, A. M., Luppi, T., Canepuccia, A., Isacch, J. P., and Iribarne, O. 2007b. Changes in rainfall pattern affect crab herbivory rates in a SW Atlantic salt marsh. *Journal of Experimental Marine Biology and Ecology*, **353**: 126–133.

Altieri, A. H., Bertness, M. D., Coverdale, T. C., Herrmann, N. C., and Angelini, C. 2012. A trophic cascade triggers collapse of a salt-marsh ecosystem with intensive recreational fishing. *Ecology*, **93**: 1402–1410.

Angelini, C., Griffin, J. N., Van de Koppel, Lamers, J. L. P. M., Smolders, A. J. P., Derksen-Hooijberg, M., Van der Heide, T. and Silliman, B. R. 2016. A keystone mutualism underpins resilience of a coastal ecosystem to drought. *Nature Communications*, **7**: 12473.

Angelini, C., and Silliman, B. R. 2012. Patch size-dependent community recovery after massive disturbance. *Ecology*, **93**: 101–110.

Armitage, A. R., and Fong, P. 2004. Upward cascading effects of nutrients: shifts in a benthic microalgal community and a negative herbivore response. *Oecologia*, **139**: 560–567.

Baldwin, A. H., and Mendelssohn, I. A. 1998. Response of two oligohaline marsh communities to lethal and nonlethal disturbance. *Oecologia*, **116**: 543–555.

Basan, P. B., and Frey, R. W. 1977. Actual-palaeontology and neoichnology of salt marshes near Sapelo Island, Georgia. *Geological Journal Special Issue*, **9**: 41–70.

Bazely, D. R., and Jefferies, R. L. 1986. Changes in the composition and standing crop of salt-marsh communities in response to the removal of a grazer. *Journal of Ecology*, **74**: 693–706.

Beeftink, W. G. 1977. The coastal salt marshes of western and northern Europe: an ecological and phytosociological approach. In V. J. Chapman, ed., *Wet Coastal Ecosystems*. Elsevier Scientific Publishing Company, Amsterdam, pp. 109–155.

Bernik, B. M., Li, H., and Blum, M. J. 2016. Genetic variability of *Spartina alterniflora* intentionally introduced to China. *Biological Invasions*, **18**: 1485–1498.

Bertness, M. D. 1984. Ribbed mussels and *Spartina alterniflora* production in a New England salt marsh. *Ecology*, **65**: 1794–1807.

Bertness, M. D. 1985. Fiddler crab regulation of *Spartina alterniflora* production on a New England salt marsh. *Ecology*, **66**: 1042–1055.

Bertness, M. D. 1991a. Interspecific interactions among high marsh perennials in a New England salt marsh. *Ecology*, **72**: 125–137.

Bertness, M. D. 1991b. Zonation of *Spartina patens* and *Spartina alterniflora* in a New England salt marsh. *Ecology*, **72**: 138–148.

Bertness, M. D., Brisson, C. P. Coverdale, T. C. Bevil, M. C. Crotty, S. M., and Suglia, E. R. 2014. Experimental predator removal causes rapid salt marsh die-off. *Ecology Letters*, **17**: 830–835.

Bertness, M. D., and Callaway, R. 1994. Positive interactions in communities. *Trends in Ecology and Evolution*, **9**: 191–193.

Bertness, M. D., Crain, C., Holdredge, C., and Sala, N. 2008. Eutrophication and consumer control of New England salt marsh primary productivity. *Conservation Biology*, **22**: 131 139.

Bertness, M. D., and Ellison, A. M. 1987. Determinants of pattern in a New England salt marsh plant community. *Ecological Monographs*, **57**. 129 147

Bertness, M. D., and Ewanchuk, P. J. 2002. Latitudinal and climate-driven variation in the strength and nature of biological interactions in New England salt marshes. *Oecologia*, **132**: 392–401.

Bertness, M. D., Gough, L., and Shumway, S. W. 1992a. Salt tolerances and the distribution of fugitive salt marsh species. *Ecology*, **73**: 1842–1851.

Bertness, M. D., and Hacker, S. D. 1994. Physical stress and positive associations among marsh plants. *American Naturalist*, **144**: 363–372.

Bertness, M. D., and Pennings, S. C. 2000. Spatial variation in process and pattern in salt marsh plant communities in Eastern North America. Pages 39–57 in M. P. Weinstein and D. A. Kreeger, eds., *Concepts and Controversies in Tidal Marsh Ecology*. Kluwer Academic Publishers, Dordrecht.

Bertness, M. D., and Shumway, S. W. 1992. Consumer driven pollen limitation of seed production in marsh grasses. *American Journal of Botany*, **79**: 288–293.

Bertness, M. D., and Shumway, S. W. 1993. Competition and facilitation in marsh plants. *American Naturalist*, **142**: 718–724.

Bertness, M. D., Wikler, K., and Chatkupt, T. 1992b. Flood tolerance and the distribution of *Iva frutescens* across New England salt marshes. *Oecologia*, **91**: 171–178.

Bilkovic, D. M., Mitchell, M. M., Isdell, R. E., Schliep, M., and Smyth, A. R. 2017. Mutualism between ribbed mussels and cordgrass enhances salt marsh nitrogen removal. *Ecosphere*, **8**: e01795.

Blakeslee, A. M. H., Altman, I., Miller, A. W., Byers, J. E., Hamer, C. E., and Ruiz, G. M. 2012. Parasites and invasions: a biogeographic examination of parasites and hosts in native and introduced ranges. *Journal of Biogeography*, **39**: 609–622.

Blum, M. J., Bando, K. J., Katz, M., and Strong, D. R. 2007. Geographic structure, genetic diversity and source tracking of *Spartina alterniflora*. *Journal of Biogeography*, **34**: 2055–2069.

Boesch, D. F., and Turner, R. E. 1984. Dependence of fishery species on salt marshes – the role of food and refuge. *Estuaries*, **7**: 460–468.

Bortolus, A., and Iribarne, O. 1999. Effects of the SW Atlantic burrowing crab *Chasmagnathus granulata* on a Spartina salt marsh. *Marine Ecology, Progress Series*, **178**: 79–88.

Bradley, P. M., and Dunn, E. L. 1989. Effects of sulfide on the growth of three salt marsh halophytes of the southeastern United States. *American Journal of Botany*, **76**: 1707–1713.

Brewer, J. S., and Bertness, M. D. 1996. Disturbance and intraspecific variation in the clonal morphology of salt marsh perennials. *Oikos*, **77**: 107–116.

Brewer, J. S., Levine, J. M., and Bertness, M. D. 1998. Interactive effects of elevation and burial with wrack on plant community structure in some Rhode Island salt marshes. *Journal of Ecology*, **86**: 125–136.

Byers, J. E. 2000. Competition between two estuarine snails: implications for invasions of exotic species. *Ecology*, **81**: 1225–1239.

Byers, J. E., Rogers, T. L., Grabowski, J. H., Hughes, A. R., Piehler, M. F., and Kimbro, D. L. 2014. Host and parasite recruitment correlated at a regional scale. *Oecologia*, **174**: 731–738.

Callaway, J. C., Sullivan, G., and Zedler, J. B. 2003. Species-rich plantings increase biomass and nitrogen accumulation in a wetland restoration experiment. *Ecological Applications*, **13**: 1626–1639.

Callaway, R. M. 1994. Facilitative and interfering effects of Arthrocnemum subterminale on winter annuals. *Ecology*, **75**: 681–686.

Callaway, R. M., and Pennings S. C. 1998. Impact of a parasitic plant on the zonation of two salt marsh perennials. *Oecologia*, **114**: 100–105.

Cavanaugh, K. C., Kellner, J. R., Forde, A. J., Gruner, D. S., Parker, J. D., Rodriguez, W., and Feller, I. C. 2014. Poleward expansion of mangroves is a threshold response to decreased frequency of extreme cold events. *Proceedings of the National Academy of Science, USA*, **111**: 723–727.

Cebrian, J. 1999. Patterns in the fate of production in plant communities. *American Naturalist*, **154**: 449–468.

Chapman, V. J. 1974. Salt marshes and salt deserts of the world. In: R. J. Reimold and W. H. Queen, editors. *Ecology of Halophytes*. Academic Press, New York, pp. 3–19.

Coverdale, T. C., Herrmann, N. C., Altieri, A. H., and Bertness, M. D.. 2013. Latent impacts: the role of historical human activity in coastal habitat loss. *Frontiers in Ecology and the Environment*, **11**: 69–74.

Craft, C. 2007. Freshwater input structures soil properties, vertical accretion, and nutrient accumulation of Georgia and U.S. tidal marshes. *Limnology and Oceanography*, **52**: 1220–1230.

Crain, C. M., and Bertness, M. D. 2006. Ecosystem engineering across environmental gradients: implications for conservation and management. *Bioscience*, **56**: 211–218.

Crain, C. M., Silliman, B. R., Bertness, S. L., and Bertness, M. D. 2004. Physical and biotic drivers of plant distribution across estuarine salinity gradients. *Ecology*, **85**: 2539–2549.

Cresswell, W., Lind, J., and Quinn, J. L. 2010. Predator-hunting success and prey vulnerability: quantifying the spatial scale over which lethal and non-lethal effects of predation occur. *Journal of Animal Ecology*, **79**: 556–562.

Crichton, O. W. 1960. Marsh crab: intertidal tunnel-maker and grass-eater. *Estuarine Bulletin*, **5**: 3–10.

Crotty, S. M., Sharp, S. J., Bersoza, A. C., Prince, K. D., Cronk, K., E. Johnson, E., and Angelini, C. 2018. Foundation species patch configuration mediates salt marsh biodiversity, stability and multifunctionality. *Ecology Letters*, **21**: 1681–1692.

Currin, C. A., Newell, S. Y., and Paerl, H. W. 1995. The role of standing dead *Spartina alterniflora* and benthic microalgae in salt marsh food webs: considerations based on multiple stable isotope analysis. *Marine Ecology Progress Series*, **121**: 99–116.

Dai, T., and Wiegert, R. G. 1996a. Estimation of the primary productivity of *Spartina alterniflora* using a canopy model. *Ecography*, **19**: 410–423.

Dai, T., and Wiegert, R. G. 1996b. Ramet population dynamics and net aerial primary productivity of *Spartina alterniflora*. *Ecology*, **77**: 276–288.

Daleo, P., Alberti, J., Bruschetti, C. M., Pascual, J., Iribarne, O., and Silliman, B. R. 2015. Physical stress modifies top-down and bottom-up forcing on plant growth and reproduction in a coastal ecosystem. *Ecology*, **96**: 2147–2156.

Daleo, P., Alberti, J., Canepuccia, A., Escapa, M., Fanjul, E., Silliman, B. R., Bertness, M. D., and Iribarne, O. 2008. Mychorrhizal fungi determine salt-marsh plant zonation depending on nutrient supply. *Journal of Ecology*, **96**: 431–437.

Daleo, P., Fanjul, E., Casariego, A. M., Silliman, B. R., Bertness, M. D., and Iribarne, O. 2007. Ecosystem engineers activate mycorrhizal mutualism in salt marshes. *Ecology Letters* **10**: 902–908.

Daleo, P., and Iribarne, O. 2009. Beyond competition: the stress-gradient hypothesis tested in plant–herbivore interactions. *Ecology*, **90**: 2368–2374.

Darley, W. M., Montague, C. L., Plumley, F. G., Sage, W. W., and Psalidas, A. T. 1981. Factors limiting edaphic algal biomass and productivity in a Georgia salt marsh. *Journal of Phycology*, **17**: 122–128.

Davidson, A., Griffin, J. N., Angelini, C., Coleman, F., Atkins, R. L., and Silliman, B. R. 2015. Non-consumptive predator effects intensify grazer-plant interactions by driving vertical habitat shifts. *Marine Ecology Progress Series*, **537**: 49–58.

de Bettencourt, A. M. M., Neves, R. J. J., Lança, M. J., Batista, P. J., and Alves, M. J. 1994. Uncertainties in import/export studies and the outwelling theory. An analysis with the support of hydrodynamic modelling. In W. J. Mitsch, ed., *Global Wetlands: Old world and new*. Elsevier Science B. V., Amsterdam, pp. 235–256.

Deegan, L. A., Johnson, D. S., Warren, R. S., Peterson, B. J., Fleeger, J. W., Fagherazzi, S., and Wollheim, W. M. 2012. Coastal eutrophication as a driver of salt marsh loss. *Nature*, **490**: 388–392.

Denno, R. F. 1980. Ecotope differentiation in a guild of sap-feeding insects on the salt marsh grass, *Spartina patens*. *Ecology*, **61**: 702–714.

Denno, R. F., Gratton, C., Dobel, H., and Finke, D. L. 2003. Predation risk affects relative strength of top-down and bottom-up impacts on insect herbivores. *Ecology*, **84**: 1032–1044.

Denno, R. F., Lewis, D., and Gratton, C. 2005. Spatial variation in the relative strength of top-down and bottom-up forces: causes and consequences for phytophagous insect populations. *Annales Zoologici Fennici*, **42**: 295–311.

Denno, R. F., Peterson, M. A., Gratton, C., Cheng, J., Langellotto, G. A., Huberty, A. F., and Finke, D. L. 2000. Feeding-induced changes in plant quality mediate interspecific competition between sap-feeding herbivores. *Ecology*, **81**: 1814–1827.

Denno, R. F., and Roderick, G. K. 1992. Density-related dispersal in planthoppers: effects of interspecific crowding. *Ecology*, **73**: 1323–1334.

Denno, R. F., Roderick, G. K., Peterson, M. A., Huberty, A. F., Dobel, H. G., Eubanks, M. D., Losey, J. E., and Langellotto, G. A. 1996. Habitat persistence underlies intraspecific variation in the dispersal strategies of planthoppers. *Ecological Monographs*, **66**: 389–408.

Diaz-Ferguson, E., Robinson, J. D., Silliman, B., and Wares, J. P. 2010. Comparative phylogeography of North American Atlantic salt marsh communities. *Estuaries and Coasts*, **33**: 828–839.

Döbel, H. G., Denno, R. F., and Coddington, J. A. 1990. Spider (Araneae) community structure in an intertidal salt marsh: effects of vegetation structure and tidal flooding. *Environmental Entomology*, **19**: 1356–1370.

Donnelly, J. P., Bryant, S. S., Butler, J., Dowling, J., Fan, L., Hausmann, N., Newby, P., et al. 2001a. 700 yr sedimentary record of intense hurricane landfalls in southern New England. *Geological Society of America Bulletin*, **113**: 714–727.

Donnelly, J. P., Roll, S., Wengren, M., Butler, J., Lederer R., and Webb, III, T. 2001b. Sedimentary evidence of intense hurricane strikes from New Jersey. *Geology*, **29**: 615–618.

Ellison, A. M. 1987. Effects of competition, disturbance, and herbivory on *Salicornia europaea*. *Ecology*, **68**: 576–586.

Ellison, A. M. 1991. Ecology of case-bearing moths (Lepidoptera: coleophoridae) in a New England salt marsh. *Environmental Entomology*, **20**: 857–864.

Elschot, K., Vermeulen, A., Vandenbruwaene, W., Bakker, J. P., Bouma, T. J., Stahl, J., Castelijns, H., and Temmerman, S. 2017. Top-down vs. bottom-up control on vegetation composition in a tidal marsh depends on scale. *PLOS ONE*, **12**: e0169960.

Engels, J. G., and Jensen, K. 2010. Role of biotic interactions and physical factors in determining the distribution of marsh species along an estuarine salinity gradient. *Oikos*, **119**: 679–685.

Escapa, M., Minkoff, D. R., Perillo, G. M. E., and Iribarne, O. 2007. Direct and indirect effects of burrowing crab *Chasmagnathus granulatus* activities on erosion of southwest Atlantic *Sarcocornia*-dominated marshes. *Limnology and Oceanography*, **52**: 2340–2349.

Ewanchuk, P. J., and Bertness, M. D. 2003. Recovery of a northern New England salt marsh plant community from winter icing. *Oecologia*, **136**: 616–626.

Fariña, J. M., He, Q., Silliman, B. R., and Bertness, M. D. 2017. Biogeography of salt marsh plant zonation on the Pacific coast of South America. *Journal of Biogeography*, **45**: 238–247.

Fariña, J. M., Silliman, B. R., and Bertness, M. D. 2009. Can conservation biologists rely on established community structure rules to manage novel systems?...Not in salt marshes. *Ecological Applications*, **19**: 413–422.

Feher, L. C., Osland, M. J., Griffith, K. T., Grace, J. B., Howard, R. J., Stagg, C. L., Enwright N. M., et al. 2017. Linear and nonlinear effects of temperature and precipitation on ecosystem properties in tidal saline wetlands. *Ecosphere*, **8**: e01956.

Finke, D. L., and Denno, R. F. 2004. Predator diversity dampens trophic cascades. *Nature*, **429**: 407–410.

Finke, D. L., and Denno, R. F. 2006. Spatial refuge from intraguild predation: implications for prey suppression and trophic cascades. *Oecologi*, **149**: 265–275.

Foster, W. A., and Treherne, J. E. 1976. Insects of marine saltmarshes: problems and adaptations. In: L. Cheng, ed., *Marine Insects*. North-Holland Publishing Company, Amsterdam, pp. 5–42.

Frey, R. W., and Basan, P. B. 1978. Coastal salt marshes. In: R. A. Davis Jr., ed., *Coastal Sedimentary Environments*. Springer-Verlag, New York, pp. 101–169.

Gabler, C. A., Osland, M. J., Grace, J. B., Stagg, C. L., Day, R. H., Hartley, S. B., Enwright, N. M., et al. 2017. Macroclimatic change expected to transform coastal wetland ecosystems this century. *Nature Climate Change*, **7**: 142–147.

Gallagher, J. L., Reimold, R. J., Linthurst, R. A., and Pfeiffer, W. J. 1980. Aerial production, mortality, and mineral accumulation-export dynamics in *Spartina alterniflora* and *Juncus roemerianus* plant stands in a Georgia salt marsh. *Ecology*, **61**: 303–312.

Ganong, W. F. 1903. The vegetation of the Bay of Fundy salt and diked marshes: an ecological study. *Botanical Gazette*, **36**: 161–186, 280–302, 350–367, 429–455.

Gedan, K. B., Crain, C. M., and Bertness, M. D. 2009. Small-mammal herbivore control of secondary succession in New England tidal marshes. *Ecology*, **90**: 430–440.

Grewell, B. J. 2008a. Hemiparasites generate environmental heterogeneity and enhance species coexistence in salt marshes. *Ecological Applications*, **18**: 1297–1306.

Grewell, B. J. 2008b. Parasite facilitates plant species coexistence in a coastal wetland. *Ecology*, **89**: 1481–1488.

Griffin, J. N., and Silliman, B. R. 2011 Predator diversity stabilizes and strengthens trophic control of a keystone grazer. *Biology Letters*, **7**: 79–82.

Grosholz, E. 2010. Avoidance by grazers facilitates spread of an invasive hybrid plant. *Ecology Letters*, **13**: 145–153.

Guo, H., and Pennings, S. C. 2012. Mechanisms mediating plant distributions across estuarine landscapes in a low-latitude tidal estuary. *Ecology*, **93**: 90–100.

Hacker, S. D., and Bertness, M. D. 1995. A herbivore paradox: why salt marsh aphids live on poor-quality plants. *American Naturalist*, **145**: 192–210.

Hacker, S. D., and Bertness M. D. 1999. Experimental evidence for factors maintaining plant species diversity in a New England salt marsh. *Ecology*, **80**: 2064–2073.

Hackney, C. T., and Bishop, T. D. 1981. A note on the relocation of marsh debris during a storm surge. *Estuarine, Coastal and Shelf Science*, **12**: 621–624.

Haines, E. B. 1976. Stable carbon isotope ratios in the biota, soils and tidal water of a Georgia salt marsh. *Estuarine and Coastal Marine Science*, **4**: 609–616.

Haines, E. B., and Montague, C. L. 1979. Food sources of estuarine invertebrates analyzed using 13C/12C ratios. *Ecology*, **60**: 48–56.

Hanley, T. C., Kimbro, D. L., and Hughes, A. R. 2017. Stress and subsidy effects of seagrass wrack duration, frequency, and magnitude on salt marsh community structure. *Ecology*, **98**: 1884–1895.

Hardwick-Witman, M. N. 1985. Biological consequences of ice rafting in a New England salt marsh community. *Journal of Experimental Marine Biology and Ecology*, **87**: 283–298.

He, Q., Altieri, A. H., and Cui, B. 2015. Herbivory drives zonation of stress-tolerant marsh plants. *Ecology*, **96**: 1318–1328.

He, Q., and Bertness, M. D. 2014. Extreme stresses, niches, and positive species interactions along stress gradients. *Ecology*, **95**: 1437–1443.

He, Q., Bertness M. D., and Altieri A. H. 2013. Global shifts towards positive species interactions with increasing environmental stress. *Ecology Letters*, **16**: 695–706.

He, Q., Bertness, M. D., Bruno, F., Li, B., Chen, G., Coverdale, T. C., Altieri, A. H., et al. 2014. Economic development and coastal ecosystem change in China. *Scientific Reports*, **4**: 5995.

He, Q., and Cui, B. 2015. Multiple mechanisms sustain a plant-animal facilitation on a coastal ecotone. *Scientific Reports*, **5**: 8612.

He, Q., Cui B., Bertness, M. D., and An, Y. 2012. Testing the importance of plant strategies on facilitation using congeners in a coastal community. *Ecology*, **93**: 2023–2029.

He, Q., and Silliman, B. R. 2015. Biogeographic consequences of nutrient enrichment for plant-herbivore interactions in coastal wetlands. *Ecology Letters*, **18**: 462–471.

He, Q., and Silliman, B. R. 2016. Consumer control as a common driver of coastal vegetation worldwide. *Ecological Monographs*, **86**: 278–294.

He, Q., Silliman, B. R., and Cui, B. 2017a. Incorporating thresholds into understanding salinity tolerance: a study using salt-tolerant plants in salt marshes. *Ecology and Evolution*, **2017**: 6326–6333.

He, Q., Silliman, B. R., Liu, Z., and Cui, B. 2017b. Natural enemies govern ecosystem resilience in the face of extreme droughts. *Ecology Letters*, **20**: 194–201.

Hensel, M. J. S., and Silliman, B. R. 2013. Consumer diversity across kingdoms supports multiple functions in a coastal ecosystem. *Proceedings of the National Academy of Science, USA*, **110**: 20621–20626.

Hilton, G. M., Ruxton, G. D., and Cresswell, W. 1999. Choice of foraging area with respect to predation risk in redshanks: the effects of weather and predator activity. *Oikos*, **87**: 295–302.

Ho, C.-K., and Pennings, S. C. 2008. Consequences of omnivory for trophic interactions on a salt marsh shrub. *Ecology*, **89**: 1714–1722.

Holdredge, C., Bertness, M. D., and Altieri, A. H. 2008. Role of crab herbivory in die-off of New England salt marshes. *Conservation Biology*, **23**: 672–679.

Hopkinson, C. S., Gosselink, J. G., and Parrondo, R. T. 1978. Aboveground production of seven marsh plant species in coastal Louisiana. *Ecology*, **59**: 760–769.

Hovel, K. A., Bartholomew A., and Lipcius R. N. 2001. Rapidly entrainable tidal vertical migrations in the salt marsh snail *Littoraria irrorata*. *Estuaries*, **24**: 808–816.

Hughes, A. R., and Lotterhos, K. E. 2014. Genotypic diversity at multiple spatial scales in the foundation marsh species, *Spartina alterniflora*. *Marine Ecology Progress Series*, **497**: 105–117.

Hughes, A. R., Moore, A. F. P., and Piehler, M. F. 2014. Independent and interactive effects of two facilitators on their habitat-providing host plant, *Spartina alterniflora*. *Oikos*, **123**: 488–499.

Jensen, A. 1985. The effect of cattle and sheep grazing on salt-marsh vegetation at Skallingen, Denmark. *Vegetatio*, **60**: 37–48.

Johnson, D. S., and Heard, R. 2017. Bottom-up control of parasites. *Ecosphere*, **8**: e01885.

Keddy, P. A. 1990. Competitive hierarchies and centrifugal organization in plant communities. In: J. B. Grace, and D. Tilman, eds., *Perspectives on Plant Competition*. Academic Press, Inc., San Diego, pp. 265–290.

Kimbro, D. L. 2012. Tidal regime dictates the cascading consumptive and nonconsumptive effects of multiple predators on a marsh plant. *Ecology*, **93**: 334–344.

Kirwan, M. L., Guntenspergen, G. R., and Morris, J. T. 2009. Latitudinal trends in *Spartina alterniflora* productivity and the response of coastal marshes to global change. *Global Change Biology*, **15**: 1982–1989.

Kirwan, M. L., Murray, A. B., and Boyd, W. S. 2008. Temporary vegetation disturbance as an explanation for permanent loss of tidal wetlands. *Geophysical Research Letters*, **35**: L05403.

Kneib, R. T. 1987. Predation risk and use of intertidal habitats by young fishes and shrimp. *Ecology*, **68**: 379–386.

Kneib, R. T. 2000. Salt marsh ecoscapes and production transfers by estuarine nekton in the southeastern United States. In: M. P. Weinstein and D. A. Kreeger, editors. *Concepts and Controversies in Tidal Marsh Ecology*. Kluwer Academic Publishers, Dordrecht.

Kuijper, D. P. J., and Bakker, J. P. 2005. Top-down control of small herbivores on salt-marsh vegetation along a productivity gradient. *Ecology*, **86**: 914–923.

Kuris, A. M., Hechinger, R. F., Shaw, J. C., Whitney, K. L., Aguirre-Macedo, L., Boch, C. A., Dobson, A. P., et al. 2008. Ecosystem energetic implications of parasite and free-living biomass in three estuaries. *Nature*, **454**: 515–518.

Kwak, T. J., and Zedler, J. B. 1997. Food web analysis of southern California coastal wetlands using multiple stable isotopes. *Oecologia*, **110**: 262–277.

Lafferty, K. D. 1993. Effects of parasitic castration on growth, reproduction and population dynamics of the marine snail Cerithidea californica. *Marine Ecology Progress Series*, **96**: 229–237.

Lafferty, K. D., Dobson A. P., and Kuris A. M. 2006. Parasites dominate food web links. *Proceedings of the National Academy of Science, USA*, **103**: 11211–11216.

Lafferty, K. D., and Morris, K. 1996. Altered behavior of parasitized killifish increases susceptibility to predation by bird final hosts. *Ecology*, **77**: 1390–1397.

Langdon, C. J., and Newell, R. I. E. 1990. Utilization of detritus and bacteria as food sources by two bivalve suspension-feeders, the oyster *Crassostrea virginica* and the mussel *Geukensia demissa*. *Marine Ecology Progress Series*, **58**: 299–310.

Lee, S. C., and Silliman, B. R. 2006. Competitive displacement of a detritivorous salt marsh snail. *Journal of Experimental Marine Biology and Ecology*, **339**: 75–85.

Levin, P. S., Ellis, J., Petrik, R., and Hay, M. E. 2002. Indirect effects of feral horses on estuarine communities. *Conservation Biology*, **16**: 1364–1371.

Levine, J. M., Brewer J. S., and Bertness, M. D. 1998. Nutrients, competition and plant zonation in a New England salt marsh. *Journal of Ecology*, **86**: 285–292.

Lewis, D. B., and Eby, L. A. 2002. Spatially heterogeneous refugia and predation risk in intertidal salt marshes. *Oikos*, **96**: 119–129.

Li, H., Zhang, X., Zheng, R., Li, X., Elmer, W. H., Wolfe, L. M., and Li, B. 2014a. Indirect effects of non-native *Spartina alterniflora* and its fungal pathogen (*Fusarium palustre*) on native saltmarsh plants in China. *Journal of Ecology*, **102**: 1112–1119.

Li, S., and Pennings, S. C. 2016. Disturbance in Georgia salt marshes: variation across space and time. *Ecosphere*, **7**: e01487

Li, S., and Pennings, S. C. 2017. Timing of disturbance affects biomass and flowering of a saltmarsh plant and attack by stem-boring herbivores. *Ecosphere*, **8**: e01675.

Li, Z., Wang W., and Zhang, Y. 2014b. Recruitment and herbivory affect spread of invasive *Spartina alterniflora* in China. *Ecology*, **95**: 1972–1980.

Linthurst, R. A. 1980. An evaluation of aeration, nitrogen, pH and salinity as factors affecting *Spartina alterniflora* growth: a summary. In: V. S. Kennedy, ed., *Estuarine Perspectives*. Academic Press, New York, pp. 235–247

Liu, W., Maung-Douglas, K., Strong, D. R., Pennings, S. C., and Zhang, Y. 2016. Geographical variation in vegetative growth and sexual reproduction of the invasive *Spartina alterniflora* in China. *Journal of Ecology*, **104**: 173–181.

Lynch, J. J., O'Neil, E., and Lay, D. W. 1947. Management significance of damage by geese and muskrats to gulf coast marshes. *Journal of Wildlife Management*, **11**: 50–76.

Malek, J. C., and Byers, J. E. 2017. The effects of tidal elevation on parasite heterogeneity and co-infection in the eastern oyster, *Crassostrea virginica*. *Journal of Experimental Marine Biology and Ecology*, **494**: 32–27.

Marczak, L. B., Więski, K. Denno, R. F., and Pennings, S. C. 2013. Importance of local vs. geographic variation in salt marsh plant quality for arthropod herbivore communities. *Journal of Ecology*, **101**: 1169–1182.

Maricle, B. R., Cobos, D. R., and Campbell, C. S. 2007. Biophysical and morphological leaf adaptations to drought and salinity in salt marsh grasses. *Environmental and Experimental Botany*, **60**: 458–467.

Maricle, B. R., Crosier, J. J., Bussiere, B. C., and Lee, R. W. 2006. Respiratory enzyme activities correlate with anoxia tolerance in salt marsh grasses. *Journal of Experimental Marine Biology and Ecology*, **337**: 30–37.

Maricle, B. R., and Lee, R. W. 2007. Root respiration and oxygen flux in salt marsh grasses from different elevational zones. *Marine Biology*, **151**: 413–423.

McKee, K. L., Mendelssohn, I. A., and Materne, M. D. 2004. Acute salt marsh dieback in the Mississippi River deltaic plain: a drought-induced phenomenon? *Global Ecology and Biogeography*, **13**: 65–73.

Mendelssohn, I. A. 1979. Nitrogen metabolism in the height forms of *Spartina alterniflora* in North Carolina. *Ecology*, **60**: 574–584.

Mendelssohn, I. A., and Morris, J. T. 2000. Eco-physiological controls on the productivity of *Spartina alterniflora* Loisel. In: M. P. Weinstein and D. A. Kreeger, eds., *Concepts and Controversies in Tidal Marsh Ecology*. Kluwer Academic Publishers, Dordrecht, pp. 59–80.

Minello, T. J., Able, K. W., Weinstein, M. P., and Hays, C. G. 2003. Salt marshes as nurseries for nekton: testing hypotheses on density, growth and survival through meta-analysis. *Marine Ecology Progress Series*, **246**: 39–59.

Minello, T. J., and Zimmerman, R. J. 1992. Utilization of natural and transplanted Texas salt marshes by fish and decapod crustaceans. *Marine Ecology Progress Series*, **90**: 273–285.

Montague, C. L. 1982. The influence of fiddler crab burrows and burrowing on metabolic processes in salt marsh sediments. In: V. S. Kennedy, ed., *Estuarine Comparisons*. Academic Press, New York, pp. 283–301.

Moon, D. C., Rossi, A. M., and Stiling, P. 2000. The effects of abiotically induced changes in host plant quality (and morphology) on a salt marsh planthopper and its parasitoid. *Ecological Entomology*, **25**: 325–331.

Moon, D. C., and Stiling, P. 2002. The effects of salinity and nutrients on a tritrophic salt-marsh system. *Ecology*, **83**: 2465–2476.

Moon, D. C., and Stiling, P. 2004. The influence of a salinity and nutrient gradient on coastal vs. upland tritrophic complexes. *Ecology*, **85**: 2709–2716.

Morris, J. T., Kjerfve. B., and Dean J. M. 1990. Dependence of estuarine productivity on anomalies in mean sea level. *Limnology and Oceanography*, **35**: 926–930.

Morris, J. T., Sundareshwar, P. V., Nietch, C. T., Kjerfve B., and Cahoon, D. R. 2002. Responses of coastal wetlands to rising sea level. *Ecology*, **83**: 2869–2877.

Morris, J. T., Sundberg, K., and Hopkinson, C. S. 2013. Salt marsh primary production and its responses to relative sea level and nutrients in estuaries at Plum Island, Massachusetts, and North Inlet, South Carolina, USA. *Oceanography*, **26**: 78–84.

Morris, R. K. A., Reach, I. S., Duffy, M. J., Collins, T. S., and Leafe, R. N. 2004. On the loss of saltmarshes in south-east England and the relationship with *Nereis diversicolor*. *Journal of Applied Ecology*, **41**: 787–791.

Naeem, S. 2002. Ecosystem consequences of biodiversity loss: the evolution of a paradigm. *Ecology*, **83**: 1537–1552.

Naeem, S., Thompson, L. J., Lawler, S. P., Lawton, J. H., and Woodfin, R. M. 1994. Declining biodiversity can alter the performance of ecosystems. *Nature*, **368**: 734–737.

Nifong, J. C., and Silliman, B. R. 2013. Impacts of a large-bodied, apex predator (*Alligator mississippiensis* Daudin 1801) on salt marsh food webs. *Journal of Experimental Marine Biology and Ecology*, **440**: 185–191.

Nyman, J. A., Crozier, C. R., and DeLaune, R. D. 1995. Roles and patterns of hurricane sedimentation in an estuarine marsh landscape. *Estuarine, Coastal and Shelf Science*, **40**: 665–679.

O'Donnell, J. P. R., and Schalles, J. F. 2016. Examination of abiotic drivers and their influence on *Spartina alterniflora* biomass over a twenty-eight year period using Landsat 5 TM satellit imagery of the central Georgia coast. *Remote Sensing*, **8**: 477.

Odum, W. E. 1988. Comparative ecology of tidal freshwater and salt marshes. *Annual Review of Ecology and Systematics*, **19**: 147–176.

Osgood, D. T., Santos, M. C. F. V., and Zieman, J. C. 1995. Sediment physico-chemistry associated with natural marsh development on a storm-deposited sand flat. *Marine Ecology Progress Series*, **120**: 271–283.

Osland, M. J., Enwright, N., Day, R. H., and Doyle, T. W. 2013. Winter climate change and coastal wetland foundation species: salt marshes vs. mangrove forests in the southeastern United States. *Global Change Biology*, **19**: 1482–1494.

Pace, M. L., Shimmel, S., and Darley, W. M. 1979. The effect of grazing by a gastropod, *Nassarius obsoletus*, on the benthic microbial community of a salt marsh mudflat. *Estuarine and Coastal Marine Science*, **9**: 121–134.

Page, H. M. 1997. Importance of vascular plant and algal production to macro-invertebrate consumers in a southern California salt marsh. *Estuarine, Coastal and Shelf Science*, **45**: 823–834.

Paramor, O. A., and Hughes, R. G. 2004. The effects of bioturbation and herbivory by the polychaete *Nereis diversicolor* on loss of saltmarsh in south-east England. *Journal of Applied Ecology*, **41**: 449–463.

Pautzke, S. M., Mather, M. E., Finn, J. T., Deegan, L. A., and Muth, R. M. 2010. Seasonal use of a New England estuary by foraging contingents of migratory striped bass. *Transactions of the American Fisheries Society*, **139**: 257–269.

Penfound, W. T., and Hathaway, E. S. 1938. Plant communities in the marshlands of southeastern Louisiana. *Ecological Monographs*, **8**: 1–56.

Pennings, S. C., and Bertness, M. D. 2001. Salt marsh communities. In M. D. Bertness, S. D. Gaines, and M. E. Hay, eds., *Marine Community Ecology*. Sinauer Associates, Sunderland, pp. 289–316.

Pennings, S. C., and Callaway, R. M. 1992. Salt marsh plant zonation: the relative importance of competition and physical factors. *Ecology*, **73**: 681–690.

Pennings, S. C., and Callaway, R. M. 1996. Impact of a parasitic plant on the structure and dynamics of salt marsh vegetation. *Ecology*, **77**: 1410–1419.

Pennings, S. C., Ho, C.-K., Salgado, C. S., Więski, K., Davé, N., Kunza, A. E., and Wason, E. L. 2009. Latitudinal variation in herbivore pressure in Atlantic Coast salt marshes. *Ecology*, **90**: 183–195.

Pennings, S. C., and Moore, D. J. 2001. Zonation of shrubs in western Atlantic salt marshes. *Oecologia*, **126**: 587–594.

Pennings, S. C., and Richards, C. L. 1998. Effects of wrack burial in salt-stressed habitats: *Batis maritima* in a southwest Atlantic salt marsh. *Ecography*, **21**: 630–638.

Pennings, S. C., Selig, E. R., Houser, L. T., and Bertness, M. D. 2003. Geographic variation in positive and negative interactions among salt marsh plants. *Ecology*, **84**: 1527–1538.

Pennings, S. C., and Silliman, B. R. 2005. Linking biogeography and community ecology: latitudinal variation in plant-herbivore interaction strength. *Ecology*, **86**: 2310–2319.

Pennings, S. C., Stanton, L. E., and Brewer, J. S. 2002. Nutrient effects on the composition of salt marsh plant communities along the southern Atlantic and Gulf Coasts of the United States. *Estuaries*, **25**: 1164–1173.

Pfeiffer, W. J., and Wiegert, R. G. 1981. Grazers on *Spartina* and their predators. In: L. R. Pomeroy and R. G. Wiegert, ed., *The Ecology of a Salt Marsh*. Springer-Verlag, New York, pp. 87–112.

Pung, O. J., Khan, R. N., Vives, S. P., and Walker, C. B. 2002. Prevalence, geographical distribution, and fitness effects of *Microphallus turgidus* (Trematoda: Microphallidae) in grass shrimp (*Palaemonetes* spp.) from coastal Georgia. *Journal of Parasitology*, **88**: 89–92.

Purer, E. A. 1942. Plant ecology of the coastal salt marshlands of San Diego county, California. *Ecological Monographs*, **12**: 82–111.

Rachlin, J. W., Stalter, R., Kincaid, D., and Warkentine, B. E. 2012. Parsimony analysis of East Coast salt marsh plant distributions. *Northeastern Naturalist*, **19**: 279–296.

Rand, T. A. 2004. Competition, facilitation, and compensation for insect herbivory in an annual salt marsh forb. *Ecology*, **85**: 2046–2052.

Ranwell, D. S. 1961. *Spartina* salt marshes in southern England. I. The effects of sheep grazing at the upper limits of *Spartina* marsh in Bridgwater Bay. *Journal of Ecology*, **49**: 325–340.

Raybould, A. F., Gray, A. J., and Clarke, R. T. 1998. The long-term epidemic of *Claviceps purpurea* on *Spartina anglica* in Poole Harbour: pattern of infection, effects on seed production and the role of *Fusarium heterosporum*. *New Phytologist*, **138**: 497–505.

Redfield, A. C. 1972. Development of a New England salt marsh. *Ecological Monographs*, **42**: 201–237.

Reidenbaugh, T. G., and Banta, W. C. 1980. Origin and effects of *Spartina* wrack in a Virginia salt marsh. *Gulf Research Reports*, **6**: 393–401.

Rejmanek, M., Sasser, C., and Peterson, G. W. 1988. Hurricane-induced sediment deposition in a Gulf Coast marsh. *Estuarine and Coastal Shelf Science*, **27**: 217–222.

Richard, G. A. 1978. Seasonal and environmental variations in sediment accretion in a Long Island salt marsh. *Estuaries*, **1**: 29–35.

Richards, C. L., Hamrick, J. L., Donovan, L. A., and Mauricio, R. 2004. Unexpectedly high clonal diversity of two salt marsh perennials across a severe environmental gradient. *Ecology Letters*, **7**: 1155–1162.

Rowcliffe, J. M., Watkinson, A. R., and Sutherland, W. J. 1998. Aggregative responses of brent geese on salt marsh and their impact on plant community dynamics. *Oecologia*, **114**: 417–426.

Rozema, J., Bijwaard, P., Prast, G., and Broekman, R. 1985. Ecophysiological adaptations of coastal halophytes from foredunes and salt marshes. *Vegetatio*, **62**: 499–521.

Schalles, J. F., Hladik, C. M., Lynes, A. A., and Pennings, S. C. 2013. Landscape estimates of habitat types, plant biomass, and invertebrate densities in a Georgia salt marsh. *Oceanography*, **26**: 88–97.

Schindler, D. E., Johnson, B. M., MacKay, N. A., Bouwes, N., and Kitchell, J. F. 1994. Crab: snail size-structured interactions and salt marsh predation gradients. *Oecologia*, **97**: 49–61.

Schoolmaster, D. R. Jr., and Stagg, C. L. 2018. Resource competition model predicts zonation and increasing nutrient use efficiency along a wetland salinity gradient. *Ecology*, **99**: 670–680.

Schrama, M., Berg M. P., and Olff H. 2012. Ecosystem assembly rules: the interplay of green and brown webs during salt marsh succession. *Ecology*, **93**: 2353–2364.

Schubauer, J. P., and Hopkinson C. S. 1984. Above- and belowground emergent macrophyte production and turnover in a coastal marsh ecosystem, Georgia. *Limnology and Oceanography*, **29**: 1052–1065.

Seiple, W. 1981. The ecological significance of the locomoter activity rhythms of *Sesarma cinereum* (Bosc) and *Sesarma reticulatum* (Say) (Decapoda, Grapsidae). *Crustaceana*, **40**: 5–15.

Seliskar, D. M., Gallagher, J. L., Burdick, D. M., and Mutz, L. A. 2002. The regulation of ecosystem functions by ecotypic variation in a dominant plant: a *Spartina alterniflora* salt-marsh case study. *Journal of Ecology*, **90**: 1–11.

Shields, J. D., and Squyars, C. M. 2000. Mortality and hematology of blue crabs, *Callinectes sapidus*, experimentally infected with the parasitic dinoflagelate *Hematodinium perezi*. *Fishery Bulletin*, **98**: 139–152.

Shumway, S. W., and Bertness, M. D. 1994. Patch size effects on marsh plant secondary succession mechanisms. *Ecology*, **75**: 564–568.

Silliman, B. R., and Bertness, M. D. 2002. A trophic cascade regulates salt marsh primary production. *Proceedings of the National Academy of Science, USA*, **99**: 10500–10505.

Silliman, B. R., and Bortolus, A. 2003. Underestimation of *Spartina* productivity in western Atlantic marshes: marsh invertebrates eat more than just detritus. *Oikos*, **101**: 549–554.

Silliman, B. R., and Newell, S. Y. 2003. Fungal farming in a snail. *Proceedings of the National Academy of Science, USA*, **100**: 15643–15648.

Silliman, B. R., Schrack, E., He, Q., Cope, R., Santoni, A., Van der Heide, T., Jacobi, R., and Van de Koppel, J. 2015. Facilitation shifts paradigms and can amplify coastal restoration efforts. *Proceedings of the National Academy of Science, USA*, **112**: 14295–14300.

Silliman, B. R., Van de Koppel, J., Bertness, M. D., Stanton, L. E., and Mendelssohn, I. A. 2005. Drought, snails, and large-scale die-off of southern U.S. salt marshes. *Science* **310**: 1803–1806.

Silliman, B. R., and Zieman, J. C. 2001. Top-down control of *Spartina alterniflora* production by periwinkle grazing in a Virginia salt marsh. *Ecology*, **82**: 2830–2845.

Srivastava, D. S., and Jefferies R. L. 1996. A positive feedback: herbivory, plant growth, salinity, and the desertification of an Arctic salt-marsh. *Journal of Ecology*, **84**: 31–42.

Steever, E. Z., Warren, R. S., and Niering, W. A. 1976. Tidal energy subsidy and standing crop production of *Spartina alterniflora*. *Estuarine and Coastal Marine Science*, **4**: 473–478.

Stiling, P., and Moon, D. C. 2005. Quality or quantity: the direct and indirect effects of host plants on herbivores and their natural enemies. *Oecologia*, **142**: 413–420.

Stiling, P. D., and Strong, D. R. 1983. Weak competition among *Spartina* stem borers by means of murder. *Ecology*, **64**: 770–778.

Stiling, P. D., and Strong, D. R. 1984. Experimental density manipulation of stem-boring insects: some evidence for interspecific competition. *Ecology*, **65**: 1683–1685.

Stiven, A. E., and Kuenzler, E. J. 1979. The response of two salt marsh molluscs, *Littorina irrorata and Geukensia demissa*, to field manipulations of density and *Spartina* litter. *Ecological Monographs*, **49**: 151–171.

Sullivan, G., Callaway J. C., and Zedler, J. B. 2007. Plant assemblage composition explains and predicts how biodiversity affects salt marsh functioning. *Ecological Monographs*, **77**: 569–590.

Taylor, K. L., Grace, J. B., Guntenspergen, G. R., and Foote, A. L. 1994. The interactive effects of herbivory and fire on an oligohaline marsh, little lake, Louisiana, USA. *Wetlands*, **14**: 82–87.

Teal, J. M. 1958. Distribution of fiddler crabs in Georgia salt marshes. *Ecology*, **39**: 185–193.

Teal, J. M. 1962. Energy flow in the salt marsh ecosystem of Georgia. *Ecology*, **43**: 614–624.

Teal, J. M., and Howes, B. L. 1996. Interannual variability of a salt-marsh ecosystem. *Limnology and Oceanography*, **41**: 802–809.

Tilman, D., Wedin, D., and Knops, J. 1996. Productivity and sustainability influenced by biodiversity in grassland ecosystems. *Nature*, **379**: 718–720.

Tobias, C., and Neubauer, S. C. 2009. Salt marsh biogeochemistry: an overview.In: G. M. E. Perillo, E. Wolanski, D. R. Cahoon, and M. M. Brinson, eds., *Coastal Wetlands: An Integrated Ecosystem Approach*. Elsevier, Amsterdam, pp. 445–492.

Travis, S. E., and Grace, J. B. 2010. Predicting performance for ecological restoration: a case study using *Spartina alterniflora*. *Ecological Applications*, **20**: 192–204.

Travis, S. E., and Hester, M. W. 2005. A space-for-time substitution reveals the long-term decline in genotypic diversity of a widespread salt marsh plant, *Spartina alterniflora*, over a span of 1500 years. *Journal of Ecology*, **93**: 417–430.

Turner, M. G. 1987. Effects of grazing by feral horses, clipping, trampling, and burning on a Georgia salt marsh. *Estuaries*, **10**: 54–60.

Turner, R. E. 1976. Geographic variations in salt marsh macrophyte production: a review. *Contributions in Marine Science*, **20**: 47–68.

Valiela, I., and Rietsma, C. S. 1995. Disturbance of salt marsh vegetation by wrack mats in Great Sippewissett Marsh. *Oecologia*, **102**: 106–112.

Valiela, I., Teal, J. M., and Deuser, W. G. 1978. The nature of growth forms in the salt marsh grass *Spartina alterniflora*. *American Naturalist*, **112**: 461–470.

van de Koppel, J., van der Wal, D., Bakker, J. P., and Herman, P. M. J. 2005. Self-organization and vegetation collapse in salt marsh ecosystems. *American Naturalist*, **165**: E1–E12.

Van Der Wal, R., Van Wijnen, H., Van Wijnen, S., Beucher, O., and Bos, D. 2000. On facilitation between herbivores: how brent geese profit from brown hares. *Ecology*, **81**: 969–980.

Voss, C. M., Christian, R. R., and Morris, J. T. 2013. Marsh macrophyte responses to inundation anticipate impacts of sea-level rise and indicate ongoing drowning of North Carolina marshes. *Marine Biology*, **160**: 181–194.

Vu, H., Więski K., and Pennings, S. C. 2017. Ecosystem engineers drive creek formation in salt marshes. *Ecology*, **98**: 162–174.

Wang, X. Y., Shen, D. W., Jiao, J., Xu, N. N. Yu, S. Zhou, X. F., Shi, M. M., and Chen, X. Y. 2012. Genotypic diversity enhances invasive ability of *Spartina alterniflora*. *Molecular Ecology*, **21**: 2542–2551.

Wiegert, R. G., Chalmers, A. G., and Randerson, P. F. 1983. Productivity gradients in salt marshes: the response of *Spartina alterniflora* to experimentally manipulated soil water movement. *Oikos*, **41**: 1–6.

Więski, K., Guo, H., Craft, C. B., and Pennings, S. C. 2010. Ecosystem functions of tidal fresh, brackish and salt marshes on the Georgia coast. *Estuaries and Coasts*, **33**: 161–169.

Więski, K., and Pennings, S. 2014a. Latitudinal variation in resistance and tolerance to herbivory of a salt marsh shrub. *Ecography*, **37**: 763–769.

Więski, K., and Pennings, S. C. 2014b. Climate drivers of *Spartina alterniflora* saltmarsh production in Georgia, USA. *Ecosystems*, **17**: 473–484.

Wilson, A. M., Evans, T., Moore, W., Schutte, C. A., Joye, S. B., Hughes, A. H., and Anderson J. L. 2015. Groundwater controls ecological zonation of salt marsh macrophytes. *Ecology*, **96**: 840–849.

Windham, L., and Lathrop, Jr., R. G. 1999. Effects of *Phragmites australis* (common reed) invasion on aboveground biomass and soil properties in brackish tidal marsh of the Mullica River, New Jersey. *Estuaries*, **22**: 927–935.

Worm, B., Barbier, E. B., Beaumont, N., Duffy, J. E., Folke, C., Halpern, B. S., Jackson, J. B. C., Lotze, H. K., et al. 2006. Impacts of biodiversity loss on ocean ecosystem services. *Science*, **314**: 787–790.

Zacheis, A. M. Y., Hupp, J. W., and Ruess, R. W. 2001. Effects of migratory geese on plant communities of an Alaskan salt marsh. *Journal of Ecology*, **89**: 57–71.

Zerebecki, R. A., Crutsinger, G. M., and Hughes, A. R. 2017. *Spartina alterniflora* genotypic identity affects plant and consumer responses in an experimental marsh community. *Journal of Ecology*, **105**: 661–673.

Zhang, L., Wang, B., and Qi, L. 2017. Phylogenetic relatedness, ecological strategy, and stress determine interspecific interactions within a salt marsh community. *Aquatic Science*, **79**: 587–595.

Zheng, S., Shao, D., Asaeda, T., Sun, T., and Luo, S. 2016. Modeling the growth dynamics of *Spartina alterniflora* and the effects of its control measures. *Ecological Engineering*, **97**: 144–156.

6

The Role of Marshes in Coastal Nutrient Dynamics and Loss

ANNE E. GIBLIN, ROBINSON W. FULWEILER, AND CHARLES S. HOPKINSON

6.1 Introduction

Sixty-five years ago, Teal's (1962) study showed that salt marsh primary production was greater than community respiration. To explain this result, he suggested that marshes exported excess organic matter either directly as organic matter, or as organisms, to coastal waters. This concept, that marshes were "outwelling" material to the adjacent estuary and coastal oceans, was soon expanded to nutrients as well. However, the actual importance of the marsh in supplying organic matter and nutrients to adjacent coastal systems has been controversial and reviews debating the importance of outwelling from marshes have regularly appeared over the decades (Nixon 1980, Childers et al. 2000, Odum 2000, Valiela et al. 2000, Boynton and Nixon 2013). It has also been argued that in some cases the coastal ocean can act as a source of nutrients to the marsh and estuary ("inwelling").

In this chapter we will review the major inputs and outputs of nitrogen (N), phosphorus (P), and silicon (Si) into marshes and discuss characteristics of internal cycling which tend to favor either retention or loss. For all three elements, surface water inputs are very important and tend to be the best studied, but groundwater and precipitation are also sources to the marsh and estuary. Biological N_2 fixation can also be an important input for N. Losses from the system include burial in marsh and estuarine sediments, and for N, gaseous losses. As will be discussed further, the boundaries for determining marsh budgets differ and some studies only consider the marsh platform, wheras others include the marsh and tidal creeks and others budget the entire marsh estuarine complex (Fig. 6.1).

Nitrogen is by far the best studied of the nutrients as it is the one most likely to limit primary production in salt marshes and estuaries (Sullivan and Daiber 1974, Valiela and Teal 1974). Marsh P dynamics received a fair about of attention in the 1970s (discussed in Whitney et al. 1981) but P cycling in salt marshes received much less attention over the next few decades as work showing the importance of N in controlling primary production accumulated. However, there is a renewed interest in P cycling in wetland systems driven in part by a desire to understand how sea-level rise may change N vs. P limitation in vegetated systems along salinity gradients, by a greater appreciation that nutrient ratios affect algal community composition, and by a desire to gain a better understanding of how nutrient limitation might be altered if mangroves occupy zones currently vegetated by salt marshes.

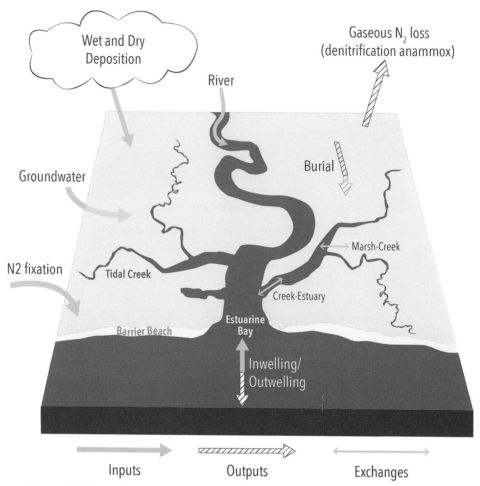

Figure 6.1 The major sources of inputs and losses of N, P, and Si to a marsh estuarine complex. The sites of exchange are also illustrated. (A black and white version of this figure will appear in some formats. For the color version, please refer to the plate section.)

Until recently, Si in marshes was rarely studied. But Si is considered a "quasi" essential nutrient for terrestrial plants, including those found in salt marshes. Better understanding of Si cycling across the land-ocean continuum is important to our broader understanding of marine primary production, higher trophic level productivity, global carbon (C) cycling, and ultimately Earth's climate (Armbrust 2009, Tréguer and De La Rocha 2013).

6.2 Key Considerations in Salt Marsh Nutrient Dynamics

Several features of salt marshes play an important role in how nutrient are cycled. One is that most vegetated marshes accrete sediments. The structure of the vegetation slows water velocities and efficiently traps organic and inorganic particles from the water column. In addition, the plants themselves produce substantial amounts of below ground material in the form of roots and rhizomes. Due to frequent tidal flooding, sediments are often waterlogged and largely anoxic within a few millimeters or centimeters below the surface. Therefore, rates of decomposition for some components of belowground plant material, especially lignin, is slowed by anaerobic conditions (reviewed in Howarth and Hobbie 1982). Depending upon the mix of organic and mineral matter from external sounces, and the production of organic matter *in-situ*, marsh sediments can range from nearly pure peat to sandy sediments with low C contents. The combination of internally produced organic matter and the deposition of sediments allows marshes to keep up with sea level rise and also leads to a significant amount of nutrient burial.

Another feature playing an important role in nutrient cycling is the chemical environment in the sediments. Due to the very high concentration of sulfate in seawater, microbial sulfate reduction makes up a major proportion of total sediment respiration below the top few millimeters of the sediment surface (Howarth & Teal 1979). This leads to highly reducing conditions and strong redox gradients. Sediments often contain significant concentrations of dissolved sulfide which readily reacts with iron to form iron sulfides. Animal burrows and drainage introduce oxygen into the sediments episodically. Soil water movement also varies over relatively short spatial scales with creekbank sediments draining much and having more water exchange than sediments further inland (Mendelssohn and Seneca 1980). As a result, the better-drained creekbank sediments are characterized by lower concentrations of nutrients, salts, and sulfides (reviewed in Mendelssohn and Morris 2000). In addition, the grasses themselves promote sediment oxidation (Howes et al. 1981) by removing water through evapotranspiration (Dacey and Howes 1984) and allowing for air entry, and by leaking some oxygen from their roots (Armstrong 1978). Seasonal changes in sulfide concentrations, the concentrations of reduced iron and manganese, and nutrients can be extremely large. These changes can be driven both by increased oxygen from higher respiration in summer driven by higher temperatures and root exudation, as well as increased sediment oxygenation from higher root radial oxygen loss and increased bioturbation. The relative balance of these production and consumption terms varies greatly between sites, in part because wetland plants differ in their ability to oxygenate the sediments (Sorrell et al. 2000). As a result, in some locations, such as the Tagus Estuary (Portugal) sediments are more oxidized during the growing season (Sundby et al. 2003) while US *Spartina alterniflora* marshes tend to be more reducing in summer (e.g., Koretsky et al. 2005).

As detailed in this chapter, these redox gradients and redox changes directly impact N cycling and indirectly alter P cycling, largely through redox driven changes in iron forms. These redox gradients can be accompanied by dynamic changes in pH. Under some conditions of enhanced draining marshes may be quite acidic while under saturated conditions marshes have soil pH values above 7 (Giblin and Howarth 1984). These dynamic pH changes can also impact nutrient cycling.

6.3 Nitrogen

6.3.1 Inputs

There are four primary sources of N to salt marshes: watershed runoff, atmospheric deposition, biological N-fixation, and the coastal ocean. The sources differ in terms of input timing, spatial distribution, magnitude, and composition (e.g., mineral vs. organic). Over longer-time periods, runoff and atmospheric deposition vary primarily as a result of anthropogenic activities. Rapid human population growth and anthropogenic activities have severely changed Earth's N cycle (Fowler et al. 2013) and the majority of US estuaries show strong signs of N-enrichment (Nixon 1995, Howarth et al. 2000a, b, Bricker et al. 2008). Quantification of inputs is challenging because source inputs are received by both aquatic and wetland portions of estuaries, and the extent to which wetlands "see" inputs to estuarine waterbodies is very difficult to quantify (Gribsholt et al. 2005, Hopkinson and Giblin 2008). There are few estimates of oceanic inputs because of the difficulty of accurately quantifying the small net balance between tidal inputs and tidal outputs from mass flux measures. Stoichiometric or metabolic approaches have proven to be successful alternatives to measuring mass fluxes (e.g., Nixon and Pilson 1984, Smith 1991, Vallino et al. 2005).

6.3.1.1 Surface and Groundwater Inputs from Adjacent Watersheds

In the early twentieth century industrial means of N fixation led to massive increases in nitrogenous fertilizer production, agricultural application of NH_4^+ fertilizer, and food production (Galloway et al. 2008), which has had major consequences on estuaries and coastal wetlands. N inputs from rivers have increased (Howarth et al. 2012) from agricultural nonpoint source runoff and from sewage effluent point sources. Fossil fuel combustion has also increased N inputs via atmospheric emissions and subsequent wet and dry deposition of N onto terrestrial ecosystems (Fowler et al. 2013), further contributing to enhanced N runoff into rivers. Rivers collect N inputs from almost the entire land surface of the Earth and deliver them into a coastal zone a fraction of 1% in total global area. Thus, N loading to estuaries is extremely high, often exceeding 40 gN m^{-2} yr^{-1}. This rate is nearly twice that typically applied to agricultural systems (Hopkinson and Giblin 2008).

Simulation models have provided great insight into what causes variability and long-term N export variability and trends over time (Bouwman et al. 2005, Seitzinger et al. 2010, Yang et al. 2015, Yu et al. 2015) Climate is typically the primary factor controlling interannual variability, while land-cover change and land-use management, such as fertilizer use, sewage treatment, and farming practices, are primarily responsible for changes over time.

The relative importance or riverine inputs of water and N varies tremendously from estuary to estuary. In a review of the N budgets of nine relatively diverse estuaries, Nixon et al. (1996) found that total riverine N loading rates varied from 1.5 to over 98 g m^{-2} y^{-1}. Direct sewage inputs increased the N loads to most of these systems, so that total river and sewage loads ranged from 1.7 to 148 g^{-2} y^{-1}. Determining importance of these inputs to the marsh requires development of a total N budget, consideration of residence time of water and N in the estuary and over adjacent tidal marshes, and lability of the various forms of N (Hopkinson and Vallino 1995, Vallino and Hopkinson 1998). External inputs of N to tidal wetlands strongly influence wetland plant N use efficiency and their dependence on retranslocated N or new N (Hopkinson 1992). Wetland plants are much more profligate in their use of N when it is readily available from outside the system.

Groundwater inputs of N are another pathway whereby watershed N enriches salt marsh estuaries. As with surface water runoff, however, the extent to which the marshes "see" groundwater N inputs is difficult to quantify and its importance is site specific (Portnoy et al. 1997). Estimates of groundwater N input range from 0.2 to 100 gN m-2 yr^{-1} (Tobias and Neubauer 2018). Several approaches are used to quantify salt marsh N inputs ranging from direct seepage measures extrapolated to the whole system (Colman and Masterson 2008) to the use of geochemical tracers (Bollinger and Moore 1993). Radium tracers can help separate the input of upland groundwater from recycled creekbank drainage (Charette et al. 2003, Charette 2007, Gardner and Gaines 2008).

6.3.1.2 Atmospheric Deposition

Atmospheric deposition of N directly onto the marsh surface varies tremendously geographically largely due to the magnitude of upwind emissions, transport time from emission sources, and vegetation height (Hopkinson and Giblin 2008, Bettez et al. 2015). Direct inputs typically range from <0.5 g N m^{-2} yr^{-1} to >1.5 gN m-2 yr^{-1} in US marshes and up to 3 gN m^{-2} yr^{-1} in European marshes (Rozema et al. 2000). Similar inputs are seen by adjacent tidal waters, so atmospheric deposition rates are actually higher when tidal flooding is also considered. As with riverine N inputs, it is difficult to determine the extent to which the tidal wetlands "see" the inputs to tidal waters, however. In systems with minor watershed inputs of N, the relative importance of atmospheric deposition increases to as high as 40% of the total N input (Nixon et al. 1996, Valigura et al. 2000).

There has been renewed interest in N deposition in the past decade, focusing on trends since the implementation of emission controls (Fowler et al. 2007) and the importance of local inputs associated with transportation and animal husbandry operations. In Europe, there was about a 28% reduction in oxidized and reduced N emissions and a similar reduction in deposition between 1980 and 2000 (Fowler et al. 2007). In North America, there was a 34% decrease in emissions and a 31% decrease in deposition from 2000 to 2018 (Lloret and Valiela 2016). The drop is particularly pronounced on the eastern side of the continent, where there has been up to a 50% decrease. The decrease results from successful governmental controls on emissions in major contributing regions and lower amounts of N oxides being transported (Lloret and Valiela 2016). However, while the data show a marked decrease in continental-scale deposition, there is a growing realization that considerable deposition is not adequately quantified by national monitoring networks, especially from agricultural and urban land in the coastal zone (Bettez and Groffman 2013). An additional 1 gN m^{-2} yr^{-1} comes from dry deposition of NOx and NH$_3$ associated with vehicular traffic on roads near coastal wetlands (Davidson et al. 2009). It has been estimated that N deposition can be underestimated by 13%–25% when gaseous N deposition from mobile sources is not considered (Bettez et al. 2013). In the city of

Baltimore (MD, USA), 77.3% of all N emissions are due to mobile sources (Bettez et al. 2013) and in general N deposition is 22% higher adjacent to suburban areas and 47% higher adjacent to urban areas, compared to nonurban areas. Elevated N deposition rates from agricultural land is primarily associated with volatilization of NH_3 from animal feeding operations and manure and fertilizer application. Coastal vegetation plays an important role in trapping dry emissions. Bettez et al. (2013) estimated continental-scale models underestimated N loading on Cape Cod (MA, USA) by about 20% because they do not account for local dry deposition. On average approximately one-third of total deposition is dry and two-thirds is wet.

Considering the high population of coastal zones worldwide, the unquantified deposition due to mobile sources could lead to serious underestimates of atmospheric deposition to the coastal zone and tidal wetlands.

6.3.1.3 N-fixation

N-fixation is typically quantified indirectly using the acetylene reduction assay and at a very small spatial scale, therefore scaling to the whole system remains challenging (Hopkinson and Giblin 2008). Fixation occurs in sediments, on the sediment surface and on standing dead shoots of *S. alterniflora* by a variety of autotrophic and heterotrophic microorganisms, but principally hetero- and nonheterocystous cyanobacteria (Newell et al. 1992, Currin and Paerl 1998a, 1998b). Important controls of N-fixation include moisture, light, and O_2 inhibition of nitrogenase (Newell et al. 1992, Joye and Paerl 1994, Currin et al. 1996). Most N-fixation rates fall in the range of <0.5–15 gN m^{-2} yr^{-1} (e.g., Rozema et al. 2000, Moisander et al. 2005; studies summarized in Hopkinson and Giblin 2008).

The magnitude and relative importance of N-fixation in meeting marsh plant N requirements varies over developmental age of wetlands (Tyler et al. 2003). In summer, when fixation rates are highest, daily N-fixation decreases with increasing marsh age. The relative importance of autotrophic or heterotrophic fixers does not appear to shift with marsh age. The reason for a decrease in the relative importance of N-fixation with age is likely a result of increasing organic N stores in marsh soils with age and thus a larger absolute magnitude of remineralization that can meet plant N growth requirements.

6.3.2 N Stocks in Vegetation and Sediments

Total stocks of N in salt marshes increase with age primarily as a result of organic N burial in conjunction with accretion and sea-level rise. Particulate organic N (PON) in sediments can therefore be orders of magnitude larger than other pools, including plant biomass, and dissolved inorganic and organic forms of N (Tobias and Neubauer 2018). Within just a 30-cm rooting zone, the mass of PON is between 500 and 1,000 gN m^{-2}. The mass of N in end-of-season plant biomass ranges between about 15 and 35 gN m^{-2} and this varies across latitudinal gradients. While tissue N concentrations are higher aboveground, larger belowground standing stock results in the total mass of N being higher belowground too. Dissolved N pools range from around 100 to 1,000 µM and are comprised primarily of organic nitrogen and ammonium (NH_4^+). NO_3^- is typically less than 10 μM. Total mass of dissolved N is very low compared to particulate N and live and or dead biomass – on the order of <1 gN m^{-2} to 20 gN m^{-2} (Hopkinson and Giblin 2008).

6.3.3 Sediment Sinks and Atmospheric Losses

Organic N burial in salt marshes is a universal characteristic of marshes that have been keeping up with sea level for the past 2,000–4,000 years. There has been considerable interest in wetland burial in the past decade or so because of the magnitude of wetland organic C burial and its impact on reducing atmospheric CO_2 levels (McLeod et al. 2011, Hopkinson et al. 2012). N burial is quantified the same way as C burial, taking into consideration accretion rates (measured across time scales of days, months, years, decades, centuries, and millennia – with settling plates, marker horizons, sediment elevation tables (SETs), ^{137}Cs, ^{210}Pb, and ^{14}C approaches), sediment bulk density, and organic N content. On average the burial rate of N in salt marsh soils is higher than almost any terrestrial ecosystem on Earth, but is comparable to that of seagrasses and mangroves. Based on global syntheses of wetland organic C burial (Hopkinson et al. 2012) and the C:N ratio of organic matter (Hopkinson and Giblin 2008, Tobias and Neubauer 2018), average annual burial of organic N in salt marshes is on the order of 1–15 gN m^{-2}yr^{-1}.

Production and loss of gaseous N in the form of N_2 is a major flux influencing the net N budget of salt marshes. Denitrification is the most important processes resulting in gaseous N_2 production. Denitrification is coupled to organic matter oxidation with NO_3^- serving as the terminal electron acceptor during respiration in the absence of O_2. Denitrification rates are controlled by O_2 concentration, NO_3^- supply, the availability of labile organic matter, and inhibitors, such as H_2S. Salt marshes are ideal environments for denitrification because of their high organic content and rate of production, tidal drainage of porewaters, especially along creekbanks, and potentially high NO_3^- availability via terrestrial runoff or high nitrification rates. Strong redox gradients and porewater drainage in salt marsh sediments removes diffusional barriers that are more common in non-bioturbated benthic environments – NH_4^+ is transported to sites favorable to nitrification and NO_3^- is transported to sites lacking O_2 but with high respiratory demand (Lettrich 2011, Koop-Jacobsen and Giblin 2010, Koch et al. 1992, Tobias et al. 2001, Tobias and Neubauer 2018). Denitrification occurs via two major pathways, direct denitrification of NO_3^- from overlying water and coupled nitrification – denitrification, where NO_3^- is produced locally by nitrification of NH_4^+. Estimated rates of denitrification are highly variable, in part due to methodological difficulties. Medium values of 14–28 mg N m^{-2} d^{-1} (compiled in Hopkinson and Giblin 2008) are roughly in balance with organic N burial (Tobias and Neubauer 2018) and N-fixation (Valiela and Teal 1979, White and Howes 1994, Anderson et al. 1997). Kaplan et al. (1979) showed that denitrification rates increase with N loading, a finding confirmed by numerous studies (reviewed in Hopkinson and Giblin 2008)

Nitrogen gas can also be produced by anammox which is a chemolithoautotrophic process where NH_4^+ is used to reduce NO_2^- to N_2. The relative balance between denitrification and anammox is controlled by the ratio of metabolizable C to NO_3^- (Algar and

Vallino 2014). At low organic C to NO_3^- ratios anammox dominates and all available C is allocated to reducing NO_3^- to NO_2, which is then further reduced by N_2. At slightly higher organic $C:NO_3^-$ ratios denitrification dominates. With high organic C content in salt marshes, anammox is not considered a favorable pathway and denitrification dominates gas flux. Anammox is typically less than 1% of total N_2 production (Koop-Jacobsen and Giblin 2009, Dalsgaard et al. 2005, Tobias and Neubauer 2018)

There have been few published studies of N_2O production in marshes. Pristine marshes can alternate between being weak sources and weak sinks of N_2O but become a source of N_2O emissions with higher N loads (Martin and Moseman-Valtierra 2015; Chmura et al. 2016; Martin et al. 2018). Although N_2O production from fertilized marshes is important from the point of greenhouse gases it is not quantitatively important in the N budget.

6.3.4 Exchanges of Inorganic and Organic N with Adjacent Systems

Tidal wetlands are almost by definition extremely open systems, as they exist in the intertidal frame and are therefore regularly flooded with estuarine water. Tidal movement of water off and on the marsh mediates the exchange of N between the marsh platform and estuarine water bodies. Particulate N (PN) moves in conjunction with sheet flow over the marsh, while dissolved forms of N move with sheet flow and with the filling and draining of creekbank porewaters. Numerous approaches are used to quantify exchange fluxes that differ according to questions being investigated. While mass balance measures through specific marsh or tidal creek cross sections show net flux, mass balance coupled with ^{15}N tracers can show tremendous internal dynamics.

Some of the earliest N studies in salt marshes examined the effect of various nutrients on macrophyte productivity. N was found to be the primary nutrient limiting plant productivity (Hopkinson and Giblin 2008). This result was interesting because of the high background concentration of NH_4^+ in salt marsh sediments. The interaction of seawater salts and NH_4^+ on root membranes decreases NH_4^+ uptake efficiency (Morris et al. 2013). Early studies also were interested in the potential of marshes to effectively process treated sewage or N-rich stormwater runoff. This interest lead to examining the fate of N applied to the marsh surface. Additions of ^{15}N as either NH_4^+ or NO_3^- revealed rapid loss via denitrification (~25%) and uptake by marsh macrophytes. After seven years, 40% of the tracer addition remained in Great Sippewissett Marsh (MA) sediments as particulate organic N with the remainder exported or denitrified. Overall one-half to three-quarters of the applied N was denitrified (White and Howes 1994). VanZomeren et al. (2011) found similar results in a brackish Louisiana marsh. With an application of $^{15}NO_3^-$ to surface water overlying the marsh, it was shown that approximately one-third of the label was incorporated into plant and sediment pools and almost two-thirds was denitrified.

Flumes stretched across the marsh platform from creekbank to the point of high water extent have been used to quantify the exchange of dissolved and particulate forms of N between tidal creeks and the marsh platform. However, since many flume studies stop near the creekbank edge, they do not always capture creekbank drainage of dissolved forms of N. Drainage is challenging to quantify because of uneven flow rates due to macro fissures and animal burrows. A promising approach is to measure porewater levels and N concentrations along transects perpendicular to tidal creek shorelines and model drainage flux, sensu Gardner and Gaines (2008).

Results of flume studies show tremendous variability between sites and over time with some exporting N and others importing it, regardless of species examined (Hopkinson and Giblin 2008). Tidal range best explains the variability (Childers et al. 1999, 2000), with low tidal range systems dominated by surface fluxes in either direction and high tidal range systems tending to show inputs via surface flowpaths and outputs via subsurface flowpaths.

With the recent interest in the ability of marshes to maintain elevation relative to an increase in the rate of sea-level rise, there has been an interest in quantifying accretion or the sedimentation of particles onto the marsh surface using marker horizons and SETs. Strong spatial gradients in total suspended solids and particulate N are found along the estuary and across the marsh platform. This translates into similarly strong spatial patterns of particle deposition. Total deposition typically decreases with distance across the marsh away from tidal creeks (Cavatorta et al. 2003) but the organic N content increases (LeMay 2007). Short-term measures of deposition are difficult to interpret because they are highly influenced by weather events, such as rainstorms, which can remove previously deposited sediments from either tiles or filter papers.

Marsh surface erosion during intense rains can also influence overall N dynamics. Using flumes, Chalmers et al. (1985) showed that 3%–20% of macrophyte N production could be exported from the marsh surface during rain events occurring when the marsh was exposed (i.e., was not flooded). Torres et al. (2003) and Chen et al. (2012) measured extremely high concentrations of suspended solids at the head of first order tidal creeks during intense rain events, sometimes exceeding 8 g L^{-1}, which is three orders of magnitude higher than typical. The particulates mobilized during these events were organic-rich with low C:N indicative of algal or vascular plant origin.

Most studies of the exchange of N between tidal marshes and estuarine waters are conducted in tidal creeks draining relatively large expanses of marsh platform. Many early studies were conducted at a time when estuarine eutrophication was observed at alarming rates and there was an interest in determining the effectiveness of marshes to filter (remove) N-rich waters thereby mitigating the effects of N-enrichment (sensu Nixon 1995). Unfortunately, it was difficult to interpret many of these studies as approaches, system scale, and species of N investigated were not uniform (Hopkinson and Giblin 2008). Valiela et al. (2000) reexamined many of the earlier results and concluded that marshes generally export total N to adjacent tidal creeks, but import NO_3^-. This may sound contradictory to marshes also sequestering organic N through burial and denitrifying most N inputs, but a marsh can still export N to adjacent tidal creeks and bury it in place, as long as inputs from other sources, such as atmospheric deposition, N-fixation, and groundwater inputs are of sufficient magnitude. In the marshes adjacent to a Mississippi River sediment diversion at Caenarvon, Louisiana, where N loading can exceed 1.5×10^6 kg NO_3^--N yr^{-1}, marshes not only bury organic N via the accumulation of undecomposed belowground plant material, they also denitrify over half the inputs (Delaune and Jugsujinda 2003) and export the remainder to Breton Sound (Hyfield et al. 2008).

Recently more information on internal N dynamics in marsh/tidal creek systems is being gained by coupling tidal creek mass flux studies with the addition of ^{15}N tracers (Gribsholt et al. 2005, Drake et al. 2009). Tracer mass balances describe the fate of a discrete mass of the added N compound, while unlabeled N compound mass balances show net processing only. For example, in the Plum Island Sound estuary in Massachusetts, ^{15}NO$_3^-$ tracer was added to both fertilized and unfertilized marsh/tidal creek systems (Drake et al. 2009) and used to trace the importance of plants and sediments as short-term sinks. Standard mass balance suggested 70% retention of NO$_3^-$ during one interval and 26% export on another. In the fertilized system 40%–50% retention was twice reported. With the ^{15}N tracer addition, however, retention was 98%–100% complete in the reference system and 50–60% in the fertilized system. The NO$_3^-$ coming into these two systems was not the same NO$_3^-$ leaving. The net balance underestimated short-term retention considerably and showed N was being sequestered in pools of short to medium duration. The addition of the tracer also made it much more apparent that the N processing capacity of a marsh ecosystem can be exceeded.

6.4 Phosphorus

6.4.1 Inputs

Phosphorus enters marshes from the atmosphere, watershed run off, and in some cases, the ocean. As is the case for N, P inputs to marshes have been greatly impacted by human activities which have accelerated the flow of P from land to aquatic systems (Howarth et al. 1995, Ruttenberg 2014). Additionally, similar to N, P inputs can vary greatly in chemical forms, magnitude and timing of the inputs, and in spatial distribution.

6.4.2 Watershed Inputs

P inputs to marshes largely come from surface waters. Because P is particle-reactive, 90% of the P in rivers globally is found in particulate forms. Organic P may comprise 20%–40% of the particulate P entering from the watershed but the majority is found in a variety of inorganic forms including P bound with iron oxides or aluminum oxyhydroxides, P in apatite, and P associated with clays (Ruttenberg 2014). Globally, iron-bound particulate P is the most important form making up 50% of the weathered P inputs to the ocean (Compton et al. 2000) and in some cases, such as in the Patuxent, Maryland, iron-bound P may be close to 80% of the total (Jordan et al. 2008). However, locally the forms of P, and their bioavailability, to estuaries vary greatly. In areas where human and animal inputs of P are abundant, dissolved P forms become increasingly important (Seitzinger et al. 2005). Particulate P inputs to marshes and estuaries are also reduced by river damming (Maavara et al. 2015).

The range of P input into estuaries is quite large. Nixon et al. (1996) was able to construct P budgets for eight of the nine estuaries where he had N budgets. The range of P inputs was similarly large as N spanning more than two orders of magnitude. River P inputs ranged from 0.1 to 24 gP m^{-2} y^{-1}, direct sewage discharge increased loading to 0.13 to 33 gP m^{-2} y^{-1}.

Groundwater P inputs from the watershed to estuaries and marshes are often small because P is adsorbed on particles during transit. Exceptions are in areas where the geology is P rich (Weston et al. 2006) or where there are high P concentrations in anaerobic groundwater. However even in areas where P concentrations are high in anaerobic groundwater, this P may not enter the marsh. Groundwater often discharges through oxic surface sediments which allows for the precipitation of iron oxyhydroxides which can scavenge P and almost completely prevent it from entering the water column, the so-called iron curtain (Charette and Sholkovitz 2002).

6.4.3 Atmospheric Deposition

Phosphorus does not have any major atmospheric gaseous forms so atmospheric P inputs largely consist of particulates which range in size from very small aerosols (<2.5 μm) to particles 10–100 μm in diameter (Redfield 2002). Graham and Duce (1979) estimated atmospheric inputs of P on land to be on the order of 27 mg m^{-2} y^{-1}. Subsequent studies have shown that P deposition can be highly variable but a recent compilation of more than five decades of data from 246 terrestrial sites found that the geometric mean of global deposition was nearly identical to Graham and Duce's (1979) estimate, 27 mg m^{-2} y^{-1} (Tipping et al. 2014). In some cases, P deposition may be underestimated because the methods used do not always include large particles (>10 μm) but these larger particles may contribute more to a local redistribution of P than long range transport (Tipping et al. 2014). Data from coastal stations are sparse but deposition in the Florida area ranged from 50 to 93 mg P mg m^{-2} y^{-1} and estimates from the Chesapeake were about 40 mg P mg m^{-2} y^{-1} (summarized in Redfield 2002), while inputs to the Baltic are assumed to less – 5–15 mg m^{-2} y^{-1} (HELCOM 2017).

Unlike the case for N, atmospheric P deposition does not appear to be changing much due to anthropogenic influences, at least over large scales. Miyazako et al. (2015) did find an increase in P deposition over the last 15 years which they attributed to sources in East Asia but Tipping et al. (2014) found no significant temporal changes in P deposition in their study. Mahowald et al. (2008) concluded that globally only 5% of total P inputs from the atmosphere and 15% of the phosphate were from anthropogenic sources.

While atmospheric P inputs are difficult to measure, they are generally the smallest source of P to salt marshes and estuaries and appear to be less impacted by human activities than watershed sources.

6.4.4 Stocks and Forms

6.4.4.1 Vegetation

There is far less data on P concentrations in salt marsh plants than N concentrations. Studies have reported P concentrations for *S. alterniflora* aboveground biomass ranging from 0.4 to nearly 4 mg g^{-1} (Table 6.1). P concentrations can show small-scale spatial variability and tend to be higher in plants from areas with higher P nutrient inputs or fertilization studies (Buresh et al. 1980, Stribling and Cornwell 2001). Most studies have found

Table 6.1 *N and P concentrations in marsh plants*

Species and Site	Month	N	P	N:P	Notes	Reference
		mg g^{-1}	mg g^{-1}	molar ratios		
S. alterniflora tall	March	20	2.6	17.0	Live	Gallagher et al.
S. alterniflora tall	November	9	1.2	16.6	Live	1980[1]
S. alterniflora short	March	12	2.5	10.6	Live	
S. alterniflora short	November	9	1.7	11.7	Live	
Juncus roemerianus	March	10	1.4	15.8	Live	
J. roemerianus	November	9	1.25	15.9	Live	
S. alterniflora streamside	June	9	1.3	15.3	Live	Buresh et al. 1980
S. alterniflora streamside	September	10	1.1	20.1	Live	
S. alterniflora 18 m inland	June	9	0.7	28.5	Live	
S. alterniflora 18 m inland	September	6.5	0.4	36.0	Live	
S. alterniflora low salinity site	February	30.7	3.8	17.9	Live	Stribling and Cornwell 2001[2]
S. alterniflora low salinity site	July	10.2	0.9	25.1	Live	
S. alterniflora low salinity site	August	7.5	0.3	55.4	Live	
S. alterniflora high salinity site	February	20	2	22.1	Live	
S. alterniflora high salinity site	July	10.3	0.8	28.5	Live	
S. alterniflora high salinity site	August	8	0.5	35.4	Live	
Phragmites australis "native"	Summer	22.2	1.03	47.7	Live	Packett and Chambers 2006[3]
Phragmites australis "invasive"	Summer	25.8	1.02	56.0	Live	
Carex lacustris		7.9	0.62	28.2	Dead	Morris and Lajtha
Typha latifolia		7.5	0.78	21.3	Dead	1986[4]
Calamagrostis canadensis		7.9	0.84	20.8	Dead	
Zizania aquatica		10.8	1.61	14.9	Dead	
Scirpus triqueter aboveground		7.6	1.1	15.3	Live	Zhou et al. 2007
Scirpus triqueter belowground		7.7	1.3	13.1	Live	

Table 6.1 (*cont.*)

Species and Site	Month	N	P	N:P	Notes	Reference
		Standing Stock g m^{-2}				
S. alterniflora aboveground		6.4	0.86	16.5	Live	Darby and Turner 2008b
S. alterniflora aboveground		2.6	0.19	30.3	Dead	
S. alterniflora belowground		11.7	0.79	32.8	Live	
S. alterniflora belowground		45.3	2.7	37.2	Dead	

[1] Plant concentrations were at their maximum in March and minimum in November.
[2] Lowest (annual average 1.3 psu) and highest salinity sites are listed.
[3] Leaves only.
[4] Collected standing dead.

P concentrations decrease faster than N concentrations at the end of growing season, and as a result N:P ratios often increase over the season and are higher in live leaves than in dead plants or litter (Table 6.1). Belowground *S. alterniflora* tissues tend to have considerably lower N/P ratios than aboveground tissues (Darby and Turner 2008) although the opposite pattern was seen in *Scirpus triqueter* (Zhou et al. 2007). The N/P ratio of *Juncus roemerianus* is similar to *Spartina* but *Phragmites australis* is considerably higher (Table 6.1). Total P stocks in vegetation vary with biomass, location, and species composition but fall within the range 1–8 g m^{-2}.

6.4.4.2 Sediments

The sediments are the major repository of P in marsh ecosystems. Total P concentrations in marsh sediments show a wide range, from 27 to over 2,000 µg P m^{-2} (<1 to over 75 umol P; Table 6.2). The N:P range in sediments shows a much wider range than the N:P range in plants, from less to 1 to over 100. Even if the data from Rhode Island is excluded, which was not a total P extraction, the range is still nearly an order of magnitude.

Some estuaries show a clear decrease in total sediment P content moving down a salinity gradient in both subtidal sediments (Jordan et al. 2008) and in salt marsh sediments (Sunwashar and Morris 1999). A comparison of sites suggests that total P concentrations in fresh and oligohaline marshes are often higher than in more saline marshes downstream reflecting diagenesis and post-depositional losses (Merrill and Cornwell 2000). Particulate P within sediments can be present in a variety of organic and chemical forms, many of which are not easily separated from one another. Many studies distinguish between five operationally defined fractions of particulate P based upon the SEDEX method (Ruttenberg 1992): loosely sorbed, bound by ferric iron, a variety of authigenic and biogenic apatite forms as well as calcium-carbonate-bound minerals, detrital apatite, and residual (largely organic P).

Table 6.2 *Concentrations of C, N, and P in marsh sediments*

Location	Site	Salinty	C mmol/g	N µmoles/g	P µmoles/g	N:P	C:P	C:N	Refs
Plum Island Marshes, MA	Reference Sites	saline	11	800	35	22.9	285.7	13.8	Buchsbaum et al. 2008[1]
Plum Island Marshes, MA	Hayed Sites	saline	10	600	25	24.0	400.0	16.7	
Narragansett, RI	Upper Bay 0–5 cm	saline	3.83	350	1.43	245.0	2,681.0	10.9	Nixon and Oviatt 1973[2]
Narragansett, RI	Mid –lower Bay, 0-5 cm	saline	3.06	171	0.86	199.5	3,570.6	17.9	
Block Island, RI	0–5 cm	saline	3.26	257	0.95	270.5	3,431.6	12.7	
Oregon Inlet, NC	natural 0–10 cm	20–35 psu	2.1	120	6.6	18.2	318.2	17.5	Craft et al. 1988[3]
	natural 10–30 cm	20–35 psu	0.6	36	2.9	12.4	206.9	16.7	
	transplanted 0–10 cm	20–35 psu	0.9	66	3.2	20.6	281.3	13.6	
	transplanted 10–30 cm	20–35 psu	0.1	8	1.1	7.3	90.9	12.5	
Snows Cut, NC	natural 0–10 cm	7–10 psu	7.2	268	18.4	14.6	391.3	26.9	
	natural 10–30 cm	7–10 psu	8.8	372	15.1	24.6	582.8	23.7	
	transplanted 0–10 cm	7–10 psu	1.5	61	6.8	9.0	220.6	24.6	
	transplanted 10–30 cm	7–10 psu	0.4	15	3.3	4.5	121.2	26.7	
Pine Knoll Shores, NC	natural 0–10 cm	25–35 psu	0.5	26	19.2	1.4	26.0	19.2	
	natural 10–30 cm	25–35 psu	0.2	11	15.5	0.7	12.9	18.2	
	transplanted 0–10 cm	25–35 psu	0.5	23	16.2	1.4	30.9	21.7	
	transplanted 10–30 cm	25–35 psu	0.2	5	12.9	0.4	15.5	40.0	
Texas Gulg, NC	natural 0–10 cm	0–15 psu	16.5	918	35.4	25.9	466.1	18.0	
	natural 10–30 cm	0–15 psu	18.3	971	18.1	53.6	1,011.0	18.8	
	natural 0–10 cm	0–15 psu	18	1156	48.1	24.0	374.2	15.6	
	natural 10–30 cm	0–15 psu	19.1	1074	21.5	50.0	888.4	17.8	

Site	Subsite	Salinity							Reference
Cooper River, SC	fresh marsh 0–10 cm	fresh	31.67	2,214.3	55.6	39.8	569.3	14.3	Sundareshwar and Morris 1999
	10–20 cm	fresh	38.17	2,071.4	31.2	66.5	1,224.6	18.4	
	brackish 0–10 cm	brackish	38.42	1,785.7	29.2	61.1	1,313.9	21.5	
	10–20	brackish	38.33	1,642.8	22.1	74.3	1,734.4	23.3	
	brackish 2 0–10 cm	brackish	33.83	1,714.3	29.0	59.1	1,166.0	19.7	
	10–20 cm	brackish	31.33	1,714.3	25.3	67.7	1,237.0	18.3	
	0–10 cm	saline	28.08	1,571.4	18.2	86.5	1,546.5	17.9	
	10–20 cm	saline	27.33	1,500.0	22.3	67.2	1,224.6	18.2	
Altamaha marshes, GA	0–30 cm	low brackish	8.5	457	18.1	25.3	469.7	18.6	Craft 2007[4]
Doboy Sound marshes, GA	0–30 cm	brackish–saline	3.5	192	12.1	15.9	290.1	18.2	
Sapelo marshes, GA	0–30 cm	saline	4.2	207	18.1	11.4	230.3	20.1	
Darss-Zingst Bodden Chain, Baltic Sea	0–2 cm basin	about 10 psu	20.83		75.7		275.2		Karstens et al. 2015[5]
	2–10 cm basin	about 10 psu	16.67		27.3		610.6		
	0–2 cm fringe	about 10 psu	2.5		12		208.3		
	2–10 cm fringe	about 10 psu	0.58		5.9		98.3		
Yangtze, China	High marsh 0–80 cm		0.45	38.6	19	2.0	23.7	11.7	Zhou et. al. 2007[6]
	Low marsh 0–80 cm		0.17	15.7	19	0.8	8.8	10.6	

[1] S. alterniflora and S. patens.

[2] Tall S. alterniflora—note this may be an incomplete extraction and not be total P.

[3] Table does not include fertilized sites.

[4] Saline marsh and brackish – mostly S. alterniflora, fresh-brackish included Zizaniopsis, Juncus, and Spartina cynosuroides.

[5] Phragmites marsh.

[6] Scirpus marsh.

127

P forms in marshes and estuaries vary significantly along salinity gradients. In general, as salinity increases sediments have lower concentrations of organic P and iron bound P and higher concentrations of P associated with calcium (Paludan and Morris 1999, Coelho et al. 2004, Hartzell et al. 2010). In most cases the P sorption capacity also decreases as salinity increases (Sundareshwar and Morris 1999). This appears to reflect a general loss of P from the iron–P phase of terrigenous sediments in saline systems where sulfide is abundant leaving behind sediments enriched in calcium-bound P and organic P (Jordan et al. 2008).

Dissolved organic and inorganic P represent a small sediment pool, ranging from nearly undetectable in low salinity sediments to several hundred micromolar. In contrast to total sediment P, dissolved phosphate concentrations are generally lowest in porewaters from sediments from low salinity sites and increase with salinity leading to a significant decrease in the N:P ratio in porewaters along estuarine salinity gradients (Paludan and Morris 1999, Sundareshwar and Morris 1999). This salinity change in marsh sediments mirrors trends observed in subtidal sediments (Jordan et al. 2008, Hartzell and Jordan 2012). The increase in dissolved P with salinity is due to the greater sorption capacity of sediments in freshwater and low salinity regions when compared to saline sediments (Sundareshwar and Morris 1999, reviewed in Ruttenberg 2014) and likely contributes to the shift in P to N limitation which is often observed moving from fresh to saline marshes.

6.4.4.3 P Cycling with the Sediments

Tobias and Neubauer (2009) estimate that for most marshes $1–8$ gP m^2 y^{-1} are mineralized within the sediments. This P recycling within sediments is critically important for macrophyte production as new P inputs from the atmosphere and watershed only supply a small fraction of the plant demand in most marshes (Paludan and Morris 1999). However, P released through microbial decomposition interacts with humics and mineral forms that vary in their bioavailability and restrict the mobility of P in porewaters.

Iron plays an important role in determining P availability and P solubility. Iron oxides have high surface areas and chemically reactive functional groups that strongly bind inorganic P (Stumm and Sulzberger 1992). In some cases, this adsorption is nearly irreversible and leads to long-term storage of P in soils and sediments as long as the oxide is present (Reddy et al. 1999). Under reducing conditions, iron can become biologically or chemically reduced and the inorganic P is liberated. However, recent data has suggested that microbial iron reduction may not be important in systems with even small amounts of sulfate (Hansel et al. 2015) and that in the absence of strong reducing agents, iron oxides are also slow to be reduced chemically. Hence, although iron is reduced in freshwater anoxic sediments, the process may be slow or incomplete, and significant amount of amounts of iron oxides, and their associated P minerals survive burial for substantial amounts of time. In systems with high concentrations of P and iron, the dissolution of iron oxides may lead to the formation of vivianite and authogenic ferrous phosphate which also sequester P over long time scales. In saline systems, sulfide produced by sulfate reduction is an effective chemical reductant for iron oxides. The reduction of the iron oxide releases P and the sulfide may also subsequently trap dissolved Fe^{+2} as iron sulfide minerals, primarily as pyrite (Howarth and Teal 1979, Giblin 1988). This difference in the fate of iron in freshwater vs. marine sediments may partially explain why P retention is greater in freshwater sediments (Caraco et al. 1990).

In spite of very high rates of sulfate reduction in salt marsh sediments (Howarth and Teal 1979), iron-associated P minerals are often still present, although in lower concentrations then in oligohaline marshes. This is because within the sediments there is a dynamic iron cycle with periods of net pyrite oxidation and net formation (Giblin and Howarth 1984, Cutter and Velinsky 1988, Gardner 1990, Koska and Luther 1995). This cycle is driven in part by the seasonal cycle of sulfate reduction (Howarth and Teal 1980) and in part by the actions of *Spartina* roots which increase sediment oxygenation. In addition, as a result of oxygen loss from roots, the roots may be covered in iron oxide plaques. This regeneration of iron oxides increases P adsorption capacity and helps retain P within the sediments (Lillebø et al. 2004). In addition to the iron cycle within the sediments, surface sediments remain oxidized for much of the year and a film of iron oxides is present in the near surface promotes not only P retention but also P adsorption from tidal waters (Scudlark and Church 1989, Chambers et al. 1992, Lillebø et al. 2007).

However, this trapping is incomplete so that in most systems, however, and especially calcareous ones, the importance of iron-bound P does decrease as salinity increases and P bound to calcium carbonate increases (Paludan and Morris 1999, Coelho et al. 2004, Ruttenberg 2014).

6.4.4.4 P Exchanges

There have been fewer studies of the exchanges of P between marsh/estuarine systems than there have been for N and many of these only quantified dissolved P fluxes. Nixon (1980) compiled studies on seven marsh systems and commented that they were all remarkably similar, showing a net export of dissolved inorganic phosphorus (DIP), and three of four sites where dissolved organic phosphorus was analyzed also show export. In a later summary, Childers et al. (2000) examined seven budgets for marsh-dominated estuarine subsystems and found that five were sources of DIP ranging from 0.1 to 0.68 g m^{-2} yr^{-2} and only two showed net imports. These and other studies would indicate that most marshes are small net exporters of DIP ranging from 0.03 to up to maximum of 2.25 g m^{-2} yr^{-1} (Tobias and Neubauer 2009). While some studies show DIP export throughout the year, others have shown a distinct seasonality with lesser export (Lillebø et al. 2004), or even DIP import during the winter months (Woodwell and Whitney 1977). It is important to point out some notable exceptions to this pattern. For example, oligohaline marsh sites in the Everglades import DIP from the oceans and, as these sites are P limited, it is the ocean not the land which supplies the limiting nutrient (Noe et al. 2001, Childers et al. 2006).

These whole systems measures of predominately DIP loss could seem to be in contrast to the flux chamber studies discussed above which concluded that the vegetated marsh largely recycled P within the sediments with dissolved losses being very small (Scudlark and Church 1989). Marsh flume studies also show little loss of dissolved P from the marsh surface and in general often find that the vegetated platform is a sink for DIP (Wolaver and Zieman 1984, Wolaver and Spurrier 1983). There are certainly cases when DIP export has been measured but these are often when the overlying water also goes anoxic. For example,

Karstens et al. (2015) found that sediments switched from a sink to a source of soluble reactive P to the water column in a *Phragmities* marsh when tidal flushing and water column mixing was low.

A comparison of flume with tidal creek export measures helps explain some of these discrepancies between the whole system flux estimates and measurements made on the marsh platform. Under oxic conditions diffusive losses of phosphate across the sediment water interface are very low as there appears to be sufficient iron to trap DIP even in saline systems. However, porewater phosphate concentrations can be quite high and some of this is advectively transported to creek water through porewater drainage in falling tides (Dame et al. 1991). This also helps explain seasonal patterns. During the warmer months, mineralization rates are at their greatest and dissolved phosphate concentrations in porewater are at their highest (Scudlark and Church 1989, Stribling and Cornwell 2001, Lillebø et al. 2004), and sediments are at their most reducing. This allows for a greater transport of DIP into the creeks in summer and fall.

Although marsh plants are not typically P limited, plants play a role in P cycling both through direct uptake and by modulating sediment oxygen conditions around the roots. Plants also remove small amounts of DIP and deposit it on the leaves as salts. This process, called guttation, was first discovered using ^{32}P tracers. Initially this was thought to be a major flux of DIP to the water column (Reimold 1972). However, subsequent studies have found that this flux is exceedingly small (Nixon 1980). A potentially larger loss is by the export of P in plant litter. Gallagher et al. (1980) examined P changes in *Spartina* and *Juncus* and estimated that if all of the aboveground senescent litter was lost it could remove between 1.5 and 2.8 g of P from the marsh per year. However, in general, marsh plants retain more P than N during senescence and litter has a higher N:P ratio than living plants so this loss may be less important than for N. In addition, as Gallagher et al. (1980) point out, there could also be considerable recycling of N and P on the marsh surface. A way to determine the importance of this loss is to compare sites where the aboveground biomass is removed by haying to nearby reference sites. Hayed sites contained statistically lower concentrations of N and P when compared to reference sites (Buchsbaum et al. 2008), suggesting that at least some nutrients in the senescent biomass are retained on site.

In the few cases where particulate P has been measured, the particulate flux dominates the P fluxes between the marsh and the estuary (e.g., Dame et al. 1991). The general pattern is the inverse of DIP with an input of particulate P with tidal inundation or little flux. More frequently flooded low marsh zones receive more of the particulate P in flooding waters than higher marsh zones following overall patterns of sediment deposition (Wolaver and Zeiman 1984). The small number of studies, and the fact that not all forms of P are measured in all of them, makes it difficult to budget particulate P fluxes. Using stoichiometric relationships and data on POC and PON fluxes, Tobias and Neubauer (2007) estimated particulate organic P fluxes and found importing systems averaged a gain of 0.6 gP m^{-2} yr^{-1}, while exporting systems lost 0.2–1.0 gP m^{-2} yr^{-1}. As pointed out by the authors, these calculations do not include mineral inputs of P which we know make up a significant part of the sediment P and in many systems rates of deposition are insufficient to keep up with sediment accretion. Large particulate P inputs can come with storms (Reed 1989, Turner et al. 2007) or be deposited by ice (Argow et al. 2011), although these are rarely quantified in salt marshes. Studies in mangroves, where hurricanes are more frequent, suggest that storm-related sediment deposition could be import for wetlands both in terms of nutrient accrual, and to help maintain accretion rates in the face of sea-level rise (Castañeda-Moya et al. 2010). There have been reports of large particulate P exports from rainfall events at low tides which scour sediments or high wave events (Dankers et al. 1984), similar to those Chen et al. (2012) report for C and N.

6.5 Silicon

6.5.1 Overview

Better understanding of silica (Si) cycling in terrestrial systems and across the land–ocean continuum is important to our broader understanding of marine primary production, higher trophic level productivity, global C cycling, and ultimately Earth's climate (Armbrust 2009, Tréguer and De La Rocha 2013). Si is considered a "quasi" essential nutrient for terrestrial plants, including those found in salt marshes. Si improves overall plant fitness by providing numerous protections from both abiotic (e.g., drought, heavy metal toxicity) and biotic (e.g., bacterial/fungal attack, grazing) stressors (Epstein 1999, Street-Perrot and Barker 2008). Si is an essential nutrient for diatoms, phytoplankton that form the base of many of our most nutritionally and economically important food webs. Diatoms also play an important role in C sequestration, and ultimately global climate. Annually, 55–113 Tmol of Si are fixed by terrestrial vegetation – an amount on par with Si fixation in the ocean (Conley 2002, Carey and Fulweiler 2012a).

Until recently, scientists described watershed Si export as a function of geology, runoff, precipitation, and temperature (e.g., Bluth and Kump 1994, Gaillardet et al. 1999). Over the last 15 years, however, a new paradigm is emerging which emphasizes the important role of biology in driving downstream Si availability. It is within this context that Si cycling in salt marshes is an area of emerging research interest, highlighting these ecosystems as key locations regulating the type, timing, and magnitude of Si availability in coastal ecosystems.

As the second most abundant element in the Earth's crust (Wedepohl 1995), Si is found in almost all parent materials and is thus a basic component of most soils (Sommer et al. 2006). Solid Si compounds found in soil can be divided into three main categories: crystalline forms such as silica minerals (e.g., quartz), primary silicates (e.g., olivine, feldspar), and secondary silicates (e.g., clay minerals); poorly crystalline and microcrystalline forms (e.g., imogolite and allophane); amorphous forms (ASi) which includes litho/pedogenic forms (e.g., silica sorbed to iron and aluminum oxi/hydroxides) and biogenic forms (BSi) (e.g., phytoliths, microorganism remains) (Sauer et al. 2006, Cornelis et al. 2010). The solubility of Si is a function of the long-range crystal order and the packing density of the Si tetrahedral (Iler 1979, Drees et al. 1989). For example, Si found in quartz is highly stable and solubility therefore low (0.10–0.25 mM) compared to the high solubility of amorphous Si (1.8–2 mM). Mineral weathering rates depend on the surface area and make-up of the mineral as well as environmental factors such as temperature, moisture, and pH (McKeague and Cline 1963).

6.5.2 Inputs

The primary source of Si to tidal salt marshes is plant mining of salt marsh soils and estuarine waters through tidal exchange (discussed below in the fluxes section). If the salt marsh is part of a river system, then the river will also provide a significant source of Si. Other, not yet well quantified sources include atmospheric deposition and potentially fresh groundwater. Unfortunately, our knowledge of salt marsh Si cycling is restricted to a limited number of studies which are summarized here. While there is still much to learn our hope is that this section provides background and motivation for future study of salt marsh Si cycling.

6.5.3 Atmospheric Deposition, Freshwater, and Groundwater

Aeolian inputs of Si to ecosystems globally are generally unconstrained and deposition to salt marshes largely unknown. Si can be delivered in both particulate (dry deposition) and dissolved (wet deposition) forms. Si associated with soil particles is transported thousands of kilometers through the atmosphere with micrometer size soil particles. The proportion of Si in this dust ultimately depends on the mineral composition of the source soil. The mean crustal Si concentration is 26.9 wt%, but mineral Si content can vary by a factor of two (Tegen and Kohfeld 2006). Estimated particulate lithogenic Si deposition to the ocean ranges from 2.8 to 4.6 Tmol Si yr^{-1} (Tegen and Kohlfeld 2006). Most of this deposition is downwind of the Saharan desert in the North Atlantic and the Gobi Desert in the North Pacific. Knowing how much Si in this deposition readily dissolves and becomes bioavailable is challenging because the solubility will depend on the lithogenic and biogenic composition of the dust. For example, Saharan dust, which is mainly quartz, has a low solubility of Si (0.02–1.1%; Baker et al. 2006) while as much as 10% of Si in feldspar dust may be soluble (Harrison 2000).

Wet deposition of dissolved silica (DSi) has been measured in a few studies. Si concentrations in rainwater from the northern Mediterranean Sea (Capo Carvallo, Corsica) ranged from 0.9 to 220 μM (Losno 1989 as found in Treguer et al.1995). In the Yellow and the East China Sea, rainwater DSi concentrations ranged from 0.5 to 15 μM (Zhang et al. 2005). Summer (July, August, September) rainwater DSi concentrations at several places in Iowa ranged from 8 to 21 μM (Anderson and Downing 2006). In a rice field in France, wet Si deposition added 107 mol Si ha^{-1} yr^{-1} (Desplanques et al. 2006).

A handful of studies have reported total deposition of Si to agricultural land in Iowa (218 mol Si ha^{-1} yr^{-1}; Anderson and Downing 2006), Lake Superior (928 × 10^6 mol Si yr^{-1}; Johnson and Eisenreich 1979), and Lake Michigan (285 × 10^6 mol Si yr^{-1}; Eisenreich et al. 1977). Several studies have reported atmospheric deposition of Si to forests which varies greatly by region (Simonson 1995), as is likely the case for salt marsh Si deposition. For temperate and tropical forests, atmospheric Si deposition ranges from 1.4 to 71 mol Si ha^{-1} yr^{-1}, respectively (Street-Perrott and Barker 2008). In the Amazon, Si inputs from rain and dust totaled 21 mol Si ha^{-1} yr^{-1} (Cornu et al. 1998). Direct deposition to the surface ocean was estimated by Tréguer and DeLaRocha (2013) to be 10 Tmol Si annually, with about 4% of that Si dissolving. Finally, Zhang et al. (2005) estimated a total aeolian input of DSi to continental margins of 0.15 (±0.05) Tmol of Si per yr^{-1}. In sum, while the atmosphere delivers Si to ecosystems globally, it appears to provide a low, although sometimes still important, input of Si (Cornelis et al. 2011).

Freshwater Si concentrations vary widely and are a function of both geological and biological watershed characteristics as well as anthropogenic activities (Meybeck 2003, Struyf et al. 2010, Carey and Fulweiler 2013b, Carey and Fulweiler 2014). Both DSi and BSi are transported to coastal ecosystems via rivers, which transport the majority (~80%) of Si to the ocean, delivering around 6 Tmol annually (Tréguer and De La Rocha 2013). The global average of river Si concentration is 150 μM (Tréguer and DeLaRocha 2013), and the total availability will depend on the flux of water entering the salt marsh. For any salt marshes influenced by freshwater this will surely be an important source of Si as concentrations will be higher than those found in estuarine waters.

Groundwater flow in salt marshes is a driven by tidal forcing, evapotranspiration, precipitation, and sometimes discharge of fresh groundwater from uplands (Harvey et al. 1987, Nuttle 1988, Gardner and Reeves 2002, Wilson and Gardner 2006). To our knowledge, no one has yet reported Si values for fresh groundwater entering salt marshes. Groundwater Si concentrations range widely depending on underlying geology and land use (Maguire 2017) and thus are too broadly unconstrained to discuss here. It is relevant to point out that groundwater DSi concentrations could be higher than estuarine waters and thus are a potentially important source needing further consideration for salt marsh Si budgets.

6.5.4 Silica Stocks in Sediment

The Si content of salt marsh sediments has been reported in a handful of studies. Some studies report the concentration of amorphous Si (ASi), which includes both pedogenic Si and BSi (e.g., Carey and Fulweiler 2014a). Some studies report BSi (e.g., Müller et al. 2013) but reading their methods it appears they also measured sediment ASi. Here forward we will use ASi to describe sediment Si concentrations.

Generally, ASi concentrations in salt marsh sediments are ~2% SiO_2 by dry weight (Norris and Hackney 1999, Hou et al. 2010, Müller et al. 2013). These concentrations are similar to reported values in tidal freshwater marshes (Struyf et al. 2005, 2006) and at the lower range of terrestrial soils (Sommer et al. 2006). Two studies report higher ASi concentrations, reaching 12.9% SiO_2 by dry weight in a Georgia salt marsh (Chen and Windom 1997) and in a temperate salt marsh in Narragansett Bay (Carey and Fulweiler 2013a). The reasons for these higher concentrations are unknown but are likely the product of varying salt marsh characteristics (e.g., bulk density, sediment type, tidal amplitude, etc.; Carey and Fulweiler 2013a). These high values argue for more thorough documentation of sediment ASi concentrations across a variety of salt marsh types and locations.

ASi concentrations are typically highest in the surface sediments of wetland sediments, including salt marshes (Hou et al. 2010, Struyf et al. 2010, Carey and Fulweiler 2013a, 2014a). These higher concentrations could be driven by several mechanisms including: diatom deposition or growth on sediment surface (Colman and Bratton 2003), import and deposition of ASi from tidal creek and estuarine waters (Jacobs et al. 2008, Vieillard et al. 2011), or the product of Si rich aboveground biomass decomposition (Struyf et al. 2010). Carey and Fulweiler (2013a) suggest that these higher surface concentrations could also be driven by plant "nutrient uplift" where plants take up Si from deep sediment layers, incorporate them into their biomass, and then this Si is incorporated into the surface sediments upon plant senescence.

6.5.5 Silica Stocks in Vegetation

Plants take up DSi as silicic acid (H_4SiO_4), the dominant form of Si soil solutions (Epstein 1994). DSi moves through the transpiration stream where it becomes precipitated as biogenic Si (BSi) or phytoliths throughout the plants – from the roots to the shoots. Si improves overall plant fitness by providing numerous protections from both abiotic (e.g., drought, heavy metal toxicity) and biotic (e.g., bacterial/fungal attack, grazing) stressors (Epstein 1999, Street-Perrot and Barker 2008).

BSi concentrations found in terrestrial vegetation often exceed those of well-known macronutrients such as N and can comprise between 1 and 10% of the dry weight of plant material (Epstein, 1994, Alexandre et al. 1997). The variability in plant Si concentrations is accounted for primarily by taxonomic order with the highest BSi concentrations found in grasses (Poaceae) and sedges (Cyperaceae) (Hodson et al. 2005). Salt marshes are dominated by these types of plants and thus contain large quantities of BSi, making salt marshes "hot spots" of Si cycling and ecosystems that play a key role in regulating downstream Si availability for diatoms.

6.5.5.1 Aboveground Vegetation

Salt marshes are characterized by halophytes, with *Spartina* grasses being one of the most common genera of salt marsh grasses globally. As such, salt marsh studies have focused on quantifying concentrations of BSi in *Spartina* sp. in a variety of locations (e.g., Lanning and Eleuterius 1981, Hou et al. 2010) and under different environmental stressors (e.g., wave action: Querné et al. 2012, excess N loading: Carey and Fulweiler 2013a).

BSi concentrations in the aboveground, living biomass of *S. alterniflora* range from 0.3% to 5.2% SiO_2 by dry weight (Table 6.3). BSi in *Spartina patens*, found on the high marsh, typically has lower BSi concentrations compared to *S. alterniflora*. We could find three other studies that report BSi concentrations for other *Spartina* species and the concentrations are comparable to those found in *S. alterniflora* and *S. patens* (Table 6.3). Generally, Si concentrations vary by geographic location and tend to increase over the growing season, with highest BSi concentrations reported in biomass collected in the fall or winter.

Table 6.3 *Silica Concentration in Marsh Plants*

Plant	% SiO$_2$	Sampling time	Location	Method	Reference
Spartina alterniflora	0.26	Spring	Block Island, Great Salt Pond, RI	alkaline digestion	Carey and Fulweiler 2014a
Spartina alterniflora	0.30	Spring	Zeek's Creek, Narragansett Bay, RI	alkaline digestion	
Spartina alterniflora	0.43	Spring	Nag Creek, Narragansett Bay, RI	alkaline digestion	
Spartina alterniflora	0.45	Summer	Block Island, Great Salt Pond, RI	alkaline digestion	
Spartina alterniflora	0.48	Spring	Little Massachuck Creek, Providence River Estuary, RI	alkaline digestion	
Spartina alterniflora	0.53	Spring	Babson Creek, ME	alkaline digestion	
Spartina alterniflora	0.60	Summer	Nag Creek, Narragansett Bay, RI	alkaline digestion	
Spartina alterniflora	0.67	Dec–Feb	Mott Creek Marsh, NC	alkaline digestion	Norris and Hackeny 1999
Spartina alterniflora	0.71	Summer	Zeek's Creek, Narragansett Bay, RI	alkaline digestion	Carey and Fulweiler 2014a
Spartina alterniflora	0.78	annual mean	Yangtze River Estuary, China	alkaline digestion	Hou et al. 2010
Spartina alterniflora	0.78	June	Bay of Brest salt marsh, France	alkaline digestion	Querné et al. 2012
Spartina alterniflora	0.83	-	Davis Bayou, MS	Ash, HCl digestion	Lanning and Eleuterius 1981
Spartina alterniflora	0.95	Summer	Babson Creek, ME	alkaline digestion	Carey and Fulweiler 2014a
Spartina alterniflora	0.96	Summer	Little Massachuck Creek, Providence River Estuary, RI	alkaline digestion	
Spartina alterniflora	1.09	annual mean	Prudence Island salt marsh, RI	alkaline digestion	Carey and Fulweiler 2013
Spartina alterniflora	1.30	annual mean	Prudence Island salt marsh, RI	alkaline digestion	
Spartina alterniflora	2.20	November	Poplar Island Restored Marsh, MD	alkaline digestion	Straver 2015
Spartina alterniflora	2.28	October	Graveline Bayou, MS	Ash, HCl digestion	Lanning and Eleuterius 1981
Spartina alterniflora	5.20	Apr-Sept	Mott Creek Marsh, NC	alkaline digestion	Norris and Hackeny 1999
Spartina angelica	2.61	September	St. Annaland salt marsh, the Netherlands	HNO3, HCl, HF digestion	de Bakker et al. 1999

Table 6.3 (*cont.*)

Plant	% SiO$_2$	Sampling time	Location	Method	Reference
Spartina cynosuroides	0.56	August	Bell Fontaine Beach Marsh, MS	Ash, HCl digestion	Lanning and Eleuterius
Spartina cynosuroides	2.52	July	Simmons Bayou, MS	Ash, HCl digestion	1981
Spartina patens	0.29	Spring	Nag Creek, Narragansett Bay, RI	alkaline digestion	Carey and Fulweiler
Spartina patens	0.31	Summer	Nag Creek, Narragansett Bay, RI	alkaline digestion	2014a
Spartina patens	0.66	annual mean	Greenwich Bay salt marsh, RI	alkaline digestion	Carey and Fulweiler
Spartina patens	0.87	annual mean	Greenwich Bay salt marsh, RI	alkaline digestion	2013
Spartina patens	0.89	July	Simmons Bayou, MS	Ash, HCl digestion	Lanning and Eleuterius 1981
Spartina patens	0.89	Summer	Babson Creek, ME	alkaline digestion	Carey and Fulweiler
Spartina patens	1.01	Spring	Babson Creek, ME	alkaline digestion	2014a
Spartina patens	2.19	October	Graveline Bayou, MS	Ash, HCl digestion	Lanning and Eleuterius 1981

6.5.5.2 Belowground Vegetation

BSi concentrations in *Spartina* sp. roots have only been reported for a few studies (Hou et al. 2010, Querné et al. 2012, Carey and Fulweiler 2013a, 2014a) and rhizome concentrations only in one (Carey and Fulweiler 2013a). Querné et al. (2012) measured BSi content in the roots of *S. alterniflora* in June, reporting a concentration (0.26% BSi) – three times less than the value they reported in mature leaves during the same time period. Similarly, in the Yangtze Estuary, *S. alterniflora* root BSi concentrations (0.44%–0.54% BSi) were much lower than aboveground tissue concentrations (Hou et al. 2010). These lower concentrations were attributed to the transfer of dissolved Si from the roots to the shoots during water transport. In contrast, root and rhizome concentrations in Narragansett Bay, Rhode Island, salt marshes varied widely from 0.9% to 4.1% and 0.1% and 4.9% BSi in *S. patens* and *S. alterniflora*, respectively (Carey and Fulweiler 2013a). In this same study, rhizome Si concentrations were between 1.1 and 15 times lower than the BSi concentrations in roots. Unlike aboveground BSi tissue concentrations, belowground BSi concentrations did not vary seasonally. Together, these studies report values similar to the range (0.22%–1.09% BSi) reported for five herbaceous wetland species in a tidal freshwater marsh (Struyf et al. 2005).

6.5.5.3 Why do BSi Concentrations in Salt Marsh Vegetation Vary?

Various mechanisms have been proposed for the variation in plant BSi concentrations including Si availability in the environment, environmental stressors, and differences in growth and/or transpiration rates (Norris and Hackney 1999, Struyf et al. 2005, Carey and Fulweiler 2013a). Carey and Fulweiler (2014a) proposed a model where Si uptake by *Spartina* is dependent on environmental conditions and plant genetic origin, highlighting a strong phenotypic plasticity for *Spartina* Si uptake. Recent experimental research supports this idea as the negative effects of high salinity (Mateos-Naranjo et al. 2013) or copper (Mateos-Naranjo et al. 2015) were ameliorated in *Spartina densiflora* plants exposed to higher concentrations of Si. Si amendments appeared to decrease sodium uptake and increase net photosynthetic rate and water use efficiency in *S. denisflora* under a high salinity treatment (Mateos-Naranjo et al. 2013). Similarly, Si additions increased net photosynthetic rate in copper-exposed *S. denisflora* and decreased copper translocation from the roots to the leaves (Mateos-Naranjo et al. 2015). These adaptive responses in *Spartina* sp. are important as we consider how salt marshes respond to change driven by human activities such as draining, ditching, excess N loading, and sea level rise as well as naturally occurring environmental variability (e.g., droughts, floods, major storms). It is possible that these types of disturbance may cause salt marsh plants to take up more Si thus altering Si fluxes across the salt marsh/estuary interface and ultimately Si availability in estuarine water columns.

6.5.6 Si Exchange with Adjacent Estuarine and Marine Systems

As detailed in the preceding sections, salt marshes are large reservoirs of Si as well as dynamic bioreactors transforming Si from one form to another, and ultimately impacting Si availability in nearby estuarine waters. The exact role of salt marshes as sources or sinks of Si to estuarine waters, however, is still an open research question needing further work. Observations in both fresh- and saltwater tidal wetlands have found them to be sinks of BSi and sources of DSi to adjacent waters (Struyf et al. 2006, Vieillard et al. 2011). Vieillard et al. (2011), working in the Great Marsh (Rowley, MA) found that the salt marsh acted as a point source of Si in summer, exporting DSi at the same magnitude as some nearby, mid-sized rivers. Also in this study, summer BSi fluxes were tidally variably with a net export of BSi during a spring tide. In contrast, Müller et al. (2013) found two Wadden Sea salt marshes to be nets sinks of Si. Here BSi fluxes were driven by large inputs of freshly deposited sediments which greatly exceeded exports of both DSi and BSi in seepage waters (Müller et al. 2013). Carey and Fulweiler (2014b) report seasonal variations in Si fluxes from two salt marshes in Narragansett Bay over six tidal cycles. Spring was dominated by BSi inputs from the estuary to the salt marsh and exports of DSi from the salt marsh to the estuary. In the summer, however, they observed a net import form the estuary to the salt marsh for both BSi and DSi. They proposed that this import was driven by enhanced BSi remineralization during the warmer summer months which increased the flood water DSi concentrations (Carey and Fulweiler 2014b). Combining the spring and summer fluxes resulted in the salt marshes being a net sink of BSi and DSi, with approximately three times more Si imported than exported. However, during the spring the salt marsh provided an order of magnitude more Si than nearby rivers (Carey and Fulweiler 2014b). Importantly, this spring export coincides with the time when temperate estuarine waters are often depleted in DSi and DSi fluxes from rivers can be at a minimum (Fulweiler and Nixon 2005, Carey and Fulweiler 2013b). Thus, this spring export of Si from the salt marsh to the adjacent estuary could help eliminate Si limitation thereby fueling spring diatom blooms.

Carey and Fulweiler (2014b) also measured N fluxes during this study in order to better constrain the stoichiometry of marsh nutrient export. These ratios are critical in understanding the role salt marshes play in driving adjacent estuarine phytoplankton production because diatoms require Si on a 1:1 molar basis with N (Redfield et al. 1963). N:Si ratios less than 1 indicate that N, not Si is limiting. In the spring, N:Si export ratios were well below 1 for the neap (0.1) and spring tide (0.01), thereby suppling enough Si and N to support diatom growth (Carey and Fulweiler 2014b). In the summer, N:Si flux ratios above and below 1 were observed, clouding this pattern. Although, they note, that ~70% of the time the marsh supplied more Si than N during the summer, and the average N:Si concentration ratios were well below 1. Finally, working in Breton Sound Estuary (LA, USA), Lane et al. (2004) reported that the Si:DIN ratio increased from 0.9 to 2.6 as diverted Mississippi river water flowed through fresh and brackish wetlands, mainly as a result of rapid nitrate uptake.

What do these studies mean for the role of salt marshes in regulating Si availability? First, it's clear that there are only a handful of published studies focused on Si fluxes from salt marshes and only one that examines Si in relation to other nutrients. Clearly, this is an area ripe for future research. Second, together these studies suggest that marshes provide an important role in recycling Si and regulating its availability to estuarine waters. Carey and Fulweiler (2014b) proposed that salt marshes are Si sponges – on the one hand they release DSi necessary for diatom growth, on the other they help capture and store BSi from diatom production, increasing the residence time of Si in northeastern salt marshes and allowing a delay in its export. They go on to highlight another key question – how has the large-scale decrease in salt marsh area impacted Si cycling and availability in estuarine systems? These questions and many others related to Si cycling in salt marshes are ultimately important to the functioning of coastal ecosystems.

6.6 Estuarine Nutrient Exchanges – Summary

Boynton and Nixon (2013) nicely summarized some the key difficulties in analyzing estuarine budgets. Conclusions may change depending upon system boundaries and this difficulty becomes particularly acute when examining marsh processes. If one restricts the definition of "marsh" to the vegetated platform alone the conclusion one draws from the data might be different than if small tidal creeks are included, or if one considers the entire marsh–estuary complex. Except in a very few cases, such as the flume work done by Wolaver and co-workers, the contribution of the vegetated marsh itself has seldom been isolated; even in those cases, short-term tidal measures focusing on dissolved exchanges did not completely capture particulate fluxes which may be both episodic and large. Flumes also miss the importance of porewater drainage.

If only the vegetated platform is considered, as long as the marsh is keeping up with sea-level rise, the marsh must be a net sink for both P and Si. These nutrients may come from the land or the ocean. A reasonable estimate of this sink would be long-term burial rates such as those given in Table 6.4, although data for P is fairly spare and there are only a

Table 6.4 *Burial rates of elements in marshes*

Location	C	N	P	Si	N/P	Reference
	$g\ m^{-2}\ y^{-1}$	$g\ m^{-2}\ y^{-1}$	$g\ m^{-2}\ y^{-1}$	$g\ m^{-2}\ y^{-1}$		
Salt Marshes						
Plum Island Marshes, MA	110	8.0	0.35	3.5–85	22.9	Forbrich et al. 2018[1]
Great Sippewissett, MA	82	6.0				Kinney and Valiela 2013[2]
Narragansett Bay, RI, upper	70	7.4	0.03	182	254	Carey et al. 2015[3]
Narragansett Bay, RI, lower	45	2.9	0.01	146	207	
Chesapeake – Patuxent, MD		12.2	2.30		5.3	Boynton et al. 2008
Salt marsh, NC		1.3–4.1				Craft et al. 1993
Three marshes, GA	40	2.4	0.30		8.0	Loomis and Craft 2010
Doboy River mashes, GA	32	2.1	0.19		11.1	12 12
Saplo River marshes, GA	21	1.3	0.29		4.5	
Oligohaline/Mesohaline						
Choptank, MD		19.2–27.1	0.18–1.96			Merrill and Cornwell 2000
Monie Bay, MD		13.6	0.01–1.30			
Upper Patuxent, MD (lower salinity)			2.30			Harzell et al. 2010
Upper Patuxent, MD (higher salinity)			1.30			
NC		6.9–10.3				Craft et al. 1993
Tidal Fresh, GA	124	8.2	1.00		8.2	Loomis and Craft 201028
Brackish, GA	93	6.5	0.70		9.3	
Fresh-Brackish, GA	108	6.9	0.69		10.0	Craft et al. 1993
Louisiana		21				DeLaune et al. 1981

[1] With N/P from Buchsbaum et al. 2008; Si Fulweiler personal communication.
[2] Note this values is higher than White and Howes 1994 which found 3.7 $g\ m^{-1}\ y^{-1}$.
[3] With N/P from Nixon and Oviatt 1973 – note P values may be low due to extraction method.

handful of studies for Si. Nitrogen is more difficult to assess because there are gaseous exchanges which are difficult to measure, and which can be large. Measurements of denitrification and N burial suggest that these two terms are greater than N fixation which would also make the marsh a sink for N. But the uncertainty between denitrification and N fixation make it difficult to know if the N burial term for marshes is an either an under- or overestimate of the role of marshes as a N sink. Studies on young and constructed marshes also suggest that sediment N accretion may increase over time (Table 6.1, Craft et al. 1988) or with the degree of external nutrient loading (Kinney and Valiela 2013).

The role of the marsh platform as a nutrient sink may be changing. Mineral inputs to marshes, especially those from rivers, have been declining due to dams while relative sea-level has been rising (Weston et al. 2013). Fagherazzi and others (e.g., Fagherazzi et al. 2012, 2013, Mariotti and Fagherazzi 2013) have drawn attention to the interactive effects of sea-level rise and declining sediment availability on the ability of salt marshes to sustain elevation gain. They find that edge erosion of tidal wetlands increases as water depth over tidal flats increases. Water depth increases not only with sea-level rise but also with increased tidal flat erosion. There is thus a positive feedback accelerating edge erosion. While some of these eroded sediments may be captured on the marsh surface some are lost. Ganju et al. (2017) have also documented net sediment loss from numerous marsh/tidal creek systems and concludes that many of the systems will completely erode away in less than a 1,000 years.

The net edge erosion and continued elevation gain of the remaining marsh platform has a significant effect on marsh nutrient budgets. While nutrients continue to accumulate belowground, nutrients accumulated since the marsh first formed are being eroded from the edges. The fate of the eroded nutrients is not clear. Some are likely deposited onto the marsh platform after it is resuspended with strong tidal currents and waves and carried onto the marsh at high tide. But some fraction is likely remineralized, solubilized, or transported offshore. Even without loss of the platform sea-level rise would be expected to decrease at least N and P storage by marshes as oligohaline and mesohaline store more N and P than saline marshes (Table 6.4).

The emphasis on the role that marshes play in overall estuarine budgets may underestimate the importance marshes play in altering nutrient forms, nutrient ratios, and the timing of nutrient availability. Although marshes are a net sink for P and Si, by changing sediment redox and pH, and by nutrient uptake and translocation, marsh plants appear to directly and indirectly enhance P and Si availability to coastal ecosystems. Marshes moderate the supply of N to coastal waters by taking up inorganic N and storing it or removing it during warm weather months and releasing it during colder times of the year. Studies on developing and constructed marshes suggest that they may be sites of net N fixation when N availability is low, but as they mature N loss through denitrification exceeds fixation (Tyler et al. 2003). As discussed above, N losses through denitrification clearly increase with increased loading. This suggests that marshes may serve as a N buffer for other coastal ecosystems and there is evidence that fringing marshes help protect seagrass meadows from land-derived N loads at moderate loading rates (Valiela and Cole 2002).

Sea-level rise is just one of the many challenges that are facing marshes. Global change will also bring changes in temperature, altered freshwater and nutrient inputs, and changes in winds and storms. Currently we are not in a good position to predict how these changes will impact marshes what impact changing marsh function will have on the surrounding ecosystems. Estuarine models seldom explicitly integrate marsh processes. Few studies include P and Si in their measurements and these elements are not included in most salt marsh models. Because of the importance of nutrient ratios to the phytoplankton community, a better understanding of marsh P and Si dynamics under changing conditions would be valuable. A more holistic approach to studying marsh imports and exports in the landscape context would help us gain a better understanding of how marsh functioning will change in the future (Statham 2012).

Acknowledgments

Support came from NSF-OCE 1238212, NSF-OCE 1637630 and NSF 1426308.

References

Alexandre, A., Meunier, J. D., Colin, F., and Koud, J. M. 1997. Plant impact on the biogeochemical cycle of silicon and related weathering processes. *Geochimica et Cosmochimica Acta*, 61: 677–682.

Algar, C., and Vallino, J. 2014. Predicting microbial nitrate reduction pathways in coastal sediments. *Aquatic Microbial Ecology*, 71: 223238. Doi: 10.3354/ame01678

Anderson, I., Tibias, C., Neikirk, B., and Wetzel, R. 1997. Development of a process-based N mass balance model for a Virginia *Spartina alterniflora* salt marsh: implication for net DIN flux. *Marine Ecology Progress Series*, 159: 13–27.

Anderson, K. A., and Downing, J. A. 2006. Dry and wet atmospheric deposition of nitrogen, phosphorus and silicon in an agricultural region. *Water, Air, & Soil Pollution*, 176: 351–374.

Argow, B., Hughes, Z., and FitzGerald, D. 2011. Ice raft formation, sediment load, and theoretical potential for ice-rafted sediment influx on northern coastal wetlands. *Continental Shelf Research*, 31: 1294–1305.

Armbrust, E. V. 2009. The life of diatoms in the world's oceans. *Nature*, 459: 185–192.

Armstrong, W. 1978. Root aeration in the wetland condition. In: D. D. Hook and E. M. M. Crawforth, eds., *Plant Life in Anaerobic Environments*. Ann Arbor Science Publishers, pp. 269–298.

Baker, A. R., French, M., and Linge, K. L. 2006. Trends in aerosol nutrient solubility along a west–east transect of the Saharan dust plume. *Geophysical Research Letters*, 33: 7.

Bettez, N., Duncan, J., Groffman, P., Band, L., O'Neil-Dynne, Haushal, J. S., Belt, K., and Law, N. 2015. Climate variation overwhelms efforts to reduce N delivery to coastal waters. *Ecosystems*, 18: 1319–1331.

Bettez, N. and Groffman, P. 2013. N deposition in and near an urban ecosystem. *Environmental Science and Technology*, 47: 6047–6051.

Bettez, N., Marino, R., Howarth R., and Davidson, E. 2013. Roads as N deposition hot spots. *Biogeochemistry*, 114: 149–163.

Bluth, G. J. and Kump, L. R. 1994. Lithologic and climatologic controls of river chemistry. *Geochimica et Cosmochimica Acta*, 58: 2341–2359.

Bollinger, M. S., and Moore, W. 1993. Evaluation of salt marsh hydrology using radium as a tracer. *Geochimica et Cosmochimica Acta*, 57: 2203–2212.

Bouwman, A., Van Drecht, G., Knoop, J., Beusen, A., and Meinardi, C. 2005. Exploring changes in river nitrogen export to the world's oceans. *Global Biogeochemical Cycles*, 19: 1–14.

Boynton, W. R., Hagy, J. D., Cornwell, J. C., Kemp, W. M., Green, S. M., Owens, M. S., Baker, J. E., and Larsen, R. K. 2008. Nutrient budgets and management actions in the Patuxent River Estuary, Maryland. *Estuaries and Coasts*, 31: 623–651.

Boynton, W. R., and Nixon S. W. 2013. Budget analysis of estuarine ecosystems. In: J. W. Day, B. C. Crump, W. M. Kemp, and A. Yanez-Arancibia, eds., *Estuarine Ecology*. 2nd edn. John Wiley and Sons, Hoboken, NJ, pp. 443–464.

Bricker, S. B., Longstaff, B., Dennison, W., Jones, A., Boicourt, K., Wicks, C. and Woerner, J. 2008. Effects of nutrient enrichment in the nation's estuaries: a decade of change. *Harmful Algae*, 8: 21–32.

Buchsbaum, R. N., Deegan, L. A., Horowitz, J., Garritt, R. H., Giblin, A. E., Ludlam, J. P., and Shull, D. H. 2008. Effects of regular salt marsh haying on marsh plants, algae, invertebrates and birds at Plum Island Sound, Massachusetts. *Wetlands Ecological Management*, 17: 469–487.

Buresh, R. J., DeLaune, R., and Patrick Jr, W. H. 1980. Nitrogen and phosphorus distribution and utilization by *Spartina alterniflora* in a Louisiana Gulf Marsh. *Estuaries*, 3: 111–121.

Caraco, N. F., Cole J. J., and Likens, G. E. 1990. A comparison of phosphorus immobilization in sediments of freshwater and coastal marine systems. *Biogeochemistry*, 9: 227–290.

Carey, J. C., and Fulweiler, R. W., 2012a. Human activities directly alter watershed dissolved silica fluxes. *Biogeochemistry*, 111: 125–138.

Carey, J. C., and Fulweiler, R. W. 2012b. The terrestrial silica pump. *PLOS ONE*, 7, p. e52932.

Carey, J. C., and Fulweiler, R. W. 2013a. Nitrogen enrichment increases net silica accumulation in a temperate salt marsh. *Limnology and Oceanography*, 58: 99–111.

Carey, J. C., and Fulweiler, R. W. 2013b. Watershed land use alters riverine silica cycling. *Biogeochemistry*, 113: 525–544.

Carey, J. C., and Fulweiler, R. W. 2014a. Silica uptake by Spartina – evidence of multiple modes of accumulation from salt marshes around the world. *Frontiers in Plant Science*, 5: 186.

Carey, J. C., and Fulweiler, R. W. 2014b. Salt marsh tidal exchange increases residence time of silica in estuaries. *Limnology and Oceanography*, 59: 1203–1212.

Carey, J. C., Moran, S. B. Kelly, R. P., Kolker, A. S., and Fulweiler, R. W. 2015. The declining role of organic matter in New England salt marshes. *Estuaries and Coasts*, 40: 626–639.

Castañeda-Moya, E., Twilley, R. R., Rivera-Monroy, V. H., Zhang, K., Davis III, S. E., and Ross, M. 2010. Sediment and nutrient deposition associated with hurricane Wilma in mangroves of the Florida Coastal Everglades. *Estuaries and Coasts*, 33: 45–58.

Cavatorta, J., Johnston, M., Hopkinson, C., and Valentine, V. 2003. Patterns of sedimentation in a salt marsh – dominated estuary. *Biological Bulletin*, 205: 239–241.

Chalmers, A., Wiegert, R., and Wolff, P. 1985. Carbon balance in a salt marsh: interactions of diffusive export, tidal deposition and rainfall-caused erosion. *Estuarine Coastal and Shelf Science*, 21: 757–771.

Chambers, M., Harvey, J. W., and Odum, W. E. 1992. Ammonium and phosphate dynamics in a Virginia salt marsh. *Estuaries*, 15: 349–359.

Charette, M. A. 2007. Hydrologic forcing of submarine groundwater discharge: insight from a seasonal study of radium isotopes in a groundwater-dominated salt marsh estuary. *Limnology and Oceanography,* 52: 230–239.

Charette, M. A., and Sholkovitz, E. R. 2002. Oxidative precipitation of groundwater-derived ferrous iron in the subterranean estuary of a coastal bay. *Geophysical Research Letters*, 29: 851–854.

Charette, M. A., Splivallo, R., Herbold, C., Bollinger, M. A., and Moore, W. S. 2003. Salt marsh submarine groundwater discharge as traced by radium isotopes. *Marine Chemistry,* 84: 113–121.

Chen, S., Torres, R. Bizimis, M., and Wirth, E. 2012. Salt marsh sediment and metal fluxes in response to rainfall. *Limnology and Oceanography: Fluids and Environments*, 2: 54–66.

Chen, Y. C., and Windom, H. L. 1997. Sediment manganese and biogenic silica as geochemical indicators in estuarine salt marshes of coastal Georgia, USA. *Environmental Geochemistry and Health*, 19: 0. https://doi.org/10.1023/A:1018434018126.

Childers, D. L., Boyer, J. N., Davis, S. E., Madden, C. J. Rudnick, D. T., and Skalar, F. H. 2006. Relating precipitation and water management to nutrient concentrations in the oligotrophic "upside-down" estuaries of the Florida Everglades. *Limnology and Oceanography*, 51: 602–16.

Childers, D. L., Davis, S., Twilley, R., and Rivera-Monroy, V. 1999. Wetland-water column interactions and the biogeochemistry of estuary-watershed coupling around the Gulf of Mexico. In: T. Bianchi, J. Pennock, R. Twilley eds. *Biogeochemistry of Gulf of Mexico Estuaries*. Wiley, Hoboken, NJ, pp. 211–235.

Childers, D. L., Day Jr., J. W., and McKellar Jr, H. N. 2000. Twenty more years of marsh and estuarine flux studies: Revisiting Nixon (1980). In: M. P Weinstein, and D. A. Kreeger, eds., *Concepts and Controversies in Tidal Marsh Ecology*. pp. 391–423.

Chmura, G. L., Kellman, L., van Ardenne, L., and Guntenspergen, G. R. 2016. Greenhouse gas fluxes from salt marshes exposed to chronic nutrient enrichment. *PLOS ONE*, 11: e0149937. https: //doi.org/10.1371/journal.pone.0149937

Coelho, J. P., Flindt, M. R., Jensen, H. S., Lillebo, A. I., and Pardal, M. A. 2004. Phosphorus speciation and availability in intertidal sediments of a temperate estuary: relationship to eutrophication and annual P fluxes. *Estuarine, Coastal and Shelf Science*, 61: 583–590.

Colman, J. A. and Masterson, J. P. 2008. Transient simulations of nitrogen load for a coastal aquifer and embayment, Cape Cod, MA. *Environmental Science and Technology*, 42: 207–213.

Colman, S. M. and Bratton, J. F. 2003. Anthropogenically induced changes in sediment and biogenic silica fluxes in Chesapeake Bay. *Geology*, 31: 71–74.

Compton, J., Mallinson, D., Glenn, C. R., Fillippelli, G., Follmi, K., Shields, G., and Zanin, Y. 2000. Variations in the global phosphorus cycle. In: C. R. Glenn, L. Prevot-Lucas, and J. Lucas, eds., *Marine Authigenesis: from Global to Microbial*, Tulsa, SEPM (Society for Sedimentary Geology), pp. 21–33.

Conley, D. J. 2002. Terrestrial ecosystems and the global biogeochemical silica cycle. *Global Biogeochemical Cycles*, 16(4). DOI: 10.1029/2002GB001894

Cornelis, J. T., Delvaux, B., Georg, R. B., Lucas, Y., Ranger, J., and Opfergelt, S. 2011. Tracing the origin of dissolved silicon transferred from various soil-plant systems towards rivers: a review. *Biogeosciences*, 8: 89–112.

Cornu, S., Lucas, Y., Ambrosi, J. P., and Desjardins, T. 1998. Transfer of dissolved Al, Fe and Si in two Amazonian forest environments in Brazil. *European Journal of Soil Science*, 49: 377–384.

Craft, C. 2007. Freshwater input structures soil properties, vertical accretion, and nutrient accumulation of Georgia and U.S. tidal marshes. *Limnology and Oceanography*, 52: 1220–1230.

Craft, C., Broome, S. W., and Seneca, E. D. 1988. Nitrogen, phosphorus, and organic carbon pools in a natural and transplanted marsh. *Estuaries*, 11: 272–280.

Craft, C., Seneca, D., and Broome, S. W. 1993. Vertical accretion in microtidal regularly and irregularly flooded estuarine marshes. *Estuarine and Coastal Shelf Science*, 37: 371–386.

Currin, C. A., Joye, S. B., and Paerl, H. W. 1996. Diel rates of N_2-fixation and denitrification in a transplanted *Spartina alterniflora* marsh: implications for N-flux dynamics. *Estuarine, Coastal, and Shelf Science*, 42: 597–616.

Currin, C. A., and Paerl, H. W. 1998a. Environmental and physiological controls on diel patterns of N_2 fixation in epiphytic cyanobacterial communities. *Microbial Ecology*, 35: 34–35.

Currin, C. A., and Paerl, H. W. 1998b. Epiphytic nitrogen fixation associated with standing dead shoots of smooth cordgrass, *Spartina alterniflora*. *Estuaries*, 21: 108–117.

Cutter, G. A. and Velinsky, D. J. 1988. Temporal variations in sedimentary sulfur in a Delaware salt marsh. *Marine Chemistry*, 23: 311–327.

Dacey, J. W. H. and Howes, B. L. 1984. Water uptake by roots controls water table movement and sediment oxidation in short Spartina marsh. *Science*, 224: 487–489.

Dalsgaard, T., Thamdrup, B., and Canfield, D. 2005. Anaerobic ammonium oxidation (anammox) in the marine environemtn. *Research in Microbiology*, 156: 457–464.

Dame, R. F., Spurrier, J. D., Williams, T. M., Kjerfve, B., Zingmark, R. G., Wolaver, T. G., Chrzanowski, T. H., McKellar, H. N., and Vernberg, F. J. 1991. Annual material processing by a salt marsh-estuarine basin in South Carolina, USA. *Marine Ecology Progress Series*, 72: 153–166.

Dankers, N., Birsbergen, M., Zegers, K., Laane, R., and van der Loeff, M. R. 1984. Transportation of water, particulate and dissolved organic and inorganic matter between a salt marsh and the Ems-Dollard estuary, the Netherlands. *Estuarine and Coastal Shelf Science*, 19: 143–165.

Darby, F. A., and Turner, R. E. 2008a. Below- and aboveground biomass of *Spartina alterniflora*: Response to nutrient addition in a Louisiana salt marsh. *Estuaries and Coasts*, 31: 326–334.

Darby, F. A., and Turner, R. E. 2008b. Below- and aboveground *Spartina alterniflora* production in a Louisiana salt marsh. *Estuaries and Coasts*, 31: 223–231.

Davidson, E., Savage, K., Bettez, N., Marino, R., and Howarth, R. 2009. N in runoff from residential roads in a coastal area. *Water, Air, and Soil Pollution*, 210: 3–13.

De Bakker, N. V. J., Hemminga, M. A., and Van Soelen, J. 1999. The relationship between silicon availability, and growth and silicon concentration of the salt marsh halophyte *Spartina anglica*. *Plant and Soil*, 215: 19–27.

DeLaune, R. D., Reddy, C., and Patrick, W. 1981. Accumulation of plant nutrients and heavy metals through sedimentation processes and accretion in a Louisana salt marsh. *Estuaries*, 4: 328–334.

Delaune, R., and Jugsujinda, A. 2003. Denitrification potential in a Louisiana wetland receiving diverted Mississippi River water. *Chemistry and Ecology*, 19: 411–418.

Desplanques, V., Cary, L., Mouret, J. C., Trolard, F., Bourrié, G., Grauby, O., and Meunier, J. D. 2006. Silicon transfers in a rice field in Camargue (France). *Journal of Geochemical Exploration*, 88: 190–193.

Drake, D., Peterson, B. J., Galván, K. A, Deegan, L. A., Hopkinson, C., Johnson, J. M., Koop-Jakobsen, K., Lemay, L. E., and Picard, C. 2009. Salt marsh ecosystem biogeochemical responses to nutrient enrichment: a paired 15N tracer study. *Ecology*, 90: 2535–46.

Drees, L. R., Wilding, L. P., Smeck, N. E., and Senkayi, A. L. 1989. Silica in soils: quartz and disorders polymorphs, in: *Minerals in Soil Environments*, J. B. Dixon, and S. B. Weed, eds., Soil Science Society of America, Madison, pp. 914–974.

Eisenreich, S. J., Emmling, P. J., and Beeton, A. M. 1977. Atmospheric loading of phosphorus and other chemicals to Lake Michigan. *Journal of Great Lakes Research*, 3: 291–304.

Epstein, E., 1994. The anomaly of silicon in plant biology. *Proceedings of the National Academy of Sciences*, 91: 11–17.

Epstein, E., 1999. Silicon. *Annual Review of Plant Biology*, 50: 641–664.

Fagherazzi, S., Kirwan, M., Mudd, S., Guntenspergen, G., Temmerman, S., D'Alapaos, A., van de Koppel, et al. 2012. Numerical models of salt marsh evolution: ecological, geomorphic and climatic factors. *Reviews of Geophysics*, Doi: 10.1029/2011RG000359.

Fagherazzi, S., Mariotti, G., Wiberg, P., and McGlathery, K. 2013. Marsh collapse does not require sea level rise. *Oceanography*, 26: 70–77.

Forbrich, I., Giblin, A. E., and Hopkinson, C. S. 2018. Constraining marsh carbon budgets using long-term C burial and contemporary atmospheric CO_2 fluxes. *Journal of Geophysical Research*, 123: 867–878.

Fowler, D., Coyle, M., Skiba, U., Sutton, M. A., Cape, J. N., Reis, S., Sheppard, L. J., et al. 2013. The global nitrogen cycle in the twenty-first century. *Philosophical Transactions of the Royal Society, Series B*. 368: 20130164. Doi: 10.1089/rstb.2013.0164.

Fowler, D., Smith, R., Muller, J., Cape, J., Sutton, M., Erishman, J., and Fagerli, H. 2007. Long term trends in sulphur and nitrogen deposition in Europe and the cause of non-linearities. *Water Air Soil Pollut: Focus*, 7: 41–47.

Fulweiler, R. W. and Nixon, S. W. 2005. Terrestrial vegetation and the seasonal cycle of dissolved silica in a southern New England coastal river. *Biogeochemistry*, 74: 115–130.

Gaillardet, J., Dupré, B., Louvat, P., and Allegre, C. J. 1999. Global silicate weathering and CO_2 consumption rates deduced from the chemistry of large rivers. *Chemical Geology*, 159: 3–30.

Gallagher, J. L., Reimhold, R. J., Linthurst, R. A., and Pfeiffer, W. T. 1980. Aerial production, mortality and mineral accumulation-export dynamics in *Spartina alterniflora* and *Juncus roemerianus* plant stands in a Georgia salt marsh. *Ecology*, 61: 303–312.

Galloway, J. N., Townsend, A. R., Erisman, J. W., Bekunda, M., Cai, Z., Freney, J. R., Martinelli, L. A., Seitzinger, S. P., and Sutton, M. A. 2008. Transformation of the nitrogen cycle: Recent trends, questions, and potential solutions. *Science*, 320: 889–892.

Ganju, N., Defne, Z., Kirwan, M., Fagherazzi, S., D'Alpaos, A., and Carniello, L. 2017. Spatially integrative metrics reveal hidden vulnerability of microtidal salt marshes. *Nature Communications*, 8: 14156, doi: 10.1038/ncomms14156.

Gardener, L. R. 1990. Simulation of the diagenesis of carbon, sulfur and dissolved oxygen in salt marsh sediments. *Ecological Monographs*, 60: 91–111.

Gardner, L. R., and Reeves, H. W., 2002. Spatial patterns in soil water fluxes along a forest-marsh transect in the southeastern United States. *Aquatic Sciences-Research Across Boundaries*, 64: 141–155.

Gardner, W., and Gaines, E. 2008. Estimating pore water drainage from marsh soils using rainfall and well records. *Estuarine Coastal and Shelf Science*, 79: 51–58.

Giblin, A. E. 1988. Pyrite formation during early diagenesis. *Geomicrobiology Journal*, 6: 77–97.

Giblin, A. E., and Howarth, R. W. 1984. Porewater evidence for a dynamic sedimentary iron cycle in salt marshes. *Limnology and Oceanography*, 29: 47–63.

Graham, W. F., and Duce, R. A. 1979. Atmospheric pathways of the phosphorus cycle. *Geochimica et Cosmochimica Acta*, 43: 1195–1208.

Gribsholt, B., Boschker, H. T. S., Struyf, E., Andersson, M., Tramper, A., De Brabandere, L., van Damme, S., Brion, N., et al. 2005. N processing in a tidal freshwater marsh: a whole ecosystem 15N labeling study. *Limnology and Oceanography*, 50: 1945–1959.

Hansel, C. M., Lentini, C. L., Tang, Y., Johnston, D. T., Wankel, S. D., and Jardine, P. M. 2015. Dominance of sulfur-fueled iron oxide reduction in low-sulfate freshwater sediments. *The ISME Journal*, 9: 2400–2412.

Hartzell, J. L. and Jordan, T. E. 2012. Shifts in the relative availability of phosphorus and nitrogen along estuarine salinity gradients. *Biogeochemistry*, 107: 489–500.

Hartzell, J. L., Jordan, T. E., and Cornwell, J. C. 2010. Phosphorus burial in sediments along the salinity gradient of the Patuxent River, a sub estuary of the Chesapeake Bay (USA). *Estuaries and Coasts*, 33: 92–106.

Hartzell, J. L., Jordan, T. E., and Cornwell, J. C. 2017. Phosphorus sequestration in sediments along the salinity gradients of Chesapeake Bay sub-estuaries. *Estuaries and Coasts*, 40: 1607–1625.

Harrison, K. G., 2000. Role of increased marine silica input on paleo-pCO2 levels. *Paleoceanography*, 15: 292–298.

Harvey, J. W., Germann, P. F., and Odum, W. E. 1987. Geomorphological control of subsurface hydrology in the creekbank zone of tidal marshes. *Estuarine, Coastal and Shelf Science*, 25: 677–691.

HELCOM 2017. Atmospheric deposition of heavy metals on the Baltic Sea. HELCOM Baltic Sea Environment Fact Sheets. Online 27.06.2018. www.helcom.fi/baltic-sea-trends/environment-fact-sheets/.

Hodson, M. J., White, P. J., Mead, A., and Broadley, M. R. 2005. Phylogenetic variation in the silicon composition of plants. *Annals of Botany*, 96: 1027–1046.

Hopkinson, C., and A. Giblin. 2008. Salt marsh N cycling. In: R. Capone, D. Bronk, M. Mulholland, and E. Carpenter, eds., *Nitrogen in the Marine Environment*, 2nd edn. Elsevier, Amsterdam, pp. 991–1036.

Hopkinson, C. S. 1992. The effects of system coupling on patterns of wetland ecosystem development. *Estuaries*, 15: 549–562.

Hopkinson, C. S., Cai, W.-J., and Hu, X. 2012. Carbon sequestration in wetland dominated coastal systems – a global sink of rapidly diminishing magnitude. *Current Opinions in Environmental Sustainability*, 4: 1–9.

Hopkinson, C. S. and Vallino, J. 1995. The nature of watershed perturbations and their influence on estuarine metabolism. *Estuaries*, 18: 598–621.

Hou, L., Liu, M., Yang, Y., Ou, D., Lin, X., and Chen, H., 2010. Biogenic silica in intertidal marsh plants and associated sediments of the Yangtze Estuary. *Journal of Environmental Sciences*, 22: 374–380.

Howarth, R. W., Anderson, D., Church, T., Greening, H., Hopkinson, C., Huber, W. Marcus, N. et al. Committee on the Causes and Management of Coastal Eutrophication. 2000. Clean Coastal Waters – Understanding and reducing the effects of nutrient pollution. Ocean Studies Board and Water Science and Technology Board, Commission on Geosciences, Environment, and Resources, National Research Council. National Academy of Sciences, Washington, DC.

Howarth, R. W., Anderson, D., Cloern, J., Elfring, C., Hopkinson, C., Lapointe, B. Malone, T. et al. 2000. Nutrient pollution of coastal rivers, bays and seas. *Issues in Ecology*, 7: 1–15.

Howarth, R. W. and Hobbie, J. E. 1982. The regulation of decomposition and heterotophic microbial activity in salt marsh soils: a review. In: *Estuarine Comparisons*, pp. 183–207. V. S. Kennedy, ed., Academic Press, Orlando, FL, pp. 183–220.

Howarth, R. W., Jensen, H. S., Marino, R., and Postma, H. 1995. Transport to and processing of P in near-shore and oceanic waters. In: H. Tiessen, ed., *Phosphorus and the Global Environment*. John Wiley and Sons Ltd, Hoboken, NJ, pp. 1–7.

Howarth, R. W., Swaney, D., Billen, G., Carnier, J., Hong, B., Humborg, C., Johnes, P., Morth, C., and Marino, R. 2012. Nitrogen fluxes from the landscape are controlled by net anthropogenic nitrogen inputs and by climate. *Frontiers in Ecology and Environment*, 10: 37–43.

Howarth, R. W. and Teal, J. 1979. Sulfate reduction in a New England salt marsh. *Limnology and Oceanography*, 24: 999–1013.

Howes, B. L., Howarth, R. W., Teal, J. M., and Valiela, I. 1981. Oxidation-reduction potentials in a salt marsh: Spatial patterns and interactions with primary production. *Limnology and Oceanography*, 26: 350–360.

Hyfield, E., Day, J., Cable, J., and Justic, D. 2008. The impacts of re-introducing Mississippi River water on the hydrologic budget and nutrient inpus of a deltaic estuary. *Ecological Engineering*, 32: 347–359.

Iler, R. J. K. 1979. *The Chemistry of Silica: Solubility, Polymerization, Colloid and Surface Properties, and Biochemistry*. Wiley, New York, N.Y.

Jacobs, S., Struyf, E., Maris, T., and Meire, P., 2008. Spatiotemporal aspects of silica buffering in restored tidal marshes. *Estuarine, Coastal and Shelf Science*, 80: 42–52.

Johnson, T. C. and Eisenreich, S. J., 1979. Silica in Lake Superior: mass balance considerations and a model for dynamic response to eutrophication. *Geochimica et Cosmochimica Acta*, 43: 77–92.

Jordan, T. E., Cornwell J. C., Boynton, W. R., and Anderson, J. T. 2008. Changes in the phosphorus biogeochemistry along and estuarine salinity gradient: The iron conveyer belt. *Limnology and Oceanography*, 53: 172–184.

Joye, S. B. and Paerl, H. W. 1994. Nitrogen cycling in microbial mats: rates and patterns of denitrification and nitrogen fixation. *Marine Biology*, 119: 285–295

Kaplan, W., Valiela, I., and Teal, J. M. 1979. Denitrification in a Massachusetts salt marsh ecosystem. *Limnology and Oceanography*, 26: 350–360.

Karstens, S., Buczko, U., and Glatzel, S. 2015. Phosphorus storage and mobilization in coastal Phragmites wetlands: Influence of local-scale hydrodynamics. *Estuarine, Coastal and Shelf Science*, 164: 124–133.

Kinney, E. L. and Valiela, I. 2013. Changes in the δ^{15}N in salt marsh sediments in a long-term fertilization study. *Marine Ecology Progress Series*, 477: 41–52.

Koch, M. Maltby, E., Oliver, G., and Bakker, S. 1992. Factors controlling denitrification rates in tidal mudflats and fringing salt marshes in South-west England. *Estuarine, Coastal and Shelf Science*, 34: 471–485.

Koop-Jacobsen, K., and Giblin, A. 2009. Anammox in tidal marsh sediments: the role of salinity, nitrogen loading, and marsh vegetation. *Estuaries and Coasts*, 32: 238–245.

Koop-Jakobsen, K., and Giblin, A. E. 2010. The effect of increased nitrate loading on nitrate reduction via denitrification and DNRA in salt marsh sediments Limnol. *Oceanography*, 55: 789–802.

Koretsky, C. M., VanCappellen, P., DiChistina, T. J., Kostka, J. E., Lowe, K. L., Moore, C. M. Roychoudhury, A. N., and Viollier, E. 2005. Salt marsh pore water chemistry does not correlate with microbial community structure. *Estuarine Coastal and Shelf Science*, 62: 233–251.

Kostka, J. E., and Luther III, G. W. 1995. Seasonal cycling of Fe in salt marsh sediments. *Biogeochemistry*, 29: 159–181.

Lane, R. R., Day, J. W., Justic, D., Reyes, E., Marx, B., Day, J. N., and Hyfield, E. 2004. Changes in stoichiometric Si, N and P ratios of Mississippi River water diverted through coastal wetlands to the Gulf of Mexico. *Estuarine, Coastal and Shelf Science*, 60: 1–10.

Lanning, F. C., and Eleuterius, L. N. 1981. Silica and ash in several marsh plants. *Gulf and Caribbean Research*, 7: 47–52.

Lanning, F. C., and Eleuterius, L. N. 1983. Silica and ash in tissues of some coastal plants. *Annals of Botany*, 51: 835–850.

Lanning, F. C., and Eleuterius, L. N. 1985. Silica and ash in tissues of some plants growing in the coastal area of Mississippi, USA. *Annals of Botany*, 56: 157–172.

Lanning, F. C. and Eleuterius, L. N. 1989. Silica deposition in some C3 and C4 species of grasses, sedges and composites in the USA. *Annals of Botany*, 64: 395–410.

LeMay, L. 2007. *The impact of drainage ditches on salt marsh flow patterns, sedimentation and morphology: Rowley River, Massachusetts.* MS thesis. The College of William and Mary, Williamsburg, Virginia, USA.

Lettrich, M. 2011. *Nitrogen advection and denitrification loss in southeastern North Carolina salt marshes.* MS thesis, University of North Carolina Wilmington.

Lillebø, A. I., Coelho, J. P., Flindt, M. R., Jensen, H. S., Marques, J. C., Pedersen, J. B., and Pardal, M. A. 2007. *Spartina maritima* influence on the dynamics of the phosphorous sedimentary cycle in a warm temperate estuary (Mondego Estuary, Portugal). *Hydrobiologia*, 587: 195–204.

Lillebø, A. I., Neto, J. M., Flindt, M. R., Jensen, H. S., Marques, J. C., and Pardal, M. A. 2004. Phosphorous dynamics in a temperate intertidal estuary. *Estuarine, Coastal and Shelf Science*, 61: 101–109.

Lloret, J., and Valiela, I. 2016. Unprecedented decrease in deposition of nitrogen oxides over North America: the relative effects of emission controls and prevailing air-mass trajectories. *Biogeochemistry*, 129: 165–180.

Loomis, M. J., and Craft, C. B. 2010. Carbon sequestration and nutrient (nitrogen, phosphorus) accumulation in river-dominated tidal marshes, Georgia, USA. *Soil Society of America Journal*, 74: 1028–1036.

Maavara, T., Parsons, C. T., Ridenour, C., Stojanovic, S., Durr, H. H., Powley, H. R. and Van Cappellen, P. 2015. Global phosphorus retention by river damming. *Proceedings of the National Academy of Sciences*, 122: 15603–15608.

Maguire, T. J. 2017. *Anthropogenic perturbations to the biogeochemical cycle of silicon.* PhD dissertation, Boston University, Boston, MA.

Mahowald, N., Jickells, T. D., Baker, A. R., Artaxo, P., Benitex-Nelson, C. R., Bergametti, G., Bond, T. C., et al. 2008. Global distribution of atmospheric phosphorus sources, concentrations and deposition rates and anthropogenic impacts. *Global Biogeochemical Cycles*, 22: 1–19.

Mariotti, G., and Fagherazzi, S., 2013. Critical width of tidal flats triggers marsh collapse in the absence of sea-level rise. *Proceedings of the National Academy of Sciences*, 110: 5353–5356.

Martin, R. M. and Moseman-Valtierra, S. 2015. Greenhouse gas fluxes vary between Phragmites alstralis and native vegetation zones in coastal wetlands along a salinity gradient. *Wetlands*, 35: 1021–1031.

Martin, R. M., Wigand, C., Elmstrom, E., Lloret, J., and Valiea, I. 2018. Long-term nutrient addition increase respiration and nitrous oxide emissions in New England salt marsh. *Ecology and Evolution*, 8: 4958–4966.

Mateos-Naranjo, E., Andrades-Moreno, L., and Davy, A. J. 2013. Silicon alleviates deleterious effects of high salinity on the halophytic grass *Spartina densiflora*. *Plant Physiology and Biochemistry*, 63: 115–121.

Mateos-Naranjo, E., Gallé, A., Florez-Sarasa, I., Perdomo, J. A., Galmés, J., Ribas-Carbó, M., and Flexas, J. 2015. Assessment of the role of silicon in the Cu-tolerance of the C 4 grass *Spartina densiflora*. *Journal of Plant Physiology*, 178: 74–83.

McKeague, J. A., and Cline, M. G. 1963. Silica in soils. *Advances in Agronomy*, 15: 339–396.

McLeod, E., Chmura, G., Bouillon, S. Salm, R. Bjork, M. Duarte, C., Lovelock, C., Schlesinger, W., and Silliman, B. 2011. A blueprint for the blue carbon: toward an improved understanding of the role of vegetated coastal habitats in sequestering CO_2. *Frontiers Ecology Environment*, 9: 552–560.

Mendelssohn, I. A., and Morris J. T. 2000. Eco-physiological controls on the productivity of *Spartina alternilflora*, Loisel. In: *Concepts and Controversies in Tidal Marsh Ecology*. M. P. Weinstein, and D. A. Kreeger, eds., Kluwer Academic Publishers, Dordrecht, the Netherlands, pp. 59–80.

Mendelssohn, I. A., and Seneca, E. D. 1980. The influence of soil drainage on the growth of salt marsh cordgrass *Spartina alterniflora* in North Carolina. *Estuarine Coastal and Marine Science*, 2: 27–40.

Merrill, J. Z. and J. C. Cornwell. 2000. The role of oligohaline marshes in estuarine nutrient cycling. In: *Concepts and Controversies in Tidal Marsh Ecology*. M. P. Weinstein, and D. A. Kreeger, eds., Kluwer Academic Publishers, Dordrecht, the Netherlands, pp 425–441.

Metson, G. S., Lin, J., Harrison, J. A., and Compton, J. E. 2017. Linking terrestrial phosphorus inputs to riverine export across the United States. *Water Research*, 124: 177–191.

Meybeck, M. 2003. Global occurrence of major elements in rivers. *Treatise on Geochemistry*, 5: 207–223.

Miyazako, T., Kamiya, H., Godo, T., Koyama, Y., Sato, S., Kishi, M., Fujihara, A., Tabayashi, Y., and Yamamuro, M. 2015. Long-term trends in nitrogen and phosphorus concentrations in the Hii River as influenced by atmospheric deposition from East Asia. *Limnology and Oceanography*, 60: 629–640.

Moisander, P. H., Piehler, M. F., and Paerl, H. W. 2005. Diversity and activity of epiphytic nitrogen-fixers on standing dead stems of the salt marsh grass *Spartina alterniflora*. *Aquatic Microbial Ecology*, 39: 271–279.

Morris, J. T., and Lajtha, K. 1986. Decomposition and nutrient dynamics of litter from four species of freshwater emergent macrophytes. *Hydrobiologia*, 131: 215–223.

Morris, J. T., Shaffer, G. P., and Nyman, J. A. 2013. Brinson Review: Perspectives on the influence of nutrients on the sustainability of coastal wetlands. *Wetlands*, 33: 975–988.

Müller, F., Struyf, E., Hartmann, J., Wanner, A., and Jensen, K. 2013. A comprehensive study of silica pools and fluxes in Wadden Sea salt marshes. *Estuaries and Coasts*, 36: 1150–1164.

Newell, S. Y., Hopkinson, C. S., and Scott, L. 1992. Patterns of nitrogenase activity (acetylene reduction) associated with standing, decaying shoots of *Spartina alterniflora*. *Estuarine, Coastal and Shelf Science*, 35: 127–140

Nixon, S. W. 1980. Between coastal marshes and coastal waters – a review of twenty years of speculation and research on the role of salt marshes in estuarine productivity and water chemistry. In: P. Hamilton and K. B. MacDonald, eds., *Estuarine Wetland Process with Emphasis on Modelling*. Plenum Publishing Corporation, New York, pp. 438–525.

Nixon, S. W. 1995. Coastal marine eutrophication: a definition, social causes, and future concerns. *Ophelia*, 41: 199–219.

Nixon, S. W., Ammerman, J. W., Atkinson, L. P., Berounsky, V. M., Billen, G., Boicourt, W. C., Boynton, W. R., et al. 1996. The fate of nitrogen and phosphorus at the land-sea margin of the North Atlantic Ocean. *Biogeochemistry*, 35: 141–180.

Nixon, S. W., and Oviatt, C. A. 1973. Analysis of local variation in the standing crop of *Spartina alterniflora*. *Botanica Marine*, IVI: 103–9.

Nixon, W., and Pilson, M. 1984. Estuarine total system metabolism and organic exchange calculated from nutrient ratios: an example from Narragansett Bay. In: *The Estuary as a Filter*. V. S. Kennedy, ed., Academic Press, New York, pp. 261–290.

Noe, G. B., Childers, D. L., and Jones, R. D. 2001. Phosphorus biogeochemistry and the impact of P enrichment: Why is the everglades so unique? *Ecosystems*, 4: 603.

Norris, A. R., and Hackney, C. T. 1999. Silica content of a mesohaline tidal marsh in North Carolina. *Estuarine, Coastal and Shelf Science*, 49: 597–605.

Nuttle, W. K. 1988. The extent of lateral water movement in the sediments of a New England salt marsh. *Water Resources Research*, 24: 2077–2085.

Odum, E. P. 2000. Tidal marshes as outwelling pulsing systems. In: M. P. Weinstein, and D. A. Kreeger, eds., *Concepts and Controversies in Tidal Marsh Ecology*. Kluwer Academic Publishers, Dordrecht, the Netherlands, pp. 3–7.

Paludan, C. and Morris, J. T. 1999. Distribution and speciation of phosphorus along a salinity gradient in intertidal marshes. *Biogeochemistry*, 45: 197–221.

Packett, C. R., and Chambers, R. M. 2006. Distribution and nutrient status of haplotypes of the marsh grass *Phragmites australis* along the Rappahannock River in Virginia. *Estuaries and Coasts*, 29: 1222–1225.

Portnoy, J. W., Nowicki, B. L., Roman, C. T., and Urish, D. W. 1997. The discharge of nitrate-contaminated groundwater from a developed shoreline to a marsh-fringed estuary. *Water Resources Research*, **34**: 3095–3104.

Querné, J., Ragueneau, O., and Poupart, N. 2012. In situ biogenic silica variations in the invasive salt marsh plant, *Spartina alterniflora*: a possible link with environmental stress. *Plant and Soil*, 352: 157–171.

Reddy, K. R., Kadlec, R. H., Flaig, E., and Gale, P. M. 1999. Phosphorus retention in streams and wetlands – a review. *Critical Reviews in Environmental Science and Technology*, 29: 86–146.

Redfield, A. C. 1963. The influence of organisms on the composition of seawater. *The Sea*, 2: 26–77.

Redfield, G. W. 2002. Atmospheric deposition of phosphorus to the Everglades: Concepts, constraints, and published deposition rates for ecosystem management. *The Scientific World Journal*, 2: 1843–1873.

Reed, D. J. 1972. Patterns of sediment deposition in subsiding coastal salt marshes. Terrebone Bay, Louisiana: the role of winter storms. *Estuaries*, 12: 222–227.

Reimold, R. G. 1972. The movement of phosphorus through the marsh cord grass, *Spartina alterniflora*. Loisel. *Limnology and Oceanography*, 17: 606–611.

Rozema, J., P. Leendertse, J. Bakker, and H. van Wijnen. 2000. Nitrogen and vegetation dynamics in European salt marshes. In: M. P. Weinstein, and D. A. Kreeger, eds., *Concepts and Controversy in Tidal Marsh Ecology*. Kluwer Academic Publishers, Dordrecht, the Netherlands, pp. 469–494.

Ruttenberg, K. 1992. Development of a sequential extraction method for different forms of phosphorus in marine sediments *Limnology and Oceanography*, 37: 1460–1482.

Ruttenberg, K. 2014.The global phosphorus cycle. In: H. D. Holland, and K. K. Turekian, eds., *Treatise on Geochemistry*, 2nd edn, vol 10. Elsevier, Oxford, pp. 499–558.

Sauer, D., Saccone, L., Conley, D. J., Herrmann, L., and Sommer, M., 2006. Review of methodologies for extracting plant-available and amorphous Si from soils and aquatic sediments. *Biogeochemistry*, 80: 89–108.

Seitzinger, S. P., Harrison, J. A., Dumont, E., Beusen, A. H. W., and Bouwman, A. F. 2005. Sources and delivery of carbon, nitrogen, and phosphorus to the coastal zone: An overview of the Global Nutrient Export from Watersheds (NEWS) models and their application. *Global Biogeochemical Cycles* 19; GB4S01, doi: 10.1029/2005GB002606.

Seitzinger, S. P., Mayorga, E., Bouwman, A., Kroeze, C., Beusen, C., Billen, van Drecht, G., et al. 2010. Global river nutrient export: a scenario analysis of past and future trends. *Global Biogeochemical Cycles*, 24: GB0A08, doi: 10.1029/2009GB003587, 2010.

Simonson, R. W. 1995. Airborne dust and its significance to soils. *Geoderma*, 65: 1–43.

Scudlark, J. R., and Church, T. M. 1989. The sedimentary flux of nutrients at a Delaware salt marsh site: A geochemical perspective. *Biogeochemistry*, 7: 55–75.

Smith, S. V. 1991. Stoichiometry of C: N: P fluxes in shallow-water marine ecosystems. In: J. Cole, J. Lovett, and S. Findlay, eds., *Comparative Analyses of Ecosystems. Patterns, mechanisms, theories*. New York, Springer-Verlag, pp. 259–286.

Sommer, M., Kaczorek, D., Kuzyakov, Y., and Breuer, J. 2006. Silicon pools and fluxes in soils and landscapes – a review. *Journal of Plant Nutrition and Soil Science*, 169: 310–329.

Sorrell, B. Mendelssohn, I. A., McKee, K., and Woods, R. A. 2000. Ecophysiology of wetland plant roots: A modeling comparison of aeration in relation to species distribution. *Annals of Botany*, 86: 675–685.

Statham, P. J. 2012. Nutrients in estuaries – An overview and the potential impacts of climate change. *Science of the Total Environment*, 434: 213–227.

Staver, L. W. 2015. *Ecosystem dynamics in tidal marshes constructed with fine grained, nutrient rich dredged material*. PhD thesis. University of Maryland, College Park, MD.

Street-Perrott, F. A., and Barker, P. A. 2008. Biogenic silica: a neglected component of the coupled global continental biogeochemical cycles of carbon and silicon. *Earth Surface Processes and Landforms*, 33: 1436–1457.

Stribling, J. M. and Cornwell, J. C. 2001. Nitrogen, phosphorus and sulfur dynamics in a low salinity marsh system dominated by *Spartina alterniflora*. *Wetlands*, 21: 629–638.

Stumm, W., and Sulzberger, B. 1992. The cycling of iron in natural environments: Considerations based on laboratory studies of heterogeneous redox processes. *Geochimica et Cosmochimica Acta*, 56: 3233–3257.

Struyf, E., Dausse, A., Van Damme, S., Bal, K., Gribsholt, B., Boschker, H. T., Middelburg, J. J., and Meire, P. 2006. Tidal marshes and biogenic silica recycling at the land-sea interface. *Limnology and Oceanography*, 51: 838–846.

Struyf, E., Mörth, C. M., Humborg, C., and Conley, D. J., 2010. An enormous amorphous silica stock in boreal wetlands. *Journal of Geophysical Research: Biogeosciences*, 115(G4): 1–8.

Struyf, E., Smis, A., Van Damme, S., Garnier, J., Govers, G., Van Wesemael, B., Conley, D. J., et al. 2010. Historical land use change has lowered terrestrial silica mobilization. *Nature Communications*, 1: 129.

Struyf, E., Van Damme, S., Gribsholt, B., Middelburg, J. J., and Meire, P. 2005. Biogenic silica in tidal freshwater marsh sediments and vegetation (Schelde estuary, Belgium). *Marine Ecology Progress Series*, 303: 51–60.

Sullivan, M., and Daiber, F. 1974. Response in production of cord grass *Spartina alterniflora*, to inorganic nitrogen and phosphorus fertilizer. *Chesapeake Science*, 15: 121–123.

Sundareshwar, P. V., and Morris, J. T. 1999. Phosphorus sorption characteristics of intertidal marsh sediments along an estuarine salinity gradient. *Limnology and Oceanography*, 44: 1693–1701.

Sundby, B., Vale, C., Caetano, M., and Luther III, G. W. 2003. Redox chemistry in the root zone of a salt marsh sediment in the Tagus Estuary, Portugal. *Aquatic Geochemistry*, 9: 257–271.

Teal, J. 1962. Energy flow in a salt marsh ecosystem of Georgia. *Ecology*, 43: 614–624.

Tegen, I., and Kohfeld, K. E. 2006. Atmospheric transport of silicon. In: V. Ittekkot, D. Unger, C. Humborg, and N. T. An, eds., *The Silicon Cycle: Human Perturbations and Impacts on Aquatic Systems*, Island Press, Washington pp. 81–91.

Tipping, E., Benham, S., Boyle, J. F., Crow, P., Davies, J., Fischer, U., Guyatt, H., et al. 2014. Atmospheric deposition of phosphorus to land and freshwater. *Environmental Science Processes and Impacts*. DOI: 10.1039/c3em00641g.

Tobias, C., Anderson, I., Canuel, E., and Macko, S. 2001. N cycling through a fringing marsh-aquifer ecotone. *Marine Ecology Progress Series*, 210: 25–39.

Tobias, C., and Neubauer, S. 2009. Salt marsh biogeochemistry – an overview. In: G. Perillo, E. Wolanski, D. R. Cahoon, and M. M. Brinson, eds., *Coastal Wetlands: An Integrated Ecosystems Approach*, Elsevier, Amsterdam, pp. 445–492.

Tobias, C., and Neubauer, S. 2018. Salt marsh biogeochemistry – an overview. In: E., Wolanski, G. Perillo, D. Cahoon, and C. Hopkinson, eds., *Coastal Wetlands: a synthesis*. 2nd edn. Elsevier, Amsterdam, pp. 539–596.

Torres, R., Mwamba, M., and Goni, M. 2003. Properties of intertidal marsh sediment mobilized by rainfall. *Limnology and Oceanography*, 48: 1245–1253.

Tréguer, P. J., and De La Rocha, C. L. 2013. The world ocean silica cycle. *Annual Review of Marine Science*, 5: 477–501.

Treguer, P., Nelson, D. M., Van Bennekom, A. J., DeMaster, D. J., Leynaert, A., and Quéguiner, B. 1995. The silica balance in the world ocean: a reestimate. *Science*: 375–375.

Turner, R. E., Swenson, E. M., Milan, C. S., and Lee, J. M. 2007. Hurricane signals in salt marsh sediments: Inorganic sources and soil volume. *Limnology and Oceanography*, 52: 1231–1238.

Tyler, C., Mastronicola, T., and McGlathery, K. 2003. N fixation and N limitation of primary production along a natural marsh chronosequence. *Oecologia*, 136: 431–438.

Valiela, I., and Teal, J. M. 1974. Nutrient limitation in salt marsh vegetation. In: R. J. Reimold, and W. H. Queen, eds., *The Ecology of Halophytes*. Academic Press, New York, pp. 547–563.

Valiela, I., and Teal, J. 1979. The N budget of a salt marsh ecosystem. *Nature*, 280: 652–656.

Valiela, I., and Cole, M. L. 2002. Comparative evidence that salt marshes and mangroves may protect seagrass meadows from land-derived nitrogen loads. *Ecosystems*, 5: 92–102.

Valiela, I., Cole, M. L., McClelland, J., Hauxwell, J., Cebrian, J., and Joye, S. B. 2000. Role of salt marshes as part of coastal landscapes. In: M. P. Weinstein, and D. A. Kreeger, eds., *Concepts and Controversies in Tidal Marsh Ecology*.Kluwer Academic Publishers, Dordrecht, the Netherlands, pp. 23–38.

Valigura, R., Alexander, R., Castro, M., Meyers, T., Paerl, H., Stacey, P., and Turner, R., eds. 2000. Nitrogen loading in coastal water bodies: an atmospheric perspective. *Coastal and Estuarine Studies* No. 57. AGU, Washington, DC.

Vallino, J. J., and Hopkinson, C. S. 1998. Estimation of dispersion and characteristics of mixing times in Plum Island Sound Estuary. *Estuarine, Coastal and Shelf Science*, 46: 333–350.

Vallino, J. J., Hopkinson, C. S., and Garritt, R. H. 2005. Estimating estuarine gross production, community respiration and net ecosystem production: A nonlinear inverse technique. *Ecological Modeling*, 187: 281–296.

VanZomeren, C. 2011. *Fate of Mississippi River diverted nitrate on vegetated and non-vegetated coastal marshes of Breton Sounds Estuary*. MS thesis. Louisiana State University.

Vieillard, A. M., Fulweiler, R. W., Hughes, Z. J., and Carey, J. C. 2011. The ebb and flood of silica: quantifying dissolved and biogenic silica fluxes from a temperate salt marsh. *Estuarine, Coastal and Shelf Science*, 95: 415–423.

Wedepohl, K. H. 1995. The composition of the continental crust. *Geochimica et Cosmochimica Acta*, 59: 1217–1232.

Weston, N. B., Porubsky, W. P., Samarkin, V. A., Erickson, M., Macavoy, S. E., and Joye, S. B. 2006. Porewater stoichiometry of terminal metabolic products, sulfate, and dissolved organic carbon and nitrogen in estuarine intertidal creek-bank sediments. *Biogeochemistry*, 77: 375–408.

Weston, N. 2013. Declining sediments and rising seas: an unfortunate convergence for tidal wetlands. *Estuaries and Coasts*, 37: 1–23.

White, D., and Howes, B. 1994. Long-term 15N-nitrogen retention in the vegetated sediments of a New England salt marsh. *Limnology and Oceanography*, 39: 133–140.

Whitney, D. M., Chalmers, A. G., Haines, E. B., Hanson, R. B., Pomeroy, L. R., and Sherr, B. 1981. The cycles of nitrogen and phosphorus. In: L. R. Pomeroy, and R. G. Wiegert, eds. *The Ecology of a Salt Marsh*. Springer Verlag. N.Y, pp. 163–181.

Wilson, A. M., and Gardner, L. R. 2006. Tidally driven groundwater flow and solute exchange in a marsh: numerical simulations. *Water Resources Research*, 42:https://doi.org/10.1029/2005WR004302.

Woodwell, G. M., and Whitney, D. E. 1977. Flax pond ecosystems study: exchanges of phosphorus between a salt marsh and the coastal waters of Long Island Sound. *Marine Biology*, 41: 1–6.

Wolaver, T. G., and Spurrier, J. D. 1988. The exchange of phosphorus between a euhaline vegetated marsh and the adjacent tidal creek. *Estuarine and Shelf Sciences*, 26: 203–214.

Wolaver, T. G., and Zieman, J. 1984. The role of tall and medium *Spartina alterniflora* zones in the processing of nutrients in tidal water. *Estuarine, Coastal and Shelf Sciences*, 19: 1–13.

Wolaver, T. G., Zieman, J. C., Wetzel, R. and Webb, K. L. 1983. Tidal exchange of nitrogen and phosphorus between a mesohaline vegetated marsh and the surrounding estuary in the lower Chesapeake Bay. *Estuarine, Coastal and Shelf Science*, 16: 321–332.

Yang, Q., Tian, H., Friedrichs, M. A. M., Hopkinson, C. S., Lu, C., and Najjar, R. G. 2015. Increased nitrogen export from eastern North America to the Atlantic Ocean due to climatic and anthropogenic changes during 1901–2008. *Journal of Geophysical Research: Biogeosciences*, 120: 1046–1068.

Yu, D., Yan, W., Chen, N., Peng, B., Hong, H., and Zhuo, G. 2015. Modeling increased riverine nitrogen export: source tracking and integrated watershed-coast management. *Marine Pollution Bulletin*, 101: 642–652.

Zhang, J., Zhang, G. S., and Liu, S. M. 2005. Dissolved silicate in coastal marine rain-waters: Comparison between the Yellow Sea and the East China Sea on the impact and potential link with primary production. *Journal of Geophysical Research: Atmospheres*, 110(D16).

Zhou, J., Wu, Y., Kang, Q., and Zhang, J. 2007. Spatial variations in carbon, nitrogen, phosphorus and sulfur in the salt marsh sediments of the Yangtze Estuary in Chine. *Estuarine, Coastal and Shelf Science*, 71: 47–59.

Part II

Marsh Dynamics

7

Marsh Equilibrium Theory

Implications for Responses to Rising Sea Level

JAMES T. MORRIS, DONALD R. CAHOON, JOHN C. CALLAWAY,
CHRISTOPHER CRAFT, SCOTT C. NEUBAUER, AND NATHANIEL B. WESTON

7.1 Introduction

The analysis presented here was motivated by an objective of describing the interactions between the physical and biological processes governing the responses of tidal wetlands to rising sea level and the ensuing equilibrium elevation. We define equilibrium here as meaning that the elevation of the vegetated surface relative to mean sea level (MSL) remains within the vertical range of tolerance of the vegetation on decadal time scales or longer. The equilibrium is dynamic, and constantly responding to short-term changes in hydrodynamics, sediment supply, and primary productivity. For equilibrium to occur, the magnitude of vertical accretion must be great enough to compensate for change in the rate of sea-level rise (SLR). SLR is defined here as meaning the local rate relative to a benchmark, typically a gauge. Equilibrium is not a given, and SLR can exceed the capacity of a wetland to accrete vertically.

The limits of vertical accretion in coastal wetlands are of interest because the valuable ecosystem services they provide are threatened by rising sea level (Craft et al. 2009). The services at risk include storm surge abatement, vertical land building and lateral prograda-tion, recreation, soil carbon storage, and provision of habitat and nutrition for many marine species, including those of commercial importance (Oliver 1925, Costanza et al. 2008; Barbier et al. 2011; Shepard et al. 2011; Drake et al. 2015; Narayan et al. 2017). The threat from SLR, including subsidence (the other primary factor contributing to local or relative SLR), depends on the balance between SLR and local rates of erosion and accretion, i.e., the net change in sediment volume per unit area.

Previous studies have considered empirical relationships among organic and inorganic sediment fractions and vertical wetland accretion rates (e.g., Nyman et al. 1993; Turner et al. 2000; Neubauer, 2008). Model estimates of the limits of vertical accretion also have been published (Kirwan et al. 2010). They show marsh survivorship is dependent upon tidal range and rate of SLR and, depending on the model, the maximum vertical accretion rates ranged from 5 to 15 mm yr^{-1} in simulated marshes having a tidal range of 1 m and a sediment concentration of 30 mg/L. The marsh equilibrium model (MEM, Morris et al. 2002) provides insight into marsh development; here we describe the logic behind the MEM and what we refer to as marsh equilibrium theory. The results provide several generalities that improve our understanding of the relative importance of biological and physical factors that affect marsh equilibrium, when they are important, and how tidal wetlands change under different conditions of SLR and tidal amplitude.

The processes that change the volume of sediment and elevation are fundamental to the adaptation of coastal wetlands to SLR. Sediment volume decreases as a consequence of erosion and compaction (processes that we do not address here), and decomposition (of organic matter). Volume increases with deposition onto the sediment surface of mineral particles and refractory organic matter and through the subsurface accumulation of organic matter from roots and rhizomes. The annual surface deposition of mineral and organic material suspended in flood-tide water creates a lamination that can be conceptualized as a sediment cohort (Morris and Bowden, 1986). The volume of a cohort will change over time as a consequence of the ingrowth and turnover of roots and rhizomes, their decay, and their preservation. The final volume of the cohort after it has reached a depth below the root zone, after years of burial by generations of younger cohorts, is achieved when its organic matter content has stabilized (Morris and Bowden, 1986; Callaway et al. 1996; Davis et al. 2015). As cohorts traverse the root zone, the ingrowth of roots and inputs of labile organic matter are balanced by death and decay. Only the refractory portion of root and rhizome production, and in some systems the input of refractory allochthonous organic matter onto the surface, including leaf litter, adds new volume to that created by the deposition of surface material. The final bulk density and mass of the cohort will determine the rate of vertical accretion when the wetland is in equilibrium with sea level. When not in equilibrium with SLR, accretion rates could be higher or lower, depending on the rate of SLR, sediment supply, relative elevation, and other factors; and under these conditions the marsh surface will gain or lose elevation over time.

The concept of a marsh equilibrium and its development can be traced back to the early twentieth century. Stapf (1907) described how the rapid accretion of mud banks in Southampton Water (a tidal estuary north of the Solent and the Isle of Wight in England) was due the arrival of a ship carrying cordgrass (*Spartina*). Its seeds became distributed on the shores, resulting in the whole of the estuary being covered with this grass. Speaking of English and French coastlines of the English Channel, Oliver (1925) wrote "much of the silt will be held by it [*Spartina*] and the level [marsh elevation] slowly rise. In other words, the mobile silt of the channels and waterways tends to be fixed at a higher level. The immediate consequence of this is a deepening and perhaps a widening of the channels... when *Spartina* has occupied fully the available ground, the amount of mud that can be assimilated will reach its limit. For as the level rises the rate of silting will decrease till it becomes inappreciable." Ranwell (1964) reported a positive correlation between accretion, marsh elevation, vegetation height, and biomass. Richard (1978) found that sedimentation rates decrease with increasing elevation because of the reduced tidal submergence time and decreased height of the overlying water column. Pethick (1981) observed that the age/ height relationship of marsh sediment describes an asymptotic curve with the asymptote lying below the level of the highest spring tides (similar to Fig. 7.7) and apparently controlled by the frequency of tidal maxima. Krone (1987) found that "The elevation of the marsh rises after plants become established at rates that depend on the availability of suspended sediment and the depth and periods of inundation by high tides. As the elevation of the marsh rises relative to sea level, however, the frequency and duration of inundation diminishes and the rate of sediment accumulation falls," and a mostly positive accretionary balance occurs in the zone of marsh vegetation due to the effect of the vegetation cover on sedimentation and erosion protection (Dijkema et al. 1990). The early equilibrium concept was clearly about mineral sediment, flood duration and depth, and the trapping and stabilization by plants.

Randerson (1979) incorporated plant growth in a simulation of marsh accretion, including a positive feedback in which vegetative growth increases accretion by encouraging deposition and stabilizing the sediment. Numerical modeling by Allen (1990) and French (1993) showed that with an adequate sediment supply, marsh elevations will adjust to changes in the rate of SLR and that organic accretion was important. Allen (1990) also discussed the importance of relative elevation and flooding on peat formation. Morris et al. (2002) were the first to describe feedbacks among vegetative growth, relative elevation, sea level, and accretion that establish an equilibrium. They presented a model known as the marsh equilibrium model (MEM) that incorporated the growth response of vegetation to relative elevation. That early model has since taken on greater complexity (Morris 2006, 2007; Mudd et al. 2004, 2009, 2010; Kirwan et al. 2010, Morris et al. 2012, Schile et al. 2014). Kirwan and Murray (2007) added flow-induced erosion in a 3D model of accretion in which disturbance/erosion can initiate channel formation. Marani et al. (2013) took this a step further by describing coupled geomorphological–biological dynamics and marsh vegetation that actively engineer the landscape, "tuning soil elevation within preferential ranges of optimal adaptation." Alizad et al. (2016) coupled MEM with a 2D hydrodynamic model that captures feedback between plant production, sedimentation, and changing hydrodynamics across an estuarine landscape. This chapter describes marsh equilibrium theory as the tendency of feedback to modify the salt marsh environment in a passive way, resulting in a relative elevation that remains within the growth range of emergent vegetation. The success of this feedback to maintain equilibrium is dictated by the availability of mineral sediment, tidal amplitude, plant productivity, and the rate of SLR.

7.2 Model Description

7.2.1 Inundation Time

Marsh elevation is strongly related to the fraction of time (ϑ) the marsh is inundated (e.g. Krone, 1987). For a marsh flooded predominantly by astronomic tidal components, fractional inundation time ($1 \geq \vartheta \geq 0$) is determined approximately by the vertical position (relative elevation Z) of the marsh landscape between the mean high (MHW) and mean low water (MLW) levels (Fig. 7.1), i.e.

$$\vartheta \cong (\text{MHW} - Z)/(\text{MHW} - \text{MLW}) \text{ for MHW} \geq Z \geq \text{MLW} \qquad (7.1)$$

Thus, ϑ is approximately equal to dimensionless depth $(\text{MHW} - Z)/(\text{MHW} - \text{MLW})$ which scales between 0 at $Z = \text{MHW}$ and 1 at $Z = \text{MLW}$. Note that in nontidal marshes where hydroperiod is controlled by wind or river flood, inundation time would depend on

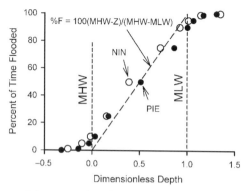

Figure 7.1 Percentage of time a marsh surface in the Plum Island, MA and North Inlet, SC is flooded as a function of dimensionless depth (MHW − Z)/(MHW − MLW). Relative mean high water (MHW) is approximately 140 and 70 cm in the two estuaries. Z is relative marsh elevation. PIE = Plum Island Estuary, MA; NIN = North Inlet, SC.

the frequency and duration of aperiodic flooding, but this would probably not scale with dimensionless depth.

7.2.2 Inorganic Sediment Deposition

Maximum sediment delivery (S, g cm^{-2} yr^{-1}) depends primarily on three factors – the concentration of sediment in suspension in nearby tidal creeks (m, g cm^{-3}), the average depth of the marsh surface below MHW (MHW − Z, cm) and the average number of flood events per year. In a semidiurnally flooded estuary, the number of tidal cycles is about 704 per year. So, for MLW<Z<MHW:

$$S = m \times (\text{MHW} - \text{Z}) \times 704 \times 0.5; \quad S = 0 \text{ for } \text{Z} \geq \text{MHW}. \tag{7.2}$$

Multiplying the depth (MHW − Z) by 0.5 gives the average depth of water during the period of inundation. So m H (MHW − Z) H 0.5 represents the suspended sediment inventory over the surface for the duration of the flood.

The deposition (W_i, g cm^{-2} yr^{-1}) or accumulation rate of inorganic sediment is given by the product of sediment delivery (S), ϑ, and capture coefficient (q):

$$W_i = \vartheta S q \tag{7.3}$$

The capture coefficient q is a dimensionless, empirically determined quantity, subject to the constraint $q\vartheta \leq 1$, i.e., in theory, the product of inundation time and capture coefficient cannot exceed the quantity of available sediment. Hence, W_i is a piece-wise function that can be applied across the vertical range of the vegetated marsh and not beyond. Across an entire vegetated marsh landscape, it must be the case that deposition cannot exceed supply ($\vartheta q \leq 1$), but locally ϑq could be greater than 1 due to resuspension and bioturbation, redistribution, and local imbalances in the supply. The value of q used here is 2.8 as determined from an analysis of SET (Sedimentation-Erosion Table) data from North Inlet,

South Carolina (www.baruch.sc.edu/geological-hydrological-databases, Morris unpublished analysis), subject to the constraint that mass across the marsh landscape is conserved. The vertical accretion rate due to mineral deposition (minerally sourced accretion) is obtained by dividing W_t by the self-packing density of inorganic sediment as discussed below.

The concept of sediment transport here is much like that of Krone (1987) who posited that suspended sediment was transported across the marsh surface at a concentration like that of the source – the nearby creeks. Further, sediment concentration across the marsh remains constant during the flooding tide, until the moment of slack high tide when flocculation occurs and particles settle. This greatly simplified the mass balance problem created by particles settling as they move, and was shown by Krone (1987) to closely simulate marsh sediment transport and deposition.

Bioturbation is another factor affecting transport. In many marshes, the density of burrowing organisms (worms and crabs, etc.) is very high. During low tide, their burrows often become hypoxic. At the moment new water floods the surface, these organisms begin to feed, irrigate their burrows, and in the process they resuspend sediment (e.g., Krüger, 1964; Vader, 1964; Torres et al. 1977; Baumfalk, 1979; Esselink and Zwarts, 1989; de Deckere et al. 2001; Webb and Eyre, 2004; Widdows et al. 2004; Ciutat et al. 2006; Volkenborn et al. 2010). This phenomenon could result in movement of sediment toward the marsh interior on successive rising tides, like a conveyor belt, resulting in a chaotic pattern of sediment deposition. Studies of the spatial pattern of deposition document its variability (Leonard, 1997; Davidson-Arnott et al. 2002; Neubauer et al. 2002; van Proosdij et al. 2006; Marion et al. 2009). However, across the entire marsh, sediment loading cannot exceed supply.

7.2.3 Biovolume Production

Vertical accretion also depends on the production of biovolume which is strongly affected by relative elevation. In salt marshes where there is little or no accumulation of surface litter, new biovolume is generated from the turnover of belowground biomass. New biovolume is the refractory portion of that turnover, and this is proportional to the standing biomass, which is a function of relative elevation. McKee and Patrick (1988) showed that *Spartina alterniflora* has a vertical growth range that spans a distance a bit greater than the tidal amplitude, centered approximately between MSL and MHW. For several species examined, including *S. alterniflora*, growth at the high end of the vertical growth range is constrained by osmotic stress (salinity, drought), while at the low end of the range, growth is constrained by hypoxia (Mendelssohn and Morris, 2000). A field bioassay experiment (marsh organ, Fig. 7.2) showed that *S. alterniflora* at North Inlet, South Carolina, grows between approximately 10 cm below MSL to about 30 cm above MHW with an optimum for productivity and standing biomass in the middle of the range (Morris et al. 2013). This seems to be a good rule of thumb that extends to estuaries differing in tidal range. We can describe the vertical biomass profile generally as a function of relative elevation (Z) of the form:

Figure 7.2 Bioassay experiment (marsh organ) at North Inlet Estuary, SC with *S. alterni-flora* growing in PVC pipes standing at different elevations, simulating differences in relative marsh elevation. (A black and white version of this figure will appear in some formats. For the color version, please refer to the plate section.)

$$B_s = a(MHW - Z) + b(MHW - Z)^2 + c \qquad (7.4)$$

where B_s is the end-of-season standing biomass (g cm^{-2}), $MHZ - Z$ is depth (D, cm), positive for depths below MHW, and the a, b, and c are empirical constants. The constants are computed after specifying the optimum depth (D_{opt}), the upper and lower limits of the growth range in terms of depth relative to MHW (D_{min} and D_{max}), and the maximum biomass at the optimum elevation (B_{max}). For calculations reported here we will assume that $B_{max} = 2000$ g m^{-2}, $D_{opt} = (D_{min} + D_{max})/2$, i.e., a symmetrical distribution, and $D_{min} = -30$ cm and $D_{max} = $ MHW $+ 10$ cm, where MSL $= 0$.

We calculated biomass distributions and subsequent accretion rates for three types of estuaries: one is microtidal and typical of US Gulf Coast with tidal amplitude of 20 cm, one is mesotidal like North Inlet marsh in South Carolina with amplitude 75 cm, and the third has an amplitude of 150 cm like Plum Island estuary, Massachusetts. We will refer to these as minor and major mesotidal sites, respectively. The resulting biomass distributions from the mesotidal systems are similar when depth is expressed as a dimensionless number, while the hypothetical microtidal estuary has a broader distribution (Fig. 7.3). The microtidal system is most dissimilar because its upper limit (MHW + 30 cm) is greater than the tidal amplitude (20 cm), which results in an optimum at MHW ($D_{opt} = 0$). More research is needed to determine how these vertical biomass distributions actually differ among sites and the degree to which they are affected by environmental variables such as rainfall and nutrients. However, for the purpose of this paper, the details are less important than the generalities. Conversion of standing biomass to new, refractory belowground production is discussed in Section 7.2.6.

7.2.4 Bulk Density

In marsh sediments, there is a relationship between soil dry density and organic matter content given by the ideal mixing rule, which states that the bulk volume of the soil

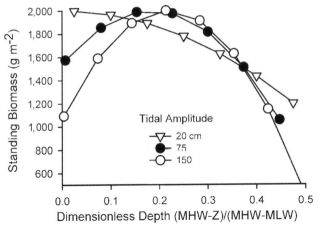

Figure 7.3 The theoretical distribution of *S. alterniflora* standing biomass from Eq. 7.4 as a function of dimensionless depth for marshes having different tidal amplitudes.

approximates the summed self-packing volumes of the organic and mineral components (Stewart et al. 1970; Morris et al. 2016), derived as follows. In a mixture, the volumes occupied by the organic and mineral components are additive (Federer et al. 1993). For example, starting with dry weights W_o and W_i of organic and inorganic matter having self-packing densities of k_1 and k_2, and volumes $V_o = W_o/k_1$ and $V_i = W_i/k_2$, when the two are mixed, the resulting bulk density is BD = $(W_o + W_i)/(W_o/k_1 + W_i/k_2)$. Substituting from loss on ignition LOI = $W_o/(W_o + W_i)$:

$$BD = 1/[LOI/k_1 + (1 - LOI)/k_2)] \qquad (7.5)$$

A fit of this model to bulk density data from 33 estuaries gave an R^2 of 0.78 and k_1 and k_2 coefficients of 0.085 ± 0.0007 g cm^{-3} and 1.99 ± 0.028 g cm^{-3} (Morris et al. 2016). The vertical accretion rate is derived by dividing the annual deposition of sediment and production of refractory organic matter, discussed above, by their respective self-packing densities, k_1 and k_2, and adding the two components (Morris et al. 2016).

7.2.5 Limits to Vertical Accretion – Mineral Sediment

It follows from the mixing rule (Eq. 5) that the relative contributions of the mineral and organic fractions to vertical accretion can be determined from their respective mass inputs. Thus, the contribution to vertical accretion from the inorganic deposition can be computed directly from the above equations and mineral bulk density k_2. Solving Eq. 2 for the inorganic loading (Fig. 7.4A) shows that it is highly sensitive to tidal amplitude, because for a given dimensionless depth, the inventory of sediment suspended over a marsh is greater for higher tidal amplitudes. For example, when the elevation is exactly mid-way between MHW and MSL (where both dimensionless depth and fractional inundation time = 0.25), a site with m = 30 mg L^{-1} and tidal amplitude of 150 cm will

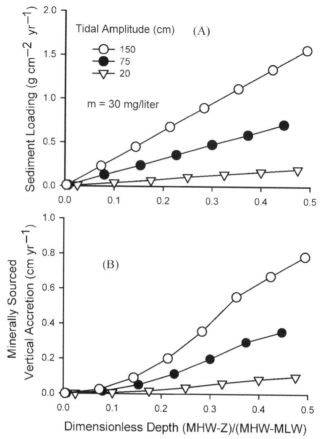

Figure 7.4 Computed sediment loading rates (A) and the corresponding vertical accretion rates (B) for marsh surfaces having tidal amplitudes of 20, 75, and 150 cm, 30 mg L^{-1} suspended sediment concentration, and dimensionless depths of 0 (at MHW) to 0.5 (MSL).

have a sediment loading of 0.79 g m^{-2} yr^{-1}, higher by virtue of the greater volume of water over the surface than a marsh with a 20 cm tidal amplitude which has a sediment loading of just 0.11 g m^{-2} yr^{-1} (Fig. 7.4A). By loading we are referring to sediment supply delivered to the marsh surface (Eq. 2).

Using the sediment deposition from Eq. 3 and dividing by mineral bulk density, we can calculate the corresponding vertical accretion rates (Fig. 7.4B). At the same mid-points between MHW and MSL ($\vartheta = 0.25$), vertical accretion rates range from 0.04 cm yr^{-1} in the case of a marsh with a 20 cm tidal amplitude, to 0.28 and 0.56 cm yr^{-1} in the cases of the 75 and 150 cm tidal amplitude estuaries, respectively, for suspended sediment concentration of 30 mg L^{-1}. Thus, all else being equal, tidal amplitude makes a big difference as doubling the amplitude should double the loading rate and contribution of mineral matter to vertical accretion. Thus, microtidal estuaries are less resilient, or less able to keep pace with SLR because of two factors: a narrower range of elevation capital

(Cahoon and Guntenspergen, 2010) and a more limited capacity for sediment loading and wetland vertical development.

We chose 30 mg L^{-1} as the sediment concentration here because it is typical of the maximum concentration seen in most East and Gulf Coast estuaries in North America (Fig. 7.5a, Weston, 2014). As a consequence of the formula used to calculate mineral sediment accumulation (Eq. 3), any change in sediment concentration should result in a proportional increase or decrease in vertical accretion rate of mineral matter. However, inputs of mineral sediment from watersheds to the coastal zone have declined in many East and Gulf Coast watersheds (Fig. 7.5b; Weston, 2014), suggesting that accelerating rates of SLR together with declining suspended sediment availability may increase marsh reliance on organic matter accumulation and/or result in the loss of tidal marsh area.

Of course, for a given suite of conditions, it is the rate of SLR that governs the vertical equilibrium as illustrated by the detail in Fig. 7.6. For the simulated virtual marshes at m = 30 mg L^{-1}, if SLR = 0.3 cm yr^{-1}, the mesotidal marshes will equilibrate at a relative elevation where the vertical accretion rate is 0.3 cm yr^{-1}. These mesotidal marshes will equilibrate at dimensionless depths (inundation times) of 0.38 and 0.26, respectively (Fig. 7.6). The microtidal marsh has no equilibrium at such a high SLR (discounting organic inputs). At SLR = 0.1 cm yr^{-1}, the microtidal marsh will equilibrate at inundation time = 0.475, equivalent to a relative elevation that is well below the optimum for the vegetation and nearing the point of drowning. At such a low SLR, the minor and major mesotidal marshes would equilibrate at dimensionless depths of 0.21 and 0.15, respectively, which both are much higher in the tidal frame than the microtidal marsh. Thus, considering only mineral inputs, for a given rate of SLR, marshes with higher tidal amplitudes will equilibrate at higher relative elevations (lower inundation times and higher elevation capital), and the range of SLR at which equilibrium is possible decreases with decreasing tidal amplitude (Fig. 7.6).

There are numerous empirical examples of the generalities discussed previously. For example, negative feedback between site elevation and the rate of vertical accretion has been described for tidal salt marshes by a number of researchers including Pethick (1981) and Williams and Orr (2002), where young, low-elevation marsh surfaces (below equilibrium) accumulate sediment quickly and asymptotically up to an equilibrium point near MHW, while high-elevation marsh surfaces accumulate matter much more slowly (Fig. 7.7). Temmerman et al. (2004) documented this negative feedback in salt, brackish, and freshwater marshes in the Scheldt Estuary, the Netherlands, varying in rates of SLR and incoming suspended sediment concentrations. They found that young, low-elevation marsh surfaces accumulate sediment quickly and asymptotically to an equilibrium level around MHW. This is consistent with the tendency of inundation time to approach zero at MHW (Fig. 7.1) and the principle behind the model of Krone (1987) as discussed earlier. Another example of the negative feedback between site elevation and vertical accretion was found in tidal mangrove forests described by Lovelock et al. (2011). They found that vertical accretion and shallow subsidence were inversely proportional to site elevation in forests located at different elevations in the tidal frame in western Moreton Bay, Australia. However, elevation change was proportional to site elevation, increasing from low to high

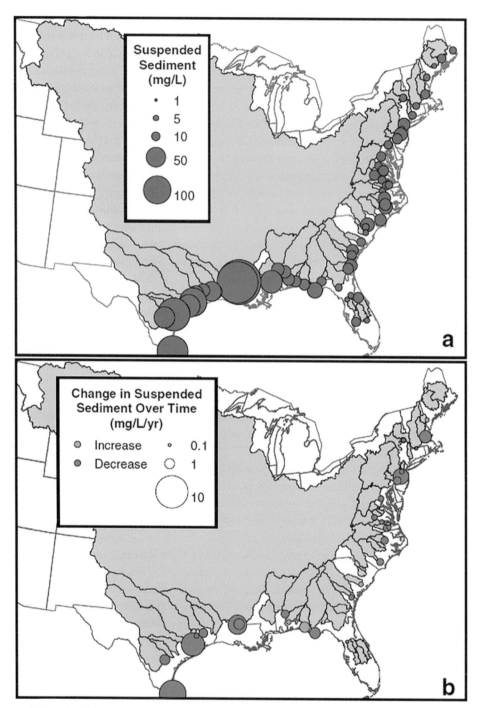

Figure 7.5 (a) Concentrations of suspended sediments in rivers, and (b) change in suspended sediment concentration over time in rivers draining to the US East and Gulf Coasts. From Weston (2014). (A black and white version of this figure will appear in some formats. For the color version, please refer to the plate section.)

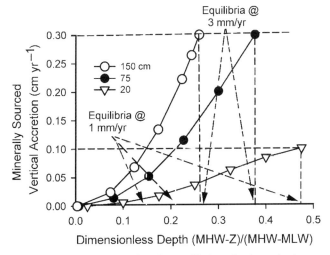

Figure 7.6 Detail from Fig. 7.4B showing the equilibrium depths at the intersection of red dashed lines [expressed as dimensionless depth (MHW−Z)/(MHW−MLW)] at rates of sea-level rise of 3 mm yr^{-1} and 1 mm yr^{-1} for virtual marshes having tidal amplitudes of 20, 75, and 150 cm, and 30 mg/L suspended mineral sediment concentration.

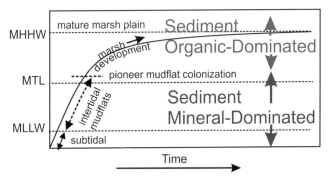

Figure 7.7 Predicted development of marsh elevation and the relative contributions of mineral and organic matter (modified from Williams and Orr 2002). MHHW: Mean Higher High Water; MTL: Mean Tide Level; MLLW: Mean Lower Low Water.

site elevations, and they attributed this to differences in root growth affecting shallow soil expansion (i.e., the opposite of shallow subsidence).

Similarly, a negative relationship was reported between elevation and vertical accretion in tidal freshwater wetlands located along a delta crevasse splay in the Mississippi River (Cahoon et al. 2011). At the Brant Pass crevasse in that delta, rates of vertical accretion, surface elevation change, and shallow subsidence were inversely proportional to soil elevation within the tidal frame of the southern splay lobe, decreasing by an order of magnitude between open water (lowest elevation), through the herbaceous marsh, and to the forested wetland (highest elevation). The elevation declined by 79 cm over a distance of about 3.25 km from the forest soil surface to the open water bottom and vertical accretion net of subsidence increased by 0.8 +/− 0.2 mm yr^{-1} for every 1 cm decrease in elevation down the full length of the delta lobe. The elevation change rate for the forest habitat was 0.07 cm yr^{-1} compared to 3.8 cm yr^{-1} for the open water habitat. The variation in sediment deposition there was driven by variation in flooding, given that suspended sediment supply is not limiting at this site. The forest site was flooded only 6% of the time whereas the open water site was permanently flooded. Lastly, there was a gradient in belowground standing crop. Belowground biomass, discussed in section 7.2.6, increased significantly with site elevation along the crevasse splay; ranging from < 80 g m^{-2} in the low-elevation open water and preemergent habitats, to 600 to 1,600 g m^{-2} in the mid-elevation herbaceous marshes, and reached a peak of >3,000 g m^{-2} in the high-elevation forest habitat (Cahoon et al. 2011).

7.2.6 Limits to Vertical Accretion – Organic Matter

Vertical accretion is ultimately limited by the mass inputs of both mineral and refractory organic material and their bulk densities. From the self-packing density k_1 and informed assumptions about the production of refractory organic matter, we can compute the corresponding vertical accretion rates resulting from organic production (Fig. 7.8). The input of the refractory organic production should be nearly equal to the production of lignin (Hodson et al. 1984; Benner and Hodson, 1985; Benner et al. 1987; Reddy et al. 2006). Using the lignin concentration for *Spartina* of 10% (Hodson et al. 1984; Buth and Voesenek, 1987; Wilson et al. 1986), a root plus rhizome to stem ratio of 2, and turnover of belowground biomass of 0.5 yr^{-1}, we calculate vertical accretion rates of 0.23 cm yr^{-1} at the elevation of optimum biomass, tailing off to zero at the upper and lower limits of

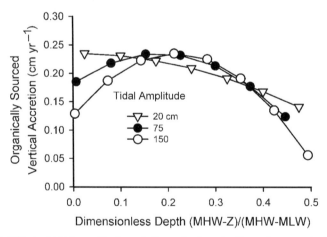

Figure 7.8 Effect of tidal amplitude and dimensionless depth on the organically sourced vertical accretion rate resulting solely from the production of refractory organic matter corresponding to the standing biomass in Fig. 7.3.

the vertical range (Fig. 7.8). This calculated rate may be conservative, as belowground turnover could be greater than 0.5 yr^{-1}, but bear in mind that the rhizome:root ratio is about 1:1 to 3:1 (Darby and Turner, 2008), and rhizomes are perennial organs with a turnover rate that is very likely much less than annual. On the other hand, this calculated rate is comparable to the 2.4 mm yr^{-1} accretion rate measured in a brackish microtidal marsh on the Pamlico River estuary, North Carolina, where soils are highly organic (70%–80%) (Craft et al. 1993).

From the inputs of mineral and organic matter as described above, the corresponding soil organic matter concentration can be readily computed (Fig 7.9A). These soil organic matter concentrations are those expected below the root zone after the inputs and decay of labile organic matter. We also observe high concentrations of organic matter and lower rates of accretion in real marshes that are equilibrated at elevations high in the tidal frame, where inundation time is greatly reduced (for example, Fig. 7.10). At lower relative elevations where inundation time is greater, organic matter concentrations are lower as the contribution of organic production declines and mineral inputs increase, and rates of vertical accretion are higher (Figs. 7.9 and 7.10). This is true irrespective of tidal amplitude. However, we also see that the sensitivity of minerally sourced accretion to inundation time or dimensionless depth increases with increasing tidal amplitude, due

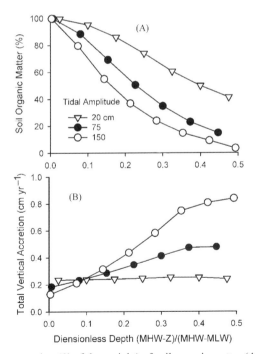

Figure 7.9 The concentration (% of dry weight) of soil organic matter (A) and total vertical accretion rate (B) from the computed inputs of mineral and refractory organic matter (see text) at different depths and tidal amplitudes when the suspended inorganic sediment concentration was 30 mg/L.

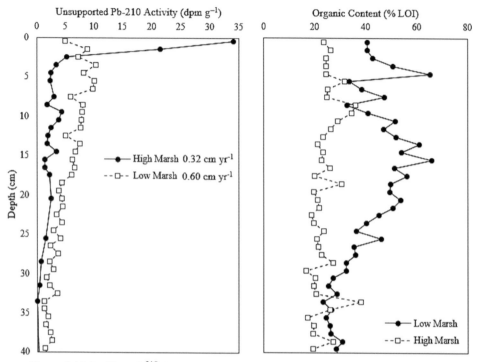

Figure 7.10 Profiles of ^{210}Pb activity (left) and soil organic matter (right) from cores collected from a high-elevation marsh site dominated by *Spartina patens* and a low-elevation marsh site dominated by *S. alterniflora* at the macrotidal Plum Island Sound, Massachusetts. The high marsh site is 32 cm higher in elevation than the low marsh.

largely to the effect of tidal amplitude on the suspended sediment inventory (Fig. 7.4). In contrast, biomass distribution (Fig. 7.3) and organic accretion are predicted to be less sensitive to amplitude (Fig. 7.8).

Total vertical accretion rate (Fig. 7.9B) is equal to the sum of the inorganic (Fig. 7.4B) and organic accretion rates (Fig. 7.8) by virtue of the ideal mixing rule. We show that at SLR less than 0.4 cm yr^{-1} and m = 30 mg L^{-1}, total accretion in mesotidal estuaries is dominated by organic matter inputs and at higher SLR by mineral inputs. At the simulated microtidal site, total accretion was dominated by organic matter at rates of SLR < 0.24 cm yr^{-1} (the upper range of simulations). Thus, highly organic soils develop at low SLR and mineral soils at high SLR. A SLR of 0.2 cm yr^{-1} is typical of the recent history of SLR (Thompson et al. 2016; Dangendorf et al. 2017) while the suspended sediment concentration used here is probably high for a majority of East and Gulf Coast estuaries today (Weston 2014).

These patterns of declining importance of mineral inputs and increasing importance of organic matter inputs are even more obvious when marshes out of equilibrium are considered, i.e., restoration sites (e.g., Callaway et al. 2013). In this case, restoration elevations typically start below the threshold at which plants can grow, and accumulation of sediment is driven entirely by mineral matter or allochthonous organic matter before plants become fully established. As the restored marsh builds elevation, mineral matter inputs would gradually decrease, and at the threshold for vegetation establishment, this would lead to the initiation of organic matter accumulation. At this point, the restored marsh would continue to build elevation with a mix of mineral and organic matter until it reaches the appropriate equilibrium elevation based on tidal range, suspended sediment, and other local factors. Bear in mind that the equation for sediment deposition used here, because of the conditions on capture efficiency, applies only to the range of possible elevations within a vegetated marsh landscape.

These predicted patterns for restored marsh development have been seen in San Francisco Bay at the South Bay Salt Pond Restoration Project, where initial rates of sedimentation in restored salt ponds are orders of magnitude greater (50–200 mm yr^{-1}; Callaway et al. 2013) than those found in natural marshes within San Francisco Bay (3–4 mm yr^{-1}; Callaway et al. 2012). These patterns also have been observed in constructed tidal marshes along the North Carolina coast where, using feldspar marker layers, the accretion rate in a 1 year-old marsh was two times greater (2.2 cm yr^{-1}) than in older constructed marshes (0.6 cm yr^{-1}) and nearby mature natural marshes (0.9–1.5 cm yr^{-1}) (C. B. Craft unpublished data). Likewise, an eight year-old natural marsh in Georgia, USA that developed from mudflat exhibited an accretion rate (3–4 cm yr^{-1}) that was an order of magnitude greater than those measured in nearby natural marshes (2–3 mm yr^{-1}), and well above the local rate of SLR (3 mm yr^{-1}) (He et al. 2016). As predicted, organic matter concentrations are also greater in natural marshes than in the restored marshes, and the rate of accretion declines as the marsh builds elevation (Craft et al. 2003). This leads to an asymptotic increase in elevation toward marsh equilibrium elevations that have been described previously, based on the Krone model (Williams and Orr, 2002).

7.3 Conclusions

Marsh equilibrium theory informs us about how marshes adjust their elevations and productivity in response to changes in rate of SLR and makes predictions about the effects of variables such as tidal amplitude and sediment supply – variables that differ among estuaries. The model predicts that mineral accretion will decline monotonically with increasing relative elevation while accretion of organic matter is distributed nonlinearly with respect to relative elevation – trends that are supported by empirical observations. Organic production spans the range of the tides above MSL to about the level of mean higher high water. For microtidal marshes, the biomass optimum elevation is probably closer to MHW. In mesotidal systems, minimal organic production occurs at the extremes of the distribution, with an optimum elevation near mid-range. The relative elevation of a marsh will be established within this range at a position that depends on the rate of SLR. As the SLR rate increases, equilibrium relative elevation will decrease. However, there are numerous examples of marshes that are not in equilibrium with SLR, and marshes in these cases can accrete vertically at rates that exceed the equilibrium rates. This would happen for a prograding marsh that is colonizing an emerging bar or a restoration site where the initial elevation is below the equilibrium elevation.

The total vertical accretion rate is equal to the sum of the inorganic and organic accretion rates by virtue of the ideal mixing rule (Morris et al. 2016). At relative elevations favorable to organic matter production and higher, soils will be dominated by organic matter and vertical accretion will be determined largely by organic matter production, because the volume occupied by organic matter is $23 \times$ greater than the volume of an equivalent dry weight of mineral sediment (Morris et al. 2016). Low SLR results in development of highly organic soils, while high SLR results in the formation of mineral soils. The equilibrium model highlights the declining importance of mineral matter accumulation and the corresponding increase in the importance of organic matter accumulation as elevation builds.

There is a rate of SLR that exceeds the capacity of a vegetated marsh to maintain its elevation within its vertical range of survival, and this depends on the tidal amplitude and mineral sediment supply. We predict that microtidal marshes cannot survive SLR greater than about 2.5 mm yr^{-1} without large inputs of mineral matter or substantial accretion of organic matter above what the model assumes. The low concentrations of suspended inorganic sediments that characterize most East Coast estuaries of the USA (Weston 2014) cannot alone indefinitely support microtidal marshes. For a fixed concentration of suspended sediment, macrotidal marshes have greater rates of sediment loading. The vegetation in a mesotidal marsh also has a larger vertical growth range, i.e., higher elevation capital. Thusly, marsh equilibrium theory predicts that microtidal marshes will drown before mesotidal marshes.

Acknowledgements

This work was supported by NSF grants 1457622, 1457435, 1654853, and 1457442. We thank Joel Carr and three anonymous reviewers for their constructive critiques of the manuscript. Use of trade, product, or firm names is for descriptive purposes only and does not imply endorsement by the US Government.

References

Alizad, K., Hagen, S. C., Morris, J. T., Medeiros, S. C., Bilskie, M. V., and Weishampel, J. F. 2016. Coastal wetland response to sea level rise in a fluvial estuarine system. *Earth's Future*, 4. doi:10.1002/2016EF000385

Allen, J. R. L. 1990. Saltmarsh growth and stratification: a numerical model with special reference to the Severn Estuary, southwest Britain. *Marine Geology*, 95: 7–96.

Barbier, E. B., Hacker, S. D. Kennedy, C., Koch, E. W., Stier, A. C., and Silliman, B. R. 2011. The value of estuarine and coastal ecosystem services. *Ecological Monographs*, 81: 169–193.

Baumfalk, Y. A. 1979. On the pumping activity of *Arenicola marina*. *Netherlands Journal of Sea Research*, 13: 422–427.

Benner, R., Fogel, M. L., Sprague, E. K., and Hodson, R. E. 1987. Depletion of ^{13}C in lignin and its implications for stable carbon isotope studies. *Nature*, 329: 708–710.

Benner, R., and Hodson, R. E. 1985. Microbial degradation of the leachable and ligno-cellulosic components of leaves and wood from *Rhizophora* mangle in a tropical mangrove swamp. *Marine Ecology Progress Series*, 23: 221–230.

Buth, G. J. C., and Voesenek, L. A. C. J. 1987. Decomposition of standing and fallen litter of halophytes in a Dutch salt marsh. In: A. H. L. Huiskes, C. W. P. M. Blom, and J. Rozema, eds., *Geobotany 11: Vegetation Between Land and Sea*. Dr. W. Junk Publishers, Dordrecht, pp. 146–165.

Cahoon, D., and Guntenspergen, G. 2010. Climate change, sea-level rise, and coastal wetlands. *National Wetlands Newsletter*, 32: 8–12.

Cahoon, D. R., White, D. A., and Lynch, J. C. 2011. Sediment infilling and wetland formation dynamics in an active crevasse splay of the Mississippi River delta. *Geomorphology*, 131: 57–68.

Callaway, J. C., Borgnis, E. L., Turner, R. E., and Milan, C. S. 2012. Carbon sequestration and sediment accretion in San Francisco Bay tidal wetlands. *Estuaries & Coasts*, 35: 1163–1181.

Callaway, J. C., Nyman, J. A., and DeLaune, R. D. 1996. Sediment accretion in coastal wetlands: a review and a simulation model of processes. *Current Topics in Wetland Biogeochemistry*, 2: 2–23.

Callaway, J. C., Schile, L. M., Borgnis, E. L., Busnardo, M., Archbald, G., and Duke, R. 2013. Sediment dynamics and vegetation recruitment in newly restored salt ponds – Final Report for Pond A6 Sediment Study: Submitted to Resources Legacy Fund and South Bay Salt Pond Restoration Project, 29 p., www.southbayrestoration .org/documents/technical/Pond%20A6%20FINAL%20report.COMBINED.08.21 .2013.pdf

Ciutat, A., Widdows, J., and Readman, J. W. 2006. Influence of cockle Cerastoderma edule bioturbation and tidal-current cycles on resuspension of sediment and polycyclic aromatic hydrocarbons. *Marine Ecology Progress Series*, 328: 51–64.

Costanza, R., Perez-Maqueo, O., Martinez, M. L., Sutton, P. Anderson, S., and Mulder, K. 2008. The value of coastal wetlands for hurricane protection. *Ambio*, 37: 241–248.

Craft, C., Clough, J. Ehman, J. Joye, S. Park, R., Pennings, S., Guo, H., and Machmuller, M. 2009. Forecasting the effects of accelerated sea-level rise on tidal marsh ecosystem services. *Frontiers in Ecology*, 7: 73–78.

Craft, C. B., Megonigal, J. P., Broome, S. W., Cornell, J., Freese, R., Stevenson, R. J., Zheng, L., and Sacco, J. 2003. The pace of ecosystem development of constructed *Spartina alterniflora* marshes. *Ecological Applications*, 13: 1417–1432.

Craft, C. B., Seneca, E. D., and Broome, S. W. 1993. Vertical accretion in regularly and irregularly flooded microtidal estuarine marshes. *Estuarine Coastal and Shelf Science*, 37: 371–386.

Dangendorf, S., Marcos, M., Wöppelmann, G., Conrad, C. P., Frederikse, T., and Riva, R. 2017. Reassessment of 20th century global mean sea level rise. *Proceedings of the National Academy of Sciences of the USA*, 114: 5946–5951.

Darby, F. A., and Turner, R. E. 2008. Below- and aboveground biomass of *Spartina alterniflora*: Response to nutrient addition in a Louisiana salt marsh. *Estuaries and Coasts*, 31: 326–334.

Davidson-Arnott, R. G. D., van Proosdij, D., Ollerhead, J., and Schostak, L. E. 2002. Hydrodynamics and sedimentation in salt marshes: examples from a macro-tidal marsh, Bay of Fundy. *Geomorphology*, 48: 209–231

Davis, J. L., Currin, C. A., O'Brien, C., Raffenburg, C., and Davis, A. 2015. Living shorelines: Coastal resilience with a blue carbon benefit. *PLOS ONE*, 10(11): e0142595. doi: 10.1371/journal.pone.0142595

de Deckere, E. M. G. T., Tolhurst, T. J., and de Brouwera, J. F. C. 2001. Destabilization of cohesive intertidal sediments by infauna. *Estuarine, Coastal and Shelf Science*, 53: 665-669.

Dijkema, K. S., Bossinade, J. H., Bouwsema, P., and de Glopper, R. J. 1990. Salt marshes in the Netherlands Wadden Sea: Rising high-tide levels and accretion enhancement. In: J. J. Beukema, W. J Wolff, and J. J. W. M. Brouns, eds., *Developments in Hydrobiology 57: Expected Effects of Climatic Change on Marine Coastal Ecosystems*. Springer, Dordrecht, pp. 173–188.

Drake, K., Halifax, H., Adamowicz, S. C., and Craft, C. 2015. Carbon sequestration in tidal salt marshes of the northeast United States. *Environmental Management*, 56: 998–1008.

Esselink, P., and Zwarts, L. 1989. Seasonal trend in burrow depth and tidal variation in feeding activity of Nereis diversicolor. *Marine Ecology Progress Series*, 56: 243–254.

Federer, C. A., Turcotte, D. E., and Smith, C. T. 1993. The organic fraction–bulk density relationship and the expression of nutrient content in forest soils. *Canadian Journal of Forest Research*, 23: 1026–1032.

French, J. R. 1993. Numerical simulation of vertical marsh growth and adjustment to accelerated sea-level rise, North Norfolk, UK. *Earth Surface Processes and Landforms*, 18: 63–81.

He, Y., Widney, S. E., Ruan, M., Herbert, E. R., Li, X., and Craft, C. 2016. Accumulation of soil carbon drives denitrification potential and laboratory-incubated greenhouse gas flux along a chronosequence of salt marsh development. *Estuarine, Coastal and Shelf Science*, 172: 72–80.

Hodson, R. E., Christian, R. R., and Maccubbin, A. E. 1984. Lignocellulose and lignin in the salt marsh grass *Spartina alterniflora*: initial concentrations and short-term, post-depositional changes in detrital matter. *Marine Biology*, 81: 1–7.

Kirwan, M. L., Guntenspergen, G. R., D'Alpaos, A., Morris, J. T., Mudd, S. M., and Temmerman, S. 2010. Limits on the adaptability of coastal marshes to rising sea level. *Geophysical Research Letters*, 37: L23401. doi: 10.1029/2010GL045489.

Kirwan, M. L., and Murray, A. B. 2007. A coupled geomorphic and ecological model of tidal marsh evolution. *Proceedings of the National Academy of Sciences of the USA*, 104: 6118–6122,

Krone, R. B. 1987. A method for simulating historic marsh elevations. In: N. C. Krause, ed., *Coastal Sediments '87*: American Society of Civil Engineers, New York, NY, pp. 316–323.

Krüger, F. 1964. Messungen der Pumptätigkeit von Arenicola marina L. im Watt. *Helgoländer wissenschaftliche Meeresuntersuchungen*, 11: 70–91.

Leonard, L. A. 1997. Controls on sediment transport and deposition in an incised mainland marsh basin, southeastern North Carolina. *Wetlands*, 17: 263–274.

Lovelock, C. E., Bennion, V., Grinham, A., and Cahoon, D. R. 2011. The role of surface and subsurface processes in keeping pace with sea level rise in intertidal wetlands of Moreton Bay, Queensland, Australia. *Ecosystems*, 14: 745–757.

Marani, M, Da Lio, C., and D'Alpaos, A. 2013. Vegetation engineers marsh morphology through multiple competing stable states. *Proceedings of the National Academy of Sciences of the USA*, 110: 3259–3263.

Marion, C., Anthony, E. J., and Trentesaux, A. 2009. Short-term (≤ 2 yrs) estuarine mudflat and saltmarsh sedimentation: High-resolution data from ultrasonic altimetry, rod surface-elevation table, and filter traps. *Estuarine, Coastal and Shelf Science*, 83: 475–484.

McKee, K. L., and Patrick, Jr, W. 1988. The relationship of smooth cordgrass (*Spartina alterniflora*) to tidal datums: a review. *Estuaries*, 11: 143–151.

Mendelssohn, I. A., and Morris, J. T. 2000. Ecophysiological controls on the growth of *Spartina alterniflora*. In: N. P. Weinstein, and D. A. Kreeger, eds., *Concepts and Controversies in Tidal Marsh Ecology*. Kluwer Academic Publishers, Dordrecht, pp. 59–80.

Morris, J. T. 2006. Competition among marsh macrophytes by means of geomorphological displacement in the intertidal zone. *Estuarine and Coastal Shelf Science*, 69: 395–402.

Morris, J. T. 2007. Ecological engineering in intertidal saltmarshes. *Hydrobiologia*, 577: 161–168.

Morris, J. T., Barber, D. C., Callaway, J.C., Chambers, R., Hagen, S. C., Hopkinson, C. S., Johnson, B. J. et al. 2016. Contributions of organic and inorganic matter to sediment volume and accretion in tidal wetlands at steady state, *Earth's Future*, 4. doi: 10.1002/2015EF000334.

Morris, J. T., and Bowden, W. B. 1986. A mechanistic, numerical model of sedimentation, mineralization, and decomposition for marsh sediments. *Soil Science Society of America Journal*, 50: 96–105.

Morris, J. T., Edwards,J., Crooks, S., and Reyes, E. 2012. Assessment of carbon sequestration potential in coastal wetlands. In R. Lal, K. Lorenz, R. Hüttl, B. U. Schneider, and J. von Braun, eds., *Recarbonization of the Biosphere: Ecosystem and Global Carbon Cycle*. Springer, Dordrecht, pp. 517–531.

Morris, J. T., Sundareshwar, P. V., Nietch, C. T., Kjerfve, B., and Cahoon, D. R. 2002. Responses of coastal wetlands to rising sea level. *Ecology*, 83: 2869–2877.

Morris, J. T., Sundberg, K., and Hopkinson, C. S. 2013. Salt marsh primary production and its responses to relative sea level and nutrients in estuaries at Plum Island, Massachusetts, and North Inlet, South Carolina, USA. *Oceanography*, 26: 78–84.

Mudd, S. M., D'Alpaos, A., and Morris, J. T. 2010. How does vegetation affect sedimentation on tidal marshes? Investigating particle capture and hydrodynamic controls on biologically mediated sedimentation. *Journal of Geophysical Research-Earth Surface*, 115, F03029, doi: 10.1029/2009JF001566.

Mudd, S. M., Fagherazzi, S., Morris, J. T., and Furbish, D. J. 2004. Flow, sedimentation, and biomass production on a vegetated salt marsh in South Carolina: toward a predictive model of marsh morphologic and ecologic evolution. In: S. Fagherazzi, A. Marani, and L. K. Blum, eds., *The Ecogeomorphology of Tidal Marshes*American Geophysical Union, Washington, DC, pp. 165–187.

Mudd, S. M., Howell, S., and Morris, J. T. 2009. Impact of the dynamic feedback between sedimentation, sea level rise, and biomass production on near surface marsh stratigraphy and carbon accumulation. *Estuarine, Coastal and Shelf Science*, 82: 377–389.

Narayan, S., Beck, M. W., Wilson, P., Thomas, C. J., Guerrero, A., Shepard, C. C., Reguero, B. G., Franco, G., Ingram, J. C., and Trespalacios, D. 2017. The value of coastal wetlands for flood damage reduction in the northeastern USA. *Scientific Reports*, 7. doi: 10.1038/s41598-017-09269-z.

Neubauer, S. C. 2008. Contributions of mineral and organic components to tidal freshwater marsh accretion. *Estuarine, Coastal and Shelf Science*, 78: 78–88.

Neubauer, S. C., Anderson, I. C., Constantine, J. A., and Kuehl, S. A. 2002. Sediment deposition and accretion in a mid-Atlantic (U.S.A.) tidal freshwater marsh. *Estuarine, Coastal and Shelf Science*, 54: 713–727.

Nyman, J. A., Delaune, R. D., Walters, R. J., and Patrick Jr, W. H. 1993. Relationship between vegetation and soil formation in a rapidly submerging coastal marsh. *Marine Ecology Progress Series*, 96: 269–279.

Oliver, F.W. 1925. *Spartina townsendii*: Its mode of establishment, economic uses and taxonomic status. *Journal of Ecology*, 13: 74–91

Pethick, J. 1981. Long-term accretion rates on tidal salt marshes. *Journal of Sedimentary Petrology*, 51: 571–577.

Randerson, P. F. 1979. A simulation model of salt-marsh development and plant ecology. In: B. Knights and A. J. Phillips, eds., *Estuarine and Coastal Reclamation and Water Storage*. Saxon House, Farnborough, pp. 48–67.

Ranwell, D. S. 1964. *Spartina* salt marshes in Southern England. II: Rate and seasonal pattern of sediment accretion. *Journal of Ecology*, 52: 79–94.

Reddy, K. R., Osborne, T. Z., Inglett, K. S., and Corstanje, R. 2006. Influence of water levels on subsidence of organic soils in the upper St. Johns River basin. Final Report SH45812, St. Johns Water Management District. Palatka, FL.

Richard, G. A. 1978. Seasonal and environmental variations in sediment accretion in a Long Island salt marsh. *Estuaries*, 1: 29–35.

Schile, L. M., Callaway, J. C., Morris, J. T., Stralberg, D., Parker, V. T., and Kell, M. 2014. Modeling tidal wetland distribution with sea-level rise: Evaluating the role of vegetation, sediment, and upland habitat in marsh resiliency. *PLOS ONE*, 9: e88760. doi: 10.1371/journal.pone.0088760

Shepard, C. C., Crain, C. M., and Beck, M. W. 2011. The protective role of coastal marshes: A systematic review and meta-analysis. *PLOS ONE*, 6: e27374.

Stapf, O. 1907. Mud binding grasses. *Bulletin of Miscellaneous Information (Royal Botanic Gardens, Kew)*, 1907: 190–197.

Stewart, V. I., Adams, W. A., and Abdulla, H. H. 1970. Quantitative pedological studies on soils derived from Silurian mudstones. *Journal of Soil Science*, 21: 248–255.

Temmerman, S., Govers, G., Wartel, S., and Meire, P. 2004. Modelling estuarine variations in tidal marsh sedimentation: response to changing sea level and suspended sediment concentrations. *Marine Geology*, 211: 1–19.

Thompson, P. R., Hamlington, B. D., Landerer, F. W., and Adhikari, S. 2016. Are long tide gauge records in the wrong place to measure global mean sea level rise? *Geophysics Research Letters*, 43: 403–410,411.

Torres, J. J., Gluck, D. L., and Childress, J. J. 1977. Activity and physiological significance of the pleopods in the respiration of *Callianassa californiensis* (Dana) (Crustacea: Thalassinidea). *Biological Bulletin*, 152: 134–146.

Turner R. E., Swenson, E. M., and Milan, C. S. 2000. Organic and inorganic contributions to vertical accretion in salt marsh sediments. In: M. P. Weinstein, and D. A. Kreeger, eds., *Concepts and Controversies in Tidal Marsh Ecology*, Kluwer Academic Publishers, Dordrecht, pp. 583–595.

Vader, W. J. M. 1964. A preliminary investigation into the reactions of the infauna of the tidal flats to tidal fluctuations in water level. *Netherlands Journal of Sea Research*, 2: 189–222.

van Proosdij, D., Ollerhead, J., and Davidson-Arnott, R. G. D. 2006. Seasonal and annual variations in the sediment mass balance of a macro-tidal salt marsh. *Marine Geology*, 225: 103–127.

Volkenborn, N., Polerecky, L., Wethey, D. S., and Woodin, S. A. 2010. Oscillatory porewater bioadvection in marine sediments induced by hydraulic activities of *Arenicola marina*. *Limnology and Oceanography*, 55: 1231–1247.

Webb, A. P., and Eyre, B. D. 2004. Effect of natural populations of burrowing thalassini-dean shrimp on sediment irrigation, benthic metabolism, nutrient fluxes and denitrifi-cation. *Marine Ecology Progress Series*, 268: 205–220.

Weston, N. B. 2014. Declining sediments and rising seas: an unfortunate convergence for tidal wetlands, *Estuaries and Coasts*, 37: 1–23,

Widdows, R. J., Blauw, A., Heip, C. H. R., Herman, P. M. J., Lucas, C. H., Middelburg, J. J., Schmidt, S., Brinsley, M. D., Twisk, F., and Verbeek, H. 2004. Role of physical and biological processes in sediment dynamics of a tidal flat in Westerschelde Estuary, SW the Netherlands. *Marine Ecology Progress Series*, 274: 41–56.

Williams, P. B., and Orr, M. K. 2002. Physical evolution of restored breached levee salt marshes in the San Francisco Bay estuary. *Restoration Ecology*, 10: 527–542

Wilson, J. O., Buchsbaum, R., Valiela, I., and Swain, T. 1986. Decomposition in salt marsh ecosystems: phenolic dynamics during decay of litter of *Spartina alterniflora*. *Marine Ecology Progress Series*, 29: 177–187.

8

Salt Marsh Ecogeomorphic Processes and Dynamics

CAROL A. WILSON, GERARDO M. E. PERILLO, AND ZOE J. HUGHES

8.1 Introduction

Salt marshes are considered some of the most biologically diverse and ecologically important regions on Earth, containing thousands of species of robust salt-tolerant plants, crabs, fish, mollusks, zooplankton, algae, and bacteria. Isolated between topographic headlands, laterally continuous behind protective barriers, or associated with extensive delta landscapes, salt marshes are regulated by a variety of physical forces such as waves, tides, rivers, and storm surges, but they are also impacted by climatic variations in temperature and precipitation, riverine flooding, local tectonics, and subsidence (i.e., a deltaic process that describes the lowering of the land surface). Biological forces also play important roles in controlling salt marsh landscapes as many species shape geomorphic development. As these landscapes form and evolve, there exist significant interactions between biology, hydrology, and geology; thus it is impossible to consider salt marsh geomorphology – i.e., how the landscape changes over time – without taking into account these principal interactions.

Ecogeomorphology is the study of the complex interactions between biological, hydrological, and geological forcings, and how they impact a particular landscape. Salt marshes are low-energy coastal landscapes comprised of halophytic vegetation that colonizes relatively flat areas subject to intermittent saline water inundation. These unique, hydrodynamically regulated areas are dissected with extensive creeks and contain unvegetated mudflats within and along the periphery of the marsh. All of these areas are teeming with life. For important reviews on the geological formation, hydrodynamic processes, and ecology of salt marshes, readers are referred to Chapters 2–6 of this book. This chapter will focus on the salt marsh landscape (the continuum from tidal channels and creeks to the vegetated marsh platform), discuss relationships and feedbacks that exist between the ecological engineers that live upon and within the salt marsh, and examine the controls that ultimately shape geomorphic development and evolution. This review is not intended to be fully exhaustive, but provides a comprehensive summary of some prominent ecogeomorphic processes at play within salt marshes around the world. For example, once an intertidal mudflat aggrades to the elevation most favorable to support vegetation, salt marsh macrophytes will colonize it and, once formed, marsh platforms are further maintained by the deposition and vertical accretion of sediment, as long as enough sediment is available. In this process, very fine-grained inorganic particles suspended within tidal waters are trapped by salt marsh vegetation that slow current velocities through the canopy, leading to enhanced particle settling. In addition, suspended sediment can be adhered to plant stems or the salt marsh substrate by sticky exopolymeric substances excreted by benthic algae and microbial mats colonizing these surfaces. Further, root production below the salt marsh surface similarly aids in the vertical maintenance of the salt marsh platform, particularly as it is regulated relative to local sea-level rise.

The aforementioned examples illustrate how biology aids in the creation and stability of salt marshes. However, biology can also impart destabilizing forces. A prime example is crab bioturbation that reworks the fine-grained sediment in marshes, redistributing it from depth to the surface, and repackaging it into more easily erodible pellets (Botto & Iribarne, 2000; Escapa et al. 2007; Wilson et al. 2012; Farron, 2018). Bioturbation can also disrupt the proper functioning of plant roots, stressing the plants, which can lead to their migration, modification of their reproduction system, and/or mortality (Escapa et al. 2015). Herbivory is another example of a destabilizing biological force within salt marshes. For instance, marshes in the Mississippi Delta have had negative impacts from grazing by rodent-like mammals called nutria that were introduced during the past century, and in northern subarctic salt marshes geese have been implicated for extensive salt marsh deterioration (Dionne, 1985). Over-consumption by grazers can result in the deflation – or lowering – of the salt marsh surface and, in cases with low or null sediment input, the system is transformed into a mudflat (Lynch et al. 1947; Dionne, 1985; Silliman et al. 2005). Finally, disruption to the underlying root structure can result in zones of weakness at depth, which dictate how erosion might take place along salt marsh fringes, especially during high energy events such as storms. From these brief examples, it is evident a combination of biological, geological, and hydrodynamic interactions create the underlying architecture of salt marshes, regulate deposition and erosion processes, and impact marsh sustainability in the long-term.

A list of the physical and biological processes that have been found to impact the geomorphology of salt marshes is featured in Table 8.1. The timescales over which these processes occur can vary considerably, ranging from seconds to millennia. While some of these processes may be short-lived, it is important to note that the geomorphic changes they impart can be much longer lasting, depending upon total impact to the salt marsh infra-structure and biological functioning, as will be discussed throughout this chapter. In addition, ecogeomorphic interactions will vary spatially within a particular salt marsh, and across global variations in climate, geology, hydrology, and biology. Thus, processes that may be prevalent in one particular salt marsh may not be universally present in salt marshes worldwide. Therefore, site-specific conditions in climate change, sea level fluctu-ation, regional tectonic movement, tidal dynamics, wave climate, and biozonation are all factors that should be considered when examining ecogeomorphology in salt marshes.

8.2 The Salt Marsh Landscape

The 3D surface structure of salt marshes is characterized by the presence of subtidal and intertidal courses – which include rills, grooves, gullies, creeks, and channels (see Perillo 2019 for definitions) – intertidal unvegetated mudflats, and intertidal vegetated marsh

Table 8.1 *Physical and biological processes common in salt marshes, and associated timescales*

	Duration								
Seconds	10	100	1000	10,000					
Days			0.01	0.12	1.16	12	120	1200	
Years						0.03	0.33	3.30	33	33,000

PHYSICAL
Waves
Precipitation events
Tides
Temperature variation
Storm
River flood
Drought
Subsidence
Tectonic activity

BIOLOGICAL
Biofilm life cycle
Crab burrow excavation
Seasonal plant productivity
Fish life cycle
Halophytic vegetation life cycle
Belowground biomass preservation

platforms that may contain depressions, hummocks, cliffs, etc. (see Fig. 8.1 for a generalized profile). Tidal courses are elongated incisions that are inundated permanently to semipermanently from tidal fluctuation. These courses are the conduits of water, sediment, nutrients, and energy between salt marsh platforms and adjacent water bodies (i.e., bay, estuary, ocean). Intertidal areas, in contrast, are inundated during isolated portions of the tide depending on local surface elevation relative to water elevation. Water levels fluctuate based on tidal and wave conditions, but are often also impacted by local meteorological and fluvial conditions, as applicable (e.g., storm surge, coastal setup, riverine flooding).

Localized elevation and tidal range define the frequency and duration of flooding, which dictates sediment transport patterns and ecotones, based on the preferences, competition, and adaptations of flora and fauna that colonize the salt marsh. Salt marshes are characterized as either ramp or platform type (see [FitzGerald et al. 2008], and references therein). In general, both types have very gentle seaward-dipping gradients, approaching 10^{-3}, though there can be microtopography across the landscape including slight hummocks, natural levees adjacent to creekbanks, and interior depressions that may be unvegetated (defined as pannes, pans, or pools; see Fig. 8.1, [Perillo, 2019]). Small changes in elevation and surface slope can have profound effects on the prevailing physical and chemical conditions, with implications for the physiological tolerances of intertidal species and, as a result, their distribution within the salt marsh (Niering & Warren, 1980; Bertness et al. 1992; Pennings & Callaway, 1992; Wilson et al. 2014; Julien, 2018). North American salt marshes are dominated by *Spartina alterniflora*, *S. patens*, *Distichlis spicata*, and *Juncus roemerianus* vegetation (Odum, 1988). At lower tidal heights and along creekbanks, tall-form *S. alterniflora* (1–2 m high) colonizes, giving way to short-form *S. alterniflora* (<0.5 m tall), *S. patens*, *Distichlis*, and *Juncus* at higher tidal heights. In New England salt marshes, *S. patens* appears to be prevalent in areas of the high salt marsh platform that are better drained, and short-form *S. alterniflora* dominates where elevations are high but drainage is poor

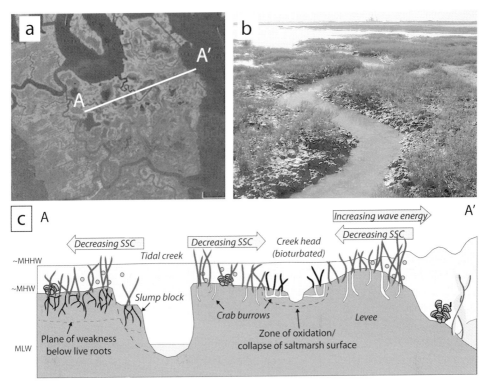

Figure 8.1 Salt marshes are comprised of vegetated marsh platforms dissected with tidal channels and smaller creeks, adjacent to extensive mudflats, shown here in aerial view for Great Bay, NJ (a) and on the ground in Bahia Blanca, Argentina (b). A few ecogeomorphic processes common in salt marshes are shown in the cross section of transect A–A′, idealized in (c): natural levees form adjacent to larger bodies of water (channel or bay) where wave and tidal current energies are attenuated by vegetation and/or offshore oyster reefs, favoring sediment deposition; crab burrowing is prevalent across the platforms, but can be concentrated by creekbanks or interior creek heads; interior lowlands, defined as pans or ponds (see Perillo, 2019), are present where waterlogging is prevalent and vegetation is stunted or absent; live macrophyte rooting can extend up to 1 m depth, below which there can be a zone of weakness that can act as a failure plane along creekbanks and channel edges. SSC: suspended sediment concentration. (A black and white version of this figure will appear in some formats. For the color version, please refer to the plate section.)

(Wilson et al. 2014). Short-form *S. alterniflora* in these marshes appears to represent physiologically stunted plants rather than a genetically different form, as they are characterized by reduced soil drainage and oxygenation (Mendelssohn & Seneca, 1980; Howes et al. 1981; Mendelssohn et al. 1981), increased soil sulfide levels (King et al. 1982), and increased root mat density and substrate hardness (Ringold, 1979; Bertness, 1985; Ellison et al. 1986). These geochemical conditions control what infauna colonize different reaches of the salt marsh, as well (Julien, 2018).

The present day geomorphology of a salt marsh ecosystem depends on emplacement location, the initial conditions of formation, and the net balance of erosion and accumulation of sediment that has subsequently occurred. Inorganic sediment sourced from rivers or the nearby coastal environment is important for salt marsh sustainability as it provides essential nutrients for growth and a stable substrate for the colonization and maintenance of vegetation and benthic communities (Reed, 1988; Pethick, 1992; French and Spencer, 1993; Nyman et al. 1995a; Reddy and Delaune, 2008; Bartholdy, 2012, and references therein). Sediment can also be organic in origin, comprised of the particles of organisms, by far the most common being the extensive roots of salt marsh macrophytes that accumulate over time (Redfield and Rubin, 1962; Ellison et al. 1986; Delaune et al. 1994; Nyman et al. 1995b). This amalgamation of organic and inorganic sediment, typically very fine-grained, is subject to compaction and decomposition over time (Kaye and Barghoorn, 1964; van Asselen et al. 2009, and references therein). It is well recognized that both organic and inorganic sediment must provide a continuous input to build the wetland both vertically and horizontally in order to keep it in equilibrium with water level fluctuation due to (1) tidal and wave energy (Schwimmer, 2001; Turner et al. 2002; Temmerman et al. 2004) and (2) sea-level variations (Redfield and Rubin, 1962; Redfield, 1972; Morris et al. 2002; Nyman et al. 2006; Kirwan and Guntenspergen, 2010). As both inorganic and organic sediment contributions are important, physical processes that control sediment delivery, as well as flora and fauna and their relative life cycles play significant roles in building salt marsh surfaces and their evolution (Table 8.1).

As Table 8.1 shows, processes impacting geomorphic processes – particularly sediment accumulation and erosion – can occur on timescales varying from seconds (waves), hours (tides), and days to years (river flooding, bioturbation, decomposition), or decades to millennia (compaction, tectonic subsidence). An important thing to note is most processes in salt marshes are cyclical in nature, displaying high spatial and temporal variability. The following sections will address how biology, or ecological engineers, impact different zones of the marsh. These engineers can impart both stabilizing and destabilizing forces on the salt marsh, and the landscape response can be quite varied.

8.3 Ecological Engineers in Marshes

Ecological engineers are organisms that alter resources such as space, substrate, habitat, nutrients, and/or food sources (Jones et al. 1994). These natural engineers impact the design, construction, operation, and maintenance of structures within the salt marsh, which have an important effect to the overall functioning of salt marsh systems. All filter-feeding, burrowing, and deposit-feeding animals of salt marshes potentially have important habitat-modifying effects that may influence marsh grass productivity and, ultimately, marsh growth (Bertness, 1984a, 1984b). Foundational studies started with single macrofaunal and -floral interactions (e.g., Bertness, 1984a, 1984b; da Cunha Lana and Guiss, 1992), while subsequent work has diversified to include how biology interacts with the salt marsh substrate and how complex interactions shape salt marsh planform evolution (e.g., sediment deposition or erosion, geo-technical properties, platform or tidal course aggradation or erosion). It is well recognized in recent studies that a multidisciplinary approach is often needed to fully understand and appreciate the dynamic changes biology imparts on the morphology of salt marshes.

The term "living shorelines" has been recently adopted to specify naturally occurring shoreline stabilization, including salt marsh vegetation or oyster reefs for wave damping (O'Donnell, 2017; Davenport et al. 2018). These nature-based methods of shoreline erosion control are implemented due to the stabilizing impact that vegetation and *in-situ* benthic organisms have on salt marsh substrate. Successful living shorelines not only provide erosion control, but also maintain ecosystem services (critical habitat for fish, shellfish, marine plants), improve water quality through groundwater infiltration, reduce surface water runoff, and decrease sediment transport (e.g., Currin et al. 2010; Onorevole et al. 2018, to name a few). In this light, living shorelines have become desirable alternatives for coastal engineering practices. The US National Oceanographic and Atmospheric Administration (NOAA) has an online "Green Infrastructure Effectiveness" database as a reference for coastal managers (NOAA, 2018). However improved understanding of the interdependency of salt marsh geomorphology, ecology, and hydrodynamics is needed, especially for coastal managers and engineers to justify reduction of traditional "hard" erosion-control structures and encourage more ecosystem-based approaches (Spalding et al. 2014). Here we will first discuss the role of various ecological stabilizers, and then examine the converse role other organisms play in destabilizing salt marsh environments.

8.3.1 Ecological Stabilizers

Ecological stabilizers predominantly include certain macroflora and -fauna species, but even microflora and -fauna can play an important role in stabilizing salt marsh environments. Stability is achieved if the *in-situ* habitat or behavior of organisms changes the ambient soil or water properties such that salt marsh surfaces (vegetated and bare) are maintained or preserved. Ecological engineers may create an environment whereby energies are reduced, sediment deposition is favored, and/or favorable habitat for the colonization of flora or fauna is produced. In this section, we focus our discussion on vegetation, macrophytobenthos, and bivalves, though other stabilizing organisms may exist in various salt marsh settings.

8.3.1.1 Vegetation

Sediment trapping – Salt marshes are historically considered low energy environments where inorganic sediment is deposited from river, tidal, and storm inundation. As mentioned previously, sediment is important for salt marsh sustainability as it provides essential nutrients for growth, a stable substrate for the colonization and maintenance of vegetation and benthic communities, and the ability of the salt marsh to vertically accrete and keep pace with fluctuations in water level from local sea-level rise. Sediment is imported to salt marshes from external sources (e.g., fluvial, marine environments), transported with varying currents through tidal courses, and deposited over mudflats and vegetated surfaces where current velocity and wave activity diminish. Vegetation plays a key role in trapping sediment as aboveground structures attenuate (1) current velocity, (2) wave action, and (3) flow turbulence when salt marsh platforms are inundated, as detailed below.

Over regular tidal cycles, most salt marshes exhibit shallow water distortion, whereby the duration of the flood tide is much shorter than the ebb (Pethick, 1980; French and Stoddart, 1992; Friedrichs and Perry, 2001, and references therein). Mean current velocity is controlled by water surface slope, which in turn depends on the spatial gradient of the tidal wave, surface roughness, and the landscape topography. Current velocities in excess of 1 m s^{-1} have been measured within tidal channels and over mudflats (e.g., Perillo et al. 1993; Wang et al. 1999, to name a few) while flow across salt marsh platforms is diminished to <1–20 cm s^{-1} (e.g., French and Stoddart, 1992; Christiansen et al. 2000, to name a few). Over unvegetated mudflat surfaces, phase asymmetry combined with friction, settling- and scour-lag provides an important mechanism for the landward transport of fine-grained sediment (Postma, 1961; French and Stoddart, 1992, and references therein). Over vegetated intertidal surfaces, however, there exists a vegetation-induced hydraulic roughness that further attenuates the flow above bankfull stage on flood tides (see Fig. 8.2; French and Stoddart, 1992; Christiansen et al. 2000; Madsen et al. 2001; Neumeier and Ciavola, 2004; Lightbody and Nepf, 2006). This has been supported by many authors, who argue that vegetation additionally decreases turbulence that keeps sediment in suspension (Lightbody and Nepf, 2006; Neumeier and Amos, 2006). Reduction in current velocity and turbulence result in net import of sediment to the salt marsh surface, revealed by tide averaged suspended sediment concentrations that are typically greater on the flood tide than ebb tide (Fig. 8.2; Reed et al. 1999; Christiansen et al. 2000; Li and Yang, 2009). This sediment is either: (1) deposited directly to the salt marsh surface (Leonard and Luther, 1995; Christiansen et al. 2000; Van Proosdij et al.

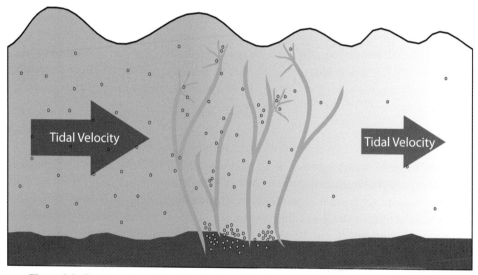

Figure 8.2 Conceptual model of the direct and indirect effects of macrophytes (halophytic vegetation, submerged aquatic vegetation common along marsh shorelines) on physical environmental parameters such as current velocity and wave energy, impacting sediment erosion and deposition in salt marshes. Modified from Madsen et al. (2001). Figure credit B. Gregory

2000) or (2) trapped on aboveground vegetation, and subsequently deposited to the surface once the plant senesces and falls (see Fig. 8.2; Li and Yang, 2009).

The capability of marsh vegetation to trap sediment is dependent on a number of factors: sediment supply, tidal range, and marsh elevation (which combined dictate inundation frequency and duration), and morphological properties of vegetation such as aboveground biomass density, height, thickness, degree of branching, and strength (Lightbody and Nepf, 2006; Li and Yang, 2009; O'Donnell, 2017; Shepard et al. 2011). In salt marshes of the Yangtze Delta, the amount of sediment adhering to plants was significant and positively correlated with plant biomass within species (including branching and flowering structures), but varied greatly between species: introduced *S. alterniflora* is significantly more efficient in trapping suspended sediment than the native *Scirpus* and *Phragmites* plants (Li and Yang, 2009). In addition, that study found more sediment was adhered to the base of plant stems as opposed to higher on the stem, supporting that the depth, inundation duration, and suspended sediment concentration are important factors (Li and Yang, 2009), as greater suspended sediment concentrations are typically found near the bed than with increasing height above the bed (Dyer, 1988; Li and Yang, 2009). Stem density also plays an important role: as can be expected, greater stem density and stem thickness are positively correlated with current attenuation (Lightbody and Nepf, 2006). Experimental studies show even with current velocities sufficient to erode non-cohesive sediment (>20 cm/s), *S. anglica* densities greater than 1,500 stems m^{-2} could generate sedimentation (Bouma et al. 2009; Addino et al. 2015). Observed stem densities of common marsh macrophytes include 50–300 m^{-2} for *S. alterniflora*, 200–3,500 m^{-2} for *S. patens*, 500–2,000 m^{-2} for *Distichlis*, 50–1,800 m^{-2} for *Juncus*, and 250 m^{-2} for *Phragmites* (see Gleason et al. 1979; Hopkinson et al. 1980; Bertness, 1984b; Robertson & Weis, 2005).

The attenuation of current velocity and enhancement of sediment deposition from vegetation impacts the geomorphic evolution of salt marshes. The deposition of sediment on the salt marsh bed due to macrophytes contributes to the vertical accretion of the marsh (Delaune et al. 1983; Stevenson et al. 1986; Reed, 1988; Day et al. 2011; Shi et al. 2012). In North American salt marshes, vertical accretion rates range from 6 to >10 mm yr^{-1} in low ramp marshes to 1–4 mm yr^{-1} on high salt marsh platforms (FitzGerald et al. 2008; and references therein; see also Kirwan et al. 2016). Rates on the order of 0.5–5 mm yr^{-1} have been recorded in UK, Australia, and South Korea salt marshes (see Reed, 1988; Howe et al. 2009; to name a few). In North Norfolk, UK, 8.5% of salt marsh accretion was attributed to sediment adhering to vegetation (French and Spencer, 1993), and in salt marshes of Delaware basin, USA, 50% of the salt marsh accretion was contributed by this mechanism (Stumpf, 1983; Li and Yang, 2009).

Sediment accumulation is typically strongly correlated to distance from tidal courses and shoreline edges, i.e., positions of the marsh most proximal to sediment source exhibit the greatest sedimentation (see Delaune et al. 1978; Letzsch and Frey, 1980; Hatton et al. 1983; Reed, 1988; French and Spencer, 1993; Chmura and Kosters, 1994; Reed et al. 1999; Christiansen et al. 2000; Temmerman et al. 2005; van Proosdij et al. 2006; Perillo, 2019). Fractionation in grain size also occurs with distance from tidal courses, as coarser grains are deposited with proximity to channels (Christiansen et al. 2000; Yang et al. 2008).

Further, even with finest grain sizes (i.e., mud), flocculation can cause a fractionation in grain size moving to the interior of marshes (French and Stoddart, 1992; Christiansen et al. 2000). These studies found reduced flow velocities associated with marsh canopies resulted in flocculation of fine grained mud particles, such that 70–80 % of the sediment deposited within 8 m of the tidal creek was in a flocculated form, while in the marsh interior some 25 m from the tidal creek only small individual grains and small flocs were deposited.

This enhanced sediment deposition from vegetation along the edges of tidal courses and shorelines results in the formation of "natural levees," or slightly higher elevation relative to interior marsh, as shown in Fig. 8.1 (Letzsch and Frey, 1980; Hatton et al. 1983; Reed, 1988). As elevation controls inundation frequency, natural levees are inundated over a smaller proportion of the tidal cycle than interior marshes (Reed et al. 1999; Christiansen et al. 2000). In addition, coarser grains impart better soil drainage, and natural levees tend to exhibit vegetation with greater aboveground biomass due to enhanced soil oxidation (Howes et al. 1986; Bertness et al. 1992; see Fig. 8.1). This enhanced aboveground biomass production provides a natural feedback for greater sediment trapping (Fig. 8.2; Reed et al. 1999; Christiansen et al. 2000; Perillo, 2019). Further, levee formation stabilizes tidal courses such as tidal creeks and channels, allowing for vertical growth (Perillo, 2019, and references therein). Of course, levees cannot grow indefinitely but must reach a kind of dynamic equilibrium based on tidal range, overflow velocities, suspended sediment concentration, and plant species dominance. Over time, natural levees tend to maintain higher elevation as the coarser grain sizes contained within are less compactable than finer grain sizes deposited in the interior of salt marshes (Kennish, 2001, and references therein).

On the other hand, there are examples where levees are not present. They seldom occur along tidal courses in erosive environments or when the concentration of suspended sediments is very low (<100 mg L^{-1}; Perillo, 2019). In some cases, it has been observed that the marsh to mudflat transition has a characteristic depression and shoulder (i.e., Tavy and Bahía Blanca estuaries, shown in Fig. 8.3). Studies by Widdows et al. (2008) and Pratolongo et al. (2010) show this morphology is a result of higher sedimentation on the mudflat. In these cases, wave activity on the mudflat induces bed liquefaction (Mehta, 1996), and the resulting fluid mud is transported by tidal currents landward. The edge of the marsh then acts as a dam for the sediment moving into the marsh, inducing a higher level on the mudflat than in the adjacent marsh (Fig. 8.3).

French and Spencer (1993) remark that sedimentation over single tidal cycles is much less than what is preserved over annual time scales, thus sedimentation from fluvial and storm events in addition to tides are contributing factors to many salt marshes worldwide (Turner, 2010; Smith et al. 2015). This is discussed further in the next section.

Decreases in wave energy – Salt marshes are living shorelines, and serve as "horizontal levees" for upland areas (Costanza et al. 2008; O'Donnell, 2017), particularly due to the ability of vegetation to decrease wave energy. The ability of marsh vegetation to attenuate small and medium wave heights is well documented in field and lab-based studies using real and artificial vegetation (Knutson et al. 1982; Kobayashi et al. 1993; Nepf, 1999; Augustin et al. 2009). Most wave attenuation occurs within the first few

Figure 8.3 Example of mud accumulation at the edge of a Spartina marsh in Bahía Blanca Estuary.

meters of the seaward edge of marshes due to friction (Knutson et al. 1982; Moller and Spencer, 2002) or the presence of some obstruction like a levee (Koch et al. 2009), and this decrease in wave energy provides a secondary control where sediment is deposited rather than remains in suspension (Fig. 8.2). Early studies by Knutson et al. (1982) revealed that wave energy decreased greater than 50% within the first 5 m of salt marsh vegetation (see Fig. 8.4a). But the variety of field conditions such as wave climate, inundation depth, and type and morphology of marsh stems and leaves (all of which vary temporally and spatially) makes it difficult to make exact determinations of the effect of marsh vegetation on wave attenuation (see O'Donnell, 2017, and references therein). It is generally accepted that wave attenuation increases with stem thickness and density and the overall width of the vegetated marsh platform (Anderson et al. 2011); however, as water depth increases the attenuation diminishes (Boorman et al. 1998; Koch et al. 2009; Gedan et al. 2011). Recent investigations have looked at the impact of vegetation on wave energy during storms (e.g., Shepard et al. 2011; Moller et al. 2014; Hu et al. 2015) and these studies maintain salt marshes provide important shoreline protection during high-energy storm events. In this manner, salt marshes serve as protective "horizontal levees" for upland areas (Costanza et al. 2008). Under hurricane conditions, the relative stem height, not density, was a bigger factor for attenuating storm surge (Hu et al. 2015). However the protective services provided by salt marsh vegetation has

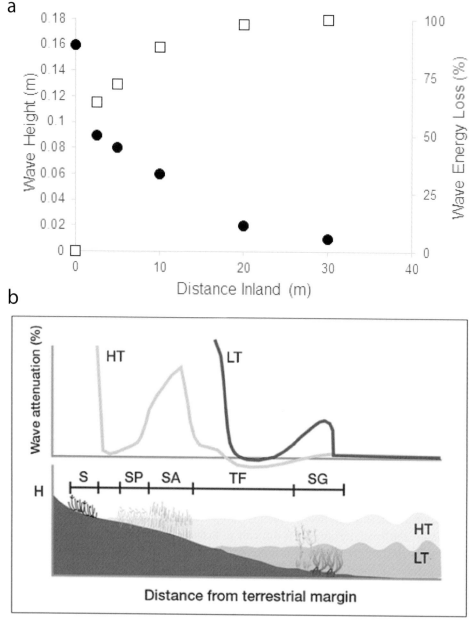

Figure 8.4 (a) Decreases in wave height (circles) and wave energy (squares) associated with friction along vegetated salt marsh platforms in Chesapeake Bay, Virginia, USA (from Knutson et al. 1982). (b) Idealized scheme of wave attenuation across the profile of a marsh ecosystem at high (HT) and low tide (LT). Waves attenuation change due to the kind of plants or bare sediments (TF). *SG = seagrass, SA = Spartina alterniflora, SP = Spartina patens, S = Sarcocornia.* (modified from Koch et al. 2009). (A black and white version of this figure will appear in some formats. For the color version, please refer to the plate section.)

limitations: with increasing wind speed and storm duration, storm surge attenuation diminishes (Hu et al. 2015).

Koch et al. (2009) provide a schematic representation of the percentage of wave attenuation along the cross-shore profile of an ideal temperate marsh ecosystem at high and low tide conditions. As shown in Fig. 8.4b, the degree of attenuation is obviously different at both stages as well as the distance inland that the waves can reach. Also the degree of attenuation is zonation-dependent due to the available water depth (further inland, smaller inundation depth) but also due to the type of plants and morphology of the marsh.

Belowground biomass production: aiding vertical accretion and stabilizing salt marsh substrates – Not only do macroflora assist in the stability of salt marshes with their aboveground biomass, belowground root and rhizome production can also impart stabilizing forces. As previously mentioned, salt marsh sediments are comprised of live and dead root and rhizome structures (collectively termed "belowground biomass") within an inorganic sediment matrix (see Fig. 8.5a). The relative fraction of each varies seasonally, spatially, and with depth (see Fig. 8.5; Ellison et al. 1986; Gross et al. 1991; Deegan et al. 2012; S. Warren, per com.). Salt marshes with predominantly inorganic sediment matrix are termed "mineralogenic," while those that have extensive belowground biomass are termed "organogenic." Early studies show that belowground biomass production can range from 300 to 1,200 g m^{-2} throughout the year (*S. alterniflora* and *S. patens* rhizome production averages 600–800 g m^{-2}, Valiela et al. 1976) and more recent use of computed tomography allows 3D visualization and quantification of roots, rhizomes, and peat (e.g., Davey et al. 2011; Blum and Davey, 2013). Studies show the accumulation of organic material helps maintain surface elevations in salt marshes (e.g., Nyman et al. 1990; Turner et al. 2002; Day et al. 2011; Wilson et al. 2014). It is also generally accepted that roots and rhizomes resist salt marsh erosion as they physically bind the underlying architecture and help stabilize it (Howes et al. 2010; Turner, 2011). Two common methods employed in determining the erosion resistance of salt marsh soils are measuring shear strength, discussed here, or the critical shear velocity, discussed in Section 8.3.1.2.

Shear strength is typically measured with a Torvane instrument, a hand-held paddle that is inserted into the soil and rotated until the substrate breaks cohesion (see Fig. 8.5b-c). This method is employed to measure the relative cohesion of the soil, which alludes to the erodibility of the substrate to ~1 m depth (for methods, see Wilson and Allison, 2008; Howes et al. 2010; Turner, 2011). Salt marsh belowground biomass production (i.e., root and rhizome density), water content, grain size, and autocompaction all affect the shear strength of the soil, and these parameters vary with depth. Root density and structure (fibrous, branching, pennate, or taproot) varies within and between species. For example, live roots of tall-form *S. alterniflora* (smooth cordgrass) can extend down to >35 cm depth, while those of short-form *S. alterniflora* typically only extend ~15–20 cm depth (Bertness, 1985; Gross et al. 1991; Wilson et al. 2014). *S. patens* roots are more fibrous in comparison and extend a maximum of 20 cm deep (Windham, 2001; Deegan et al. 2012; Wilson et al. 2014). When plants die, their roots lose turgor pressure and dewater, however autocompaction and decreased decomposition under loading and reducing conditions, respectively, forms peaty

Figure 8.5 (a) An example of the extensive root network of belowground biomass of *Spartina alterniflora* (photo credit Amanda Davis). (b) The strength of the root network is measured using a hand-held shear vane, shown here in a *S. alterniflora* marsh in South Carolina. (c) Data from *S. alterniflora* marshes of Louisiana show a decrease in strength with depth, indicative of the limit of live rooting (Howes et al. 2010; Turner, 2011; Valentine and Mariotti, 2019). An increase in shear strength at deeper depths is from compaction and consolidation of sediment. (d) Zones of weakness at depth can be responsible for creekbank failures, shown here from a nutrient-enriched *S. patens* marsh in Plum Island (Deegan et al. 2012). (A black and white version of this figure will appear in some formats. For the color version, please refer to the plate section.)

substrates composed predominantly of plant lignin (Kaye and Barghoorn, 1964; Niering and Warren, 1980; Benner et al. 1991; Mitsch and Gosselink, 2003).

Marsh edges versus interior high platforms have different vegetation composition, age, and belowground redox potential, which effects the total belowground biomass accumulation (Bertness, 1985) and thus shear strength of the soil. In general, the more live belowground biomass accumulation, the more robust the salt marsh substrate and greater shear strength (see Fig. 8.5; Bertness, 1985; Howes et al. 2010; Turner et al. 2011; Hughes and Wilson, unpub data). Additionally, as salt marsh substrate dewaters with burial and consolidation, bulk density and strength increases while porosity decreases (Hazelden and Boorman, 2001). Several studies in salt marshes of Louisiana found the greatest scatter in shear strength is observed near the surface, which is attributed to variations in both rooting density and the strength of individual roots (see Fig. 8.5; Wilson and Allison, 2008; Howes et al. 2010; Turner, 2011, Valentine and Mariotti, 2019). In general, shear strength decreases with depth as live root density decreases, then increases due to soil consolidation and compaction (Fig. 8.5; van Eerdt, 1985; Howes et al. 2010; Turner, 2011; Valentine and Mariotti, 2019). In a study of a restored marsh in UK, Watts et al. (2003) found marshes situated higher than mean high water (inundated <15 % of time) had less water content, greater bulk density (thus greater consolidation), and greater shear strength (up to 35× more) compared to sites situated lower in the tidal frame and inundated more frequently. In the same study, Watts et al. (2003) found that while control salt marsh soils displayed higher water content than the restored marshes, they were better structured, with high plant root density and moderate soil organic matter content (5%–10%). The shear strength of these soils was ~2× greater than the restored marsh, further supporting the stabilizing role of vegetation in salt marshes.

Salt marshes, in general, have lower organic content at depth than freshwater marshes due to either greater sediment delivery or less preservation of organic material at depth (Odom, 1988; Nyman et al. 1990). However, the rooting characteristics and strength of the soils along salinity gradients varies considerably. Studies in Louisiana found that salt marshes in Breton Sound were more resistant to erosion from the passage of Hurricane Katrina compared to interior freshwater marshes due to the increased strength of these soils (Howes et al. 2010; Turner, 2011). However, nutrient loading may also have been a factor. Studies there and in many freshwater, brackish, and saltwater coastal wetlands show root and rhizome biomass declines with increased nutrient loading (Holm 2006; Darby and Turner 2008a, b; Langley et al. 2009; Deegan et al. 2012).

Nutrient additions to salt marshes increase the aboveground biomass (Darby and Turner 2008b), lower belowground biomass (Darby and Turner 2008a, 2008b), and increase decomposition rates and reduce soil strength (Wigand et al. 2009; Turner, 2011). In a New England salt marsh, creekbank failures were more readily observed in a nutrient-enriched marsh as a result of reduced belowground biomass (Deegan et al. 2012, see Fig. 8.5d), further supporting the stabilizing role of plant roots and rhizomes.

8.3.1.2 Biofilms and Macrophytobenthos

Biofilms are the initial stage of more complex structures called microbial mats that can form both within the salt marsh substrate (endobenthos) or on top of the marsh substrate (epibenthos; Stolz, 2000; Cuadrado et al. 2014). The first few millimeters of intertidal sediment can contain varied microbial and microphytobenthos biomass by organisms such as diatoms and cyanobacteria. These organisms typically form secretions of a carbohydrate-rich exopolymer, commonly called extracellular polymeric substance , which surrounds the sediment matrix and may result in an increased interparticle cohesion that increases the stability of the sediment (Grant et al. 1986; Madsen et al. 1993; Underwood et al. 1995; Yallop et al. 2000; Pan et al. 2013).

Normally microbial mats are better developed in mudflats than in marshes because vascular plants and crab bioturbation somewhat prevent extensive mat formation (see Fig. 8.6; Underwood et al. 1995). Microbial mats can generate chemical changes in the sediment structure by trapping nutrients and forming minerals (i.e., apatite, carbonates) that are not regularly found in mudflats and marshes (Soudry, 2000; Stolz, 2000; Dupraz and Visscher, 2005). More importantly, however, extracellular polymeric substance secreted by macrophytobentos sticks to sediment particles, making them more cohesive (Grant et al. 1986; Underwood et al. 1995). This has been measured using a Cohesive Strength Meter (CSM) or EROMES (Tolhurst et al. 1999; Lanuru, 2008). A CSM uses a vertical jet of water at increasing velocities and an optical sensor to detect sediment erosion (indicated by a reduction in light transmission; see Tolhurst et al. 1999 for methods). In an evaluation of tidal flat sediments in the UK, Tolhurst et al. (1999) found the critical erosion threshold in sediments covered by biofilms was greater than those without (horizontal bed shear stress increased from 0.37 N m^{-2} to 2.31 N m^{-2} when biofilms present). Other studies have yielded similar results in the Wadden Sea (Austen et al. 1999; Lanuru, 2008), Puerto Rosales (Argentina; Cuadrado et al. 2014), and in flume experiments (Widdows et al. 1998; Widdows and Brinsley, 2002). Figure 8.6 shows results from Puerto Rosales, where the authors found microbial mats, both endo- and epibenthic, significantly increased the critical shear stress needed to erode sediment (indicated by the symbol θ) by either currents or waves (Cuadrado et al. 2014).

Since biofilms reduce the erosion of sediments on salt marshes and adjacent mudflats, they impart a stabilizing effect on the marsh. In fact, as will be discussed later in section 8.3.2.3, disturbance of microphytobenthos by feeding white croaker fish (*Micropogonias furnieri*) along channel margins in Bahia Blanca Estuary (Argentina) has been implicated in enhanced erosion of sediment, evidenced by reduction in bed shear stress and chlorophyll-a concentration in disturbed areas (Molina et al. 2017).

8.3.1.3 Oysters/Clams/Mussels

Beds of benthic infaunal organisms such as oysters, clams, and mussels are common features in the intertidal and shallow subtidal zone. Extensive work has shown infauna can strengthen soil networks, armor beds, and prevent erosion, aid in biodeposition, and/ or affect species distribution, production, and abundance due to habitat creation or alteration (Kraeuter, 1976; Bertness, 1984b; Gutiérrez and Iribarne, 1999; Murray et al. 2002, and references therein; Gutiérrez et al. 2003, and references therein; Altieri et al. 2007; Angelini et al. 2015; Bayne, 2017). As all of these processes exert geomorphological control on the salt marsh, many bivalve species are considered ecological engineers. In salt marshes, common infaunal species include *Geukensia demissa*, *Polymesoda* sp., *Crassostrea virginica*, *Mercenaria*, and *Mytilus edulis*, which grow as bioherms, or aggradations of individuals that can either be laterally extensive or isolated plots (Bertness, 1984b; Julien, 2018, and references therein; see Fig. 8.7). In general, these

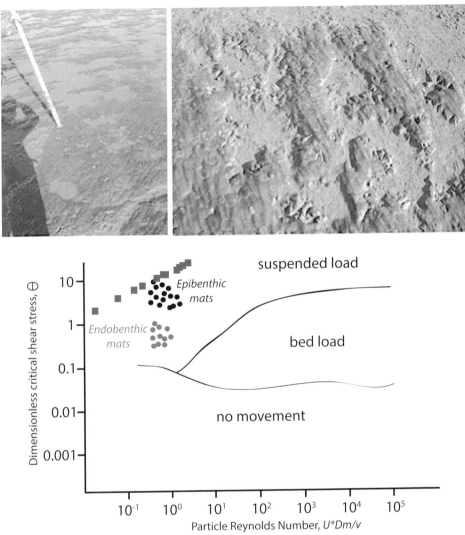

Figure 8.6 (*top*) Examples of biofilms on a tidal flat of Bahía Blanca Estuary, Argentina. The patches covered by the biofilms protect wave ripples formed previously, which makes the substrate more resistant to erosion. In the patches where the biofilm was removed, sediment is eroded forming small depressions called pans. (Photo credit Lea Olsen) (bottom) Field measurements for initiation of grain sediment transport from Puerto Rosales, where Cuadrado et al. (2014) found microbial mats, both endo- and epi-benthic, significantly increased the critical shear stress needed to erode sediment. The squares are the maximum values of critical shear measured on the flat over epibenthic microbial mats.

different infaunal groups exhibit strong affinity to different types of substrata (i.e., sediment-specific distribution patterns) and hydrodynamic conditions; however, these preferences can vary by latitude, local climate conditions, and biological forcings (e.g., predation, food availability).

Figure 8.7 A bioherm of the eastern oyster, (*Crassostrea virginica*) armoring a *Spartina alterniflora* salt marsh shoreline in St Helena Sound, SC. The common ribbed mussel, *Geukensia demissa*, is typically found in small bioherms in salt marshes (b–c), and recent studies show *Geukensia* associations with *S. alterniflora* are beneficial for productivity (b). Photo credits W. Doar and A. Julien

In North Atlantic and Gulf of Mexico salt marshes, intertidal populations of the Eastern oyster (*Crassostrea virginica*) form armored bioherms along unvegetated muddy shorefaces and within tidal creeks (see Fig. 8.7). These bioherms, comprised of both live individuals and shell hash, actively baffle water flow and stabilize soft sediments associated with mudflat and creek beds, effectively armoring the shoreline in low energy environments (Meyer et al. 1997; Piazza et al. 2005; Scyphers et al. 2011; see Figs. 8.1 and 8.7). In an experiment in Mobile Bay, Scyphers et al. (2011) found that oyster reefs situated normal to the marsh shoreline served as natural breakwaters and decreased shoreline retreat. Results from a retreating marsh in Louisiana supported decreased shoreline retreat when armored with oyster reefs, but only in low wave energy environments; higher energy shorelines, particularly those impacted by storm events, were less impacted (Piazza et al. 2005).

As oyster reefs are increasingly utilized as a nature-based method of shoreline erosion control ("living shorelines," O'Donnell, 2017, and references therein), understanding their impact on salt marsh geomorphology requires further investigation. For example, shell hash that collects along mudflats and salt marsh surfaces after the death of individuals armors the shoreline and may even influence salt marsh elevation: in an experimental study, Meyer et al. (1997) found salt marsh areas immediately inland of cultch sites (shell fragments that juvenile oysters can colonize) exhibited gain in elevation compared to control sites. In this study, it was not revealed if the gain in elevation was due to increase in sedimentation or belowground biomass production; however, it was evident the oyster cultch dampened wave energy, which enhanced the ability of the marsh vegetation to stabilize sediment (Meyer et al. 1997; see also Section 8.3.1.1). In some cases, however, shell hash that has washed ashore from storms may be detrimental to the salt marsh: in a study of marsh shorelines in Louisiana, Crawford (2018) found the storm overwash of shell material smothered live *S. alterniflora* plants, reducing aboveground biomass up to 40 m inland; it remains unclear if this process ultimately enhances shoreline erosion or marsh submergence.

The common ribbed mussel, *Geukensia demissa*, is predominantly found at salt marsh edges of smooth cordgrass (*S. alterniflora*) in northern US marshes, but more discrete aggregations in the high marsh platform and near the mouths of tidal creeks in more southern US salt marshes (Bertness, 1984b; Julien, 2018, and references therein; see Fig. 8.9). *Geukensia* has been termed a secondary foundation species for its ability to create new habitat, particularly within stands of smooth cordgrass (Altieri et al. 2007; Angelini et al. 2015), and there appears to be facultative mutualism between *Geukensia* and *S. alterniflora*, whereby neither species is dependent on the other, but their associations are beneficial (Bertness, 1984b). *Geukensia* colonize within the salt marsh sediment, attaching themselves with proteinaceous byssal threads, and where mussel densities are great, substrate binding serves to prevent soil erosion and physical disturbance of aboveground vegetation (Bertness, 1984b; Julien, 2018, and references therein). At the landscape scale, studies show *Geukensia* increase *S. alterniflora* productivity and drought resistance (see Fig. 8.7; Bertness, 1984b; Angelini et al. 2015, 2016; Julien, 2018). While some studies show invertebrate abundance and diversity is enhanced by the presence of ribbed mussels (Altieri et al. 2007; Angelini et al. 2015), recent research shows transient nekton abundance may not be affected (Julien, 2018). Finally, biodeposition by *Geukensia* is common, as found in salt marshes in Massachusetts and Georgia: through filter feeding, mussels transfer and concentrate sediment and nutrients from the water column to the soil surface through the production of feces and pseudofeces (Bertness, 1984b; Smith and Frey, 1985). Production of fecal pellets enhances sedimentation rates, which can stabilize marsh growth, and deposits high in nutrients and concentrated minerals may alleviate nutrient limitation (Jordan and Valiela, 1982; Bilkovic et al. 2017).

While clams are similar to mussels, they have a more advanced internal structure, including a digestive and circulatory system. Certain species have byssal threads that similarly stabilize surface sediment, and Gutiérrez and Iribarne (1999) found removal of *Tagelus plebeius* clams due to harvesting in a southwest Atlantic salt marsh (Mar Chiquita lagoon) destabilized the substrate and enhanced erosion. Similar to mussels, clams also produce nutrient-rich feces/pseudofeces to intertidal surfaces and the water column, affecting biogeochemical cycles (see Kraeuter, 1976, and references therein). Interestingly, fecal pellets produced by infauna may also impart a destabilizing effect on salt marsh surfaces, as these particles may be more easily eroded (see Section 8.3.2.1).

In a recent study by Addino et al. (2015), ecosystem engineering by *S. alterniflora* on sediment type influenced *T. plebeius* clam somatic growth in a similar marsh in Bahia Blanca. The authors found *S. alterniflora* stem density affected the turbulence and thus, grain size deposited across a salt marsh platform, and *T. plebeius* exhibited longer shells in more cohesive mudflat substrate where energy allocation to burrow maintenance was lower (Addino et al. 2015). Future research that investigates whether larger organisms impart more stabilization, or how anthropogenic harvesting has impacted the salt marsh stability is needed (see also Gutiérrez et al. 2003, and references therein).

8.3.2 Ecological Destabilizers

While some ecological engineers impart stabilizing forces within salt marshes, others create destabilization with their inherent behavior and living practices. Destabilization includes processes that decrease plant productivity or surface elevation, weaken substrate architecture, or alter hydrodynamic flow paths, leading to erosion or deflation of the salt marsh surface. These processes may or may not affect the overall sustainability of the salt marsh, so this is a heavily studied and debated subject of salt marsh ecogeomorphology.

Destabilization may occur when biology *directly* disturbs vegetation or substrate, which is most commonly reached when high populations of destabilizing organisms colonize a particular region of the salt marsh (Montague, 1980). Destabilization may also be achieved via feedback mechanisms where vegetation, substrate, or hydrodynamic flow are disturbed *indirectly* by destabilizing organisms. In this section, we focus our discussion on the most prominent destabilizers in salt marsh systems including crabs, snails, fish, and mammals, though other destabilizing organisms may exist in various salt marsh settings.

8.3.2.1 Crab Bioturbation

Crabs have been implicated for destabilizing salt marshes through all of the aforementioned effects (decreasing plant productivity and surface elevation, affecting substrate architecture, and altering hydrodynamic flow paths) from the formation of crab burrows and bioturbation – physical mixing of the sediment surface. It is not uncommon to find populations of crabs in salt marshes worldwide in excess of 200 m^{-2}, depending on species and location. Physical factors such as tidal variation, slope, sediment composition, and ease of excavation impact the ability of crabs to make and maintain their protective burrows in salt marshes, and biological factors such as food availability, competition, and predation are also important factors dictating crab abundance and distribution (Teal, 1958; Montague, 1980; Bertness and Miller, 1984; Bertness, 1985; Weissburg, 1992; Botto and Iribarne, 2000; McCraith et al. 2003; Bertness et al. 2009; Wang et al. 2008; 2010; Vu, 2018). Tidal variation and slope control the hydroperiod, which affects the wetting and drying of the marsh substrate, and there are evident species-specific preferences where infaunal habitation takes place. Early studies in northern Atlantic salt marshes found that deposit feeding fiddler crabs *Uca pugilator* and *U. pugnax* prefer sandy and muddy substrates, respectively (Teal, 1958), while the herbivorous crab *Sesarma* sp. preferentially colonizes peaty substrates along salt marsh edges (Smith, 2009; Bertness et al. 2009).

Marsh substrate affects the day-to-day maintenance required by crabs, which in turn affects burrow morphology and distribution. For example, most crabs can readily form burrows in soft muddy substrates, but these burrows require higher maintenance due to poor structural support (Bertness and Miller 1984; Bertness et al. 2009). Marsh peat comprised of dense plant roots and rhizomes is more difficult to burrow into, but these burrows are not as prone to collapse and require lower maintenance, thus can be structurally more complex (Bertness and Miller 1984; Bertness et al. 2009; Wang et al. 2014); however, very high densities of salt marsh roots and rhizomes can actually impede burrow activity (e.g., Wang et al. 2014). Crab burrow densities typically range from 20 to 300 burrows m^{-2}, with the highest densities most commonly observed near creekbanks (Ringold, 1979; Katz, 1980; Montague, 1980, 1982; Bertness and Miller, 1984; Bertness, 1985; Jaramillo and Lunecke, 1988; Genoni, 1991; Mouton and Felder, 1996; McCraith et al. 2003).

Crab burrow morphology such as average width, depth, structure, and connectivity varies among species (see Allen and Curran, 1974; Montague, 1980; Sharma et al. 1987;

Figure 8.8 Crab burrows in salt marshes can have various morphology and structure, both aboveground and belowground: (a) mudflat in *a Salicornia ambigua* salt marsh in Bahia Blanca (Perillo and Iribarne, 2003b); (b) Australian salt marsh male *Uca capricornis* and chimney structure (Slatyer et al. 2008); (c–d) resins of crab burrows in a *Sueda japonica* salt marsh in Korea (Koo et al. 2005).

Escapa et al. 2008; Katrak et al. 2008; Wang et al. 2010; 2014), and advances in understanding burrow morphology and structure have been achieved from molds created using various resins, featured in Figure 8.8 (Allen and Curran, 1974; Genoni, 1991; McCraith et al. 2003; Katrak et al. 2008; Wang et al. 2010; Wang et al. 2014). Burrows belowground can be J, L, Y, U, H, or more complex shapes, and aboveground structures and connectivity also vary, but hood or chimney structures are common (Fig. 8.8; Allen and Curran, 1974; Genoni, 1991; Perillo et al. 2005; Wang et al. 2010; Wang et al. 2014). The aforementioned fiddler crab species, *Uca pugnax* and *pugilator* create and compete for individually occupied J-shaped burrows, 1–2 cm wide extending to ~10–25 cm depth (Allen and Curran, 1974; McCraith et al. 2003). *Sesarma* spp. appear to be more social: they create an interconnected burrow network 1–2 cm wide to ~20 cm depth occupied by many individuals (Allen and Curran, 1974; Wilson et al. 2012; Vu et al. 2017). South Atlantic deposit-feeding crabs *Neohelice* (formerly *Chasmagnathus) granulata* and *Uca uruguayensis* occupy similar muddy regions of the salt marsh, however *N. granulata* burrows have large funnel-shaped entrances (up to 14 cm diameter) that remain open at high tide, while *U. uruguayensis* burrows are much smaller ($\leq$2 cm wide) and the individuals prefer to close them at high tide (Botto and Iribarne, 2000, and references therein; Perillo et al. 2005). Some crab species have variable burrow morphology, for example, the aforementioned *N. granulata* crabs exhibit both tubular and funnel-shaped burrow entrances, and the chambers can be isolated or connected (Escapa et al. 2008; Perillo, 2019).

Species-specific burrowing activity dictates the effect of bioturbation, and this is heavily studied in salt marshes. Some of the most pressing questions actively investigated include: what is the sediment turnover rate resulting from burrowing by individual species? How does this effect sediment transport, particularly at the marsh surface? What is the effect of bioturbation on sediment structure and character belowground, including biochemical and physical changes that affect decomposition and/or erosion? Foundational and cutting edge work that address these questions is detailed in the following sections.

Physical mixing from crab bioturbation – Bioturbation and sediment turnover by crabs is one type of non-diffusive mixing referred to as bioadvection (Teal, 1958; Rice, 1986; Boudreau and Imboden, 1987). Sediment turnover is typically analyzed from field and mesocosm experiments that measure quantity and character of the material advected onto the surface outside of crab burrows (typically in the form of feeding, fecal, and burrow pellets) relative to what is reworked back into burrows using burrow mimics (see Fig. 8.9; McCraith et al. 2003; Escapa et al. 2008; Wang et al. 2010; Vu and Pennings, 2018). Feeding pellets are small (mm-scale), irregular-shaped balls deposited from non-ingested particles of deposit feeders, while fecal pellets are more regularly shaped particles that have passed through the gut of an organism. Burrow pellets are larger (up to 10 cm), more regularly shaped balls comprised of sediment particles extracted from burrow chambers during maintenance operations. Once particles (feeding, fecal, or burrow) are placed outside of burrow entrances, they are typically reworked by tides, currents, or organisms at the surface over time. Sediment from the surface can replace sediment at depth (Fig. 8.9; Sharma et al. 1987; Botto and Iribarne, 2000; Wang et al. 2010), and this material typically has high water and organic content as measured by sediment collected within burrow mimics (Botto and Iribarne, 2000; Botto et al. 2006; Wang et al. 2010).

Sediment turnover rates vary among species, and is typically dictated by individual – and thus burrow – size, sediment cohesion factors, and population density (see Katz 1980; Takeda and Kurihara 1987; Iribarne et al. 1997; Botto and Iribarne 2000; McCraith et al. 2003; Gutiérrez et al. 2006; Fanjul et al. 2007; Escapa et al. 2008; Wang et al. 2010). Botto and Iribarne (2000) found in the salt marshes of Bahia Samborombón relatively large-sized *Neohelice granulata* rework 2,234.6 g m^{-2} d^{-1} while smaller fiddler crabs *U. uruguayensis* rework 678.9 g m^{-2} d^{-1}. Meanwhile fiddler crabs *U. pugnax*, *U. pugilator*, and *U. minax* in the salt marshes of North Inlet, South Carolina, rework as much as 171 g m^{-2} d^{-1} (McCraith et al. 2003, using bulk density for salt marsh soil of 1.1 g cm^{-3}). Due to routine maintenance, more sediment is excavated by crabs on a daily basis than infills their burrows

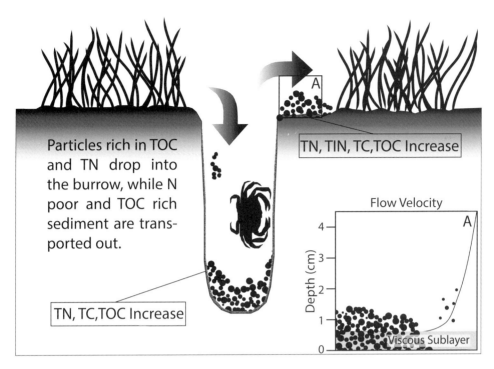

Figure 8.9 Due to the cycling of water, sediment, and nutrients, crab burrows can be considered extensions of the sediment–water interface. Note sediment is cycled into and outside of burrows, and sediment mounds placed outside of burrows increase the surface roughness of the marsh, enhancing sediment transport if particles project above the viscous sublayer (Inset A). TC: total carbon; TOC: total organic carbon; TN: total nitrogen; TIN: total inorganic nitrogen. Modified from Wang et al. (2010). Figure credit B. Gregory

(see Escapa et al. 2008; Wang et al. 2010, and references therein). McCraith et al. (2003) reported that burrows in North Inlet typically last 50–100 days, reinforcing that bioturbation and sediment turnover is a persistent process where crabs live in salt marshes (see also Takeda and Kurihara, 1987; Wang et al. 2010).

Bioturbation enhanced erodibility of salt marsh substrate – While salt marsh sediment is typically very fine-grained and difficult to erode once deposited (see Sections 8.3.1.1 and 8.3.1.2), several authors argue that crab bioturbation can enhance the erodibility of the soil (Botto and Iribarne, 2000; Cadee, 2001; Escapa et al. 2007; Wilson et al. 2012). This is achieved by reworking what is normally cohesive fine-grained sediment into larger pellets that increase the surface roughness of the salt marsh bed. Surface roughness has a profound impact on flow structure and sediment transport when tides and currents pass over the salt marsh: a surface where the particles are entrained in the sub-cm bottom boundary layer – called the "viscous sublayer" – is considered smooth, while if particles project above the viscous sublayer, it is considered rough because the particles create turbulence (see Fig. 8.9; Kirchner et al. 1990; van Rijn et al. 1990; Leeder, 1998). Particles that project above the viscous sublayer can experience enhanced lift force and thus become entrained in the flow and erode (Fig. 8.9; Murray et al. 2002).

Particles that are eroded from the bed must then be balanced with how much is deposited back inside the burrows, but this can be quite complex and vary according to hydrodynamic flow conditions (see Murray et al. 2002; Escapa et al. 2008; Wang et al. 2010, and references therein). In habitats with low flow energy, burrowing enhances sediment trapping within burrows, whereas in habitats with high flow energy, erosion of sediment could be increased: Escapa et al. (2008) report enhanced erosion of sediment from the crab *N. granulata,* but only in regions of the salt marsh where current velocities were sufficient to initiate transport (creek bottoms and creek basins on the salt marsh platform). In addition, the authors note that desiccation of the sediment excavated is an important factor: desiccated mounds produced cohesive blocks that were then transported by currents. Botto and Iribarne (2000) argue while *N. granulata* and *U. uruguayensis* are important bioturbators in soutwest Atlantic salt marshes, they have contrasting effects on sediment transport: *N. granulata* stabilizes the sediment by placing fine-grained, cohesive sediment on the surface that persists over several tidal cycles; *U. uruguayensis* disrupts the sediment by pelletizing it, increasing surface roughness, and increasing erosion.

Biogeochemical cycles & decomposition resulting from crab bioturbation – Burrows increase the surface area of a salt marsh (as much as 60% according to Katz, 1980) and increase sediment porosity and permeability (Mendelssohn and Seneca 1980; Widdows et al. 1998; Perillo et al. 2005; Xin et al. 2009), therefore it is readily recognized that crab bioturbation may significantly affect the composition and chemistry of salt marsh sediments, a few of which are shown in Figure 8.9 (Gribsholt et al. 2003; Wang et al. 2010). Foundational studies show that crab burrow waters typically have greater nutrients and oxygen, and lower salinity and sulfide as a result of more exchange of water (Montague, 1982; Perillo and Iribarne, 2003a). This creates a microzone of oxidized conditions in salt marsh substrate surrounding the burrows (few mm, Howes et al. 1981; Koo et al. 2005). Some studies show that this has improved salt marsh productivity, as seen in the standing stocks of short-form *Spartina alterniflora* on Sapelo Island, Georgia (western Atlantic; Montague, 1982; see also Bertness, 1985). Plant and microbial growth may also be stimulated from organic matter brought to the sediment surface by crab bioturbation (e.g., Genoni, 1991; Bertness, 1985), and rates of carbon transfer have been documented as much as $10 \text{ g m}^{-2} \text{ d}^{-1}$ (Gutiérrez et al. 2006; Thomas and Blum, 2010; Wang et al. 2010).

However, crab bioturbation has also been implicated for enhancing decomposition of belowground biomass, under both oxidized and reduced conditions. Under oxidized (oxygenated) conditions, organic matter can be remineralized or decomposed by reacting with oxygen, producing carbon dioxide. In wetlands, including salt marshes, oxidized conditions typically exist only in the water column and within a few millimetres of the sediment–water surface. Under reducing conditions (where oxygen is limited or not present, i.e., most of the salt marsh substrate), other electron acceptors such as iron, manganese, and sulfate are utilized in the decomposition of organic matter. This is typically biologically mediated by microorganisms (Reddy and Delaune, 2008), though recent studies show crab bioturbation can enhance this process (Bertness, 1985; Gribsholt et al. 2003; Wilson et al. 2012). In South Carolina and Georgia salt marshes, researchers showed *Sesarma* burrowing near creek heads enhanced the oxygenation of the substrate, facilitating the decomposition of *S. alterniflora* belowground biomass (reducing the belowground biomass as much as 50 %; Wilson et al. 2012; Vu et al. 2017). Others have documented that burrowing from both *Sesarma* and *U. pugnax* have enhanced Fe(III) reduction in different regions of the salt marsh, accounting for 54–86 % of the total carbon oxidation within 4 cm distance of burrows (Kostka et al. 2002; Gribsholt et al. 2003) and 3 cm depth near bioturbated creek heads (Hyun et al. 2007). These studies highlight how crab bioturbation can enhance decomposition rates in salt marshes, particularly in certain regions. Section 8.4.2 addresses geomorphological implications of these processes on a larger (planform) scale.

Salt marsh destabilization from herbivory –Certain crab species are omnivorous, feeding on live plant material, dead leaf litter, other crabs, terrestrial fungi, and marsh sediment (deposit feeders; Pennings et al. 1998). Prominent examples of those that forage on salt marsh macrophytes include *Armases cinereum*, *Sesarma reticulatum*, and *N. granulata* (see Pennings et al. 1998; Alberti et al. 2007, 2008; Holdredge et al. 2008; Bertness et al. 2009). Herbivory can take place predominantly aboveground with the crabs foraging on plant stems and leaves (as in the case of *Neohelice*; Alberti et al. 2007, 2008), or also belowground with crabs foraging on plant roots and rhizomes (e.g., *Sesarma*; Wilson et al. 2012; Coverdale et al. 2012; Vu et al. 2018). Alberti et al. (2007) found *Spartina densiflora* and *S. alterniflora* productivity in southwest Atlantic salt marshes (measured as stem number, proportion of live stems, and stem height) was diminished by herbivory from *N. granulata*. Increases in herbivorous crab populations (*Sesarma*, in particular) have been associated with die-off and conversion from vegetated marsh platforms to mudflat in New England and mid-Atlantic salt marshes (Holdredge et al. 2008; Smith, 2009; Bertness et al. 2009; Wilson et al. 2012; Vu et al. 2017). Since *S. alterniflora* is responsible for sediment binding, peat deposition, and the vertical accretion of New England marshes (Redfield, 1965), Bertness et al. (2009) and Coverdale et al. (2012) maintain cordgrass die-off from crab herbivory has the potential to compromise the ability of salt marshes to keep pace with sea-level rise. Other authors argue that the alteration of tidal flows that result from *S. alterniflora* dieback has the potential to increase the sustainability of the salt marsh platform (Hughes et al. 2009; Wilson et al. 2012; Vu et al. 2017). Regardless, it is well recognized that transitions from vegetated to unvegetated surfaces alters the species richness and biodiversity in the salt marsh (*sensu* van Wieren and Bakker, 2008). Certainly this subject matter remains heavily debated in salt marsh ecogeomorphology and deserves further investigation.

8.3.2.2 Snail Disturbance

In the past, it was widely accepted that grazers play a relatively unimportant role in the salt marsh community. However, recent investigations into plant stress and die-off have shifted opinions, and an increasing number of studies are revealing the important role that herbivores and bioturbators play in salt marsh dynamics (Pennings and Silliman, 2005;

Silliman et al. 2005; Vu et al. 2018). Recent work by Silliman and Zieman (2001) and Silliman et al. (2005) highlight that marsh and mud snails are primary drivers of *Spartina alterniflora* productivity in mid-Atlantic salt marshes. The salt marsh snail, *Littorina littorea,* is an invasive species in North America, with populations along the western Atlantic increasing since introduction in the 1840s (Bertness, 1984a, and references therein). Populations of adult *L. littorea* are rare above the mean high water line, densities can be up to 400–800 snails m^{-2} in the middle intertidal zone, and distribution abruptly drops below mean low water, likely due to predation (Montague, 1982, and references therein; Bertness, 1984a; Silliman and Zieman, 2001). Exclusion experiments show salt marsh snails can decrease *S. alterniflora* productivity >50 % (in one case from 1490 to 274 g m^{-2}; Silliman and Zieman, 2001). Snails consume the diatoms, algae, and fungus from surface of the *S. alterniflora* plant. During grazing, however, the radular tongue can induce damage to plant cell walls, enhancing or facilitating fungal infection and senescence (Silliman et al. 2005, and references therein). In an investigation of die-off areas in the southeast USA (GA and LA), Silliman et al. (2005) reported very high densities of *Littorina* snails along the fringes of the dieback areas (at 11 of the 12 sites observed; see Fig. 8.10). They reported that while drought was the initiating cause of plant dieback, snail grazing was a significant contributing factor to the expansion of dieback areas (as much as 30 m in one year in Georgia where snail densities were ≥2,000 m^{-2}, and as much as 10 m in one year in Louisiana where snail densities were >1,000 m^{-2}).

8.3.2.3 Fish Disturbance

Salt marsh fishes are among the most highly valued animals of the marsh because of their commercial and recreational importance (Costanza et al. 2008, and references therein) and connections to ocean carbon and nutrient cycles (e.g., Nelson et al. 2012). Depending on region, permanent residents are present as well as migratory species during breeding seasons. Species dominance varies latitudinally and, of course, these are constantly in flux due to food web dynamics and anthropogenic overfishing pressure. In general, fishes do not spend a great deal of time within vegetated salt marsh areas due to limited periods of inundation (some areas in Australia, for instance, are only inundated a few tides a year; Laegdsgaard, 2006), but during flooding both transient and resident species can migrate to the interior in search of food (Peterson and Turner, 1994, and references therein). Fish are most common within tidal creeks and channels, and salt marsh edges inundated more frequently (Peterson and Turner, 1994; West and Zedler, 2000).

Bioturbation by fish within salt marshes is relatively rare; however, it has been reported in the literature, and in some cases with major geomorphic response. Perillo et al. (2005) report that the white croaker fish (*Micropogon opercularis*) in Bahía Blanca Estuary create circular depressions 1–3 cm deep along channel margins where they forage on crabs housed within their burrows (*N. granulata*). These depressions facilitate the percolation of groundwater into the burrows and indirectly enhance the incision and headward erosion of marsh grooves that may later become larger tidal courses over time (creeks or channels). Experiments reveal that the critical shear stress of sediments bioturbated by the white croaker is less than areas where the fish are excluded, thus leading researchers to argue that

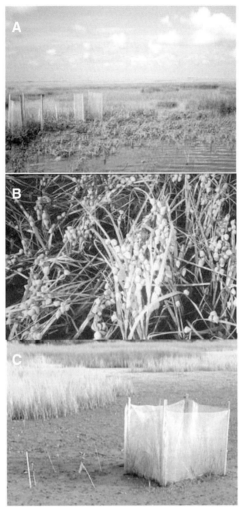

Figure 8.10 Figures from Silliman et al. (2005) showing a dieoff border within a *S. alterniflora* marsh in their study area (Louisiana and Georgia, USA); the hardware and cloth in the background is a snail exclusion cage (a). Very high densities of salt marsh snails *Littorina littorea* were found at the frontline of the dieback area (b). Exclusion cages effectively preserved patches of *S. alterniflora* salt marsh, indicating that grazing by *L. littorea* was a major contributing factor to the growth of the dieback areas (c).

fish bioturbation can enhance the erodibility of salt marsh sediments, particularly along creekbanks and mudflats where bioturbation is prevalent (see Molina et al. 2017, for further details).

8.3.2.4 Waterfowl Disturbance

Although rare, disturbance by waterfowl has been reported in salt marshes, ranging from areas around the Gulf of Mexico to the St. Lawrence Estuary (Lynch et al. 1947, and references therein; Dionne, 1985; Gauthier et al. 2005). This is most extensively described for the greater snow goose (*Chen caerulescens atlanticus*) by Dionne (1985), where he reports that the geese dig many thousands of small holes in the substrate (6–12 cm depth, 10–25 cm in diameter) in search of clean rhizomes of *Scirpus pungens*. The geese tear out the roots of the plants and disturb the substrate, which is only ~30 cm thick muddy alluvium overlying a denser glaciomarine clay. Dionne (1985) further reports that the sediment extracted is easily resuspended by waves and currents, causing on average lowering of the lower tidal marsh of the St. Lawrence Estuary each year in the area impacted by geese, deflating the surface up to 10 cm yr^{-1}. In this particular instance, lowering of the marsh surface also enhances the wave energy impacting the adjacent higher marsh platform, and retreat rates on the order of 5–6 m yr^{-1} and 1 m yr^{-1} have been measured for the lower and higher tidal marshes, respectively (Dionne, 1985). Interestingly, hunting may regulate this process, as the author reports regions of the marsh where geese hunting is prohibited exhibit more severe erosion.

It is estimated that feeding snow geese, in addition to other waterfowl, pull up and discard at least ten times the amount of plant material they actually consume (Lynch et al. 1947, and references therein). Trampling of salt marshes by waterfowl has also been mentioned in the literature, and in the case of the greater snow goose in the St. Lawrence Estuary, trampling damages the vegetation and produces depressions on the surface that can later enhance the erosion in the winter (i.e., freeze thaw of pools; Dionne, 1985). Populations of the snow goose have grown 9% yr^{-1} since the mid-1970s, and a recent study shows peak aboveground production of *Scirpus pungens* has declined in some regions of the estuary since the early 1980s (Gauthier et al. 2005). However, in recent decades marsh foraging by the geese has been on the decline in some regions (as much as 40%) as foraging in nearby agricultural fields is on the rise (Gauthier et al. 2005).

In another example that waterfowl affect sediment stability, in the Minas Basin (Bay of Fundy, Canada) extracellular polymeric substance secretion of diatoms is normally regulated by the concentration of a foraging amphipod, *Corophium volutator*. The arrival of large flocks of the migratory semipalmated sandpiper *Calidris pusilla* completely reduces the abundance of amphipods due to predation, resulting in an increase of the production of extracellular polymeric substance and strengthening of the surface sediments (Daborn et al. 1993; see also Section 8.3.1.2). These conditions appear to be temporary, however, as when the birds leave the grazing area, the system is quickly restored to previous conditions within a few days (Daborn et al. 1993).

8.3.2.5 Disturbance by Mammals

Phenomena, popularly called "eatout" events, have been reported in salt marshes world-wide, most commonly by the semiaquatic muskrat (*Ondatra zibethica*) and nutria (*Myocastor corpus*). Lynch et al. (1947) reported that trappers captured 3–9 million muskrat per year in Louisiana at the time, and that eatouts from muskrat lowered the marsh elevation and formed swale-like ponds on the surface that were then susceptible to erosion ("loosely anchored soils and vegetation are liable to be torn loose and washed away"; Lynch et al. 1947). Many researchers have since argued that extreme herbivory by a larger cousin to the muskrat, the nutria (which range in size up to 10 kg), has been a contributing factor for degrading salt marshes, particularly in the USA. Nutria were introduced to North America in the past century, and recent research shows herbivory by the species decreases aboveground salt marsh production $\geq$20 % (*S. alterniflora* from $\geq$1,000 to $\leq$800 g m^{-2}; *S. patens* from $\geq$1,150 to $\leq$950 g m^{-2}, Taylor and Grace, 1995). Although nutria appear to have species-specific preferences for foraging (see Shaffer et al. 1992, for Atchafalaya delta), it is evident that overly grazed regions of salt marshes exhibit decreased soil vertical accretion, less gain in elevation, and greater shallow subsidence, presumably from the loss in aboveground vegetation to trap sediment (see Ford and Grace, 1998 for further information).

Finally trampling by larger mammals such as deer have been known to create paths in salt marshes, which can affect the hydrologic flow directions across the marsh (e.g., Keusenkothen, 2002; Keusenkothen and Christian, 2004; Hannaford et al. 2006); however, major permanent geomorphic alterations are not apparent.

8.4 Landscape Evolution and Dynamics

Since we have established some prominent relationships and feedbacks that exist between ecological engineers that that live upon and within the salt marsh realm, we will now focus on the salt marsh landscape (tidal channel, creek network, and vegetated marsh platform), and examine how biology controls and shapes geomorphic development and evolution in salt marshes (i.e., ecogeomorphology). Salt marsh landscapes, like all coastal environments, are continuously evolving as they are actively interacting with oceanic and continental conditions, modulated by climate and local geological and biological processes (Marani et al. 2007). In a recent assessment of the sustainability of salt marshes, Kirwan et al. (2016) reported that marshes in North America and Europe are generally building at rates similar to or exceeding historical sea level rise, but knowledge gaps exist and the biophysical feedback processes known to accelerate soil building with sea level rise are understated in many of the assessment models used today. These authors stress more research quantifying the biophysical feedbacks is much needed. Connections between the salt marsh platform, tidal network, and the biology contained within is an integral part of the sustainability of the system, as the nexus provides nourishment (e.g., sediment and nutrients), protection for local fauna, a place for reproduction, and maintenance and growth of many ecologically and commercially important species.

For extensive discussion how marshes develop over geologic timescales, particularly with respect to rising sea levels, readers are referred to the following references: Redfield and Rubin, 1962; Redfield, 1972; French and Spencer, 1993; Roman et al. 1997; Allen, 2000; Temmerman et al. 2003; Marani et al. 2007; Kirwan and Temmerman, 2009; Kirwan et al. 2010; Millette et al. 2010; Fagherazzi et al. 2013a; Kirwan et al. 2016. Additionally, salt marsh edge geomorphic evolution is discussed by Schwimmer and Pizutto (2000), Moller and Spencer (2002), Mariotti and Fagherazzi, (2010; 2013), Priestas and Fagherazzi (2011), Fagherazzi et al. (2013b), Deegan et al. (2012), and Mariotti et al. (2016). We focus the remainder of the chapter on ecogeomorphic processes related to tidal course establishment and evolution. We briefly highlight here the complexity and dynamic balance between biology, hydrology, and geology. For more information, readers are referred to references at the end of the chapter.

8.4.1 Influence of Vegetation on Creek Formation and Dynamics

Salt marsh courses including tidal channels and creeks, together with tidal depressions, are the first features that appear in the formation of a coastal wetland, either by modification of earlier fluvial networks or by the direct action of the tides, groundwater, and precipitation. Tidal courses and depressions are regions that are incisions or negative elevations in an otherwise level or slightly seaward-inclined surface. Perillo (2019) presents a systematic classification scheme for tidal courses and depressions based on width, depth, and hydro-dynamic considerations such as inundation duration. Tidal course features include tidal rills, grooves, and gullies, which are relatively small and do not retain water during low tide, and larger sized tidal creeks and channels that retain water during low tide. A tidal course head is defined as upstream limit of observable erosion and concentration of flow within definable banks (Montgomery and Dietrich, 1988; Perillo, 2019). Tidal course morphology ranges from linear, to a combination of linear-meandering, to meandering and more complex (Pye and French, 1993).

Early models of channel development are described by the pioneer works of Yapp et al. (1917) and Pestrong (1965), among others. Even these early models highlight how salt marsh vegetation controls creek morphology and development. Yapp et al. (1917) described a possible mechanism for the origin of courses in salt marshes of the Dyfi Estuary, UK, based on the presence of hummocks formed by sediment retention due to colonization by *Glyceria* (*Puccinellia*). As the water retreats after high tide, it drains from between the macrophyte-colonized hummocks and the depressions coalesce, deepen, and widen over time as the marsh aggrades around them (Yapp et al. 1917): "As the whole marsh increases in surface extent, the main channels lengthen, new tributaries arise, and drainage systems are established" (Yapp, 1917; from Allen, 2000). The authors further describe that channels that were once shifting become more stabilized in position as the salt marsh develops, and the spreading of vegetation confines and stabilizes the channels (Yapp et al. 1917; see Allen, 2000 for further reading). Vegetation has a tendency to channelize flow; in other words, the probability of larger unchanneled flow lengths is small in the vegetated tidal marsh compared to the open unvegetated tidal flat (Vandenbruwaene et al. 2012). Recent research using both observations and modeled hydro-and sediment dynamics in the Westerschelde estuary, the Netherlands, shows that once plant establishment starts, flow velocities and bed shear stresses are reduced within and behind vegetation due to friction exerted by the vegetation (Temmerman et al. 2007; see also Section 8.3.1.1). However, vegetation patches obstruct the flow, and increased flow velocity and bed shear stress is observed between vegetation patches, causing incision and channel formation between patches as they grow laterally (Temmerman et al. 2007).

Drainage density of a salt marsh is the total channelized flow length divided by drainage area, and early work has shown that this is a function of marsh age, where creek networks

 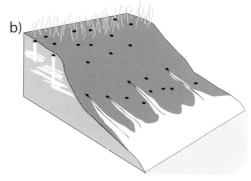

Figure 8.11 (a) Depressions in a bioturbated and unvegetated section of the Bahia Blanca salt marsh coalesce to form creek networks (Photo credit G. Perillo). (b) Percolation of groundwater and tidal flows through burrows created by *Neohelice granulata* crabs creates rill channels on mudflats that can extend into the adjacent vegetated salt marsh (modified from Perillo et al. 2005)

grow rapidly as salt marshes are first established, then contract over time as platforms elevate and the creeks widen and deepen, accommodating tidal flows (Steel and Pye, 1997; Allen, 2000). Vegetation die-off can affect the drainage density of salt marshes, as die-off areas typically deflate in elevation and become depressions in the salt marsh that can concentrate flows, forming new creeks or extending existing ones (see Fig. 8.11a; Alber et al. 2008; Hughes et al. 2009; K. Wilson et al. 2009; 2010; C. Wilson et al. 2012; Perillo, 2019). Over time, however, if flow velocities in creeks and channels are insufficient to remain erosive, plant colonization can infill channels (Wilson et al. 2014; Perillo, 2019; and references therein). Several authors have reported that channels can infill through such processes as bridging by overhanging vegetation or obstruction by collapsed sediment blocks (which may result from weak layers in the substrate; see Fig. 8.1, Section 8.3.1.1; Yapp et al. 1917; Chapman, 1960; Pestrong, 1972; Collins et al. 1987; Allen, 2000; Deegan et al. 2012; Wilson et al. 2014; Perillo, 2019). Thus the balance between erosion and deposition within the salt marsh–tidal channel continuum dictates whether the creek network will expand or contract.

8.4.2 Influence of Fauna on Creek Morphology and Dynamics

One of the most surprising and interesting findings in recent years is that creek formation and evolution can, in some cases, be a product of infaunal activity. In the high salt marshes of the Bahía Blanca Estuary (Argentina), intense burrowing by the crab *N. granulata* is focused in regions of glasswort *Sarcocornia perennis* colonization, creating holes in the salt marsh 70–100 cm deep (burrow density reaches 40–60 m^{-2}). Burrowing activity not only concentrates groundwater and tidal flows (Xin et al. 2009; Perillo, 2019), it also facilitates the erosion of sediment (researchers have documented 1,380 m^3 exported from 270 ha marsh to the estuary; Minkoff et al. 2005, 2006; Escapa et al. 2007, 2015; see also Section 8.3.2.1).

Crab activity results in depressions as large as 8 m diameter on the salt marsh surface, and water is retained at low tide or during rainfall events (see Fig. 8.11a; Perillo, 2019). Belowground connections among adjacent depressions act as avenues for groundwater and tidal flow, leading to the formation of tidal courses when the surface of the marsh collapses along elongated depressions (Fig. 8.11a). Minkoff et al. (2006) and Escapa et al. (2007) demonstrated that crab activity increases during late spring and summer, coincident with a marked increase in depressions and headward erosion of the creeks.

In another example, smaller tidal grooves formed along channel margins of the Bahía Blanca Estuary are generated from dense *N. granulata* crab burrows that penetrate the marsh surface and extend laterally into adjacent creeks (Fig. 8.11b; Perillo et al. 2005). In this case, groundwater sourced from the filling of crab burrows during high tides on the vegetated marsh platform slowly drains into the adjacent channel during low tide, producing an indentation and sill that concentrates flow and generates a tidal groove (Perillo and Iribarne, 2003a,b). As these crabs are an important food source for the white croaker fish, *Micropogon opercularis*, occasionally the mouth of the burrow is enlarged as the fish forage (producing a crater 1–3 cm deep), resulting in greater groundwater concentration and larger infiltration and, consequently, wider and deeper grooves. In high density *N. granulata* burrow patches, the lateral burrows made by the crabs generate an interconnection among burrows which otherwise will not occur (Fig. 8.11; Perillo et al. 2005). This interconnection allows for high groundwater hydraulic conductivity and eventual disturbance of the soil, which, in turn, generates the collapse of the marsh surface (Minkoff et al. 2005).

Tidal creek extension may also be a result of bioturbation activity. Hughes et al. (2009) documented that creek headward erosion in a South Carolina *S. alterniflora* salt marsh (rates approaching 2 m yr^{-1}) was an indirect result of bioturbation from the herbivorous crab, *Sesarma reticulatum* (see Fig. 8.12). Wilson et al. (2012) and Vu and Pennings (2018) collectively highlight the loss in belowground biomass from extensive burrowing from these crabs creates depressions in the salt marsh (20–40 cm lower than the surrounding salt marsh) which concentrates tidal flows, facilitating sediment erosion in bioturbated areas and creek headward incision (Fig. 8.12). In South Carolina, drainage density increased 23% between 1958 and 2006 with the headward erosion of new creeks (Wilson et al. 2012). These authors argue the increase in creek lengths observed has the potential to increase the sustainability of the salt marsh platform, as creeks facilitate (1) the delivery of sediment to the salt marsh platform and (2) the removal of toxins that build up in the soil (Hughes et al. 2009; Wilson et al. 2012; Vu et al. 2017; see also Delaune et al. 1978; Letzsch and Frey, 1980; Hatton et al. 1983; Howes et al. 1986; Reed, 1988; Bertness et al. 1992; French and Spencer, 1993; Chmura and Kosters, 1994; Reed et al. 1999; Christiansen et al. 2000; Temmerman et al. 2005; van Proosdij et al. 2006; Perillo, 2019).

8.5 Concluding Remarks

As salt marsh landscapes form and evolve, there exist significant interactions between biology, hydrology, and geology, thus it is extremely important to consider these interactions when studying salt marsh geomorphology – how the landscape changes over time.

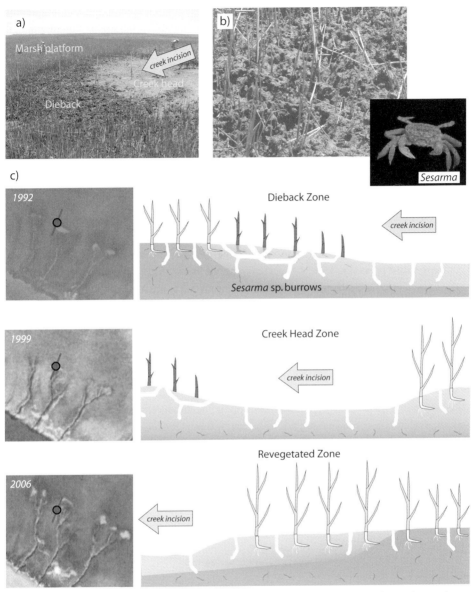

Figure 8.12 Bioturbation on salt marsh surfaces can facilitate tidal creek erosion and incision into the marsh, as seen here from a *Sesarma* bioturbated marsh in South Carolina (from Hughes et al. 2009; Wilson et al. 2012). (a) Ground view of a bioturbated creek head where dieback of *Spartina alterniflora* is observed. (b) Close up of dieback region and *Sesarma* sp. crabs. (c) Extensive burrowing near creek heads by *Sesarma* crabs results in *S. alterniflora* dieback, loss in above and belowground vegetation, and lowering of the salt marsh surface, facilitating creek headward erosion (modified from Wilson et al. 2012)

Throughout the discussions presented in this chapter, we stress that while it appears that individual species may operate on very small spatial scales within a salt marsh (<1 m^2 for some), the collective as a whole can have great impact on much larger-scale landscape evolution of salt marshes (>1 km^2). For example, recent work shows that increased erosion due to crab bioturbation in the Bahia Blanca Estuary has converted >600 ha original high marsh (*Sarcocornia perennis*, flooded about 40 times a year) into a middle-low unvegetated tidal flat (Pratolongo et al. 2013). In some areas of the low tidal flat, pioneer *S. alterniflora* plants are colonizing the region (expansion from 215 to 773 ha; Pratolongo et al. 2013). Shifting elevations and species dominance – exemplified here – has major implications for the biodiversity and functioning of the salt marsh, as highlighted throughout this chapter. Furthermore, ecogeomorphic interactions within salt marshes have feedbacks that can be very short- or long lived, stretching from seconds to geological timescales (see Table 8.1 for some examples).

As salt marsh areas are challenged to keep pace with rising sea-levels globally, understanding the relationships and feedbacks between biology, hydrology, and geology is becoming more pressing. This is especially true in those places where the sediment input to the coastal areas are reduced progressively. For instance, most marshes along the coast of Argentina are starved of sediment input due to reductions in water and sediment discharge from anthropogenic activity over the last few centuries (e.g., dam construction, irrigation, diversion; Melo et al. 2003, 2013; Kokot, 2004; Scordo et al. 2018). This is a common handicap plaguing salt marshes worldwide (Kesel et al. 1992; Kennish, 2001, and references therein; Yang et al. 2006; Syvitsky et al. 2009; Kirwan and Megonigal, 2013; Giosan et al. 2014; to name a few). Chapters 7 and 16 of this volume specifically addresses salt marsh response to sea-level rise and anthropogenic modification, respectively. Without sediment accretion (both organic and inorganic), the retention and stabilization capability of salt marshes are not enough to combat marine forces (wave and tidal currents) and keep pace with rising sea level. Though some authors report that marshes in North America and Europe are generally building at rates similar to or exceeding historical sea level rise (Kirwan et al. 2016), significant knowledge gaps remain and it is readily acknowledged more research quantifying the biophysical feedbacks, as highlighted here, is greatly needed.

Acknowledgements

Partial support for work dealing to this review was provided by grants to GMEP by CONICET, ANPCyT and Universidad Nacional del Sur.

References

Addino, M. S., Montemayor, D. I., Escapa, M., Alvarez, M. F., Valiñas, M. S., Lomovasky, B. J., and Iribarne, O. 2015. Effect of *Spartina alterniflora* Loisel, 1807 on growth of the stout razor clam *Tagelus plebeius* (Lightfoot, 1786) in a SW Atlantic estuary. *Journal of Experimental Marine Biology and Ecology*, 463: 135–142.

Alber, M., Swenson, E. M., Adamowicz, S. C., and Mendelssohn, I. 2008. Salt marsh dieback: An overview of recent events in the US. *Estuarine, Coastal and Shelf Science*, 80: 1–11.

Alberti, J., Escapa, M., Daleo, P., Iribarne, O., Silliman, B., and Bertness, M. 2007. Local and geographic variation in grazing intensity by herbivorous crabs in SW Atlantic salt marshes. *Marine Ecology Progress Series*, 349: 235–243.

Alberti, J., Escapa, M., Iribarne, O., Silliman, B., and Bertness, M. 2008. Crab herbivory regulates plant facilitative and competitive processes in Argentinean marshes. *Ecology*, 89: 155–164.

Allen, E., and Curran, H. A., 1974. Biogenic sedimentary structures produced by crabs in lagoon margin and salt marsh environments near Beaufort, North Carolina. *Journal of Sedimentary Research*, 44: 538–548.

Allen, J. R. L. 2000. Morphodynamics of Holocene salt marshes: A review sketch from the Atlantic and Southern North Sea coasts of Europe. *Quaternary Science Reviews*, 19: 1155–1231.

Altieri, A. H., Silliman, B. R., and Bertness, M. D. 2007. Hierarchical organization via a facilitation cascade in intertidal cordgrass bed communities. *The American Naturalist*, 169: 195–206

Anderson, M. E., Smith, J. M., and McKay, S. K. 2011. Wave dissipation by vegetation. USACOE Technical Report AD1003881.

Angelini, C., Griffin, J. N., Van de Koppel, J., Lamers, L. P. M., Smolders, A. J. P., Derksen-Hooijberg, M., van der Heide, T., and Silliman, B. R. 2016. A keystone mutualism underpins resilience of a coastal ecosystem to drought. *Nature Communications*, doi:10.1038/ncomms12473

Angelini, C., Heide, T., Griffin, J. N., Morton, J. P., Derksen-Hooijberg, M., Lamers, L. P. M., Smolders, A. J. P., and Silliman, B. R. 2015. Foundation species' overlap enhances biodiversity and multifunctionality from the patch to landscape scale in southeastern United States salt marshes. *Proceedings of the Royal Society of London B*, 282: 20150421.

Augustin, L. N., Irish, J. L., and Lynett, P. 2009. Laboratory and numerical studies of wave damping by emergent and near-emergent wetland vegetation. *Coastal Engineering*, 56: 332–340.

Austen, I., Andersen, T. J., and Edelvang, K. 1999. The influence of benthic diatoms and invertebrates on the erodibility of an intertidal mudflat, the Danish Wadden Sea. *Estuarine Coastal and Shelf Science*, 49: 99–111.

Bartholdy, J. 2012. Salt marsh sedimentation. In: R. A. Davis, and R. W. Dalrymple, eds, *Principals of Tidal Sedimentology*, pp. 151–185.

Bayne, B. L. 2017. Biology of oysters. *Developments in Aquaculture and Fisheries Science*. Vol. 41: 2–844.

Benner, R., Fogel, M. L., and Sprague, E. K. 1991. Diagenesis of belowground biomass of *Spartina alterniflora* in salt-marsh sediments. *Limnology and Oceanography*, 36: 1358–1374.

Bertness, M. D. 1984a. Habitat and community modification by an introduced herbivorous snail. *Ecology*, 65: 370–381.

Bertness, M. D. 1984b. Ribbed mussels and *Spartina alterniflora* production in a New England salt marsh. *Ecology*, 65: 1794–1807.

Bertness, M. D. 1985. Fiddler crab regulation of *Spartina alterniflora* production on a New England salt marsh. *Ecology*, 66: 1042–1055.

Bertness, M. D., Gough, L., and Shumway, S. 1992. Salt tolerances and the distribution of fugitive salt marsh plants. *Ecology*, 73: 1842–1851.

Bertness, M. D., Holdredge, C., and Altieri, A. H. 2009. Substrate mediates consumer control of salt marsh cordgrass on Cape Cod, New England. *Ecology*, 90: 2108–2117.

Bertness, M. D., and Miller, T. 1984. The distribution and dynamics of *Uca pugnax* (Smith) burrows in a New England salt marsh. *Journal of Experimental Marine Biology and Ecology*, 83: 211–237.

Bilkovic, D. M., Mitchell, M.M., Isdell, R. E., Schliep, M., and Smyth, A. R. 2017. Mutualism between ribbed mussels and cordgrass enhances salt marsh nitrogen removal. *Ecosphere*, 8: e01795. 10.1002/ecs2.1795

Blum, L. K., and Davey, E. 2013. Below the saltmarsh surface: visualization of plant roots by computer-aided tomography. *Oceanography*, 26: 85–87.

Boorman, L. A, Garbutt, A., and Barratt, D. 1998. The role of vegetation in determining patterns of the accretion of salt marsh sediment. In: K. S. Black, D. M. Paterson, and A. Cramp, eds, *Sedimentary Processes in the Intertidal Zone*. Geological Society (London), Special Publication No 139, pp. 389–399.

Botto, F., and Iribarne, O. 2000. Contrasting effects of two burrowing crabs (*Chasmagnathus granulata* and *Uca uruguayensis*) on sediment composition and transport in estuarine environments. *Estuarine, Coastal and Shelf Science*, 51: 141–151.

Botto, F., Iribarne, O., Gutierres, J., Bava, J., Gagliardini, A., and Valiela, I. 2006. Ecological importance of passive deposition of organic matter into burrows of the SW Atlantic crab *Chasmagnathus granulatus*. *Marine Ecology Progress Series*, 312: 201–210.

Boudreau, B. P., and Imboden, D. M. 1987. Mathematics of tracer mixing in sediments III: The theory of nonlocal mixing within sediments. *American Journal of Science*, 287: 693–719.

Bouma, T. J., Friedrichs, M., Van Wesenbeeck, B. K., Temmerman, S., Graf, G., and Herman, P. M. J. 2009. Density-dependent linkage of scale-dependent feedbacks : A flume study on the intertidal macrophyte *Spartina anglica*. *Oikos*, 118: 260–268.

Cadee, G. C. 2001. Sediment dynamics by bioturbating organisms. In: *Ecological Comparisons of Sedimentary Shores*, Springer-Verlag, Berlin, pp. 127–147.

Chapman, V. J. 1960. *Salt Marshes and Salt Deserts of the World*. Leonard Hill: London.

Chmura, G. L., and Kosters, E. C. 1994. Storm deposition and ^{137}Cs accumulation in fine-grained marsh sediments of the Mississippi Delta Plain. *Estuarine Coastal and Shelf Science*, 39: 33–44.

Christiansen, T., Wiberg, P. L., and Milligan, T. G. 2000. Flow and sediment transport on a tidal salt marsh surface. *Estuarine, Coastal and Shelf Science*, 50: 315–331.

Coverdale, T. C., Altieri, A. H. and Bertness, M. D. 2012. Belowground herbivory increases vulnerability of New England salt marshes to die-off. *Ecology*, 93: 2085–2094.

Collins, L. M., Collins, J. N., and Leopold, L. B. 1987. Geomorphic processes of an estuarine marsh: preliminary hypotheses. In: V. Gardiner, ed., *International Geomorphology, Part 1*, Wiley, Chichester, pp. 1049–1072.

Costanza, R., Pérez-Maqueo, O., Martinez, M. L., Sutton, P., Anderson, S. J., and Mulder, K. 2008. The value of coastal wetlands for hurricane protection. *AMBIO: A Journal of the Human Environment*, 37: 241–248.

Crawford, F. 2018. *Geomorphology of shell ridges and their effect on the stabilization of Biloxi marshes, east Louisiana*. MS thesis, University of New Orleans.

Cuadrado, D. G., Perillo, G. M. E. and Vitale, A. J. 2014. Modern microbial mats in siliciclastic tidal flats: Evolution, structure and the role of hydrodynamics. *Marine Geology*, 352: 367–380

Currin, C. A., Chappell, W. S., and Deaton, A. 2010. Developing alternative shoreline armoring strategies: The living shoreline approach in North Carolina. In: H. Shipman,

M. N. Dethier, G. Gelfenbaum, K. L. Fresh, and R. S. Dinicola, eds., *Puget Sound Shorelines and the Impacts of Armoring – Proceedings of a State of the Science Workshop*, May 2009, Reston, Virginia: U.S. Geological Survey, Scientific Investigations Report 2010-5254, pp. 91–102.

Daborn, G. R., Amos, C. L., Brylinsky, M., Christian, H., Drapeau, G, Faas, R. W., Grant, J., et al. 1993. An ecological cascade effect: migratory birds affect stability of intertidal sediments. *Limnology and Oceanography*, 38: 225–231.

Da Cunha Lana, P., and Guiss, C. 1992. Macrofauna-plant-biomass interactions in a euhaline salt marsh in Paranagua Bay (SE Brazil). *Marine Ecology Progress Series*, 80: 57–64.

Darby, F. A., and Turner, R. E. 2008a. Effects of eutrophication on salt marsh root and rhizome biomass accumulation. *Marine Ecology Progress Series*, 363: 63–70.

Darby, F. A., and Turner, R. E. 2008b. Below- and aboveground biomass of *Spartina alterniflora*: response to nutrient addition in a Louisiana Salt Marsh. *Estuaries and Coasts*, 31: 326–334.

Davenport, T. M., Seitz R. D., Knick,K. E., and Jackson, N. 2018. Living shorelines support nearshore benthic communities in Upper and Lower Chesapeake Bay. *Estuaries and Coasts*, 41: 197–206.

Davey, E., Wigand, C., Johnson, R., Sundberg, K., Morris, J., Roman, C. T., Davey, E. et al. 2011. Use of computed tomography imaging for quantifying coarse roots, rhizomes, peat, and particle densities in marsh soils. *Ecological Applications*, 21: 2156–2171.

Day, J. W., Jr, Kemp, G. P., Reed, D. J., Cahoon, D. R., Boumans, R. M., Suhayda, J. M., and R. Gambrell. 2011. Vegetation death and rapid loss of surface elevation in two contrasting Mississippi delta salt marshes: the role of sedimentation, autocompaction and sea level rise. *Ecological Engineering*, 37: 228–240.

Deegan, L. A., Johnson, D. S., Warren, R. S., Peterson, B. J., Fleeger, J. W., Fagherazzi, S., and Wollheim, W. 2012. Coastal eutrophication as a driver of salt marsh loss. *Nature*, 490: 388–394.

Delaune, R. D., Baumann, R. H., and Gosselink, J. G. 1983. Relationships among vertical accretion, coastal submergence, and erosion in a Louisiana Gulf Coast marsh. *Journal of Sedimentary Research*, 53: 147–157.

Delaune, R. D., Nyman, J. A., and Patrick, W. H. 1994. Peat collapse, ponding and wetland loss in a rapidly submerging coastal marsh. *Journal of Coastal Research*, 10: 1021–1030.

Delaune, R. D., Patrick, W. H., and Buresh, R. J. 1978. Sedimentation rates determined by ^{137}Cs dating in a rapidly accreting salt marsh. *Nature*, 275: 532–533.

Dionne, J.-C. 1985. Tidal marsh erosion by geese, St. Lawrence Estuary, Québec. *Géographie Physique et Quaternaire*, 39: 99–105.

Dupraz, C. and Visscher, P. T. 2005. Microbial lithification in marine stromatolites and hypersaline mats. *Trends in Microbiology*, 13: 429–438.

Dyer, K. R. 1988. Fine sediment particle transport in estuaries. In: J. Dronkers, and W. van Leussen, eds, *Physical Processes in Estuaries*. Springer, Berlin, Heidelberg, pp. 295–310.

Ellison, A. M., Bertness, M. D., and Miller, T. 1986. Seasonal patterns in the belowground biomass of *Spartina alterniflora* (Gramineae) across a tidal gradient. *American Journal of Botany*, 73: 1548–1554.

Escapa, M., Minkoff, D. R., Perillo, G. M. E., and Iribarne, O. 2007. Direct and indirect effects of burrowing crab *Chasmagnathus granulatus* activities on erosion of Southwest Atlantic *Sarcocornia*-dominated marshes. *Limnology and Oceanography*, 52: 2340–2349.

Escapa, M., Perillo, G. M. E., and Iribarne, O. 2008. Sediment dynamics modulated by burrowing crab activities in contrasting SW Atlantic intertidal habitats. *Estuarine, Coastal and Shelf Science*, 80: 365–373.

Escapa, C. M., Perillo, G. M. E., Iribarne, O. 2015. Biogeomorphically driven salt pan formation in *Sarcocornia*-dominated salt-marshes. *Geomorphology*, 228: 147–157.

Fagherazzi, S., FitzGerald, D. M., Fulweiler, R. W., Hughes, Z., Wiberg, P. L., McGlathery, K. J., Morris, J. T., Tolhurst, T. J., Deegan, L. A. and Johnson, D. S. 2013a. Ecogeomorphology of salt marshes. *Treatise on Geomorphology*, 12: 182–212.

Fagherazzi, S., Wiberg, P. L., Temmerman, S., Struyf, E., Zhao, Y., and Raymond, P. A. 2013b. Fluxes of water, sediments, and biogeochemical compounds in salt marshes. *Ecological Processes*, 2: 1–16.

Fanjul, E., Grela, M. A., and Iribarne, O. 2007. Effects of the dominant SW Atlantic intertidal burrowing crab *Chasmagnathus granulatus* on sediment chemistry and nutrient distribution. *Marine Ecology Progress Series*, 341: 177–190.

Farron, S. J. 2018. Morphodynamic responses of salt marshes to sea-level rise: upland expansion, drainage evolution, and biological feedbacks. PhD Dissertation, Boston University, 159 p.

FitzGerald, Duncan M., Fenster, M.S., Argow, B.A., and Buynevich, I.V. 2008. Coastal impacts due to sea-level rise. *Annual Review of Earth and Planetary Sciences*, 36: 601–647.

Ford, Mark A., and James B. Grace. 1998. Effects of vertebrate herbivores on soil processes, plant biomass, litter accumulation and soil elevation changes in a coastal marsh. *Journal of Ecology*, 86: 974–982.

French, J. R., and Spencer, T. 1993. Dynamics of sedimentation in a tide-dominated backbarrier salt marsh, Norfolk, UK. *Marine Geology*, 110: 315–331.

French, J. R., and Stoddart, D. R. 1992. Hydrodynamics of Salt Marsh Creek Systems: Implications for Marsh Morphological Development and Material Exchange. *Earth Surface Processes and Landforms* 17: 235–252.

Friedrichs, C. T., and Perry, J. E. 2001. Tidal salt marsh morphodynamics: A synthesis. *Journal of Coastal Research*, Special issue, no. 27: 7–37.

Gauthier, G., Giroux, J. F., Reed, Béchet, A., and Bélanger, L. 2005. Interactions between land use, habitat use, and population increase in Greater Snow Geese: What are the consequences for natural wetlands? *Global Change Biology*, 11: 856–868.

Gedan, K. B., Kirwan, M. L., Wolanski, E., Barbier, E. B., and Silliman, B. R. 2011. The present and future role of coastal wetland vegetation in protecting shorelines: answering recent challenges to the paradigm. *Climatic Change*, 106: 7–29.

Genoni, G. P. 1991. Increased burrowing by fiddler crabs *Uca rapax* (Smith) (Decapoda : Ocypodidae) in response to low food supply. *Journal of Experimental Marine Biology and Ecology*, 147: 267–285.

Giosan, L., Syvitski, J., Constantinescu, S., and Day, J. 2014. Climate change: protect the world's deltas. *Nature*, 516: 31–33.

Gleason, M. L., Elmer, D. A., Pien, N. C., Fisher, J. S. 1979. Effects of stem density upon sediment retention by salt marsh cord grass, *Spartina alterniflora* Loisel. *Estuaries*, 2: 271–273.

Grant, J., Bathmann, U. V., and Mills, E. L. 1986. The interaction between benthic diatom films and sediment transport. *Estuarine Coastal Shelf Science*, 23: 225–238.

Gribsholt, B., Kostka, J. E., and Kristensen E. 2003. Impact of fiddler crabs and plant roots on sediment biogeochemistry in a Georgia saltmarsh. *Marine Ecology Progress Series*, 259: 237–251.

Gross, M. F., Hardisky, M. A., Wolf, P. L., and Klemas, V. 1991. Relationship between aboveground and belowground biomass of *Spartina alterniflora* (Smooth Cordgrass). *Estuaries*, 14: 180–191.

Gutiérrez, J. L., and Iribarne, O. 1999. Role of Holocene beds of the stout razor clam Tagelus plebeius in structuring present benthic communities. *Marine Ecology Progress Series*, 185: 213–228.

Gutiérrez, J. L., Jones, C. G., Groffman, P. M., Findlay, S. E. G, Iribarne, O., Ribeiro, P. D., and Bruschetti, C. M. 2006. The contribution of crab burrow excavation to carbon availability in surficial salt marsh sediments. *Ecosystems*, 9: 647–658.

Gutiérrez, J. L., Jones, C. G., Strayer, D. L., and Iribarne, O. 2003. Mollusks as ecosystem engineers : The role of shell production in aquatic habitats. *Oikos*, 101: 79–90.

Hannaford, J., Pinn, E. H., and Diaz, A. 2006. The impact of sika deer grazing on the vegetation and infauna of Arne saltmarsh. *Marine Pollution Bulletin*, 53: 56–62.

Hatton, R. S., DeLaune, R. D., and Patrick Jr., W. H. 1983. Sedimentation, accretion, and subsidence in marshes of Barataria Basin, Louisiana. *Limnology and Oceanography*, 28: 494–502.

Hazelden, J., and Boorman, L. A. 2001. Soils and "managed retreat" in South East England. *Soil Use and Management*, 17: 150–154.

Holdredge, C., Bertness, M. D., and Altieri, A. H. 2008. Role of crab herbivory in die-off of new england salt marshes. *Conservation Biology*, 23: 672–679.

Holm, G. O. 2006. *Nutrient constraints on plant community production and organic matter accumulation of subtropical floating marshes*. PhD dissertation, Louisiana State University, Baton Rouge, Louisiana.

Hopkinson, C. S., Gosselink, J. G., and Parrondo, R. T. 1980. Production of coastal louisiana marsh plants calculated from phenometric techniques *Ecology*, 61: 1091–1098.

Howe, A. J., Rodriguez, J. F., and Saco, P. M. 2009. Surface evolution and carbon sequestration in disturbed and undisturbed wetland soils of the Hunter estuary, southeast Australia. *Estuarine Coastal Shelf Science*, 84: 75–83.

Howes, B. L., Goehringer, D. D., and Macey, J. W. H. 1986 Factors controlling the growth form of *Spartina alterniflora*: feedbacks between above-ground production, sediment oxidation, nitrogen and salinity. *Journal of Ecology*, 74: 881–898.

Howes, B. L., Howarth, R. W., Teal, J. M., and Valiela, I. 1981. Oxidation-reduction potentials in a saltmarsh: spatial patterns and interactions with primary production. *Limnology and Oceanography*, 26: 350–360.

Howes, N. C., FitzGerald, D. M., Hughes, Z. J., Georgiou, I. Y., Kulp, M. A., Miner, M. D., Smith, J. M., and Barras, J. A. 2010. Hurricane-induced failure of low salinity wetlands. *Proceedings of the National Academy of Sciences of the USA*, 107: 14014–14019.

Hu, K., Chen, Q., and Wang, H. 2015. A numerical study of vegetation impact on reducing storm surge by wetlands in a semi-enclosed estuary. *Coastal Engineering*, 95: 66–76.

Hughes, Z. J., FitzGerald, D. M., Wilson, C. A., Pennings, S. C. Wiçski, K., and Mahadevan, A. 2009. Rapid headward erosion of marsh creeks in response to relative sea level rise. *Geophysical Research Letters*, 36: 1–5.

Hyun, J., Smith, A. C., and Kostka, J. E. 2007. Relative contributions of sulfate- and iron (III) reduction to organic matter mineralization and process controls in contrasting habitats of the Georgia saltmarsh. *Applied Geochemistry*, 22: 2637–2651.

Iribarne, O., Bortolus, A., Botto, F. 1997. Between-habitat differences in burrow characteristics and trophic modes in the south western Atlantic burrowing crab *Chasmagnathus granulata*. *Marine Ecology Progress Series*, 155: 137–145.

Jaramillo, E., and Lunecke, K. 1988. The role of sediments in the distribution of *Uca pugilator* (Bosc) and *Uca pugnax* (Smith) (Crustacea, Brachyura) in a salt marsh at Cape Cod. *Meeresforschung*, 32: 46–52.

Jones, C. G., Lawton, J. H., and Shachak, M. 1994. Organisms as ecosystem engineers. *Oikos* 69: 373–386.

Jordan, T., and Valiela, I. 1982. A nitrogen budget of the ribbed mussel, *Geukensia demissa*, and its significance in nitrogen flow in a New England salt marsh. *Limnology and Oceanography*, 27: 75–90.

Julien, A. 2018. *Quantifying the demographics, habitat characteristics, and foundation species role of the ribbed mussel (Geukensia demissa) in South Carolina salt marshes*. Masters thesis, Dept of Biology, College of Charleston, South Carolina.

Katrak, G., Dittmann, S., and Seurant, L. 2008. Spatial variation in burrow morphology of the mud shore crab *Helograpsus haswellianus* (Brachyura, Grapsidae) in South Australian saltmarshes. *Marine and Freshwater Research*, 59, 902–911.

Katz, L. C. 1980. Effects of burrowing by the fiddler crab, *Uca pugnax* (Smith). *Estuarine Coastal and Marine Science*, 11: 233–237.

Kaye, C. A., and Barghoorn, E. S. 1964. Late Quaternary sea-level change and crustal rise at Boston Massachusetts, with notes on autocompaction of peat. *GSA Bulletin*, 75: 63–80.

Kennish, M. J. 2001. Coastal salt marsh systems in the US: a review of anthropogenic impacts. *Journal of Coastal Research*, 17: 731–748.

Kesel, R. H., Yodis, E. G., and McCraw, D. J. 1992. An approximation of the sediment budget of the Lower Mississippi River prior to major human modification. *Earth Surface Processes and Landforms*, 17: 711–722.

Keusenkothen, M. A. 2002. *The effects of deer trampling in a salt marsh*. MS thesis, East Carolina University.

Keusenkothen, M. A., and Christian, R. R. 2004. Responses of salt marshes to disturbance in an ecogeomorphological context, with a case study of trampling by deer. In: S. Fagherazzi, M. Marani, and L. Blum, eds., *The Ecogeomorphology of Tidal Marshes*, Volume 59. John Wiley, Hoboken, NJ, pp. 203–230.

King, G. M., Klug, M. J., Wiegert, R. G., and Chalmers, A. G. 1982. Relation of soil water movement and sulfide concentration of *Spartina alterniflora* production in a Georgia Salt Marsh. *Science*, 218: 61–63.

Kirchner, J. W., Dietrich, W. E., Iseya, F., and Ikeda, H. 1990. The variability of critical shear stress, friction angle, and grain protrusion in water-worked sediments. *Sedimentology*, 37: 647–672.

Kirwan, M. L., and Guntenspergen, G. R. 2010. The influence of tidal range on the stability of coastal marshland. *Journal of Geophysical Research*, 115: F02009, doi:10.1029/2009JF001400.

Kirwan, M. L., Guntenspergen, G. R., D'Alpaos, A., Morris, J., Mudd, S. M., and Temmerman, S. 2010. Limits on the adaptability of coastal marshes to rising sea level. *Geophysical Research Letters*, 37(23). https://doi.org/10.1029/2010GL045489

Kirwan, M. L., and Megonigal, J. P. 2013. Tidal wetland stability in the face of human impacts and sea-level rise. *Nature*, 504: 53–60.

Kirwan, M. L., and Temmerman, S. 2009. Coastal marsh response to historical and future sea-level acceleration. *Quaternary Science Reviews*, 28: 1801–1808.

Kirwan, M. L., Temmerman, S., Skeehan, E. E., Guntenspergen, G. R., and Fagherazzi, S. 2016. Overestimation of marsh vulnerability to sea level rise. *Nature Climate Change*, 6: 253–260.

Knutson, P. L., Brochu, R. A., and See, W. N. 1982. Wave damping in *Spartina alterniflora* marshes. *Wetlands*, 2: 87–104.

Kobayashi, N., Raichle, A., and Asano, T. 1993. Wave attenuation by vegetation. *Journal of Waterway, Port, Coastal, and Ocean Engineering Technical Report*, 119(1). https://doi.org/10.1061/(ASCE)0733-950X(1993)119:1(30)

Koch, E. W., Barbier, E. D., Silliman, B. R., Reed, D. J., Perillo, G. M. E., Hacker, S. D., Granek, E. F., et al. 2009. Non-linearity in ecosystem services: temporal and spatial variability in coastal protection. *Frontiers in Ecology and the Environment*, 7: 29–37.

Kokot, R. R. 2004. Erosión en la costa patagónica por cambio climático. *Revista de la Asociación Geológica Argentina*, 59: 715–726.

Koo, B. J., Kwon, K. K., and Hyun, J. H. 2005. The sediment-water interface increment due to the complex burrows of macrofauna in a tidal flat. *Ocean Science Journal*, 40: 221–227.

Kostka, J. E., Gribsholt, B., Petrie, E., Dalton, D., Skelton, H., and Kristensen, E. 2002. The rates and pathways of carbon oxidation in bioturbated saltmarsh sediments. *Limnology and Oceanography*, 47: 230–240.

Kraeuter, J. N. 1976. Biodeposition by salt-marsh invertebrates. *Marine Biology*, 35: 215–223.

Laegdsgaard, P. 2006. Ecology, disturbance and restoration of coastal saltmarsh in Australia: A Review. *Wetlands Ecology and Management*, 14: 379–399.

Langley, J. A., McKee, K. L., Cahoon, D. R., Cherry, J. A. and Megonigal, P. 2009. Elevated CO2 stimulates marsh elevation gain, counterbalancing sea-level rise. *Proceedings of the National Academy of Sciences of the USA*, 106: 6182–6186.

Lanuru, M. 2008. Measuring critical erosion shear stress of intertidal sediments with EROMES erosion device. *Torani Journal of Marine Science and Fisheries*, 18: 390–397.

Leeder, M. R. 1998. Lyell's Principles of Geology: Foundations of sedimentology. *Geological Society of London, Special Publication*, 143: 95–110.

Leonard, L. A., and Luther, M. E. 1995. Flow hydrodynamics in tidal marsh cano-pies. *Limnology and Oceanography* 18: 1474–1484.

Letzsch, W. S., and Frey, R. W. 1980. Deposition and erosion in a Holocene salt marsh, Sapelo Island, Georgia. *Journal of Sedimentary Research*, 50: 529–542.

Li, H., and Yang, S. L. 2009. Trapping effect of tidal marsh vegetation on suspended sediment, Yangtze Delta. *Journal of Coastal Research*, 25: 915–924.

Lightbody, A. F., and Nepf, H. M. 2006. Prediction of velocity profiles and longitudinal dispersion in salt marsh vegetation. *Limnology and Oceanography*, 51: 218–228.

Lynch, J. J., O'Neil, E., and Lay, D. W. 1947. Management significance of damage by geese and muskrats to Gulf Coast marshes. *Journal of Wildlife Management*, 11: 50–76.

Madsen, K. N., Nilsson, P., and Sunback, K. 1993. The influence of benthic microalgae on the stability of a subtidal sediment. *Journal of Experimental Marine Biology and Ecology*, 170: 159–177.

Madsen, J. D., Chambers, P. A., James, W. F., Koch, E. W., and Westlake, D. F. 2001. The interaction between water movement, sediment dynamics and submersed macrophytes. *Hydrobiologia*, 444: 71–84.

Marani, M., D'Alpaos, A., Lanzoni, S., Carniello, L., and Rinaldo, A. 2007. Biologically-controlled multiple equilibria of tidal landforms and the fate of the Venice lagoon. *Geophysical Research Letters*, 34: 1–5.

Mariotti, G., and Carr, J. 2014. Dual role of salt marsh retreat: Long-term loss and short-term resilience. *Water Resources Research*, 50: 2963–2974.

Mariotti, G., and Fagherazzi, S. 2010. A numerical model for the coupled long-term evolution of salt marshes and tidal flats. *Journal of Geophysical Research Earth Surface*, 115 (F1). https://doi.org/10.1029/2009JF00132.

Mariotti, G., and Fagherazzi, S. 2013. A two-point dynamic model for the coupled evolution of channels and tidal flats. *Journal of Geophysical Research*, 118: 1387–1399.

Mariotti, G., Kearney, W., and Fagherazzi, S. 2016. Soil creep in salt marshes. *Geology*, 44: 459–462.

McCraith, B. J., Gardner, L. R., Wethey, D. S., and Moore, W. S. 2003. The effect of fiddler crab burrowing on sediment mixing and radionuclide profiles along a topographic gradient in a southeastern salt marsh. *Journal of Marine Research*, 61: 359–390.

Mehta, A. J. 1996. Interaction between fluid mud and water waves. In: V. P., Singh, and W. H. Hager, *Environmental Hydraulics*. Kluwer, Dordretcht, pp. 153–187.

Melo, W. D., Perillo, G. M. E., Perillo, M. M., Schilizzi, R., and Piccolo, M. C. 2013. Late Pleistocene-Holocene deltas in the southern Buenos Aires Province, Argentina. In: G. Young, and G. M. E. Perillo, eds., *Deltas: Landforms, Ecosystems and Human Activities*. IAHS Press, Wallingford, UK. 358: 187–195.

Melo, W. D., Schillizzi, R., Perillo, G. M. E. y Piccolo, M. C. 2003. Influencia del área continental pampeana sobre el origen y la morfología del estuario de Bahía Blanca. *Revista de la Asociación Argentina de Sedimentología*, 10: 65–72.

Mendelssohn, I. A., McKee, K. L., and Patrick Jr., W. H. 1981. Oxygen deficiency in *Spartina alterniflora* roots: metabolic adaptation to anoxia. *Science*, 214: 439–441.

Mendelssohn, I. A., and Seneca, E. D. 1980. The influence of soil drainage on the growth of salt marsh cordgrass *Spartina alterniflora* in North Carolina. *Estuarine and Coastal Marine Science*, 11: 27–40.

Meyer, D. L., Townsend, E. C., and Thayer, G. W. 1997. Stabilization and erosion control value of oyster cultch for intertidal marsh. *Restoration Ecology*, 5: 93–99.

Millette, T., Argow, B., Marcano, E., Hayward, C., Hopkinson, C., and Valentine, V. 2010. Salt marsh geomorphological analyses via integration of multitemporal multispectralremote sensing with LIDAR and GIS. *Journal of Coastal Research*, 26: 809–816.

Minkoff, D. R., Escapa, M., Ferramola, F. E., Maraschín, S. D., Pierini, J. O., Perillo, G. M. E., and Delrieux, C. 2006. Effects of crab–halophytic plant interactions on creek growth in a S. W. Atlantic salt marsh: a cellular automata model. *Estuarine, Coastal and Shelf Science*, 69: 403–413.

Minkoff, D. R., Escapa, C. M., Ferramola, F. E., and Perillo, G. M. E. 2005. Erosive processes due to physical – biological interactions based in a cellular automata model. *Latin American Journal of Sedimentology and Basin Analysis*, 12: 25–34.

Mitsch, W. J., and Gosselink, J. G. 2000. *Wetlands*. 3rd edition. John Wiley, New York.

Molina, L. M., Valiñas, M. S., Pratolongo, P., Elias, R., and Perillo, G. M. E. 2017. Effect of *Micropogonias furnieri* on the stability of the sediment of salt marshes – an issue to be resolved. *Estuaries and Coasts*, 40: 1795–1807.

Moller, I., Kudella, M., Rupprecht, F., Spencer, T., Paul, M., van Wesenbeeck, B. K., Wolters, G., et al. 2014. Wave attenuation over coastal salt marshes under storm surge conditions. *Nature Geoscience*, 7: 727–731.

Moller, I., and Spencer, T. 2002. Wave dissipation over macro-tidal saltmarshes: Effects of marsh edge typology and vegetation change. *Journal of Coastal Research*: Special Issue 36 – International Coastal Symposium (ICS 2002): 506–521.

Montague, C. L. 1980. A natural history of temperate Western Atlantic fiddler crabs (Genus *Uca*) with reference to their impact on the salt marsh. *Contributions in Marine Science*, 23: 25–55.

Montague, C. L. 1982. The influence of fiddler crab burrowing on metabolic processes in saltmarsh sediments. In: V. S. Kennedy, ed., *Estuarine Comparisons*. Academic Press, San Francisco, pp. 283–301.

Montgomery, D. R., and Dietrich, W. E. 1988. Where do channels begin? *Nature*, 336: 232–234.

Morris, J. M., Sundareshwar, P. V., Nietch, C. T., Kjerfve, B., and Cahoon, D. R. 2002. Responses of coastal wetlands to rising sea level. *Ecology*, 83: 2869–2877.

Mouton, E C., and Felder, D. L. 1996. Burrow distributions and population estimates for the fiddler crabs *Uca spinicarpa* and *Uca longisignalis* in a Gulf of Mexico salt marsh. *Estuaries and Coasts*, 19: 51–61.

Murray, J. M. H., Meadows, A., and Meadows, P. S. 2002. Biogeomorphological implications of microscale interactions between sediment geotechnics and marine benthos: A review. *Geomorphology*, 47: 15–30.

Nelson, J., Wilson, R., Coleman, F., Koenig, C., DeVries, D., Gardner, C., and Chanton, J. 2012. Flux by fin: fish-mediated carbon and nutrient flux in the northeastern Gulf of Mexico. *Marine Biology*, 159: 365–372.

Nepf, H. M. 1999. Drag, turbulence, and diffusion in flow through emergent vegetation. *Water Resources Research*, 35: 479–489.

Neumeier, U., and Amos, C. L. 2006. The influence of vegetation on turbulence and flow velocities in European salt-marshes. *Sedimentology*, 53: 259–277.

Neumeier, U., and Ciavola, P. 2004. Flow resistance and associated sedimentary processes in a *Spartina maritima* salt-marsh. *Journal of Coastal Research*, 20: 435–447.

Niering, W., and Warren, R. S. 1980. Vegetation patterns and processes in New England Salt Marshes. *BioScience*, 30: 301–307.

NOAA. 2018. Green Infrastructure Effectiveness Database: https://coast.noaa.gov/digital coast/training/gi-database.html

Nyman, J. A., Crozier, C., and DeLaune, R. D. 1995a. Roles and patterns of hurricane sedimentation in an estuarine marsh landscape. *Estuaries, Coastal and Shelf Science*, 40: 665–679.

Nyman, J. A., Delaune, R. D., Patrick Jr., W. H. 1990. Wetland soil formation in the rapidly subsiding Mississippi River Deltaic Plain: mineral and organic matter relationships. *Estuarine, Coastal and Shelf Science*, 31: 57–69.

Nyman, J. A., DeLaune, R. D., Pezeshki, S. R., and Patrick Jr., W. H. 1995b. Organic matter cycling and marsh stability in a rapidly submerging estuarine marsh. *Estuaries*, 18: 207–218.

Nyman, J. A., Walters, R., Delaune, R. D., Patrick Jr., W. H. 2006. Marsh vertical accretion via vegetative growth. *Estuaries Coastal and Shelf Science*, 69: 370–380.

O'Donnell, J. E. D. 2017. Living shorelines: A review of literature relevant to New England coasts. *Journal of Coastal Research*, 332: 435–451.

Odum, W. 1988. Comparative ecology of tidal freshwater and salt marshes. *Annual Review of Ecology and Systematics*, 19: 147–176.

Onorevole, K. M., Thompson, S. P., and Piehler, M. F. 2018. Living shorelines enhance nitrogen removal capacity over time. *Ecological Engineering*, 120: 238–248.

Pan, J., Bournod, C. N., Pizani, N. V., Cuadrado, D. G., and Carmona, N. B. 2013. Characterization of microbial mats from a siliciclastic tidal flat (Bahía Blanca Estuary, Argentina). *Geomicrobiology Journal*, 30: 665–674.

Pennings, S. C., and Callaway, R. 1992. Salt marsh plant zonation: the relative importance of competition and physical factors. *Ecology*, 73: 681–690.

Pennings, S. C., Carefoot, T. H., Siska, E. L., Chase, M.G., and Page, T. A. 1998. Feeding preferences of a generalist salt-marsh crab: relative importance of multiple plant traits. *Ecology*, 79: 1968–1979.

Pennings, S. C., and Silliman, B. R. 2005. Linking biogeography and community ecology : Latitudinal variation in plant-herbivore interaction strength. *Ecology*, 86: 2310–2319.

Perillo, G. M. E. 2019. Geomorphology of tidal courses and depressions. In: G. M. E. Perillo, E. Wolanski, D. R. Cahoon, and C. Hopkinson, eds., *Coastal Wetlands: An Integrated Ecosystem Approach*. Elsevier, Amsterdam, pp. 185–210.

Perillo, G. M. E., Drapeau, G., Piccolo, M. C., and Chaouq, N. 1993. Tidal circulation pattern on a tidal flat, Minas Basin, Canada. *Marine Geology*, 112: 219–236

Perillo, G. M. E., and Iribarne, O. 2003a. New mechanisms studied for creek formation in tidal flats: from crabs to tidal channels. *EOS American Geophysical Union Transactions*, 84: 1–5.

Perillo, G. M. E., and Iribarne, O. 2003b. Processes of tidal channels develop in salt and freshwater marshes. *Earth Surface Processes and Landforms*, 28: 1473–1482.

Perillo, G. M. E., Minkoff, D. R., and Piccolo, M. C. 2005. Novel mechanism of stream formation in coastal wetlands by crab–fish–groundwater interaction. *Geo-Marine Letters*, 25: 214–220.

Pestrong, R. 1965. The development of drainage patterns on tidal marshes. Stanford University Publications. *Earth Science*, 10(2): 1–87.

Pestrong, R. 1972. Tidal-flat sedimentation at cooley landing, Southwest San Francisco bay. *Sedimentary Geology*, 8: 251–288.

Peterson, G. W., and Turner, R. E. 1994. The value of salt marsh edge vs interior as a habitat for fish and decapod crustaceans in a Louisiana tidal marsh. *Estuaries*, 17: 235–262.

Pethick, J. S. 1980. Velocity surges and asymmetry in tidal channels. *Estuarine Coastal and Marine Science*, 11: 331–345.

Pethick, J. S. 1992. Saltmarsh geomorphology. In: J. R. L. Allen, and K. Pye, eds, *Saltmarshes: Morphodynamics, Conservation and Engineering Significance*, Cambridge University Press, Cambridge, UK. pp. 41–62.

Piazza, B. P., Banks, P. D., and La Peyre, M. K. 2005. The potential for created oyster shell reefs as a sustainable shoreline protection strategy in Louisiana. *Restoration Ecology*, 13: 499–506.

Postma, H. 1961. Transport and accumulation of suspended matter in the Dutch Wadden Sea. *Netherlands Journal of Sea Research*, 1: 148–180.

Pratolongo, P. D., Mazzon, C., Zapperi, G., Piovan, M. J., and Brinson, M. M. 2013. Land cover changes in tidal salt marshes of the Bahía Blanca estuary (Argentina) during the past 40 years. *Estuarine, Coastal and Shelf Science*, 133: 23–31.

Pratolongo, P. D., Perillo, G. M. E., and Piccolo, M. C. 2010. Combined effects of waves and marsh plants on mud deposition events at a mudflat-saltmarsh edge. *Estuarine, Coastal and Shelf Sciences*, 87: 207–212.

Priestas, A. M., and Fagherazzi, S. 2011. Morphology and hydrodynamics of wave-cut gullies. *Geomorphology*, 131: 1–13.

Pye, K., and French, P. W. 1993. Erosion and Accretion Processes on British Saltmarshes. Volume One. Introduction: Saltmarsh Processes and Morphology. Report No. ES19. Ministry of Agriculture, Fisheries and Food. Cambridge Environmental Research Consultants, Cambridge.

Reddy, K. R., and Delaune, R. D. 2008. *Biogeochemistry of Wetlands: Science and Applications*. CRC Press, Boca Raton, FL.

Redfield, A. C. 1965. Ontogeny of a saltmarsh estuary. *Science*, 147: 50–55.

Redfield, A. C. 1972. Development of a New England Salt Marsh. *Ecological Monographs*, 42: 201–237.

Redfield, A. C., and Rubin M. 1962. The age of salt marsh peat and its relation to recent changes in sea level at Barnstable, Massachusetts. *Proceeding of the National Academy of Science of the United States of America*, 48: 1728–1735.

Reed, D. J. 1988. Sediment dynamics and deposition in a retreating coastal salt marsh. *Estuarine, Coastal and Shelf Science*, 26: 67–69.

Reed, D. J., Spencer, T., Murray, A., French, J. R., and Leonard, L. 1999. Marsh surface sediment deposition and the role of tidal creeks: implications for created and managed coastal marshes. *Journal of Coastal Conservation*, 5: 81–90.

Rice, D. L. 1986. Early diagenesis in bioadvective sediments: Relationships between the diagenesis of beryllium-7, sediment reworking rates, and the abundance of conveyor-belt deposit-feeders. *Journal of Marine Research*, 44: 149–184.

Ringold, P. 1979. Burrowing, root mat density, and the distribution of fiddler crabs in the Eastern United States. *Journal of Experimental Marine Biology and Ecology*, 36: 11–21.

Robertson, T. L., and Weis, J. S. 2005. A comparison of epifaunal communities associated with the stems of salt marsh grasses *Phragmites australis* and *Spartina alterniflora*. *Wetlands*, 25: 1–7.

Roman, C. T., Peck, J. A., Allen, J. R., King, J. W., and Appleby, P. G. 1997. Accretion of a New England (U.S.A.) salt marsh in response to inlet migration, storms, and sea-level rise. *Estuarine Coastal and Shelf Science*, 46: 717–727.

Schwimmer, R. 2001, Rates and processes of marsh shoreline erosion in Rehoboth Bay, Delaware, U.S.A. *Journal of Coastal Research*, 17: 672–683.

Schwimmer, R., and Pizzuto, J. 2000. A model for the evolution of marsh shorelines. *Journal of Sedimentary Research*, 70: 1026–1035.

Scordo, F., Bohn, V., Piccolo, M. C., and Perillo, G. M. 2018. Mapping and monitoring lakes intra-annual variability in semi-arid regions: a case study in Patagonian Plains (Argentina). *Water*, 10: 889.

Scyphers, S. B., Powers, S. P., Heck, K. L., and Byron, D. 2011. Oyster reefs as natural breakwaters mitigate shoreline loss and facilitate fisheries. *PLOS ONE*. 6(8). doi:10.1371/journal.pone.0022396.

Shaffer, G. P., Sasser, C. E., Gosselink, J. G., and Rejmanek, M. 1992. Vegetation dynamics in the emerging Atchafalaya Delta, Louisiana, USA. *Journal of Ecology*, 80: 677–687.

Sharma, P., Gardner L. R., Moore, W. S., and Bollinger, M. S. 1987. Sedimentation and bioturbation in a salt marsh as revealed by 210Pb, 137Cs, and 7Be studies. *Limnology and Oceanography*, 32: 313–326.

Shepard, C. C., Crain, C. M., and Beck, M. W. 2011. The protective role of coastal marshes: a systematic review and meta-analysis. *PLOS ONE*, 6(11): e27374. https://doi.org/10.1371/journal.pone.0027374

Shi, B. W., Yang, S. L., Wang, Y. P., Bouma, T. J., and Zhu, Q. 2012. Relating accretion and erosion at an exposed tidal wetland to the bottom shear stress of combined current-wave action. *Geomorphology*, 138: 380–389.

Silliman, B. R., Van De Koppel, J., Bertness, M. D., Stanton, L. E., and Mendelssohn, I. A. 2005. Drought, snails, and large-scale die-off of southern U.S. salt marshes. *Science*, 310: 1803–1807.

Silliman, B. R., and Zieman, J. 2001. Top-down control of *Spartina alterniflora* production by periwinkle grazing in a Virginia salt marsh. *Ecology*, 82: 2830–2845.

Slatyer, R. A., Fok, E. S. Y., Hocking, R., and Backwell, P. R.Y. 2008. Why do fiddler crabs build chimneys? *Biology Letters of the Royal Society*, 4: 616–618.

Smith, S. M. 2009. Multi-decadal changes in salt marshes of Cape Cod, Massachusetts: a photographic analysis of vegetation loss, species shifts, and geomorphic change. *Northeastern Naturalist*, 16: 183–208.

Smith, J. E., Bentley, S. J., Snedden, G. A., and White, C. 2015. What role do hurricanes play in sediment delivery to subsiding river deltas? *Scientific Reports*, 5, Article number: 17582.

Smith, J. M., and Frey, R. W. 1985. Biodeposition by the ribbed mussel *Geukensia demissa* in a salt marsh, Sapelo Island, Georgia. *Journal of Sedimentary Research*, 55: 817–828.

Soudry, D. 2000. Microbial phosphate sediment. In: R. E. Riding, and S. M. Awramik, eds., *Microbial Sediments*. Springer-Verlag, Berlin, pp. 127–136.

Spalding, M. D., Ruffo, S., Lacambra, C., Meliane, I., Zeitlin Hale, L., Shepard, C. C., and Beck. M. W. 2014. The role of ecosystems in coastal protection: Adapting to climate change and coastal hazards. *Ocean and Coastal Management*, 90: 50–57.

Steel, T. J., and Pye, K. 1997. The development of saltmarsh tidal creek networks: Evidence from the U.K. Proceedings of the Canadian Coastal Conference, pp. 267–280.

Stevenson, J. C., Ward, L. G., and Kearney, M. S. 1986. Vertical accretion in marshes with varying rates of sea-level rise. In: D. A. Wolfe, ed., *Estuarine Variability*, Academic Press, New York, pp. 241–259.

Stolz, J. F. 2000. Structure of microbial mats and biofilms. In: R. E. Riding, and S. M. Awramik, eds., *Microbial Sediments*. Springer-Verlag, Berlin, pp. 1–8.

Stumpf, R. P. 1983. The process of sedimentation on the surface of a salt marsh. *Estuarine, Coastal and Shelf Science*, 17: 495–508.

Syvitsky, J., Kettner, A. J., Overeem, I., Hutton, E. W., Hannon, M. T., Brakenridge G. R., Day, J., et al. 2009. Sinking deltas due to human activities. *Nature Geoscience*, 2: 681–686.

Takeda, S., and Kurihara, Y. 1987. The effects of burrowing of *Helice tridens* (De Haan) on the soil of a salt-marsh habitat. *Journal of Experimental Marine Biology and Ecology*, 113: 79–89.

Taylor, K. L., and Grace, J. B. 1995. The effects of vertebrate herbivory on plant community structure in the coastal marshes of the Pearl River, Louisiana, USA. *Wetlands*, 15: 68–73.

Teal, J. M. 1958. Distribution of fiddler crabs in Georgia salt marshes. *Ecology*, 39: 186–193.

Temmerman, S., Bouma, T. J., Govers, G., Wang, Z. B., de Vries, M. B., and Herman, P. M. J. 2005. Impact of vegetation on flow routing and sedimentation patterns: Three dimensional modelling for a tidal marsh. *Journal of Geophysical Research*, 110: F04019, doi: 10.1029/2005JF000301.

Temmerman, S., Bouma, T. J., Van de Koppel, J., Van der Wal, De Vries, D. M. B., and Herman, P. M. J. 2007. Vegetation causes channel erosion in a tidal landscape. *Geology*, 35: 631–634.

Temmerman, S., Govers, G., Meire P., and Wartel, S. 2003. Modelling long-term tidal marsh growth under changing tidal conditions and suspended sediment concentrations, Scheldt Estuary, Belgium. *Marine Geology*, 193: 151–169.

Temmerman, S., Govers, G., Wartel, S. and Meire, P. 2004. Modelling estuarine variations in tidal marsh sedimentation: response to changing sea levels and suspended sediment concentrations. *Marine Geology*, 212: 1–19.

Thomas, C. R., and Blum, L. K. 2010. Importance of the fiddler crab *Uca pugnax* to salt marsh soil organic matter accumulation. *Marine Ecology Progress Series*, 414: 167–177.

Tolhurst, T. J., Black, K. S., Shayler, S. A., Mather, S., Black, I., Baker, K., and Paterson, D. M. 1999. Measuring the in Situ erosion shear stress of intertidal sediments with the cohesive strength meter (CSM). *Estuarine, Coastal and Shelf Science*, 49: 281–294.

Turner, R. E. 2010. Doubt and the values of an ignorance-based world view for wetland restoration: Coastal Louisiana. *Estuaries and Coasts*, 32: 1054–1068.

Turner, R. E. 2011. Beneath the salt marsh canopy: loss of soil strength with increasing nutrient loads. *Estuaries and Coasts*, 34: 1084–1093.

Turner, R. E., Swenson, E. M., and Milan, C. S. 2002. Organic and inorganic contributions to vertical accretion in salt marsh sediments. In: M. P. Weinstein, and D. A. Kreeger, eds., *Concepts and Controversies in Tidal Marsh Ecology.* Springer, Dordrecht, pp. 583–595.

Underwood, G. J. C, Paterson, D. M., and Parkes, R. J. 1995. The measurement of microbial carbohydrate exopolymers from intertidal sediments. *Limnology and Oceanography*, 40: 1243–1453.

Valentine, K., and Mariotti, G. 2019. Wind-driven water level fluctuations drive marsh edge erosion variability in microtidal coastal bays. *Continental Shelf Research*, 176: 76–89.

Valiela, I., Teal, J. M., and Persson, N. Y. 1976. Production and dynamics of experimentally enriched salt marsh vegetation: belowground biomass. *Limnology and Oceanography*, 21: 245–252.

van Asselen, S., Stouthamer, E., and van Asch, Th. W. J. 2009. Effects of peat compaction on delta evolution: a review on processes, responses, measuring and modeling. *Earth-Science Reviews*, 92: 35–51.

van Eerdt, M. 1986. The influence of basic soil and vegetation parameters on salt marsh cliff strength. In: V. Gardiner, ed., *International Geomorphology, Part 1*, Wiley, Chichester, pp. 1073–1086.

van Proosdij, D., Davidson-Arnott, R. G. D., and Ollerhead, J. 2006. Controls on spatial patterns of sediment deposition across a macro-tidal salt marsh surface over single tidal cycles. *Estuarine, Coastal and Shelf Science*, 69: 64–86

van Rijn, L. C., van Rossum, H., and Termes, P. 1990. Field verification of 2–D and 3–D suspended-sediment models. *Journal of Hydraulic Engineering*, 116: 1270–1288.

van Wieren, S. E., and Bakker, J. P. 2008. The impact of browsing and grazing herbivores on biodiversity. In: I. J. Gordon, and H. H. T. Prins, eds. *The Ecology of Browsing and Grazing*. Springer, Berlin, pp. 236–292.

Vandenbruwaene, W., Meire, P., and Temmerman, S. 2012. Formation and evolution of a tidal channel network within a constructed tidal marsh. *Geomorphology*, 151–152: 114–125.

Vu, H. D., and Pennings, S. C. 2018. Predators mediate above- vs. belowground herbivory in a salt marsh crab. *Ecosphere*, 9(2). doi:10.1002/ecs2.2107.

Vu, H. D., Wieski, K., and Pennings, S. C. 2017. Ecosystem engineers drive creek formation in salt marshes. *Ecology*, 98: 162–174.

Wang, J. Q., Zhang, X. D., Jiang, L. F., Bertness, M. D., Fang, C. M., Chen, J. K., Hara, T., and Li, B. 2010. Bioturbation of burrowing crabs promotes sediment turnover and carbon and nitrogen movements in an estuarine salt marsh. *Ecosystems*, 13: 586–599.

Wang, J. Q., Zhang, Nie, M., Fu, C. Z., Chen, J. K., and Li, B. 2008. Exotic *Spartina alterniflora* provides compatible habitats for native estuarine crab *Sesarma dehaani* in the Yangtze River Estuary. *Ecological Engineering*, 34: 57–64.

Wang, M., Gao, X., and Wang, W. 2014. Differences in burrow morphology of crabs between *Spartina alterniflora* marsh and mangrove habitats. *Ecological Engineering*, 69: 213–219.

Wang, Y. P., Zhang, R., and Gao, S. 1999. Velocity variations in salt marsh creeks, Jiangsu, China. *Journal of Coastal Research*, 15: 471–477.

Watts, C. W., Tolhurst, T. J., Black, K. S., and Whitmore, A. P. 2003. In Situ measurements of erosion shear stress and geotechnical shear strength of the intertidal sediments of the experimental managed realignment scheme at Tollesbury, Essex, UK. *Estuarine, Coastal and Shelf Science*, 58: 611–20.

Weissburg, M. 1992. Functional analysis of fiddler crab foraging: sex-specific mechanics and constraints in *Uca pugnax* (Smith). *Journal of Experimental Marine Biology and Ecology*, 156: 105–124.

West, J. M., and Zedler, J. B. 2000. Marsh-creek connectivity: fish use of a tidal salt marsh in Southern California. *Estuaries*, 23: 699–710.

Widdows, J., and Brinsley, M. 2002. Impact of biotic and abiotic processes on sediment dynamics and the consequences to the structure and functioning of the intertidal zone. *Journal of Sea Research*, 48: 143–156.

Widdows, J., Brinsley, M. D., Bowley, N., and Barrett, C. 1998. A benthic annular flume for in situ measurement of suspension feeding/biodeposition rates and erosion potential of intertidal cohesive sediments. *Estuarine Coastal and Shelf Science*, 46: 27–38.

Widdows, J., Pope, N., and Brinsley, M. 2008. Effect of *Spartina anglica* stems on nearbed hydrodynamics, sediment erodability and morphological changes on an intertidal mudflat. *Marine Ecology Progress Series*, 362: 45–57.

Wigand, C., Brennan, P., Stolt, M., Holt, M., and Ryba, S. 2009. Soil respiration rates in coastal marshes subject to increasing watershed nitrogen loads in southern New England, US. *Wetlands*, 29: 952–963.

Wilson, K., Kelley, J., Croitoru, A., Dionne, M., Belknap, D., and Steneck, R. 2009. Stratigraphic and ecophysical characterizations of salt pools: dynamic landforms of the Webhannet Salt Marsh, Wells, ME, USA. *Estuaries and Coasts*, 32: 855–870.

Wilson, K., Kelley, J. T., Tanner, B. R., and Belknap, D. F. 2010. Probing the origins and stratigraphic signature of salt pools from north-temperate marshes in Maine, U.S.A. *Journal of Coastal Research*, 26: 1007–1026.

Wilson, C., and Allison, M. 2008. An equilibrium profile model for retreating marsh shorelines in southeast Louisiana. *Estuarine, Coastal and Shelf Science*, 80, 483–494.

Wilson, C. A., Hughes, Z. J., and FitzGerald, D. M. 2012. The effects of crab bioturbation on mid-Atlantic saltmarsh tidal creek extension: geotechnical and geochemical changes. *Estuarine, Coastal and Shelf Science*, 106: 33–44.

Wilson, C. A., Hughes, Z. J., FitzGerald, D. M., Hopkinson, C. S., Valentine, V., and Kolker, A. S. 2014. Saltmarsh pool and tidal creek morphodynamics: dynamic equilibrium of northern latitude saltmarshes? *Geomorphology*, 213: 99–115.

Windham, L. 2001. Comparison of biomass production and decomposition between *Phragmites australis* (Common Reed) and *Spartina patens* (Salt Hay Grass) in brackish tidal marshes of New Jersey, USA. *Wetlands*, 21: 179–188.

Xin, P., Jin, G., Li, L. and Barry, D. A. 2009. Effects of crab burrows on pore water flows in salt marshes. *Advances in Water Resources*, 32: 439–449.

Yallop, M. L., Paterson, D. M., and Wellsbury, P. 2000. Interrelationships between rates of microbial production, exopolymer production, microbial biomass, and sediment stability in biofilms of intertidal sediments. *Microbial Ecology*, 39: 116–127.

Yang, S. L., Li, M., Dai, S. B., Liu, Z., Zhang, J. and Ding, P. X. 2006. Drastic decrease in sediment supply from the Yangtze River and its challenge to coastal wetland management. *Geophysical Research Letters*, 33: 4–7.

Yang, S. L., Li, H., Ysebaert, T., Bouma, T. J., Zhang, W. X., Wang, Y. Y., Li, P., Li, M., and Ding, P. X. 2008. Spatial and temporal variations in sediment grain size in tidal wetlands, Yangtze Delta: On the role of physical and biotic controls. *Estuarine Coastal and Shelf Science*, 77: 657–671.

Yapp, R. H., Johns, D., and Jones, O. T. 1917. The salt marshes of the Dovey Estuary. *Journal of Ecology*, 5: 65–103.

9

Salt Marsh Sediments as Recorders of Holocene Relative Sea-Level Change

W. ROLAND GEHRELS AND ANDREW C. KEMP

9.1 Introduction

Early geoscientists recognized that salt marsh sediment overlying terrestrial deposits (e.g., soil containing the preserved, *in-situ* stumps of freshwater trees) represented submergence of the older, buried landscape at a time in the past (e.g., Bartram, 1791; Lyell, 1849). After the development of radiocarbon dating in the late 1940s (e.g., Libby 1961) it became possible to determine when salt marsh sediment was deposited and sea-level research began to focus first on building and then on interpreting Holocene relative sea level (RSL) curves (e.g., Bradley, 1953; Redfield and Rubin, 1962; Bloom and Stuiver, 1963; van de Plassche et al. 1989; Gehrels et al. 1996; Shennan and Horton, 2002). Conceptually relative sea level (RSL) is the elevation of the sea surface relative to the land surface at a specific location and averaged over a period of time to negate the influence of tides and seasonal to annual variability. For example, RSL measured by tide gauges is often expressed as a monthly or annual average, while RSL reconstructions from coastal sediment are inherently time averaged over several years to decades. A variety of physical processes acting on local-to-global spatial scales and on temporal scales from minutes to millennia can cause RSL to change across space and through time. Therefore, measured or reconstructed RSL is specific to a time and place and is often the net outcome of multiple processes acting simultaneously. Proxy-based RSL reconstructions generated by interrogation of salt marsh sediment preserved in the coastal stratigraphic record are valuable in advancing our understanding of Holocene climate (e.g., Kemp et al. 2011), the structure of Earth's interior (e.g., the viscosity and structure of the mantle; e.g., Shennan and Horton, 2002; Engelhart et al. 2011a), and of physical driving mechanisms of past, present, and future sea-level change (e.g., Kopp et al. 2016).

In this chapter, we discuss the relationship between RSL change and sediment accumulation and then use a conceptual framework supported by case studies to (1) illustrate how RSL reconstructions are generated from salt marsh sediments, and to (2) provide examples of how RSL reconstructions can be used to address research questions and test hypotheses across a range of spatial and temporal scales.

9.2 Sedimentation vs. Relative Sea-level Change

9.2.1 Sedimentation in Salt Marshes

At face value, it may seem illogical that sediment deposited in and around dynamic salt marshes can provide accurate and precise information on past RSL. Reconstructing RSL changes from salt marsh sediments requires an understanding of sedimentation patterns and salt marsh stratigraphy. This enables researchers to identify sections of salt marshes that are (or indeed are not) sensitive to RSL change and thus provide a sedimentary archive of these changes. The highest rates of sedimentation in salt marshes occur at low elevations and close to channels and creeks. In contrast, the lowest sedimentation rates occur at high elevations close to the landward edge of the salt marsh. This pattern occurs because low elevations are inundated by sediment-bearing tidal water more frequently and for longer periods of time than high elevations. For example, French et al. (1995) measured accretion rates between 1986 and 1991 in a salt marsh on the North Sea coast of England (Fig. 9.1). The measured sedimentation rates exceeded the decadal-scale rate of regional RSL rise (~1.8 mm/yr; Shennan and Horton, 2002) for much of the salt marsh area. If the rate of sedimentation exceeds the rate of local RSL rise, accommodation space is filled and the salt marsh experiences emergence (the surface moving upward to a higher elevation in the tidal frame through time). Conversely, if the rate of local RSL rise exceeds the rate of sedimentation, accommodation space is created and the salt marsh experiences submergence (the surface moving downward to a lower elevation in the tidal frame through time). Over time the gradient from the upper marsh to the low marsh becomes less steep as the low marsh accretes more quickly than the high marsh (Fig. 9.1c). Therefore, sedimentation rates in salt marshes are not always a direct equivalent to the rate of RSL. Fortunately, RSL can be reconstructed from salt marsh sediments even when sedimentation rates were less/greater than the rate of RSL rise and the salt marsh was submerging/emerging through the

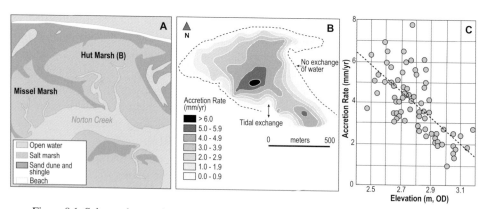

Figure 9.1 Salt marsh accretion rates measured over a 5-year period in Hut salt marsh, Scolt Head Island, in Norfolk, United Kingdom. (A) Geomorphological setting. (B) Accretion rates in Hut Marsh. (C) Relationship between accretion rates and elevation relative to Ordnance Datum (OD). Dashed line is a linear regression. Modified and redrawn from French et al. (1995)

appropriate use of sea-level indicators to reconstruct changes in paleomarsh elevation (Section 9.3). Although salt marsh sediment is most commonly used to reconstruct RSL in regions where accommodation space was created by RSL rise, many locations (particularly in the southern hemisphere) experienced RSL fall. To explore how salt marsh sediment is used to reconstruct RSL in these fundamentally different settings we use a case study from New Zealand (Section 9.4.3)

9.2.2 Salt Marsh Evolution

The relationship between sedimentation and RSL change can vary through the depositional history of a salt marsh on timescales from years to millennia (Allen 1990). To understand the capability of salt marshes to record Holocene RSL rise over centennial to millennial timescales it is useful to consider models of salt marsh evolution (Fig. 9.2). Redfield (1959, 1972) synthesized earlier hypotheses about the colonization of tidal flats by salt marsh vegetation (Shaler, 1886) and the long-term encroachment of salt marshes against an upland (Mudge, 1858; Fig. 9.2a). The Shaler model describes immature salt marshes where the rate of sedimentation exceeds the rate of RSL. This results in a regressive (decreasing marine influence caused by emergence) stratigraphic sequence in which accommodation space is filled with sediment until the surface reaches a sufficiently high tidal elevation to allow colonization by halophytic plants. In a Shaler-type marsh, sedimentation can occur rapidly and independently of sea-level rise. For example, in macrotidal settings with high sediment availability, accretion rates in low marshes can reach ~20 mm/yr (Kirwan et al. 2016).

The Mudge model describes sedimentation on mature salt marshes that have reached the upper elevation at which salt marsh sedimentation can occur, after which further sedimentation can only occur in response to local RSL rise. Redfield (1972) tacitly assumed that the upper limit of sedimentation is mean high water, although highest astronomical tide is in most cases the upper limit of salt marsh sedimentation in settings where astronomical tides are more important than wind-driven water levels. These mature salt marshes typically have a flat, platform-like geomorphology. Further RSL rise causes the salt marshes to retreat landward by colonizing adjacent upland environments and to vertically accrete more sediment on the salt marsh platform. If the rate of sedimentation is in equilibrium with sea-level rise, then the salt marsh surface maintains its elevation in the tidal frame. In the stratigraphic record this evolution is represented by a transgressive (increasing marine influence caused by submergence) stratigraphic sequence near the landward portion of the marsh and a continuous sequence of high salt marsh sediment on the platform. The salt marsh surface therefore approaches an equilibrium profile determined by the simultaneous interactions of sea-level rise, sediment compaction, and sedimentation (Allen, 1990). The rate of sediment accumulation is not a "pure" sea-level signal, but at higher elevations closer to highest astronomical tide, the contribution of sea-level rise to marsh accretion becomes increasingly dominant, making high salt marsh facies more suitable (but not "ideal") recorders of RSL rise compared to low salt marsh facies. It is these stratigraphic sequences that are commonly targeted for reconstructing RSL.

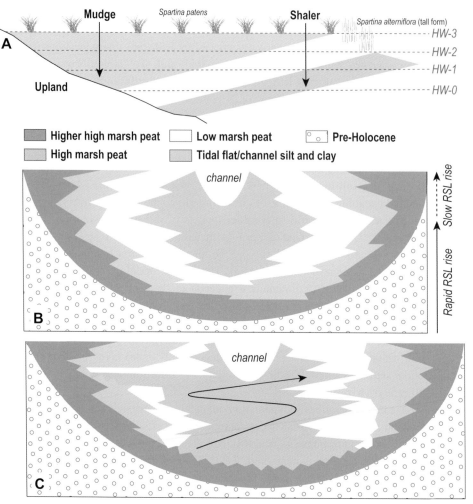

Figure 9.2 (a) Redfield's (1972) model of salt marsh evolution. The landward section is characterized by a transgressive sequence, according to the Mudge (1858) model. The Shaler (1886) model explains the regressive sequence in the seaward portion of the marsh. HW–0 is the position of high water at the beginning of marsh development. HW1–3 are various high water positions through time. Plant symbols are from the Integration and Application Network, University of Maryland Center for Environmental Science (ian. umces.edu/imagelibrary/). (b) Modification of the Redfield model showing how the rate of relative sea-level rise determines marsh evolution. Regressive sequences occur when the rate of relative sea-level rise exceeds the sedimentation rate. When relative sea-level rise slows down, the marsh progrades seaward over the tidal flat and a regressive sequence is formed. (c) Meandering tidal channels also play a role in salt marsh evolution. Only in sheltered places away from tidal channels and creeks are facies preserved that are controlled by relative sea-level rise and not disturbed by erosion. (A black and white version of this figure will appear in some formats. For the color version, please refer to the plate section.)

The Redfield model assumes that sal marsh development takes place without disturbance. Detailed stratigraphic surveys are necessary to identify if (and preferably how) a salt marsh experienced disturbance during its lifetime. Such surveys should be undertaken prior to selecting specific cores or sections to use in reconstructing RSL. This step is crucial to determine whether salt marsh evolution was controlled primarily by the rate of RSL rise, or whether disturbance events played a significant role. For example, meandering tidal channels and creeks (van der Wal and Pye, 2002), storms (Nikitina et al. 2014), or even hurricane strikes (van de Plassche et al. 2004), can erode salt marsh sediment resulting in a stratigraphic unconformity that can be recognized as salt marsh facies that is directly overlain by minerogenic deposits (Fig. 9.2a). Similarly, lateral erosion could produce displaced blocks of peaty sediments that may complicate the stratigraphy. In very dynamic settings, it is possible that RSL was the primary control on sedimentation only in sheltered locations near the upland boundary. Detailed stratigraphic exploration is needed to confirm this. We illustrate the importance of recognizing the stratigraphic model of salt marsh evolution that is applicable to the site under investigation with a case study from southwestern Florida (Section 9.4.2; Gerlach et al. 2017), in which the early part of the RSL reconstruction likely reflects local-scale sedimentation processes that resulted in a regressive sequence until accommodation space was filled after which rates of RSL rise and sedimentation became closely coupled.

9.3 Reconstructing Relative Sea Level

The conceptual framework that underpins efforts to reconstruct RSL arose from a sequence of sea-level themed International Geoscience Programs (e.g., Gehrels and Shennan, 2015; see references therein for examples of specific programs). Although the specific methods used in the field, laboratory, and data-processing evolved through time, the conceptual framework remains largely unchanged. This framework is summarized below:

$$\text{RSL}_{\text{time } i,\text{place}=j} = \text{Elevation}_{\text{time}=\text{modern},\text{place}=j} - \text{RWL}_{\text{time } i,\text{place}=j} \quad (9.1)$$

where RSL at time i and place j is calculated by subtracting a reference water level (RWL) that was reconstructed (with uncertainty) using a sea-level indicator from the measured height (elevation with uncertainty) of the sample being used to reconstruct RSL. Elevation and RWL must be expressed relative to the same datum (e.g., mean tide level, or mean high water spring tides). For a modern sample, the elevation and RWL terms are equal and thus present RSL is zero. To complete the RSL reconstruction it is necessary to estimate the age (time = i) of the sea-level indicator being used either through direct dating, or through stratigraphic correlation. In the following sections, we provide more detail on how each component of this framework is generated with the goal of providing some guidelines on how to produce reliable RSL reconstructions from salt marsh sediments and, perhaps more importantly, how to avoid potential pitfalls.

9.3.1 Salt Marsh Settings and Field Methods

Efforts to reconstruct RSL should begin with a research question and identification of the data that will be required to answer the question. Typically, multiple possible sites are identified from existing literature and/or geomorphic interpretation of maps as candidates for providing the required data, and each site is investigated in turn. At each site, systematic exploration and description of the sub-surface stratigraphy are achieved through coring at regular intervals along multiple, intersecting transects. The number and type of cores used (e.g., hand-driven gouge cores, vibracores, or Geoprobe cores) depend on the availability of resources and the nature of the sediment that must be penetrated. Following interpretation of these stratigraphies, one or more cores are selected for detailed laboratory analyses to reconstruct RSL. Best practice dictates that cores collected for analyses are stored in refrigerated conditions before, between, and after laboratory analyses to minimize the potential for degradation. Replicate cores are usually collected and are preserved as an archival copy, or are used in unforeseen, future analysis. At some locations, the stratigraphic record can be viewed in natural or excavated cross sections, in which case sampling is done from the exposed section after cleaning to remove surface contamination.

Salt marshes are variable in terms of their suitability to record RSL changes. Allen (2000) described seven geomorphological types of salt marshes (an updated rendition of these marsh types is presented in Chapter 1, Fig. 1.1). Open coast marshes (Chapter 1, Fig. 1.1c) are relatively rare but occur in macro-tidal settings where wave power is somewhat limited (relative to tidal force). They tend to be dissected by tidal creeks and are usually poorly suited for reconstructing RSL because of stratigraphic unconformities. More complete and undisturbed stratigraphic sequences in macrotidal settings are found in protected settings in sheltered open embayment (Chapter 1; Fig. 1.1e). The usefulness of this type of setting is illustrated in our case study from Maine (section 9.4.1), where uninterrupted salt marsh facies spans the period since ~6,000 cal. yr BP (Gehrels 1999). Backbarrier salt marshes are influenced by the dynamic nature of inlets (Chapter 1, Fig. 1.1), which can cause changes in tidal range and sediment supply through opening and closing. For example, breaching of the Fire Island barrier during superstorm Sandy in 2012 opened a new tidal inlet and caused tidal range in the back-barrier lagoon to increase (Fig. 9.3; Aretxabaleta et al. 2017). Similarly, small changes in embayment configuration in macrotidal estuarine marsh settings can generate a change in tidal range, whereas deltaic marshes can be strongly affected by channel meanderings (Chapter 1, Fig. 1.1f). Consequently, RSL reconstructions from these environments can be strongly influenced by local-scale processes. It is therefore important to recognize that in some geomorphological settings, RSL reconstructions derived from salt marsh sediment may include substantial contributions from physical processes that operate at local spatial scales. While the resulting reconstructions of local RSL trends can be accurate, they may be inappropriate for answering research questions that are focused on the role of processes operating on larger (regional to global) spatial scales. Alternatively, the resulting reconstructions may be well suited to answering research questions focused on understanding local-scale processes such as the time evolving distribution of barrier island inlets in a specific coastal system.

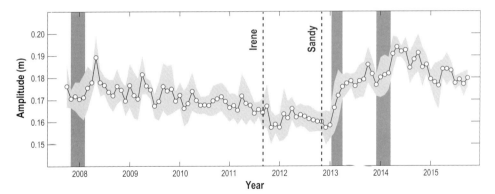

Figure 9.3 Change in M2 tidal amplitude in Great South Bay (Long Island, NY) following Hurricane Sandy (28–30 October 2012) which produced a new tidal inlet by breaching the Fire Island barrier. Shadings indicate episodes of dredging which also may have contributed to geomorphological changes in the tidal inlet. Hurricane Irene did not produce any barrier breaches. Redrawn from Aretxabaleta et al. (2017)

9.3.2 Sample Elevation

The elevation of the sample being used to reconstruct RSL is measured directly and expressed relative to modern (current) tidal datums at the same site. Modern surveying equipment (e.g., total stations and real time kinematic – RTK – satellite navigation) makes the process of levelling accurate, precise, and efficient. In most instances, the core-top elevations are surveyed relative to one another and to a control point of known elevation such as a benchmark, temporary benchmark, or RTK base station. The elevation of the sample is then simply calculated by measuring depth in core using a tape measure. For an exposed section, the ground surface can be surveyed, or stratigraphic levels can be surveyed directly. The uncertainty in this exercise can be estimated by closing a levelling loop, or through data processing in the case of RTK systems. In some instances, uncertainties for the elevation of benchmarks are reported.

While establishing relative elevations is straightforward, it is often more challenging to express absolute elevations with respect to local tidal datums because most benchmarks and RTK systems use orthometric datums such as the North American Vertical Datum of 1988 (NAVD88). Where tidal benchmarks are not available to use as a control point, tidal elevations can be established in several ways. Most simply, water levels during a day (or better still several days) are periodically surveyed and then compared to water-level changes measured by a nearby tide gauge to determine the similarity of tides between the two locations. An extension of this approach is to deploy an automated logger at the study site to make frequent (e.g., every five minutes) water-level measurements over an extended period (days to years), which allows for a more robust comparison to verified measurements at tide-gauge stations. Loggers should be affixed to a stable structure and deployment at low tide helps to ensure that the full tidal cycle is measured. In both cases, the reported datums for the tide-gauge site are transferred to the study site either unchanged, or following scaling as appropriate. Where tide gauges are not available, observations can be compared instead to predictions from a tidal model with the caveat that actual and predicted water levels can differ significantly due to meteorological conditions (wind and pressure) that themselves can have a strong seasonal bias. In this circumstance, a longer period of observation is likely less prone to being inaccurate than a shorter period of observation. At some locations, tidal models (e.g., VDatum in the USA) allow for direct conversion of measured orthometric elevations into tidal elevations. In our experience, it is good practice to establish a pseudo-permanent benchmark at each study site to quickly and easily relate future and previous sampling campaigns to one another (including repeated deployment of water-level loggers). These benchmarks can be, for example, survey nails driven into rock/concrete, or rods driven into the pre-Holocene substrate.

Through time the accumulation of salt marsh sediment causes the elevation term in equation 9.1 to increase, although the rate of sediment accumulation may not always match the rate of RSL exactly (Section 9.2; e.g., Allen, 1990). This can lead to sediment successions that are either transgressive (submergence) or regressive (emergence). The degree of disequilibrium between rates of sediment accumulation and sea-level rise can be subtle (small, short-lived differences) or extreme (large, persistent differences). The latter occurs,for example, along some seismically active coasts, where megathrust earthquakes cause sudden (seconds to hours) and large (greater than 1 m) vertical land movements (see case study in Section 9.4.4). To unravel sedimentation from RSL rise it is necessary to establish the indicative meaning (Section 9.3.3) of the sediment samples. In equation 1, this adjustment is accommodated in the RWL term. In the (desirable) circumstance that all samples in a core (i.e., through time) had a constant RWL, the resulting RSL curve would be identical to the history of sediment accumulation for the core. In reality however, few (and perhaps no) cores exhibit a reconstructed RWL that is truly constant through time even if the techniques used to estimate RWL (e.g., transfer functions, classification) suggest the changes are much smaller than their associated uncertainties (e.g., Kemp et al. 2014, 2017b).

9.3.3 Indicative Meaning

Reconstructing RSL is grounded in reasoning by analogy and requires use of a proxy, which is commonly termed a *sea-level indicator*. Physical features, chemical signatures, and biological assemblages that have a systematic and quantifiable relationship to tides can all be sea-level indicators (in practice, RSL reconstructions from salt marsh sediment almost exclusively use biological and/or chemical sea-level indicators). The relationship between a sea-level indicator and tidal elevations is quantified by its *indicative meaning* that is comprised of two parts. The *indicative range* is the range of tidal elevations between which the sea-level indicator is observed to exist or form in modern environments. The *reference water level* (RWL) is the mid-point of the indicative range. Due to differences in tidal range among sites, the indicative range is commonly expressed as an upper and a lower boundary that are both tidal datums. For example, the indicative range for high salt marsh plants (e.g., *Juncus roemerianus* or *Spartina patens*) on the US Atlantic coast is commonly taken to be mean high water to highest astronomical tide. There is no consensus on the confidence interval that should be represented by the indicative range and in many cases no confidence interval is reported. The *Handbook of Sea-Level Research* (Shennan et al. 2015) recommends that the 95% confidence interval is reported, and best practice more broadly dictates that it is helpful and transparent to explicitly provide a confidence interval (sigma or percentiles) to aid other workers. The absolute (e.g., in centimeters or meterse) indicative range of a sea-level indicator is proportional to tidal range, i.e. a larger tidal range results in a correspondingly large indicative meaning. Therefore, more precise RSL reconstructions are generally derived from sites with small tidal ranges. However, in some areas where astronomical tides are small (e.g., coastal North Carolina; Kemp et al. 2009, 2017b), wind-driven water levels can extend, sometimes considerably, the indicative range of sea-level indicators.

The indicative meaning for a sea-level indicator is established empirically through documenting its observable distribution in modern environments that are similar to those likely to be encountered in the stratigraphic record. Field-based research should focus on systematic sampling of the environmental gradient (e.g., spacing of samples at regular increments of elevation change) and emphasize capturing the uppermost and lowermost limits at which a sea-level indicator exists to avoid biasing subsequent RSL reconstructions. In the sections below we describe some of the most commonly used sea-level indicators from salt marshes.

9.3.3.1 Plants as Sea-Level Indicators

Living salt marsh plants are vertically zoned because species have varying preferences and tolerances for the balance between subaerial exposure and inundation by saltwater, which is well approximated by tidal elevation. Plants around the landward edge of a salt marsh typically have ecological preferences for slightly brackish conditions that arise from infrequent and short-lived inundation by saltwater during unusually high tides and/or storms. In contrast, plants on the salt marsh are halophytic and adapted to more frequent and prolonged marine inundation. This distribution has long been recognized by coastal ecologists (e.g., Wells, 1928; Chapman, 1940); indeed up until the early twentieth century surveyors in Boston used the informal "Marsh Datum" to establish local elevations with respect to mean high water by leveling to particular species of plants on nearby mature salt marshes recognized by their wide and flat geomorphology[1]. Although the aboveground parts of salt marsh plants are often decomposed, some shallow surface structures such as rhizomes are

---

[1] John Ripley Freeman and Herbert Shedd were prominent civil engineers working in Boston during the late nineteeth and early twentieth centuries. In a letter dated 30th June 1903, Shedd wrote to Freeman that:

"In isolated cases, I often used marsh level as representing mean high tide... Whenever I took those levels, I was surprised to find how accurately the surface of the marsh for considerable areas came to the same level plane. Whenever there appeared to be a slight depression or elevation varying from this plane the character of the grass was different and it seemed to be necessary only select a spot that standard marsh grass or sod to find the real marsh level."

In 1909 Freeman testified before court in a land ownership case that:

"the marsh level is remarkably persistent at the plane of mean high water, or very nearly to that, and one of the stumbling blocks in the way of the theory of subsidence would be that fact, of the persistence of the marsh level at mean high water, unless it is also proved that the deposition goes on at substantially the same rate as the subsidence"

This correspondence shows that the relationship between plants, salt-marsh geomorphology, sedimentation, and tidal elevation have long been recognized. The original letter and a copy of the court transcript are held in the papers of John Ripley Freeman at the Massachusetts Institute of Technology.

often well preserved and can be identified to the species level using existing guides and/or through collection of representative examples while in the field. If plants are used as sea-level indicators it is important to use the plant parts that grew close to the marsh surface since this horizon is the implicit focus of equation 1. A limitation of using plants as sea-level indicators is that few identifiable remains are present in a single sample (example.g., a short section of core), which prohibits statistical analysis and increases the possibility that the plant macro-fossil under examination is not representative of the plant community that was present at the time of sediment deposition. For example, salt marshes in North Carolina are primarily vegetated by *J. roemerianus* (up to 77% by area; Eleuterius, 1976), but the rhizomes of *Distichlis spicata* are disproportionally common in cores of high salt marsh peat (e.g., Kemp et al. 2017). This difference between the apparent paleo and modern abundance of the two plant species probably reflects the preferential preservation of *D. spicata* over *J. roemerianus* rather than a change in the dominant plant community through time.

9.3.3.2 *Microfossils as Sea-Level Indicators*

Like plants, the vertical zonation of microfossils (e.g., foraminifera, diatoms, and testate amoebae) in and around salt marshes is controlled directly by the differing species tolerance to subaerial exposure and, indirectly, by other environmental variables (e.g., grain size, organic content, salinity) that have a strong correlation with tidal elevation. A primary advantage of using microfossils as sea-level indicators is that they respond rapidly to environmental change (e.g., Engelhart et al. 2013; Horton et al. 2017) and they form assemblages with high numbers of individuals from multiple species, which makes them suitable for statistical analysis. The primary disadvantage of using microfossils to reconstruct RSL is the time needed to process sediment samples in the laboratory to concentrate tests, count large numbers of individuals, and the training required to become familiar with the taxonomy of different groups (easiest/quickest for foraminifera and most difficult/time consuming for diatoms in our experience).

Foraminifera are single-celled marine organisms that live primarily on, or just beneath, the salt marsh surface. The pioneering work by David Scott (e.g., Scott and Medioli, 1978) and subsequent work around the world (e.g., Gehrels, 1994; Horton and Edwards, 2004; Hawkes et al. 2010; Callard et al. 2011; Wright et al. 2011; Strachan et al. 2015; Avnaim-Katav et al. 2017; Kemp et al. 2017a) demonstrates that salt marsh foraminifera meet the criteria to be sea-level indicators (Fig. 9.4). Their vertical zonation is controlled by the differing tolerances of species to subaerial exposure, such that the uppermost limits of a salt marsh where subaerial exposure is greatest (and tidal flooding lowest) is an extreme environment for foraminifera and often only the most hardy species are able to live here (e.g., *Jadammina macrescens.* Note that this taxon was recently renamed *Entzia macrescens*; LeCove and Hayward, 2017; but we use the old name because it is entrenched in the literature). Shallow subtidal and tidal-flat foraminiferal assemblages are often dominated by taxa with calcareous tests, while in the salt marsh agglutinated taxa are dominant. Low salt marsh environments at locations throughout the world are typically dominated by *Miliammina fusca*. High salt marsh environments can be dominated by various taxa, many of which are common at sites around the world and there is commonly a high degree of

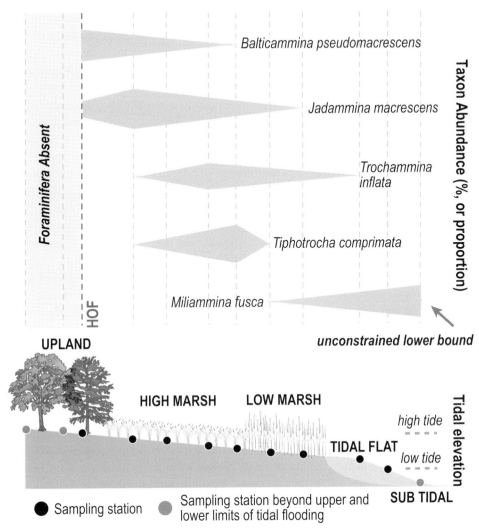

Figure 9.4 An example of intertidal zonation displayed by foraminifera in samples collected along the sloping surface of a salt marsh in eastern Maine. Plant zones are also shown. Redrawn and modified from Gehrels (1994). (A black and white version of this figure will appear in some formats. For the color version, please refer to the plate section.)

analogy between modern assemblages and those found in the Holocene sedimentary record. At mid and high latitudes (e.g., Maine, Canada, northern Europe, and the US Pacific Northwest), high salt marsh zones often contain near mono-specific assemblages of *J. macrescens* and/or *Balticammina pseudomacrescens* (e.g., Gehrels, 2002; Wright et al. 2011; Barnett et al. 2016; Kemp et al. 2017a). In extreme cases, salt marsh foraminifera can form two binary and monospecific assemblages where *M. fusca* is the only species at lower elevations and *J. macrescens* and/or *B. pseudomacrescens* is the only taxon present at higher elevations. In this instance the utility of foraminifera to reconstruct RSL is diminished (e.g., Gehrels et al. 2006; Kemp et al. 2013, 2017) because reconstructions become binary (similar to classification of discrete environments using plants rather than reconstruction of a continuous variable) and transfer functions are rendered of little use.

Documenting the distribution of foraminifera is typically done by collecting surface (0–1 cm) sediment samples along transects. The tidal elevation of each sample is measured at the time of collection and the assemblage of foraminifera is established by counting (and identifying to the species level) individual tests. Debate remains as to how many individuals should be counted to adequately describe the assemblage in a single sample (Patterson and Fishbein, 1989; Fatela and Taborda, 2002; Edwards and Wright, 2015), although most studies seek to count 100 or more. This dilemma also applies to other microfossil groups (e.g., Payne and Mitchell, 2009). The addition of a stain (e.g., rose Bengal) allows living and dead individuals to be differentiated. Some early literature (e.g., Scott and Medioli, 1980; Murray, 1982) debated the relative merits of using the live, dead, or total (live + dead) assemblage, although most recent studies use the dead assemblage as a sea-level indicator because it is naturally time and taphonomy averaged (e.g., Horton, 1999). Sampling in the field often fails (for practical reasons) to define the lower limit of tidal-flat taxa, which makes it difficult to establish their indicative meaning. Similarly, it is imperative to establish the uppermost limit at which foraminifera exist (termed the highest occurrence of foraminifera; Wright et al. 2011) and we note from experience that this can lie above highest astronomical tide and in soil-forming environments vegetated by fresh-water upland vegetation. This is particularly pronounced in regions where wind-driven water levels exceed the astronomical tidal range (e.g., Kemp et al. 2009, 2017b).

Diatoms are single-celled, photosynthezing algae that are found in all aquatic environments and that form a siliceous test. On salt marshes, their distributions are controlled by several environmental (co)variables, including salinity, grain size, temperature, and nutrient supply. Many of these variables are directly, or indirectly, controlled by tidal inundation and, like foraminifera, distinct diatom assemblages occur within narrow vertical zones in many salt marshes around the world and are therefore suitable sea-level indicators (Watcham et al. 2013; Barlow et al. 2014; Long et al. 2014; Saher et al. 2015). An advantage of diatoms over foraminifera is that their distribution extends from salt marshes into adjacent freshwater upland environments. This makes them particularly useful in reconstructing RSL changes from stratigraphic settings that include supratidal sediments (e.g., the case study about earthquake-driven RSL change in Section 9.4.4). A unique problem for using diatoms as sea-level indicators is their high diversity within and between samples and sites (often an order of magnitude greater than foraminifera). For this reason, many studies aim to count 300 or more individual diatoms in a single sample. High diversity within samples means that modern datasets may have more taxa than samples, which is a barrier to statistical analysis. In some cases, species may be consolidated into generic or ecological (e.g., fresh, brackish, marine) groups to reduce diversity. The high degree of variability from one site to another is reflected in the frequent lack of analogy between modern and fossil assemblages (Watcham et al. 2013) and likely reflects the influence of variables other than tidal elevation in controlling assemblages of diatoms.

Testate amoebae are single-celled organisms that live in many terrestrial and aquatic habitats, including peat bogs, soils, lakes, and salt marshes. They are also referred to as rhizopods, thecamoebians, and/or arcellaceans in some literature. Their vertical intertidal niches appear to be narrower than for foraminifera and diatoms and assemblage compositions are quite similar on many coastlines (Barnett et al. 2017). Relative intolerance of salinity appears to limit their occurrence to the uppermost part of salt marshes, making them rare at lower elevations and usually absent below mean high water (Charman et al. 2002). Important taxa with widespread occurrences in the upper salt marsh include *Tracheleuglypha dentata*, *Trinema lineare*, *Euglypha rotunda*, *Centropyxis cassis*, *Centropyxis platystoma*, *Difflugia pristis*, and *Cyphoderia ampulla*. The diversity of testate amoebae assemblages is typically greater than foraminifera and less than diatoms (Gehrels et al. 2001). Therefore, it is common for studies to aim to count approximately 100 or more individuals (Charman et al. 2002; Barnett et al. 2017). The preservation potential of testate amoebae, however, is less than for other microfossil groups and some taxa tend to disintegrate, particularly in air-dry sediments (Roe et al. 2002; Barnett et al. 2017). Specifically, idiosomic genera (tests composed of proteinaceous secretion and siliceous plates) are preferentially removed from the sedimentary record compared to xenosomic genera (tests formed from agglutinated particles), which potentially results in a systematic bias and no-modern-analogue-outcomes (e.g., Kemp et al. 2017c) in RSL reconstructions generated using testate amoebae.

9.3.3.3 Transfer Functions

Transfer functions are empirically derived equations for reconstruction past environmental conditions from paleontological data (Sachs, 1977). The term refers not to a specific mathematical technique, but more broadly to a quantitative approach. Although transfer functions were first used in paleoceanographic studies in the early 1970s, their adoption by the sea-level research community occurred later and they have become increasingly common place since the mid-1990s. Transfer functions use the modern distribution of microfossils to generate quantitative reconstructions of RWL. The regression models used for this purpose describe the distribution of taxa across the surface of salt marshes from lower to higher tidal elevations. These models are then used to calculate the tidal elevation at which a fossil analogue preserved in the stratigraphic record was originally deposited. A key characteristic of transfer functions is that they treat tidal elevation as a continuous variable, which enables them to identify subtle disequilibria between the rate of sediment accumulation and RSL change. This stands in contrast to classification-based approaches (e.g., using plants or discriminant functions) where core samples are placed into bins (e.g., high marsh) and the treatment of elevation as a discrete variable only allows recognition of pronounced disequili-bria. Sample-specific (~1σ) uncertainties for RWL reconstructions are generated by transfer functions. The typical precision achieved using foraminifera is ±10%–20% of tidal range. For detailed explanations of the use of transfer functions in salt-marsh-based RSL recon-structions using microfossils we refer the reader to Barlow et al. (2013) and Kemp and Telford (2015). While the vast majority of RSL reconstruction that employed transfer functions used regression models (e.g., Weighted Averaging, Weighted Averaging Partial Least Squares), a new generation of Bayesian transfer functions is emerging that is underpinned by fundamentally different numerical approaches and can incorporate other proxies (e.g., bulk-sediment geochemistry, Section 9.3.3.4) in a multi-proxy framework (e.g., Cahill et al. 2016; Kemp et al. 2018). An appealing characteristic of Bayesian transfer functions is that they enable each species of microfossil to have a relationship to tidal elevation that is unique in both form (i.e., not a Gaussian response curve for all species) and specific parameter values. It is likely that these models will become more commonplace.

9.3.3.4 Sediment Biogeochemistry as a Sea-Level Indicator

Recent work investigated the utility of biogeochemical measurements as sea-level indicators (see Wilson, 2017 for a review). In most studies, analysis is performed on bulk-sediment samples (i.e. undifferentiated slices of surface or core sediment that are not separated into their constituent parts prior to analysis), although in some instances specific size fractions of the sediment, or specific chemical compounds (e.g., n-alkanes) are analyzed. The most commonly used measurements characterize the quantity and isotopic composition of carbon and nitrogen in sediment, including total organic carbon (TOC), the ratio of carbon to nitrogen (C:N, calculated from measured total nitrogen) and $\delta^{13}C$. These parameters are anticipated to have a quantifiable relationship to tidal elevation for two reasons. Firstly, higher elevations receive relatively high input of terrestrial carbon, while lower elevations receive relatively high input of marine carbon. Since terrestrial and marine carbon can be differentiated measured parameters (e.g., TOC, C:N, $\delta^{13}C$), the relative inputs from these two sources are likely to be strongly correlated to elevation (Lamb et al. 2006; Wilson, 2017; Goslin et al. 2017). Secondly, at some locations (e.g., the mid-Atlantic and Northeastern USA) salt marsh plants, the remains of which may be difficult to identify, follow a different photosynthetic pathway (C_3 vs. C_4) than plants that live in adjacent freshwater environments (e.g., Kemp et al. 2012). The $\delta^{13}C$ difference between these groups of plants is substantial and preserved in bulk sediment, although differences between modern and fossil values are found due to plant decomposition and the ^{13}C Suess effect (the introduction of ^{12}C-enriched CO_2 into the atmosphere due to fossil fuel burning and deforestation; e.g., Wilson, 2017). The key strength of using TOC, C:N, and $\delta^{13}C$ as a sea-level indicator is that measurements can be made relatively simply, cheaply, and quickly on instruments that are now widely available in academic settings. Although TOC, C:N, and $\delta^{13}C$ are (to date) the most commonly used geochemical proxy for reconstructing RSL from salt marsh sediment, other approaches have been used, including Rock-Eval pyrolysis (e.g., Kemp et al. 2017).

9.3.4 Dating Salt Marsh Sediment

The age of salt marsh sediments older than ~300 years is established by radiocarbon dating. The preferred materials for dating are plant fragments, horizontally embedded within the sediment sequence (e.g., Gehrels et al. 2005) and/or subsurface rhizomes and stalks with welldefined relationship to the surface (e.g., van de Plassche et al. 2000). These materials are typically separated from the surrounding sediment matrix, cleaned carefully under a binocular microscope to remove older adhered sediment and younger in-growing roots, and then oven dried (at low temperature, ~40°C to avoid temperature-dependent fractionation of carbon isotopes). The prepared samples are submitted to a national, academic, or commercial radiocarbon dating lab where the actual age measurement is performed (usually through accelerator mass spectrometry). The reported radiocarbon age (and uncertainty), is converted to a calendar age through calibration (Reimer et al. 2013). Radiocarbon dating can also be performed on bulk sediment, although this material commonly contains carbon spanning a range of ages and is therefore likely to return a less precise (or inaccurate) age. In rare cases, optically stimulated luminescence dating has been successfully applied to salt marsh sediments (e.g., in Denmark; Szkornik et al. 2008).

Radiocarbon dates on material that is less than ~300 years old typically yield multiple possible calendar ages and are often associated with increased chronological uncertainty that arises from the calibration. Therefore, studies that seek to reconstruct recent changes in RSL using salt marsh sediment must employ alternative methods if it is necessary to achieve multi-decadal precision to answer the research question under investigation. Some specialized radiocarbon methods (e.g., bomb spike and high-precision dating of multiple aliquots; Marshall et al. 2007) can provide improved precision. Methods that measure sediment accumulation rates (e.g., ^{210}Pb), or identify horizons of known age from downcore elemental, isotopic, and/or pollen profiles, provide additional chronological constraints for recent salt marsh sediments (Gehrels et al. 2006, 2008; Kemp et al. 2012). For example, the onset of industrialization which occurred at ~1850–1900 CE in North America can be recognized by marked increases in the concentration of heavy metals such as Pb that were emitted to the atmosphere as pollutants (Kemp et al. 2012). Similarly, the isotope ^{137}Cs was introduced to the atmosphere by aboveground testing of nuclear weapons. Its first occurrence therefore corresponds to the onset of widespread testing in the 1950s and its peak activity in a downcore profile is interpreted as representing 1963–1965 CE when aboveground testing was banned. Changes in the composition of pollen assemblages may also record historic events such as land clearance by European settlers in eastern North America, Australia, and New Zealand, the introduction of invasive/exotic species, and other disturbances such as fire or disease (e.g., chestnut blight; Donnelly et al. 2004; Gehrels et al. 2005; Kemp et al. 2011). In many cases marker horizons represent environmental or industrial changes that were regional in scale. Therefore, specific horizons are not present at all locations, or must be ascribed an age that is appropriate for the region under examination. The ^{137}Cs horizons are typically considered to be hemispherically widespread and synchronous. Our case studies from Florida (Section 9.4.2) and New Zealand (Section 9.4.3) illustrate how several of these dating techniques are used to reconstruct late Holocene and historic RSL change from salt marsh sediments.

9.3.5 Types of Relative Sea-Level Reconstruction

There are broadly two types of RSL reconstructions that are generated from salt-sediment. The first are sea-level index points, which provide discrete RSL reconstructions from a single time and place, usually on multi-centennial to millennial time scales. Each sea-level index point has a vertical and chronological uncertainty (Fig. 9.5). A suite of sea-level index points from a site provides a RSL history for that site, while efforts to compile and standardize sea-level index points provide regional- to continental-scale RSL histories (e.g., Shennan and Horton, 2002; Engelhart et al. 2011b; Engelhart et al. 2015; Vacchi et al. 2016, 2018). Discrete sea-level index points are independent of one another because each sample is dated directly. Where these samples directly overlie an incompressible substrate they are termed "basal" (Jelgersma, 1961; Kaye and Barghoorn, 1964; Streif, 1979; Gehrels, 1999; Shennan and Horton, 2002) and considered unaffected by compaction (Fig. 9.5). Therefore, comparison of basal and non-basal sea-level index points provides a means to test for the influence of postdepositional lowering caused by compaction. The case study from Maine (Section 9.4.1) provides an example where the use of basal sea-level index points reveals considerable compaction in a 6000-yr long section of salt marsh sediment.

The second type of RSL reconstructions are near-continuous records generated from a single core or section of salt marsh sediment. These studies typically seek to provide precise RSL reconstructions to investigate physical processes acting over relatively short (multi-decadal to centennial time scales) time periods. In this approach, multiple depths in the core are directly dated and an age-depth model is used to estimate the age (with uncertainty) of all depths in the core including those that were not originally dated. Specialized software packages (e.g., BACON, Bchron) are important tools to generate age-depth models (see summary and review in Parnell and Gehrels, 2015). Data points that estimate the height and timing of RSL are generated from each level in the core where sea-level indicators are available (e.g., depths where a transfer function was applied to a microfossil assemblage). The use of an age–depth model means individual data points in this type of reconstruction are not independent of one another, which has important statistical implications for quantifying RSL trends. The typical resolution of the resulting RSL reconstructions is on the order of one data point per decade and ±5–20 cm, with most precise values obtained along microtidal coasts (where the indicative meaning of the sea-level indicators is small) with high sedimentation rates (single samples represent relatively little time). This precision has enabled RSL reconstructions from salt marshes to be meaningfully compared to, or combined with, nearby tide-gauge records and thus to bridge the gap between geological and instrumental records. The case studies from Maine, Florida, and New Zealand in Section 9.4 are examples of how a single core can be used to produce a near-continuous RSL reconstruction.

9.4 Case Studies

In this section we provide examples of Holocene RSL reconstructions generated from salt marsh sediment. The case studies were selected to provide examples of various types of research questions that require different types of analyses to answer those questions. Our first two examples are from North America and highlight the need to consider salt marsh geomorphology and stratigraphy to understand how the site evolved and its capability to

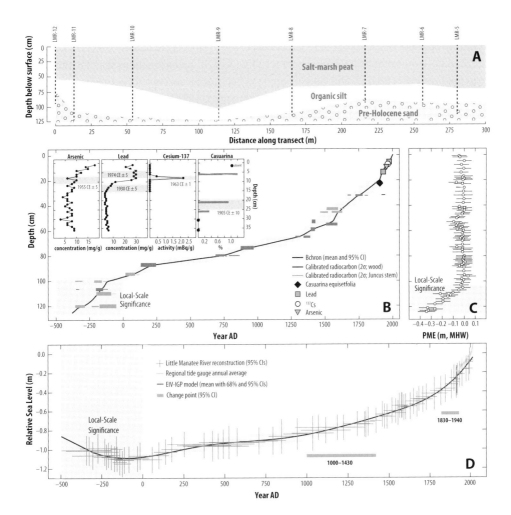

Figure 9.5 Relative sea-level reconstruction from salt marsh sediments in the Little Manatee River estuary, Gulf coast of Florida. (a) Stratigraphy underlying the site described from cores. Core LMR-9 was selected for detailed analysis. The lower part of the sequence represents a Shaler-type (colonizing) salt marsh when local-scale processes were inferred as the dominant cause of relative sea-level change. (b) Chronological data and age-depth model for LMR-9. Inset panels show downcore profiles of pollutants, radioisotopes, and pollen from the exotic genus *Casuarina*. Radiocarbon ages and chronological markers were combined using a Bchron age-depth model. (c) Reconstructed paleo-marsh elevations, or indicative meaning based on a regional-scale transfer function. (d) Relative sea-level reconstruction from LMR-9 combined with tide gauge measurements from Key West, Naples, St. Petersburg, and Fort Meyers. Modified from Gerlach et al. (2017). (A black and white version of this figure will appear in some formats. For the color version, please refer to the plate section.)

record regional-scale RSL trends (see Section 9.2.2). We also include a southern hemisphere example, because salt marshes here experienced late Holocene RSL fall, which is unfavorable to long-term salt marsh development resulting in RSL reconstructions that are often only span the last few centuries. Our final example is one from a seismically active coast (the US Pacific Northwest along the Cascadia subduction zone). Here, many of the same principles for reconstructing RSL apply as along passive margins, but RSL changes are often abrupt as a consequence of rapid vertical land movements caused by (M_w>8) megathrust earthquakes.

9.4.1 Maine, USA

The RSL reconstruction from Sanborn Cove marsh in eastern Maine (USA) addresses the question of compaction in thick salt marsh sequences. The area is important due to perceived along-coast differential isostatic and neotectonic crustal movements (Gehrels and Belknap, 1993; Gehrels et al. 2006). To detect such motion, it is important that differential compaction can be ruled out as an explanation. Another methodological problem that is addressed in this study is whether or not it is possible to detect sub-centennial and decimeter-scale sea-level fluctuations from foraminifera preserved in salt marsh sediment.

Sanborn Cove is a unique salt marsh, because it contains a thick section of salt marsh deposits that spans the past ca. 6000 years, which places it among the longest known sequences of uninterrupted salt marsh facies along the US East Coast (Fig. 9.6). It is an open-bay type marsh with a diurnal tidal range of ~6 m. The geomorphological setting makes it a suitable location to record RSL fluctuations that are not influenced by dynamic features such as barrier spits and tidal inlets. On the other hand, as a consequence of the macro-tidal setting the uncertainties associated with the indicative meaning of the sea-level indicators are correspondingly large (see Section 9.3.3), illustrating how sometimes site selection requires compromises if the "ideal" site is not present. Such compromises should consider the research questions that are posed.

Gehrels (1999) reconstructed long-term and short-term RSL changes from the sediments, the former from a series of basal sea-level index points, the latter from detailed analyses of a single core. By comparing ages of in-core samples with those from basal samples at approximately the same elevation, it is possible to assess whether compaction occurred (Fig. 9.6a), although it should be taken into account that basal samples are generally formed in a higher setting than in-core samples. The indicative meaning was established using foraminifera and a regional-scale transfer function calibrated using the modern distribution of foraminifera in Maine from a previous study (Gehrels, 1994). The resulting RSL reconstructions are shown in Figure 9.6b. The important message of this case study is the demonstrable influence of compaction in long salt marsh sequences (see also Brain et al. 2017 for geotechnical assessments of compaction). Although not all RSL reconstructions along the US East Coast are significantly affected by compaction (e.g., Brain et al. 2015, 2017; Zoccarato and Teatini, 2017), in Sanborn Cove most basal sea-level index points plot higher than the RSL reconstruction from the core. This pattern indicates that the sediment in the core experienced postdepositional lowering through sediment compaction, which serves to lower the elevation term in equation 9.1 and results in the amount and rate of RSL rise being overestimated. The core reconstruction based on

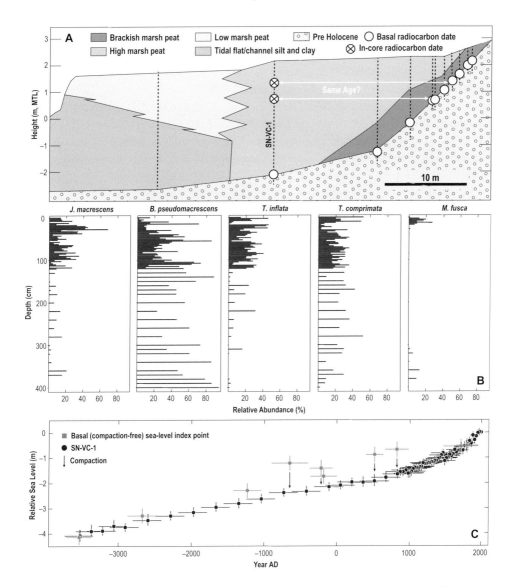

Figure 9.6 Relative sea-level reconstructions from salt marsh sediments at Sanborn Cove, eastern ME (Gehrels 1999). (a) Stratigraphy and radiocarbon dates. Core SN-VC-1 was used to produce a near-continuous reconstruction. (b) Foraminifera in core SN-VC-1. (c) Relative sea-level reconstructions from basal sea-level index points (blue squares) and from core SN-VC-1 (red circles). The offset between the two records is assumed to be the result of compaction (vertical arrows) of sediment in SN-VC-1. (A black and white version of this figure will appear in some formats. For the color version, please refer to the plate section.)

foraminiferal analyses also exhibits details not resolvable by the basal record. For example, the occurrence of *M. fusca* in the top of the core is a direct result of the modern acceleration of RSL rise that started at the end of the nineteenth century and caused relative submergence of the salt marsh.

9.4.2 Gulf of Mexico, Florida, USA

Gerlach et al. (2017) sought to understand how and why RSL changed in the Gulf of Mexico and western Atlantic Ocean during the past ~2,000 years. The reconstruction filled an important spatial gap between Louisiana (e.g., Gonzalez and Tornqvist, 2009) and the southeastern Atlantic coast of North America (e.g., Kemp et al. 2011). The diurnal tidal range in this region is small (0.60 m at Little Manatee River), which makes it well suited to producing precise RSL reconstructions (Fig. 9.5). Foraminifera were used as sea-level indicators and a transfer function was calibrated using modern distributions from five sites in the study region.

The Little Manatee River site is in an estuarine setting and preserves ~1 m of high salt marsh sediment overlying a ~0.2-m unit of tidal flat or low salt marsh sediment, and consolidated pre-Holocene sediment. The stratigraphy indicated that early in the site's history it was a Shaler-type marsh where sedimentation exceeded sea-level rise resulting in emergence, after which a Mudge-type marsh vegetated by *J. roemerianus* evolved and took on its characteristic flat geomorphology and in which rates of sedimentation likely tracked sea-level rise closely.

The history of sediment accumulation was established by using a Bchron age–depth model constrained by 16 radiocarbon dates on identifiable plant macrofossils that grew close to the salt marsh surface and pieces of wood recognized as having fallen onto the salt marsh surface. The upper part of the core was dated by recognition of chronological markers from downcore pollen, elemental, and isotopic profiles because the study sought to generate a RSL reconstruction that spanned the transition from the late Holocene to historical periods. The chronological markers provide a useful illustration of the need to consider regional setting and environmental histories. Activity of ^{137}Cs displayed a clear peak that was interpreted to represent aboveground nuclear weapons testing prior to its prohibition in 1963, which is consistent with the presence and interpretation of this marker throughout the Northern Hemisphere. The abundance of lead in the core increased above low and stable background concentrations, and reached a peak before decreasing at the top of the core. This profile is typical of salt marsh stratigraphies elsewhere in eastern North America (e.g., Kemp et al. 2011) where the onset of lead pollution is interpreted as being synchronous with (and a consequence of) the onset of industrialization in the mid to late nineteenth century. However, this interpretation is not applicable in Florida because prevailing trade winds rather than westerlies mean that pollution was not transported to Little Manatee River from the centres of industrialization in the mid-west and on the Atlantic Coast. Recognizing the regional nature of atmospheric pollution, Gerlach et al. (2017) instead attributed the onset of lead pollution in the core to the widespread and nearby use of leaded gasoline that began in the 1930s. They also identified region-specific chronological horizons associated with the introduction of ornamental trees (increase in *Casuarina* pollen) and the use of arsenic-based herbicides in citrus groves and golf courses. The resulting age–depth model showed that the unit of tidal-flat and low-marsh sediment accumulated rapidly, after which sedimentation rates were largely constant at ~0.3 mm/yr until a historic increase beginning at ~1900 CE.

The near-continuous RSL reconstruction was combined with instrumental measure-ments of RSL from nearby tide gauges in order to utilize all available information about how RSL varied through time. The primary driver of multi-centennial to millennial-scale RSL change in this region was ongoing glacio-isostatic adjustment (e.g., Peltier, 2004). After removing the contribution from this process (0.3 mm/yr) the sea-level reconstruc-tion represents regional -scale processes that are non-linear through time and local-scale processes. An apparent sea-level fall from ~500 BCE to 0 CE was interpreted as reflecting local-scale processes as the Shaler-type marsh rapidly filled accommodation space. After 0 CE, the sea-level record from Little Manatee River was indistinguishable (within uncertainty) in the amount and rate of change from similar records in Louisiana (Gonzalez and Tornqvist, 2009) and northeastern Florida (Kemp et al. 2014). Gerlach et al. (2017) used this similarity to conclude that the Little Manatee reconstruction primarily represented regional- rather than local-scale processes after 0 CE, until the onset of modern sea-level rise, which was correlated with global scale trends in tide-gauge compilations and other RSL reconstructions. The location of the site suggests that the sea-level reconstruction could be prone to the influence of evolving river discharge, so it is important to consider this factor when interpreting the RSL reconstruction. For this region, Piecuch et al. (2018) subsequently showed that river discharge affects regional rather than just local RSL and that its influence is weak on timescales longer than a few years. Since the 1-cm thick sediment slices used to reconstruct RSL are inherently time averaged over years to decades the salt marsh record of RSL changes likely filters out any short-lived influence of river discharge.

The case study from Florida illustrates how pollution and pollen profiles must be interpreted as representing histories and events that act on local to global scales and that it is often inappropriate to apply an interpretation from one location to another if conditions (e.g., prevailing winds) are markedly different. It also shows how the causes of RSL can be difficult to untangle from a single reconstruction, but that comparison with instrumental data and reconstructions from other sites can help to resolve the relative importance of local-, regional-, and global-scale processes acting across different timescales, particularly when site geomorphology is taken into consideration.

9.4.3 New Zealand

In contrast to the North Atlantic region, salt marshes in the southern hemisphere are generally characterized by a thin (less than ~0.5 m) surface unit of salt marsh sediment overlying older tidal flat deposits (or a pre-Holocene substrate). This stratigraphy is a consequence of the RSL history and the availability of accommodation space. Late Holocene RSL has been falling in much of the southern hemisphere but most tide gauges now show rising RSL, signalling a reversal at some time in the past centuries. Gehrels et al. (2011) reconstructed RSL changes at Pounawea on the south coast of New Zealand's South Island from a thin sequence of salt marsh sediments (Fig. 9.7) to investigate the late Holocene RSL history and capture this reversal. The chronology of the reconstruction was based on a combination of radiocarbon and marker techniques. The latter are especially

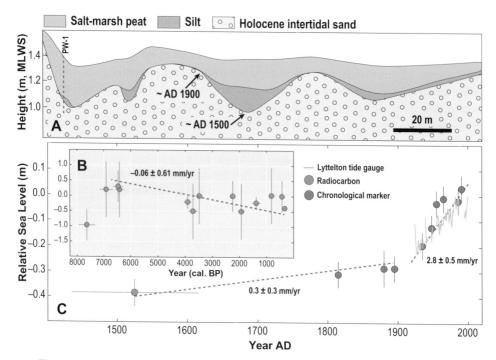

Figure 9.7 Relative sea-level reconstruction from salt marsh sediments at Pounawea, Catlins coast of the South Island, New Zealand. (a) Stratigraphy. (b) Holocene sea-level reconstruction derived from radiocarbon-dated sediment. (c) Recent sea-level reconstruction where sample age was determined by recognizing pollution trends and events of known age. Comparison to the nearby Lyttelton tide gauge (solid line) indicates that relative sea level is accurately reconstructed from salt-mars sediment. Linear rates of relative sea-level rise are shown. Redrawn and modified from Gehrels et al. (2008)

useful for the past 200 years and include analyses of Pb isotopes, which demonstrated that outfall of the 1815 Tambora volcanic eruption had been deposited on the marsh surface. Pb isotopes also reflected the pollution history of the area, including the introduction of unleaded gasoline in the 1980s. Pollen markers showed the clearing of forests, first by Polynesians in the early fourteenth century and since the 1850s by European settlers who also introduced exotic plant species 20 years later. A charcoal peak was correlated with the extensive forest fire of 1935, while the 1965 level was detected by analyses of ^{137}Cs.

The salt marsh at Pounawea is underlain by intertidal sands and has developed during the past ca. 500 years. The foraminiferal assemblage preserved in the oldest salt marsh sediments is dominated by *Haplophragmoides wilberti* which is evidence that these sediments were deposited in an upper salt marsh environment, close to the limit of the high spring tides. Therefore, the marsh represents a Mudge-type (encroaching) marsh, rather than a Shaler-type (flat-colonizing) marsh. Without the foraminiferal analyses this would have been difficult to determine. The stratigraphic succession shows that the intertidal sands must have fallen dry following the late Holocene RSL drop, before RSL began rising, which allowed upper salt marsh grasses to colonize the sandy substrate and provided the accommodation space for the salt marsh deposits to accumulate. For the first 400 years, salt marsh accretion was slow under conditions of very slow RSL rise, producing a silty salt marsh facies in which organics were poorly preserved. Around 1900 the RSL accelerated and the nature of deposition changed from silty to highly organic, but the marsh kept pace with RSL rise, as demonstrated by the appearance of the foraminifer *Trochamminita salsa*, also an upper salt marsh species. After 1950 the foraminiferal assemblage is more diverse and also includes the common salt marsh species *M. fusca*, *Trochammina inflata*, and *J. macrescens*, an assemblage indicative of middle marsh conditions. The microfossils and chronology show that 30 cm of RSL rise occurred in the past 100 years. The acceleration of RSL rise from the nineteenth into the twentieth century was very rapid and is reflected in the RSL reconstruction by a sharp inflexion around 1900. This case study from New Zealand demonstrates how RSL changes affect the stratigraphy and sediment types of the coastal zone and provide the vertical accommodation space in which sediments can accumulate.

9.4.4 Oregon, USA

The Cascadia subduction zone extends from Cape Mendocino in Northern California to Vancouver Island in British Columbia. Along this active margin the earthquake deformation cycle drives a characteristic and unusual pattern of RSL change that is recorded in, and can be reconstructed from, stratigraphy preserved beneath salt marshes. During the relatively protracted (lasting 100s–1,000s of years) interseismic phase of Cascadia's deformation cycle, strain accumulates along the offshore subduction zone. Part of this strain is expressed as uplift of the coast (a process causing RSL fall). When the strain is released during a great ($M_w >8$) megathrust earthquake, subsidence of the coast causes an instantaneous RSL rise. Co-seismic RSL changes are an example of pronounced disequilibrium between the rate of sedimentation and the rate of RSL rise. The RSL change caused by a megathrust earthquake on the Cascadia subduction is on the order of tens of centimeters to 1–2 m. Because this change occurs over a period of seconds to minutes, the rate of RSL rise is extreme even though the sediment filling the newly created accommodation space is deposited rapidly at rates of several millimetres to centimetres each year.

Nelson et al. (1995) sought to establish a co-seismic origin for an abrupt RSL change that occurred in 1700 along the Cascadia subduction zone (Fig. 9.8). This research aim necessitated a sampling regime that investigated a single event at multiple sites along the coastline. The stratigraphic record at each site was investigated to ensure that it met criteria to attribute RSL change to a great earthquake (Atwater et al. 1995, Nelson et al. 1996). Among these criteria were evidence of: (1) submergence; (2) sudden RSL rise; (3) a geographically widespread change; (4) synchroneity of changes among sites; and (5) presence of a tsunami deposit. Nelson et al. (1995) demonstrated that nine sites from Vancouver Island to northern California met these criteria for the 1700 event

We examine results from one site (Salmon River, Oregon) where Nelson et al. (1995) confirmed a coseismic origin for an abrupt RSL rise and Hawkes et al. (2011) later used

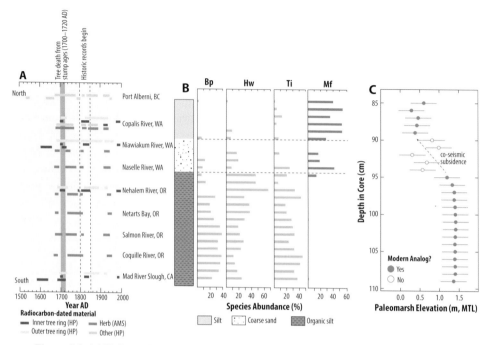

Figure 9.8 (a) Estimated timing of coseismic subsidence at nine salt marsh sites (ordered from north to south) along the Cascadia subduction zone. Bars represent possible calibrated age, where shading denotes the material that was dated. Regional-scale submergence of tree stumps at 1700–20 was caused by earthquake-driven subsidence (RSL rise). Historic records begin in ~1800 and do not include a great earthquake, therefore the event likely took place before this time. Modified from Nelson et al. (1995) (b) Foraminifera preserved in a core of coastal sediment collected from the Salmon River site in Oregon (after Hawkes et al. 2011). The pre-earthquake sediment is organic silt and includes high proportions of high salt marsh foraminifera (blue bars; Bp = *Balticammina pseudomacrescens*, Jm = *Jadammina macrescens*, Hw = *Haplophragmoides wilberti*). The sand overlying this unit was likely deposited by a tsunami that accompanied the earthquake. The uppermost unit of silt is dominated by low salt marsh foraminifera (red bars; Mf = *Miliammina fusca*). The abrupt change in sediment and foraminifera indicates rapid relative sea level rise. (c) Paleo-tidal elevation reconstructed using a regional-scale transfer function (errors are ~1σ). Open symbols represent samples that lacked an analog in the modern training set of Hawkes et al. (2010). These samples are from the tsunami-deposited sand, which includes a mixed assemblage. The reconstructed change in elevation indicates that coseismic subsidence caused by a great megathrust earthquake resulted in ~0.9 m of relative sea-level rise.

foraminifera to more precisely estimate the magnitude of subsidence that occurred in 1700. Hundreds of cores across multiple salt marshes in this estuary revealed that a unit of high salt marsh peat was abruptly overlain by silt and clay. Nelson et al. (1995) radiocarbon dated the leaves and stems of herbaceous plants that were rooted in the underlying unit of high salt marsh peat to estimate that the abrupt change in RSL occurred at ~1700. The choice and careful selection of material for radiocarbon dating was key (although there are issues related to plateaus in the recent part of the radiocarbon calibration curve). Leaves and stems of herbaceous plants are short-lived and delicate, meaning that an *in situ* date (as evidenced by being found in rooted position) likely records the timing of submergence accurately and precisely. Hawkes et al. (2011) counted foraminifera preserved in a single core at Salmon River and used a regional-scale transfer function (Hawkes et al. 2010) to reconstruct RSL. The underlying unit of high salt marsh peat was dominated by *Balticammina pseudomacrescns* and *H. wilberti*, which are characteristic of modern high salt marsh environments in Oregon. A 4-cm thick unit of sand containing a mixed assemblage of foraminifera and lacking a modern analog was interpreted as a tsunami deposit that transported individual tests from lower to higher elevation and also entrained tests eroded from high elevations. The overlying unit of silt and clay was dominated by *M. fusca*, which is analogous to the assemblage that occupies low salt marsh environments in Oregon today. The difference in reconstructed RWL between the top of the pre-earthquake peat unit and the bottom of the post-earthquake clastic unit was 0.60 ± 0.29 m, which was used as an estimate for the amount of subsidence that took place.

The study from Salmon River shows how methodologies for reconstructing RSL changes from salt marsh sediments are not only applicable to passive margin coastal settings, but equally useful along active coastlines and in circumstances where there is an extreme disequilibrium between sedimentation and sea-level change. Stratigraphic surveys can demonstrate abrupt changes in depositional environments. Microfossils can establish the magnitudes of co-seismic submergence (or emergence) and dating methods can establish synchronicity of stratigraphic contacts.

9.5 Concluding Remarks

We described how salt marsh sediments can be used to reconstruct Holocene RSL changes. The stratigraphy in some salt marshes will pose problems for the sea-level researcher, in particular in dynamic settings such as barrier marshes, or marshes that are heavily dissected by tidal creeks and channels. It is important that these complexities are recognized, and that an understanding is obtained of the long-term evolution of the marsh, be it the colonization of a tidal flat (Shaler-type marsh), or the transgressive onlap of the marsh against an upland (Mudge-type marsh) – the latter is more useful for RSL reconstructions. Marshes can change their "type" throughout their history and gain (or lose) their usefulness for RSL reconstructions. Microfossils (e.g., foraminifera, diatoms, testate amoebae) preserved in the sediments are accurate and precise sea-level indicators, especially in micro-tidal settings, while biogeochemical analyses (e.g., $\delta^{13}C$) provide additional constraints on indicative meaning of salt marsh sediment. Radiocarbon dating of plant macrofossils underpins the chronology of salt marsh based RSL reconstructions. For the past three to four centuries radiocarbon dating is supplemented by ^{210}Pb dating and identification of chrono-horizons from pollen and chemical analyses, or from short-lived radionuclides (^{137}Cs). Sophisticated techniques are commonly used to improve the statistical robustness of RSL reconstructions, including the application of transfer functions and age–depth modelling.

References

Allen, J. R. L. 1990. Constraints on measurements of sea-level movements from salt-marsh accretion rates. *Journal of the Geological Society of London*, 147: 5–7.

Allen, J. R. L. 2000. Morphodynamics of Holocene salt marshes: a review sketch from the Atlantic and Southern North Sea coasts of Europe. *Quaternary Science Reviews*, 19: 1155–1231.

Aretxabaleta, A. L., Ganju, N. K., Butman, B., and Signell, R. P. 2017. Observations and a linear model of water level in an interconnected inlet-bay system. *Journal of Geophysical Research Oceans*, 122: 2760–2780.

Atwater, B. F., Nelson, A. R., Clague, J. J., Carver, G. A., Bobrowsky, P. T., Bourgeois, J., and Darienzo, M. E. 1995. Summary of coastal geologic evidence for past great earthquakes at the Cascadia subduction zone. *Earthquake Spectra*, 11: 1–18.

Avnaim-Katav, S., Gehrels, W. R., Brown, L. N., Fard, E., and MacDonald, G. M. 2017. Distributions of salt-marsh foraminifera along the coast of SW California, USA: implications for sea-level reconstructions. *Marine Micropalaeontology*, 131: 25–43.

Barlow, N. L. M., Long, A. J., Saher, M. H., Gehrels, W. R., Garnett, M. H., and Scaife, R. G. 2014. Salt-marsh reconstructions of relative sea-level change in the North Atlantic during the last 2000 years. *Quaternary Science Reviews*, 99: 1–16.

Barlow, N. L. M., Shennan, I., Long, A. J., Gehrels, W. R., Saher, M., Woodroffe, S. A., and Hillier, C. 2013. Salt marshes as late Holocene tide gauges. *Global and Planetary Change*, 106: 90–110.

Barnett, R. L., Garneau, M., and Bernatchez, P. 2016. Salt-marsh sea-level indicators and transfer function development for the Magdalen Islands in the Gulf of St. Lawrence, Canada. *Marine Micropaleontology*, 122: 13–26.

Barnett, R. L., Gehrels, W. R., Charman, D. J., Saher, M. H., and Marshall, W. A. 2015. Late Holocene sea-level change in Arctic Norway. *Quaternary Science Reviews*, 107: 214–230.

Barnett, R. L., Newton, T. L., Charman, D. J., and Gehrels, W. R. 2017. Salt-marsh testate amoebae as precise and widespread indicators of sea-level change. *Earth Science Reviews*, 164: 193–207.

Bartram, W. 1791. *Travels through North and South Carolina, Georgia, East and West Florida*. Philadelphia, James and Johnson.

Bloom, A. L., and Stuiver, M. 1963. Submergence of the Connecticut coast. *Science*, 139: 332–334.

Bradley, W. H. 1953. Age of intertidal tree stumps at Robinhood, Maine. *American Journal of Science*, 251: 543–546.

Brain, M. J., Kemp, A. C., Hawkes, A. D., Engelhart, S. E., Vane, C. H., Cahill, N., Hill, T. D., Donnelly, J. P., and Horton, B. P. 2017. Exploring mechanisms of compaction in salt-marsh sediments using Common Era relative sea-level reconstructions. *Quaternary Science Reviews*, 167: 96–111.

Brain, M. J., Kemp, A. C., Horton, B. P., Culver, S. J., Parnell, A. C., and Cahill, N. 2015. Quantifying the contribution of sediment compaction to late Holocene salt-marsh sea-level reconstructions, North Carolina, USA. *Quaternary Research*, 83: 41–51.

Cahill, N., Kemp, A. C., Horton, B. P., and Parnell, A. C. 2016. A Bayesian hierarchical model for reconstructing relative sea level: from raw data to rates of change. *Climate of the Past*, 12: 525–542.

Callard, S. L., Gehrels, W. R., Morrison, B. V. and Grenfell, H. R. 2011. Suitability of salt-marsh foraminifera as proxy indicators of sea level in Tasmania. *Marine Micropaleontology*, 79: 121–131.

Chapman, V. J. 1940. Succession on the New England salt marshes. *Ecology*, 21: 279–282.

Charman, D. J., Roe, H. M., and Gehrels, W. R. 2002. Modern distribution of saltmarsh testate amoebae: regional variability of zonation and response to environmental variables. *Journal of Quaternary Science*, 17: 387–409.

Donnelly, J. P., Cleary, P., Newby, P., and Ettinger, R. 2004. Coupling instrumental and geological records of sea-level change: evidence from southern New England of an increase in the rate of sea-level rise in the 19th century. *Geophysical Research Letters*, 30, doi:10.1029/2003GL017801.

Edwards, R., and Wright, A. 2015. Foraminifera. In: I. Shennan, A. J. Long, and B. P. Horton, eds., *Handbook of Sea-Level Research*, Wiley, Chichester, pp. 191–217.

Edwards, R. J., van de Plassche, O., Gehrels, W. R., and Wright, A. J. 2004. Assessing sea-level data from Connecticut, USA, using a foraminiferal transfer function for tide level. *Marine Micropalaeontology*, 51: 239–255.

Eleuterius, L. N. 1976. The distribution of Juncus roemerianus in the salt marshes of North America. *Chesapeake Science*, 17: 289–292.

Engelhart, S. E., Horton, B. P., and Kemp, A. C. 2011b. Holocene sea-level changes along the United States' Atlantic Coast. *Oceanography*, 24: 70–79.

Engelhart, S. E., Horton, B. P., Nelson, A. R., Hawkes, A. D., Witter, R. C., Wang, K., Wang, P.-L., and Vane, C. H. 2013. Testing the use of microfossils to reconstruct great earthquakes at Cascadia. *Geology*, 41: 1067–1070.

Engelhart, S. E., Peltier W. R., and Horton, B. P. 2011a. Holocene relative sea- level changes and glacial isostatic adjustment of the U.S. Atlantic coast. *Geology*, 39: 751–754.

Engelhart, S. E., Vacchi, M., Horton, B. P., Nelson, A. R., and Kopp, R. E. 2015. A sea-level database for the Pacific coast of central North America. *Quaternary Science Reviews*, 113: 78–92.

Fatela, F., Taborda, R. 2002. Confidence limits of species proportions in microfossil assemblages. *Marine Micropaleontology*, 45: 169–174.

French, J. R., Spencer, T., Murray, A. L., and Arnold, N. S. 1995. Geostatistical analysis of sediment deposition in two small tidal wetlands. *Journal of Coastal Research*, 11: 308–321.

Gehrels, W. R. 1994. Determining relative sea-level change from salt-marsh foraminifera and plant zones on the coast of Maine, U.S.A. *Journal of Coastal Research*, 10: 990–1009.

Gehrels, W. R. 1999. Middle and late Holocene sea-level changes in eastern Maine reconstructed from foraminiferal saltmarsh stratigraphy and AMS 14C dates on basal peat. *Quaternary Research*, 52: 350–359.

Gehrels, W. R. 2000. Using foraminiferal transfer functions to produce high-resolution sea-level records from saltmarsh deposits, Maine, USA. *The Holocene*, 10: 367–376.

Gehrels, W. R. 2002. Intertidal foraminifera as palaeoenvironmental indicators. In: S. K. Haslett, ed., *Quaternary Environmental Micropalaeontology*, Arnold Publishers, New York, pp. 91–114.

Gehrels, W. R., and Belknap, D. F. 1993. Neotectonic history of eastern Maine evaluated from historic sea-level data and 14C dates on salt-marsh peats. *Geology*, 21: 615–618.

Gehrels, W. R., Belknap, D. F., and Kelley, J. T. 1996. Integrated high-precision analyses of Holocene relative sea-level changes: Lessons from the coast of Maine. *Geological Society of America Bulletin*, 108: 1073–1088.

Gehrels, W. R., Callard, S. L., Moss, P. T., Marshall, W. A., Blaauw, M., Hunter, J., Milton, J. A., and Garnett, M. H. 2012. Nineteenth and twentieth century sea-level changes in Tasmania and New Zealand. *Earth and Planetary Science Letters*, 315–316: 94–102.

Gehrels, W. R., Kirby, J. R., Prokoph, A., Newnham, R. M., Achterberg, E. P., Evans, E. H., Black, S., and Scott, D. B. 2005. Onset of recent rapid sea-level rise in the western Atlantic Ocean. *Quaternary Science Reviews*, 24: 2083–2100.

Gehrels, W. R., Marshall, W. A., Gehrels, M. J., Larsen, G., Kirby, J. R., Eiriksson, J., Heinemeier, J., and Shimmield, T. 2006. Rapid sea-level rise in the North Atlantic Ocean since the first half of the 19th century. *The Holocene*, 16: 948–964. Erratum, The Holocene 17: 419-420.

Gehrels, W. R., Roe, H. M., and Charman, D. J. 2001. Foraminifera, testate amoebae and diatoms as sea-level indicators in UK saltmarshes: a quantitative multiproxy approach. *Journal of Quaternary Science*, 16: 201–220.

Gehrels, W. R., and Shennan, I. 2015. Sea level in time and space: revolutions and inconvenient truths. *Journal of Quaternary Science*, 30: 131–143.

Gerlach, M. J., Engelhart, S. E., Kemp, A. C., Moyer, R. P., Smoak, J. M., Bernhardt, C. E., and Cahill, N. 2017. Reconstructing Common Era relative sea-level change on the Gulf Coast of Florida. *Marine Geology*, 390: 254–269.

González, J. L. and Törnqvist, T. E. 2009. A new Late Holocene sea-level record from the Mississippi Delta: Evidence for a climate/sea level connection? *Quaternary Science Reviews*, 28, 1737–1749.

Goslin, J., Sansjofre, P., Van Vliet-Lanoë, B., and Delacourt, C. 2017. Carbon stable isotope δ13C) and elemental TOC, TN) geochemistry in saltmarsh surface sediments Western Brittany, France): a useful tool for reconstructing Holocene relative sea-level. *Journal of Quaternary Science*, 32: 989–1007.

Guilbault, J. -P., Clague, J. J., and Lapointe, M. 1995. Amount of subsidence during a late Holocene earthquake – evidence from fossil tidal marsh foraminifera at Vancouver Island, west coast of Canada. *Palaeogeography, Palaeoclimatology, Palaeoecology*, 118: 49–71.

Hawkes, A. D., Horton, B. P., Nelson, A. R., and Hill, D. F. 2010. The application of intertidal foraminifera to reconstruct coastal subsidence during the giant Cascadia earthquake of AD 1700 in Oregon, USA. *Quaternary International*, 221: 116–140.

Hawkes, A. D., Horton, B. P., Nelson, A. R., Vane, C. H., and Sawai, Y. 2011. Coastal subsidence in Oregon, USA, during the giant Cascadia earthquake of AD 1700. *Quaternary Science Reviews*, 30: 364–376.

Horton, B. P. 1999. The distribution of contemporary intertidal foraminifera at Cowpen Marsh, Tees Estuary,UK: implications for studies of Holocene sea-level changes. *Palaeogeography, Palaeoclimatology, Palaeoecology*, 149: 127–149.

Horton, B. P., and Edwards, R. J. 2006. Quantifying Holocene sea level change using intertidal foraminifera: lessons from the British Isles. Retrieved from http://repository.upenn.edu/ees_papers/50.

Horton, B. P., Edwards, R. J., and Lloyd, J. M. 1999. A foraminiferal-based transfer function: Implications for sea- level studies. *Journal of Foraminiferal Research*, 29: 117–129.

Horton, B. P., Milker, Y., Dura, T., Wang, K., Bridgeland, W. T., Brophy, L., Ewald, M., et al. 2017. Microfossil measures of rapid sea-level rise: timing of response of two microfossil groups to a sudden tidal-flooding experiment in Cascadia. *Geology*, 45: 535–538.

Imbrie, J., and Kipp, N. G. 1971. A new micropaleontological method for quantitative paleoclimatology: Application to a late Pleistocene Caribbean core. In: K. K. Turekian, ed., *The Late Cenozoic Glacial Ages*, Yale University Press, New Haven, pp. 71–181.

Jelgersma, S. 1961. Holocene sea-level changes in the Netherlands. *Mededelingen Geologische Stichting C-IV*, 7: 1–100.

Kaye, C. A., and Barghoorn, E. S. 1964. Late Quaternary sea-level change and crustal rise at Boston, Massachusetts, with notes on the autocompaction of peat. *Geological Society of America Bulletin*, 75: 63–80.

Kemp, A. C., Cahill, N., Engelhart, S. E., Hawkes, A. E., and Wang, K. 2018. Revising estimates of spatially variable subsidence during the A.D. 1700 Cascadia earthquake using a Bayesian foraminiferal transfer function. *Bulletin of the Seismological Society of America*, 108: 654–673.

Kemp, A. C., Engelhart, S. E., Culver, S. J., Nelson, A., Briggs, R. W., and Haeussler, P. J. 2013. Modern salt-marsh and tidal-flat foraminifera from Sitkinak and Simeonof Islands, southwestern Alaska. *Journal of Foraminiferal Research*, 43: 88–98.

Kemp, A. C., Horton, B. P., Culver, S. J., Corbett, D. R., van de Plassche, O., Gehrels, W. R., Douglas, B. C., and Parnell, A. C. 2009. Timing and magnitude of recent accelerated sea-level rise (North Carolina, United States). *Geology*, 37: 1035–1038.

Kemp, A. C., Horton, B. P., Donnelly, J. P., Mann, M. E., Vermeer, M., and Rahmstorf, S. 2011. Climate related sea-level variations over the past two millennia. *Proceedings of the National Academy of Sciences of the USA*, 108: 11017–11022.

Kemp, A. C., Horton, B. P., Nikitina, D., Vane, C. H., Potapova, M., Weber-Bruya, E., Culver, S. J., Repkina, T., and Hill, D. F. 2017a. The distribution and utility of sea-level indicators in Eurasian sub-Arctic salt marshes White Sea, Russia *Boreas*, 46: 562–584.

Kemp, A. C., Kegel, J. J., Culver, S. J., Barber, D. C., Mallinson, D. J., Leorri, E., Bernhardt, C. E., Cahill, N., et al. 2017b. Extended late Holocene relative sea-level histories for North Carolina, USA. *Quaternary Science Reviews*, 160: 13–30.

Kemp, A. C., Sommerfield, C. K., Vane, C. H., Horton, B. P., Chenery, S., Anisfeld, S., and Nikitina, D. 2012. Use of lead isotopes for developing chronologies in recent salt-marsh sediments. *Quaternary Geochronology*, 12: 40–49.

Kemp, A. C., Telford, R. J. 2015. Transfer functions. In: I. Shennan, A. J. Long, and B. P. Horton, eds. *Handbook of Sea-Level Research*, Wiley, Chichester, pp. 470–499.

Kemp, A. C., Vane, C. H., Horton, B. P., Engelhart, S. E., and Nikitina, D. 2012. Application of stable carbon isotopes for reconstructing salt-marsh floral zones and relative sea level, New Jersey, USA. *Journal of Quaternary Science*, 27: 404–414.

Kemp, A. C., Wright, A. J., Barnett, R. L., Hawkes, A. D., Charman, D. J., Sameshima, C., King, A. N., et al. 2017c. Utilty of salt-marsh foraminifera, testate amoebae and bulk-sediment $\delta^{13}C$ values as sea-level indicators in Newfoundland, Canada. *Marine Micropaleontology*, 130: 43–59.

Kemp, A. C., Wright, A. J., Edwards, R. J., Barnett, R. L., Brain, M. J., Kopp, R. E., Cahill, N., et al. 2018. Relative sea-level change in Newfoundland, Canada during the past ~3000 years. *Quaternary Science Reviews*, 201: 89–110.

Kirwan, M. L., Temmerman, S., Skeehan, E. E., Guntenspergen, G. R., and Fagherazzi, S. 2016. Overestimation of marsh vulnerability to sea level rise. *Nature Climate Change*, 6: 253–260.

Kopp, R. E., Kemp, A. C., Bittermann, K., Horton, B. P., Donnelly, J. P., Gehrels, W. R., Hay, et al. 2016. Temperature-driven global sea-level variability in the Common Era. *Proceedings of the Natural Academy of Sciences of the United States of America*, 113: E1434–E1441.

Lamb, A. L., Wilson, G. P., and Leng, M. L. 2006. A review of coastal palaeoclimate and relative sea-level reconstructions using δ13C and C/N ratios in organic material. *Earth-Science Reviews*, 75: 29–57.

Le Coze, F., and Hayward, B. 2017. *Entzia macrescens Brady*, 1870 In: B. W. Hayward, F. Le Coze, and O. Gross World Foraminifera Database. Accessed at www .marinespecies.org/foraminifera/aphia.php?p=taxdetails&id=742429 on 2017-10-26.

Libby, W. F. 1961. Radiocarbon dating. *Science*, 133: 621–629.

Long, A. J., Barlow, N. L. M., Gehrels, W. R., Saher, M. H., Woodworth, P. L., Scaife, R. G., Brain, M. J., and Cahill, N. 2014. Contrasting records of sea-level change in the eastern and western North Atlantic during the last 300 years. *Earth and Planetary Science Letters*, 388: 110–122.

Lyell, C. 1849. A Second Visit to the United States of North America, in Two Volumes. Volume 1. New York, Harper and Brothers.

Marshall, W. A., Gehrels, W. R., Garnett, M. H., Freeman, S. P. H. T., Maden, C., and Xu, S. 2007. The use of "bomb spike" calibration and high-precision AMS 14C analyses to date salt-marsh sediments deposited during the past three centuries. *Quaternary Research*, 68: 325–337.

Mudge, B. F. 1858. The salt marsh formations of Lynn. *Proceedings of the Essex Institute*, 2: 117–119.

Murray, J. W. 1982. Benthic foraminifera: the variability of living, dead or total assemblages in the interpretation of palaeoecology. *Journal of Micropalaeontology*, 1: 137–140.

Nelson, A. R., Shennan, I., and Long, A. J. 1995. Identifying coseismic subsidence in tidal-wetland stratigraphic sequences at the Cascadia subduction zone of western North America. *Journal of Geophysical Research*, 101: 6115–6135.

Nikitina, D. L., Kemp, A. C., Horton, B. P., Vane, C. H., van de Plassche, O., and Engelhart, S. E. 2014. Storm erosion during the past 2000 years along the north shore of Delaware Bay, USA. *Geomorphology*, 208: 160–172.

Parnell, A. C., and Gehrels, W. R. 2015. Using chronological models in late Holocene sea level reconstructions from saltmarsh sediments. In: I. Shennan, A. J. Long, and B. P. Horton, eds. *Handbook of Sea-Level Research*, Wiley, Chichester, pp. 500–513.

Patterson, R. T., and Fishbein, E. 1989. Re-examination of the statistical methods used to determine the number of point counts needed for micropaleontological quantitative research. *Journal of Palaeontology*, 63: 245–248.

Payne, R. J., and Mitchell, E. A. D. 2009. How many is enough? Determining optimal count totals for ecological and palaeoecological studies of testate amoebae. *Journal of Paleolimnology*, 42: 483–495.

Piecuch, C. G., Bittermann, K., Kemp, A. C., Ponte, R. M., Little C. M., Engelhart, S. E., and Lentz, S. J. 2018. River-discharge effects on United States Atlantic and Gulf coast sea-level changes. *Proceedings of the National Academy of Sciences of the USA*, 30: 7729–7734.

Redfield, A. C. 1959. The Barnstable marsh. In R. A. Ragotzkie, L. R. Pomeroy, J. M. Teal, and D. C. Scott, eds., *Proceedings, Salt marsh Conference*, May 25–28, 1958, Sapelo Island, Athens, Georgia, University of Georgia, pp. 37–42.

Redfield, A. C. 1972. Development of a New England salt marsh. *Ecological Monographs*, 42: 210–237.

Redfield, A. C., and Rubin, M. 1962. The age of salt marsh peat and its relations to recent change in sea level at Barnstable, Massachusetts. *Proceedings of the National Academy of Sciences of the USA*, 48: 1728–1735.

Reimer, P. J., Bard, E., Bayliss, A., Beck, J. W., Blackwell, P. G., Bronk Ramsey, C., Buck, C. E., et al. 2013. IntCal13 and Marine13 radiocarbon age calibration curves 0–50,000 years cal BP. *Radiocarbon*, 55: 1869–1887.

Roe, H. M., Charman, D. J., and Gehrels, W. R. 2002. Fossil testate amoebae in coastal deposits in the UK: implications for studies of sea-level change. *Journal of Quaternary Science*, 17: 411–429.

Sachs, H. M. 1977. Paleoecological transfer functions. *Annual Review of Earth and Planetary Sciences*, 5: 159–178.

Saher, M. H., Gehrels, W. R., Barlow, N. L. M., Long, A. J., Haigh, I. D., and Blaauw, M. 2015. A 600-year multiproxy record of sea-level change and the influence of the North Atlantic Oscillation. *Quaternary Science Reviews*, 108: 23–36.

Scott, D. B., and Medioli, F. S. 1978. Vertical zonations of marsh foraminifera as accurate indicators of former sea-levels. *Nature*, 272: 528–531.

Scott, D. B., and Medioli, F. S. 1980a. Quantitative studies of marsh foraminifera distribution in Nova Scotia: implications for sea-level studies. *Cushman Foundation for Foraminiferal Research Special Publication*, 17: 1–58.

Shaler, N. S. 1886. Preliminary report on sea-coast swamps of the eastern United States. US Geological Survey 6th Annual Report, pp. 353–398.

Shennan, I., and Horton, B. 2002. Holocene land- and sea-level changes in Great Britain. *Journal of Quaternary Science*, 17: 511–526.

Shennan, I., Long, A.J., and Horton, B.P., eds. 2015. *Handbook of Sea-Level Research*, Wiley, Chichester.

Streif H. 1979. Cyclic formation of coastal deposits and their indications of vertical sea-level changes. *Oceanus*, 5: 303–306.

Strachan, K. L., Hill, T. R., Finch, J. M., and Barnett, R. L. 2015. Vertical zonation of foraminifera assemblages in Galpins Salt Marsh, South Africa. *Journal of Foraminiferal Research*, 45: 29–41.

Szkornik, K., Gehrels, W. R., and Murray, A. S. 2008. Aeolian sand movement and relative sea-level rise in Ho Bugt, western Denmark, during the Little Ice Age. *The Holocene*, 18: 951–965.

Törnqvist, T. E., van Ree, M. H. M., van 't Veer, R., and van Geel, B. 1998. Improving methodology for high-resolution reconstruction of sea-level rise and neotectonics by paleoecological analysis and AMS 14C dating of basal peats. *Quaternary Research*, 49: 72–85.

Vacchi, M., Engelhart, S. E., Nikitina, D., Ashe, E. L., Peltier, W. R., Roy, K., Kopp, R. E., and Horton, B. P. 2018. Postglacial relative sea-level histories along the eastern Canadian coastline. *Quaternary Science Reviews*, 201: 124–146.

Vacchi, M., Marriner, N., Morhange, C., Spada, G., Fontana, A., and Rovere, A. 2016. Multiproxy assessment of Holocene relative sea-level changes in the western Mediterranean: sea-level variability and improvements in the definition of the iso-static signal. *Earth-Science Reviews*, 155: 172–197.

Van de Plassche, O. 2000. North Atlantic climate-ocean variations and sea level in Long Island Sounds, Connecticut, since 500 cal yr AD. *Quaternary Research*, 53: 89–97.

Van de Plassche, O., Mook, W. G., and Bloom, A. L. 1989. Submergence of coastal Connecticut 6000-3000 14C) years B.P. *Marine Geology*, 86: 349–354.

Van de Plassche, O., Wright, A. J., van der Borg, K., and de Jong, A. F. M. 2004. On the erosive trail of a 14th and 15th century hurricane in Connecticut USA) salt marshes. *Radiocarbon*, 46: 775–784.

Van der Wal, D., and Pye, K. 2002. Patterns, rates and possible causes of saltmarsh erosion in the Greater Thames area UK. *Geomorphology*, 61: 373–391.

Watcham, E. P., Shennan, I., and Barlow, N. L. M. 2013. Scale considerations in using diatoms as indicators of sea-level change: lessons from Alaska. *Journal of Quaternary Science*, 28: 165–179.

Wells, B. W. 1928. Plant communities of the coastal plain of North Carolina and their successional relations. *Ecology*, 9: 230–242.

Wilson, G. P. 2017. On the application of contemporary bulk sediment organic carbon isotope and geochemical datasets for Holocene sea-level reconstruction in NW Europe. *Geochimica et Cosmochimica Acta*, 214: 191–208.

Woodroffe, S. A. 2009. Testing models of mid to late Holocene sea-level change, North Queensland, Australia. *Quaternary Science Reviews*, 28: 2474–2488.

Wright, A. J., Edwards, R. J., and van de Plassche, O. 2011. Reassessing transfer-function performance in sea-level reconstruction based on benthic salt-marsh foraminifera from the Atlantic coast of NE North America. *Marine Micropaleontology*, 81: 43–62.

Zoccorato, C., and Teatini, P. 2017. Numerical simulations of Holocene salt-marsh dynamics under the hypothesis of large soil deformations. *Advances in Water Resources*, 110: 107–119.

10

Storm Processes and Salt Marsh Dynamics

KATHERINE A. CASTAGNO, JEFFREY P. DONNELLY, AND JONATHAN D. WOODRUFF

10.1 Introduction

Marshes have long been considered useful for their ecosystem service of coastal protection. Their roles in protection from storms and floods are seen as necessary and important to many coastal communities (Barbier et al. 2011; Costanza et al. 1997; Millennium Ecosystem Assessment. 2005; Morgan et al. 2009). Understanding the impacts that storms have on coastal ecosystems and adjacent coastal communities is imperative to increasing coastal resilience in the face of future increases in coastal flooding and associated damage (Mendelsohn et al. 2012; Pielke et al. 2008). Salt marshes have been lauded as buffers to storm surges, wind-generated waves, and elevated water levels (French 2006; Möller 2012). The ecological restoration economy, which includes salt marsh restoration, in the USA alone generates $9.5 billion in annual economic output and employs an estimated 126,000 workers (BenDor et al. 2015). After Hurricane Sandy, the US Fish and Wildlife Service spent more than $40 million on salt marsh restoration projects in response to this single event, including $11 million toward restoring a series of salt marshes along Long Island.

Recent research, reviewed in this chapter, explores the impacts of storms on salt marsh erosion and accumulation, with an eye toward the utility of salt marshes for coastal protection in the face of rising sea levels and increasing storminess. Our review is primarily focused to the East and Gulf Coast of the USA as well as Europe. This is in part due to the concentration of researchers and the prevalence of marsh systems in these regions, compared to mangrove wetlands that tend to dominate the tropics (Fig. 10.1). Our review highlights the processes that occur in marshes resulting from storms (Fig. 10.2), and provides insight into the future response of marshes to changes in storminess (a discussion that complements that provided in Chapter 18 on future sea level rise impacts on marshes). The variability in storm impacts to marshes is largely based on both intensity and circumstance, with the potential for a spectrum of effects to be seen at a single marsh (Fig. 10.3; Cahoon 2006; Morton & Barras 2011). Though marshes may be generally resilient to coastal flooding on larger spatial and temporal scales, storms may have more subtle impacts, causing cascading effects on a variety of smaller scales. A single storm may result in overwash that provides sediment to build the marsh in some sections, while wave activity and higher channel velocities increases marsh-edge erosion/retreat in other sections

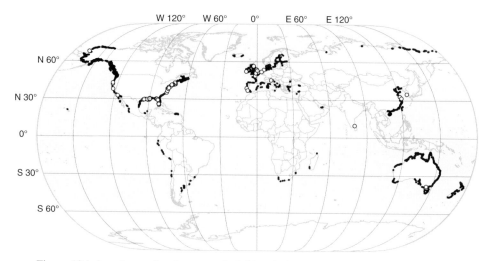

Figure 10.1 Locations of major research (white circles) on the effects of storms on salt marshes. Global marsh extent is indicated in black (from Mcowen et al. 2017). For a full list of research locations, please see http://arcg.is/1CCuXj. (A black and white version of this figure will appear in some formats. For the color version, please refer to the plate section.)

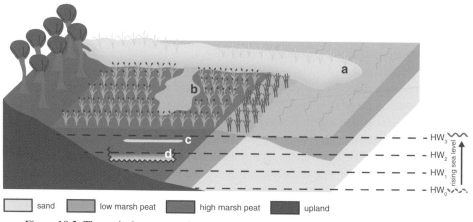

sand   low marsh peat   high marsh peat   upland

Figure 10.2 Theoretical structure of a salt marsh based on Redfield's bidirectional model of salt marsh evolution (Redfield 1972), including a coastal barrier (a). As sea level rises, the salt marsh spreads over both sand and upland through accumulation of sediment. Storm waves can overtop the barrier, leaving both present (b) and past (c) sandy overwash deposits in the marsh. Evidence of erosion (d) may also be seen. HW: High water.

due to scarping, undercutting, and slumping (Fig. 10.3). Given the prevalence of marshes worldwide and the likelihood of changes in storminess as climate changes (Walsh et al. 2016), it is important to understand the effectiveness of marshes in flood mitigation, as well as the role of storms in controlling marsh morphology. Our review begins with a discussion on the depositional record of storms in marshes, followed by an overview of modern storm impacts on marshes, and marsh attenuation of waves and storm surge.

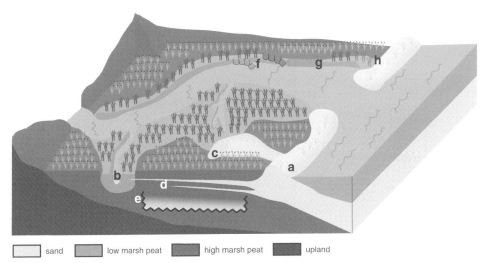

| | sand | | low marsh peat | | high marsh peat | | upland |

Figure 10.3 Theoretical structure of a salt marsh, including a coastal barrier (a) and channels (b). Impacts of storms on marshes exist along a spectrum based on a variety of factors, including intensity, circumstance, storm track, and marsh structure. Depositional features, such as present (c) and past (d) sandy overwash deposits, can occur concurrently with erosional features, such as stratigraphic evidence (e), edge erosion (f), slumping (g), and vegetation plucking (h).

10.2 Stratigraphic Evidence of Storms in Marshes

Marsh stratigraphy and sediments provide both accretional and erosional evidence of paleo-storms. With respect to accretion, storm-induced overwash deposits in salt marshes have been used to augment our understanding of hurricane strikes (e.g., Boldt et al. 2010; Donnelly et al. 2001a, 2001b). Elevated water levels from storm surge and waves can be significant enough to overtop sandy barriers, transporting and depositing coarse sediment on top of salt marsh peat. During the years following the storm, organic material accumulates/grows over the associated overwash layer, thus preserving the storm event as a stratigraphically distinct allochthonous layer. Donnelly et al. (2001b) described this sedimentary pattern in a salt marsh in southern Rhode Island (Fig. 10.4a). At this salt marsh, aerial photos after major hurricanes in 1938 and 1954 confirm overwash fans deposited by the events. These overwash fans, along with four older deposits, were confirmed in a series of 14 sediment cores, suggesting that several intense hurricanes had impacted the area since 1635. Other studies have also identified storm-induced deposits in marsh, including in the Gulf of Mexico (Williams 2012), South Carolina (Hippensteel 2008; Hippensteel & Martin 1999), and Georgia (Kiage et al. 2011). For a complete datebase of paleo-storm reconstructions from marshes and other back-barrier environments please see recent reviews by Oliva et al. (2018) and Muller et al. (2017).

van de Plassche et al. (1999, 2004, 2006) proposed that intense storms (under a specific, yet unknown, combination of conditions) may be the cause of preserved stratigraphic

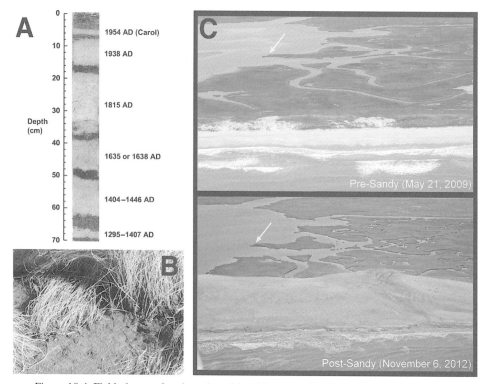

Figure 10.4 Field photos of various depositional impacts of storms on marshes. (a) Storm-induced overwash deposits from Succotash Marsh, East Matunuck, RI (modified from Cheung et al. 2007). (b) Example of sediment deposited on the marsh after Hurricanes Katrina and Rita (modified from Turner et al. 2006). (c) Aerial photos of overwash deposits left by Hurricane Sandy in Assateague Island, VA (modified from Sopkin et al. 2014). (A black and white version of this figure will appear in some formats. For the color version, please refer to the plate section.)

evidence of widespread erosional events in a Connecticut salt marsh. At Pattagansett River Marsh, two distinct erosional events were discovered (van de Plassche et al. 2006), where the portion of the marsh that had been eroded appeared to have been quickly infilled with tidal mud, which then transitioned to low and then high marsh peat. This was interpreted to represent a small-scale complete transition from marsh to tidal creek/mudflat system then back to a low-to-high marsh again. Though storms have been shown to cause smaller, localized erosional events, the large-scale unconformities observed by van de Plassche et al. (on the order of $>100\text{m}^2$) were surprising. Erosion of this scale has not been documented in modern salt marshes and did not appear to have deposited large blocks of excavated peat elsewhere. van de Plassche et al. (2006) found that the dates of their two identified erosional events corresponded with the age ranges of known hurricane deposits found at another salt marsh approximately 60 kilometers away (Donnelly et al. 2001a), including a hurricane in 1635, which is considered by many to be the most intense hurricane in the historical record to impact southern New England (Boose et al. 2001; Ludlum 1963).

Nikitina et al. (2014) expanded on the van de Plassche et al. (2006) research with sedimentary evidence of potential storm erosion from more than 200 gouge cores along seven transects in a salt marsh in New Jersey. The authors documented similar depositional sequences as those seen by van de Plassche et al. (2006) across great swathes of marsh: in at least seven sequences, there were abrupt contacts between salt marsh peat and overlying intertidal mud, suggesting that the underlying peat was eroded and then rapidly infilled by tidal mud (Fig. 10.5f). Though the authors suggested several different

Figure 10.5 Field photos of various erosional impacts of storms on marshes. (a) Eroding cliffs in the Westerschelde, the Netherlands (modified from van de Koppel et al. 2005). (b) Example of *Salicornia* spp. causing massive bank erosion in Tillingham, Essex, UK (modified from Möller 2012). c–e) Different kinds of marsh erosion (slumping, undercutting, and root scalping, respectively) in the Virginia Coast Reserve (modified from Fagherazzi et al. 2013). (f) Stratigraphic evidence of salt marsh peat overlain by tidal mud after a proposed erosional event in Sea Breeze Marsh, New Jersey (modified from Nikitina et al. 2014). (A black and white version of this figure will appear in some formats. For the color version, please refer to the plate section.)

processes that may produce these sequences, they developed a chronology that suggested that these events may have correlated with historic and prehistoric tropical cyclone events observed previously in the northeast (Donnelly et al. 2001b). The authors proposed that the most recent episodes of marsh erosion may correlate with tropical cyclones in 1903 AD, 1821/1788 AD, and 1635 AD.

Erosional features in marsh stratigraphy are attributed to storms largely through the discrediting of other potential mechanisms (Nikitina et al. 2014). Alternative explanations for the intertidal mud deposits overlying high marsh peat include an increase in tidal range (Long et al. 2006), gradual migration of tidal channels (Stumpf 1983), background rates of marsh-cliff retreat (McLoughlin et al. 2015), marsh pond formation (Wilson et al. 2009), and changes in sea level (Schwimmer & Pizzuto 2000). Migration of tidal channels was ruled out in the Nikitina et al. (2014) study since tidal creeks are generally considered stable over long timescales (Redfield 1972), and the process of tidal creek migration in the case of stabilizing vegetation loss is a slow time-transgressive process that would not result in the observed sharp contacts (Stumpf 1983). Though many marshes have been extensively ditched for mosquito control since the 1800s and the erosive contacts may represent infilling of ditches, the authors suggest that would limit those contacts to post-1800, which is not observed in the record. Such erosive sequences would then also be common across many marshes – an observation also not seen as yet. Salt pans, which form either from marsh growing around a tidal depression or after significant vegetation disturbance (Wilson et al. 2009), have a similar stratigraphic signature to that seen by Nikitina et al. (2014) and van de Plassche et al. (2006). The authors assert, however, that the stratigraphic sequences seen in their study site appear to be laterally continuous across large areas of the marsh. This observation is more consistent with a major erosional event than a localized salt pan. Additionally, while most of these ruled-out processes occur over longer timescales, higher-energy events could also be responsible for more rapid changes in tidal range, channel geometry, or cliff erosion, among others.

It is important to consider why the Nikitina et al. (2014) and van de Plassche et al. (2006) studies observed evidence of widespread marsh erosion that they attribute to hurricanes when other marshes in the vicinity do not (e.g. Donnelly et al. 2001a; Donnelly 2004; Miller et al. 2009). A variety of factors – including sampling biases, marsh geomorphology, marsh composition, storm track, and storm intensity – may play roles in observing eroding or depositing sediment in different locations or at different times. Indeed, sediment cores from marshes in New England suggest that a given location can experience both depositional and erosional events (Fig. 10.6). While the extensive erosion of marsh platforms by intense hurricanes remains an explanation for the features mapped in these marshes, the physical processes of erosion and the character and magnitude of the storms potentially responsible remains elusive. Smaller-scale erosional processes have been observed historically and therefore the mechanisms responsible for them are better understood.

Analysis of paleorecords shows that salt marshes and intense storm events have coexisted for thousands of years. Boldt et al. (2010) identified 30 distinct storm events (including seven historical severe landfalling hurricanes) over the past 2,000 years in a sediment

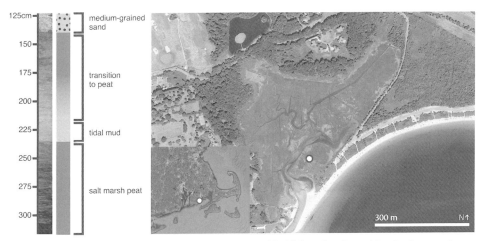

Figure 10.6 Evidence of erosional (sharp contact with tidal mud and transition back to peat) and depositional (medium-grained sand overwash) features in the same core from Round Hill Beach, Dartmouth, MA.

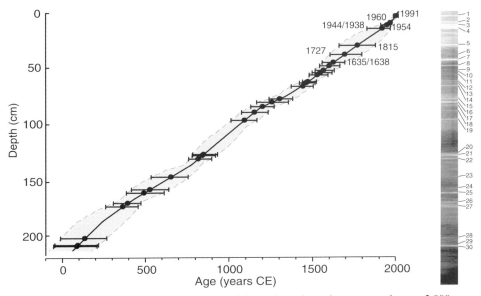

Figure 10.7 Evidence of continued storm activity and marsh persistence over the past 2,000 years from a salt marsh in Mattapoisett, MA (modified from Boldt et al. 2010). Left, radiocarbon-derived age model (gray) for 30 identified storm-induced overwash events (black circles, with uncertainties). Right, radiograph imagery identifying storm-induced overwash events.

core collected from a salt marsh in Mattapoisett, Massachussetts (Fig. 10.7). Though there have been periods of increased hurricane activity and quiescence in the last two millennia (Donnelly et al. 2015), cores from Mattapoisett suggest continuous marsh development. Despite significant variability in both storminess and storm impact over the past millennia

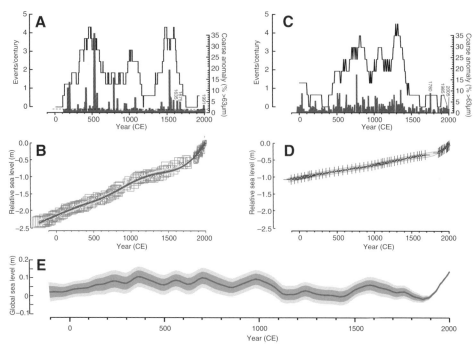

Figure 10.8 Though storminess has fluctuated over the past 2,000 years, sea level remained relatively stable prior to the Industrial Revolution. (a) Storm record for Salt Pond, Falmouth, MA, where peaks in coarse anomaly indicate a storm deposit (modified from Donnelly et al. 2015). (b) Relative sea-level reconstruction from East River Marsh, Guilford, CT, from salt marsh sediment dates and tide-gauge records (modified from Kemp et al. 2015). (c) Storm record for Mullet Pond, Apalachee Bay, FL (modified from Lane et al. 2011). (d) Relative sea-level reconstruction from Nassau Landing, FL, from salt marsh sediment dates and tide-gauge records (modified from Kemp et al. 2014). (e) Global sea-level curve for the past 2,000 years from statistical meta-analysis of proxy relative sea-level reconstructions and tide-gauge data. (Modified from Kopp et al. 2016)

(Fig. 10.8a and 10.8c), sea levels have remained relatively stable (Fig. 10.8b and 10.8d), with the most notable increase in global sea levels occurring in the last 150 years (Donnelly et al. 2015; Kemp et al. 2014, 2015; Kopp et al. 2016; Lane et al. 2011; Fig. 10.8e). Tandem increases in sea levels (as discussed in Chapter 17) and storminess may serve to multiply marsh vulnerability. The paleorecord shows that patterns of and mechanisms for erosion or deposition from intense storms are complex. These patterns and mechanisms will be explored in more detail in the following section on modern processes.

10.3 Modern Storm Impacts on Marshes

To determine the modern impacts of storm events on marshes, one must first consider the factors that contribute to marsh stability (Fig. 10.9). A detailed review on the topic of marsh stability is provided in Chapter 7. Here, we provide an abbreviated review on the topic

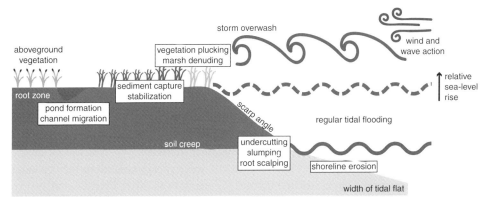

Figure 10.9 Diagram of different processes discussed in text related to storms that impact marsh stability. Several other processes not discussed in the text (e.g., ice rafting, vegetation loss from crabs and snails, availability of sediment, organic degradation rates, among many others) and not included in this diagram are discussed in Chapters 8 and 15.

specific to storm impacts. A series of biological and physical feedbacks allow marshes to vertically keep pace with sea-level rise. Following Redfield's bidirectional model of salt marsh evolution (Redfield 1965), as sea level rises and organic material and clastic sediment accumulates on the marsh, the marsh accretes vertically over basement material or mudflats (Fig. 10.2). Following the model described by Stumpf (1983), deposition of sediment on the marsh occurs predominantly with tidal flooding. As water from tidal currents floods the marsh surface, marsh vegetation slows the currents, trapping suspended sediment (Postma 1961). In addition to regular tidal flooding, storms can be a major source of sediment for the marsh (Donnelly 2004; Donnelly et al. 2001a, 2001b; Turner et al. 2006; Walters et al. 2014; Walters & Kirwan 2016; Fig. 10.4). Large storms can produce storm surge and wave heights significant enough to overtop sandy barriers, transporting and depositing coarse sediment on top of salt marsh peat. Highly turbid water from storms can also be carried into the marsh via creek networks along the back side of the marsh (Schuerch et al. 2013). Kolker et al. (2009) found a strong correlation between short-term sea-level change (e.g., storms) and increased accretion in marshes in Long Island, New York. Stumpf (1983) proposed that total sedimentation is from a combination of tidal and storm sources, with storms controlling sediment supply and movement on smaller time scales than daily tidal flooding, when considered over decades or centuries.

Though flooding during storms has documented impacts on marshes, in some cases very high storm surges over a marsh may actually help protect its surface from both erosion and accumulation of sediment. Elsey-Quirk (2016) measured the impacts of Hurricane Sandy (2012) on salt marshes in New Jersey, finding only localized and temporary marsh elevation changes and disturbances, despite Sandy making direct landfall. Similar conditions were seen after Hurricane Hugo (1989) in a salt marsh in South Carolina (Gardner et al. 1992). In both cases, the perpendicular coastal approach by both storms resulted in significant storm surges. The surge completely inundated the marsh, such that sediment associated with the surge was deposited more inland, being transported over and bypassing much of the marsh surface. Storm waves on top of the deep surge caused minimal erosion along the marsh surface or edge.

The marsh edge is subjected to constant stress from wind, waves, and currents. Field and model observations confirm that how a marsh edge erodes depends largely on the type of waves to which it is exposed. Leonardi and Fagherazzi (2014) found that, on small spatial scales, high storm wave energy conditions erode marsh boundaries uniformly, whereas low wave energy conditions result in a jagged erosion pattern – largely due to the influence of local marsh resistance. On larger, whole-marsh scales, boundaries of rapidly eroding marshes are significantly smoother than those of sheltered or slowly eroding marshes, though marshes experiencing low-wave energy conditions may be more susceptible to increased large, isolated failures (Leonardi et al. 2016; Priestas et al. 2015). There does not appear to be a threshold of wave power over which marsh erosion accelerates drastically. For example, Leonardi et al. (2016) identified a linear relationship between salt marsh erosion and wave power and determined that wind speeds associated with moderate storms ($1.6–10.7$ m s^{-1}) are associated with the greatest amounts of marsh deterioration due to their higher frequency of occurrence. In contrast, extreme wind speeds associated with rarer and more violent storms and hurricanes (>28.5 m s^{-1}) contribute less than 1% of long-term marsh deterioration along the lateral plane. Due to short observational periods, however, studies such as Leonardi et al. (2016) may miss more intense (e.g., >50 m s^{-1}), relatively rare events.

Morton and Barras (2011) explored patterns of wetland erosion and deposition following major hurricanes in southern Louisiana. The authors determined – through analysis of aerial photography, satellite imagery, and on the ground mapping – several depositional features associated with major storms, including wrack zones, interior-marsh deposits, and shoreline deposits. If a storm does not deposit an extremely thick wrack or sediment layer, which would kill supportive vegetation, depositional events generally favor marsh resilience. Overwash fans associated with Hurricanes Audrey (1957), Andrew (1992), Lili (2002), Rita (2005), Gustav (2008), and Ike (2008) often exceeded 50 meters in width, with washover terraces from Gustav as wide as 150 meters. These overwash deposits, however, were coupled with massive erosional signatures throughout the marshes, suggesting a more complicated relationship between marsh resilience and major storms. Following Hurricanes Audrey (1957), Hilda (1964), Andrew (1992), Katrina (2005), and Rita (2005), erosional features (including pond formation or expansion, marginal incised damage, braided channels, plucked or denunded marsh, floating-marsh redistribution, and shoreline erosion) were substantial. Hurricanes Katrina and Rita, for example, increased the water area in coastal Louisiana by 230 km^2 and 295 km^2, respectively (Barras 2007). Most substantially, ponds expanded or created by hurricane erosion tended to become permanent features – increasing the open water area substantially across southern Louisiana. Contrary to the previously mentioned findings of Leonardi et al. (2016), Barras (2007) asserts that storm-related erosion is the primary natural process by which marshes in southern Louisiana degrade, especially since the self-healing capabilities of marsh systems using sediment and nutrients from the Mississippi delta system have been hindered since land development in the late ninetheenth century. Barras (2007) suggests that increased wetland loss will increase overall vulnerability of southern Louisiana marshes to extreme storms, particularly in seasons with multiple storms, but the difficulty in predicting where and when extreme landfalls will occur makes it challenging to predict loss in specific areas at specific rates. Many of these marshes experiencing losses were fresh to brackish marshes, also emphasizing the importance of salinity for storm resilience (Howes et al. 2010).

Howes et al. (2010) determined that 2005 Hurricanes Katrina and Rita eroded more than 500 km² of wetlands within coastal Louisiana. Higher-salinity marshes were found to be more resilient to erosion than lower-salinity marshes, with lower-salinity marshes preferentially failing. Using field and lab measurements of soil strength, lower-salinity marshes were found to have a weak zone approximately 30 cm below the marsh surface (with shear strengths between 500 and 1,450 Pa). Higher-salinity marshes had no documented weak zones, with shear strengths greater than 4,500 Pa throughout the entire soil profile. Though controlled by a variety of factors, the difference is likely due to vegetation – *Spartina alterniflora* marshes are more tolerant to increased salinities, with more substantial, deeper root systems than less-saline marshes dominated by *Spartina patens*. The authors suggest that this research may have broader implications for current freshwater diversion plans. For example, the introduction of freshwater to marsh systems may decrease their resilience to large storms by creating weak zones within the marsh soil more susceptible to erosion.

As seen in the Howes et al. (2010) study, variations in vegetation assemblage influence both sediment capture and platform stabilization, and such differences in vegetation are controlled by several factors that include salinity, tidal range, wave energy, and climatic setting (Frey and Basan 1978; see Chapter 1 for a more detailed review on marsh vegetation and Chapter 8 for a more detailed review of ecogeomorphic feedbacks). In addition to the findings of Howes et al. (2010), Snedden et al. (2014) found a negative relationship between the duration of flood inundation during an experimental mesocosm study of above- and belowground *Spartina* biomass in the Mississippi deltaic plain, as did Watson et al. (2017) in southern New England marshes. Möller (2012) explored how individual plants can act as erosional agents, discovering that a stand of *Salicornia* spp. within a *Spartina* marsh may be responsible for large-scale bank failure in Tillingham, Essex, UK (Fig. 10.5b). These individual plants can interact with waves, causing small-scale erosion at the plant scale, which, when extended to each individual plant within a stand, can lead to erosion on massive scales. Goodbred and Hine (1995) explored the impacts of the March 1993 "Storm of the Century" in a *Juncus roemerianus*-dominated marsh in west-central Florida, where despite storm surges nearing 3 m and deposition of up to 2 cm of sediment on the marsh, there was minimal marsh shoreline erosion. The authors attribute this to increased sediment stabilization from dense *J. roemeranius* root mats, as well as decreased surface erosion from a reduction in near-bed flow velocities by its canopy (Leonard et al. 1995). Tate and Battaglia (2013), however, found *J. roemeranius* to be vulnerable to storm surge and wrack deposition in an experimental setup in a marsh on the Florida panhandle.

Using a three-dimensional hydrodynamic and sediment transport model for a salt marsh in the Netherlands, Temmerman et al. (2005) found that vegetative impacts on flow velocity may influence marsh erosion. Erosion of sediment occurs as flow velocity increases over unvegetated portions of the marsh. Accumulation of sediment occurs as flow velocity decreases at the vegetation edge and sediment is trapped. The water level of the flow in relation to the height of the vegetation is also important – when the water level exceeds the vegetation height, flow shifts from a more regional deposition–erosion pattern to larger-scale, homogenous sheet flow. A certain threshold, however, may exist above which stabilizing vegetation cannot recover from destruction by wave action, as explored by a numerical model of salt marsh development by van de Koppel et al. (2005). Storm-induced disturbances, therefore, may trigger cascading vegetation loss and marsh collapse.

The positive feedback between vegetation growth and sediment accumulation, such as that described by Temmerman et al. (2005) among others, continues until the marsh reaches a critical state where the marsh edge grows so steep that it is particularly vulnerable to erosion by storm waves. As the marsh collapses, it forms a scarp, which can retreat inland even in the absence of storm stress. The composition of the marsh sediment also plays a role in scarp retreat. Allen (1989) explored how sandy and muddy marsh scarp systems in estuaries in western Britain responded to wave attack. They found that muddy marsh systems in the Severn Estuary were more resilient to storm waves due to increased cohesion, often reach 5–10m in height, and only experience small cantilever failures after strong storms. Sandier marsh systems in Solway Firth and Morecambe Bay more frequently experience toppling and cantilever failures, with much of the soil strength limited to the first 10–20 cm of root-dominated sediment.

Baustian and Mendelssohn (2015) analyzed the recovery rate of plant cover in coastal Louisiana salt marshes following Hurricanes Gustav and Ike (2008). To determine marsh resilience in the absence of stabilizing vegetation, the study followed control and experimentally disturbed plots (treated with herbicide), and quarterly monitored rate of recovery – both prior to and after the impact of the 2008 hurricanes. The authors found that hurricane-induced sedimentation was highly correlated with aboveground primary vegetative production and increased recovery rates after the disturbance. The sedimentation from Hurricanes Gustav and Ike was found to increase the vigor of the vegetation in the experimental plots and, overall, increase marsh resilience. Though the authors acknowledge marsh destruction from the two hurricanes studied, they suggest that increased sedimentation may be beneficial in locally increasing marsh resilience to long-term sea-level rise and future storms.

A marsh requires sediment to accrete fast enough to ensure resilience to rising sea level. Walters and Kirwan (2016) suggest that overwash from hurricane waves actually may be responsible for increasing marsh resilience, citing an optimal thickness of deposit to maximize vegetation productivity, above which vegetation death begins to occur. Using a series of mesocosms to simulate a range of burial scenarios of *S. alterniflora* in a marsh in the Virginia Coast Reserve, the authors found that, though major overwash events (>10 cm) may cause marsh loss, smaller, more regular storm events (between 5 and 10 cm of overwash) contribute to the continued resilience of these marshes.

Turner et al. (2006) investigated the accumulation of inorganic sediment on coastal wetlands of Louisiana from Hurricanes Katrina and Rita and found that accumulation exceeded 131×10^6 metric tons over more than 38,500 km^2, with an average thickness of 5.18 cm. Sediment was not equally distributed throughout the marshes, with almost 5 m of storm surge depositing >10 cm of inorganic mud in some regions. Analyzing the spatial distribution and amount of sediment, the authors determine that major storm action is the dominant pathway by which offshore inorganic sediment moves inshore onto the coastal marshes of Louisiana. Burkett et al. (2007), however, suggest that this study fails to take into consideration the significant erosion that also occurred during these major storms, and did not adequately discern the true fluvial source of sediment sampled. Regardless, sedimentation from large storms in Louisiana (McKee & Cherry 2009; Tweel & Turner 2012) and elsewhere (Allison & Kepple 2001; Hu et al. 2018) remains well documented and modeled, and attempts at developing a sediment budget of a marsh system over the course of a storm are necessary to understand the dynamics and impacts of major storms on these systems (Ganju et al. 2015).

Storm sediment dynamics play a large role in the elevation of a marsh, which ultimately determines its resilience. Marsh elevation is a controlling factor in several ecogeomorphic feedbacks, including vegetation establishment, growth, and survival (Cahoon 2006). Cahoon (2006) reviewed major storm impacts on marsh elevations from Louisiana, Florida, North Carolina, Maryland, California, and Republic of Honduras. Storm impacts on elevation vary based on both storm and marsh properties. Storms impacted marsh elevation positively in a variety of ways, including sediment deposition, root growth, soil swelling, and lateral folding of the root mat. Storms impacted marsh elevation negatively through sediment erosion, sediment compaction, soil shrinkage, and root decomposition. For a more comprehensive review of ecogeomorphic feedbacks, please see Chapter 8.

10.4 Marsh Attenuation of Storm Waves and Surge

Salt marshes have recently been considered a critical resource in coastal protection (e.g. Gedan et al. 2011; Temmerman et al. 2013). Hurricanes Katrina and Rita both devastated southern Louisiana in 2005, but despite being of similar size and intensity, Hurricane Rita traveled over 30–50 km of wetland before reaching a main population center. In contrast, Hurricane Katrina traveled over a series of large lagoons, artificial channels, and highly degraded wetlands before reaching a main population center. In the case of the Rita event, marshes were able to accommodate extra surge, reducing the impact when the storm reached populated areas, while the surge for Katrina was far less attenuated (Day et al. 2007). Though the marshes in both storm paths experienced significant destruction (Day et al. 2007; Tweel & Turner 2012), the large extent of marshes in Hurricane Rita's path provided substantial protection. Wamsley et al. (2009) modeled the effects of Katrina and Rita on southern Louisiana marshes, finding similar results – increased restoration of marshes results in decreased storm surge and wave heights, though the amount of attenuation was highly variable among different marshes. Modeling results by Barbier et al. (2013) indicate that sea-to-land storm surge in Louisiana decreased as wetland continuity (i.e., presence of wetlands) and vegetation roughness (i.e., presence of vegetation in these wetlands) increased, and that a 10% increase in wetland continuity along a 6 km transect could save 3–5 homes/properties per storm. Some research has indicated that increased water depth associated with marsh erosion and conversion to a mudflat or open-water system may reduce storm surge levels (Loder et al. 2009). Decreased marsh continuity, however, is also associated with wetland loss increases storm surge levels and helps to highlight potential benefits of marsh restoration for coastal protection in many cases.

Stark et al. (2015) studied tidal and storm surge attenuation in a marsh along the Western Scheldt estuary, the Netherlands. The study site was a 4-km intertidal channel with a surrounding marsh platform of varying width. The authors found that maximum attenuation occurred along channels with wider marsh platforms. Interestingly, channels amplified smaller neap tides and attenuated larger spring tides. Optimal storm surge attenuation occurred when inundation heights were between 0.5 and 1 m. Inundation heights outside of this range were not attenuated as greatly, perhaps due to reduced bottom or vegetation friction. The authors propose that storm surge attenuation is both location- and event-specific, with storm surge reductions in the literature ranging from as much as 25 cm km^{-1} length of marsh, down to -2 cm km^{-1} where this negative value denotes actual amplification. This suggests that coastal management plans should consider the height and geometry of an individual marsh and the storm intensity threshold of interest on a case-by-case basis when considering their utility in mitigating storm surge.

Orton et al. (2015) ran hydrodynamic models for Jamaica Bay, New York City, to determine the utility of the site's marsh system for coastal protection. The Jamaica Bay marsh, though undergoing restoration beginning in 2006, has deep dredged channels throughout. Orton et al. (2015) found that the presence of these deep channels effectively short circuited the ability of the marsh to reduce storm surge, and that channel infilling would be far more beneficial for storm surge attenuation than the restoration of surrounding marshes, particularly for fast-moving events. Marsooli et al. (2016), however, showed that the Jamaica Bay marsh system did substantially reduce flow velocity of waves simulated for a tropical storm impacting the region in 2011 (Irene). Numerical modeling of morphological change caused by Hurricane Sandy (2012) found that the presence of marsh vegetation reduced erosion during coastal storms as well (Hu et al. 2018). Thus, though marshes may not be effective in all cases to mitigate storm surge, they may still prove to be highly effective as a means to attenuate storm waves.

Experimentally, lab-based research into the utility of marshes for attenuation of storm waves has involved the use of flume tanks or other in-lab procedures with natural or simulated marsh. Möller et al. (2014) used a 300-m flume tank to simulate a series of storm surge conditions on a transplanted natural marsh from the German Wadden Sea. The authors found that up to 60% of observed wave reduction in the marsh was due to the presence of vegetation. As the waves grew in intensity (heights up to 0.9 m in 2 m of water), the marsh continued to resist surface erosion, even as the waves began to flatten and break down vegetation. This suggests that wave attenuation by marsh vegetation is particularly important to continued resilience of marshes – especially since many field studies focus primarily on relatively low-energy conditions. Möller et al. (2014) built on the meta-analysis of Shepard et al. (2011), which found that smaller, more frequent waves are easily attenuated by marsh vegetation. Across all 30 studies considered by Shepard et al.

(2011), marsh vegetation was observed to play a significant role in shoreline stabilization. Although many studies focus on the utility of vegetation to attenuate high wave energy, a study by Feagin et al. (2009) suggests that soil type may be more influential than vegetation cover, with humic soils richer in fine-grained organics more resistant to erosion than soils rich in coarse-grained organics like roots or other plant debris. This is also consistent with the earlier mentioned study by Allen (1989), which showed a higher degree of stability for mud-dominated compared to sand-dominated marshes. Vegetation, however, does play a major role in these soil parameters, so it is important to consider how marsh ecosystem maintenance, restoration, and creation can continue to develop sustainable and ecologically sound coastal protection (Temmerman et al. 2013).

10.5 Regional Setting

Storms are a natural component of most ecological systems. Though paleorecords show significant variability in both storminess and storm impact, coastal salt marshes have thrived since their development over the Holocene (Engelhart et al. 2009; Shennan & Horton 2002). Intense storms have the capacity to dynamically impact marshes – particularly at meso- and microscales – but marshes also have the capacity to adapt and adjust to storm-induced changes, provided they have room to migrate. These dynamic impacts may be erosion of the marsh, accumulation of sediment on the marsh, or, more likely, a combination of both. Through the last several millennia, storm climate has varied considerably (Brandon et al. 2013; Donnelly et al. 2015), sea level has remained relatively stable (Kemp et al. 2014, 2015), and marshes have persisted (Figs. 10.7 and 10.8). Only in the last 150 years has sea-level risen at a significant rate (Fig. 10.8e), potentially leading marshes to become more vulnerable to storm impacts. Even so, a recent preliminary assessment of aerial photography before and after Hurricanes Michael (2018; category 4 landfall in Florida) and Florence (2018; category 1 landfall in North Carolina) shows only limited accumulation and erosion effects. This speaks to the ongoing resilience of salt marshes in the face of intense storms.

It is important to consider that marshes vary widely in terms of their geologic, geomorphic, geographic, and climatic setting. *J. roemeranius* marshes, primarily located in the southeastern USA, are often more directly in the path of tropical cyclones. South-facing *Spartina*-dominated marshes in the northeastern USA may be more susceptible to tropical cyclones tracking up the east coast, whereas north-facing New England marshes may be more susceptible to damage from nor'easters. More protected marshes (e.g., those located up drowned river valleys) may be less susceptible to wave attack and surge associated with less intense storms, whereas coastal marshes are exposed to wave attack both daily and from storms of varying intensities. Storm impacts also are often highly variable on smaller scales, dependent on both storm characteristics (proximity to marsh, angle of approach, tide levels, rainfall levels, wind speeds, etc.; Resio & Westerink 2008) and marsh properties (marsh health, soil strength, local bathymetry, groundwater levels, etc.; Cahoon 2006). The interplay among all of these properties dictate the impact of a particular storm on a given marsh, and different impacts from the same storm may be seen at the same marsh or proximal marshes.

10.6 Summary

The appreciation of marshes for their storm protection and erosion control services continues to grow and is largely responsible for over an order of magnitude increase in the appraised value of tidal marshes in recent decades (e.g. from $14k to $194k ha^{-1}yr^{-1} between 1997 and 2014; Costanza et al. 2014). Critical to this increase in value is the perceived effectiveness of marshes in storm protection. Marsh sediments preserve evidence of storm-induced erosion and deposition that has and should continue to be utilized to gain more insight on both early historic and pre-historic storm events. This sedimentological evidence of past storm events also exhibits a high degree of spatial variability, both in terms of deposition and erosion. Studies of modern marsh processes confirm the spatial temporal variance in marsh response to storms. When depositional conditions occur, recent research indicates that storms can be a valuable mechanism for marsh resilience to sea-level rise. However, erosion primarily on the marsh edge causes systems in open water to be in a continued state of flux. Minor and moderate storms have been found to largely be responsible for most marsh edge erosion, although catastrophic loss from more intense events has the potential to significantly modify a marsh system. The diversity of responses and rates of erosion and accretion of different marshes during storms also highlight the importance that vegetative species and sediment composition play.

The effectiveness of marshes as a form of flood mitigation is an area of active research. It is clear that marshes provide an effective means of wave attenuation; however, the capacity of these systems to reduce storm surge varies significantly from site to site. As investments in ecosystem-defenses continue to grow, so should research on marsh–storm relationships. Of particular importance are the spatial and temporal factors that determine a marsh's effectiveness in storm flood mitigation, as it is likely that costs and benefits will vary widely depending on the specific setting.

References

Allen, J. R. L. 1989. Evolution of salt-marsh cliffs in muddy and sandy systems: A qualitative comparison of British West-Coast estuaries. *Earth Surface Processes and Landforms*, 14: 85–92.

Allison, M. A., and Kepple, E. 2001. Modern sediment supply to the lower delta plain of the Ganges-Brahmaputra River in Bangladesh. *Geo-Marine Letters*, 21: 66–74.

Barbier, E. B., Georgiou, I. Y., Enchelmeyer, B., and Reed, D. J. 2013. The value of wetlands in protecting Southeast Louisiana from hurricane storm surges. *PLOS ONE*, 8: 1–6.

Barbier, E. B., Hacker, S. D., Kennedy, C., Koch, E. W., Stier, A. C., and Silliman, B. R. 2011. The value of estuarine and coastal ecosystem services. *Ecological Monographs*, 81: 169–193.

Barras, J. A. 2007. Land area changes in coastal Louisiana after Hurricanes Katrina and Rita. In G. S. Farris, G. J. Smith, M. P. Crane, C. R. Demas, L. L. Robbins, and D. L. Lavoie, eds., *Science and the Storms: The USGS Response to the Hurricanes of 2005* pp. 97–112. U.S. Geological Survey Circular 1306.

Baustian, J. J., and Mendelssohn, I. A. 2015. Hurricane-induced sedimentation improves marsh resilience and vegetation vigor under high rates of relative sea level rise. *Wetlands*, 35: 795–802.

BenDor, T., Lester, T. W., Livengood, A., Davis, A., and Yonavjak, L. 2015. Estimating the size and impact of the ecological restoration economy. *PLOS ONE*, 10: 1–15. https://doi.org/10.1371/journal.pone.0128339

Boldt, K. V., Lane, P., Woodruff, J. D., and Donnelly, J. P. 2010. Calibrating a sedimentary record of overwash from Southeastern New England using modeled historic hurricane surges. *Marine Geology*, 275: 127–139.

Boose, E. R., Chamberlin, K. E., and Foster, D. R. 2001. Landscape and regional impacts of hurricanes in New England. *Ecological Monographs*, 71: 27–48.

Brandon, C. M., Woodruff, J. D., Lane, D. P., and Donnelly, J. P. 2013. Tropical cyclone wind speed constraints from resultant storm surge deposition: A 2500 year reconstruction of hurricane activity from St. Marks, FL. *Geochemistry, Geophysics, Geosystems*, 14: 2993–3008.

Burkett, V., Groat, C. G., and Reed, D. 2007. Hurricanes not the key to a sustainable coast. *Science*, 315: 1366–1367.

Cahoon, D. R. 2006. A review of major storm impacts on coastal wetland elevations. *Estuaries and Coasts*, 29: 889–898.

Costanza, R., d'Arge, R., de Groot, R., Farber, S., Grasso, M., Hannon, B., van den Belt, M.. Limburg, K. et al. 1997. The value of the world's ecosystem services and natural capital. *Nature*, 387: 253–260.

Costanza, R., de Groot, R., Sutton, P., van der Ploeg, S., Anderson, S. J., Kubiszewski, I., Farber, S., and Turner, R. K. 2014. Changes in the global value of ecosystem services. *Global Environmental Change*, 26: 152–158.

Craft, C., Clough, J., Ehman, J., Jove, S., Park, R., Pennings, S., Guo, H. and Machmuller, M. 2009. Forecasting the effects of accelerated sea-level rise on tidal marsh ecosystem services. *Frontiers in Ecology and the Environment*, 7: 73–78.

Day, J. W., Boesch, D. F., Clairain, E. J., Kemp, G. P., Laska, S. B., Mitsch, W. J., Orth K. et al. 2007. Restoration of the Mississippi Delta: Lessons from Hurricanes Katrina and Rita. *Science*, 315: 1679–1684.

Donnelly, J. P. 2004. Coupling instrumental and geological records of sea-level change: Evidence from southern New England of an increase in the rate of sea-level rise in the late 19th century. *Geophysical Research Letters*, 31: 2–5.

Donnelly, J. P., Hawkes, A. D., Lane, P., Macdonald, D., Shuman, B. N., Toomey, M. R., van Hengstum, P. J., and Woodruff, J. D. 2015. Climate forcing of unprecedented intense-hurricane activity in the last 2000 years. *Earth's Future*, 3: 49–65.

Donnelly, J. P., Roll, S., Wengren, M., Butler, J., Lederer, R., and Webb, T. 2001. Sedimentary evidence of intense hurricane strikes from New Jersey. *Geology*, 29: 615–618.

Donnelly, J. P., Smith Bryant, S., Butler, J., Dowling, J., Fan, L., Hausmann, N., Newby, P., et al. 2001. 700 yr sedimentary record of intense hurricane landfalls in southern New England. *Geological Society of America Bulletin*, 113: 714–727.

Elsey-Quirk, T. 2016. Impact of Hurricane Sandy on salt marshes of New Jersey. *Estuarine, Coastal and Shelf Science*, 183: 235–248.

Engelhart, S. E., Horton, B. P., Douglas, B. C., Peltier, W. R., and Törnqvist, T. E. 2009. Spatial variability of late Holocene and 20th century sea-level rise along the Atlantic coast of the United States. *Geology*, 37: 1115–1118.

Fagherazzi, S., Mariotti, G., Wiberg, P. L., and McGlathery, K. J. 2013. Marsh collapse does not require sea level rise. *Oceanography*, 26: 70–77.

Feagin, R. A., Lozada-Bernard, S. M., Ravens, T. M., Moller, I., Yeager, K. M., and Baird, A. H. 2009. Does vegetation prevent wave erosion of salt marsh edges? *Proceedings of the National Academy of Sciences of the USA*, 106: 10109–10113.

French, J. 2006. Tidal marsh sedimentation and resilience to environmental change: Exploratory modelling of tidal, sea-level and sediment supply forcing in predominantly allochthonous systems. *Marine Geology*, 235: 119–136.

Frey, R. W., and Basan, P. B. 1978. Coastal salt marshes. In R. A. Davis Jr., ed., *Coastal Sedimentary Environments*. New York: Springer, pp. 225–302.

Ganju, N. K., Kirwan, M. L., Dickhudt, P. J., Guntenspergen, G. R., Cahoon, D. R., and Kroeger, K. D. 2015. Sediment transport-based metrics of wetland stability. *Geophysical Research Letters*, 42: 7992–8000.

Gardner, L. R., Michener, W. K., Kjerve, B., and Lipscomb, D. J. 1992. Disturbance effects of Hurricane Hugo on a pristine coastal landscape: North Inlet, South Carolina, USA. *Netherlands Journal of Sea Reasearch*, 30: 249–263.

Gedan, K. B., Altieri, A. H., and Bertness, M. D. 2011. Uncertain future of New England salt marshes. *Marine Ecology Progress Series*, 434: 229–237.

Goodbred, S. L., and Hine, A. C. 1995. Coastal storm deposition: Salt-marsh response to a severe extratropical storm, March 1993, west-central Florida. *Geology*, 23: 679–682.

Hippensteel, S. P. 2008. Preservation potential of storm deposits in South Carolina back-barrier marshes. *Journal of Coastal Research*, 243: 594–601.

Hippensteel, S. P., and Martin, R. E. 1999. Foraminifera as an indicator of overwash deposits, Barrier Island sediment supply, and Barrier Island evolution: Folly Island, South Carolina. *Palaeogeography, Palaeoclimatology, Palaeoecology*, 149: 115–125.

Howes, N. C., FitzGerald, D. M., Hughes, Z. J., Georgiou, I. Y., Kulp, M. A., Miner, M. D., Smith, J. M., Barras, J. A. 2010. Hurricane-induced failure of low salinity wetlands. *Proceedings of the National Academy of Sciences of the USA*, 107: 14014–14019.

Hu, K., Chen, Q., Wang, H., Hartig, E. K., and Orton, P. M. 2018. Numerical modeling of salt marsh morphological change induced by Hurricane Sandy. *Coastal Engineering*, 132: 63–81.

Kemp, A. C., Bernhardt, C. E., Horton, B. P., Kopp, R. E., Vane, C. H., Peltier, W. R., Hawkes, A. D., et al. 2014. Late Holocene sea- and land-level change on the U.S. southeastern Atlantic coast. *Marine Geology*, 357: 90–100.

Kemp, A. C., Hawkes, A. D., Donnelly, J. P., Vane, C. H., Horton, B. P., Hill, T. D., Anisfeld, S. C. et al. 2015. Relative sea-level change in Connecticut USA) during the last 2200 yrs. *Earth and Planetary Science Letters*, 428: 217–229.

Kiage, L., Deocampo, D., Mccloskey, T. A., Bianchette T. A., and Hursey, M. 2011. A 1900-year paleohurricane record from Wassaw Island, Georgia, USA. *Journal of Quaterary Science*, 26: 714–722.

Kirwan, M. L., Temmerman, S., Skeehan, E. E., Guntenspergen, G. R., and Faghe, S. 2016. Overestimation of marsh vulnerability to sea level rise. *Nature Climate Change*, 6: 253–260.

Kolker, A. S., Goodbred, S. L., Hameed, S., and Cochran, J. K. 2009. High-resolution records of the response of coastal wetland systems to long-term and short-term sea-level variability. *Estuarine, Coastal and Shelf Science*, 84: 493–508.

Kopp, R. E., Kemp, A. C., Bittermann, K., Horton, B. P., Donnelly, J. P., Gehrels, W. R., Hay, C. C., et al. 2016. Temperature-driven global sea-level variability in the Common Era. *Proceedings of the National Academy of Sciences of the USA*, 113: 1–8.

van de Koppel, J., van der Wal, D., Bakker, J. P., and Herman, P. M. J. 2005. Self-organization and vegetation collapse in salt marsh ecosystems. *The American Naturalist*, 165: E1–E12.

Lane, P., Donnelly, J. P., Woodruff, J. D., and Hawkes, A. D. 2011. A decadally-resolved paleohurricane record archived in the late Holocene sediments of a Florida sinkhole. *Marine Geology*, 287: 14–30.

Leonard, L. A., Hine, A. C., and Luther, M. E. 1995. Surficial sediment transport and deposition processes in a *Juncus roemerianus* marsh. *Journal of Coastal Research*, 11: 322–336.

Leonardi, N., Defne, Z., Ganju, N. K., and Fagherazzi, S. 2016. Salt marsh erosion rates and boundary features in a shallow bay. *Journal of Geophysical Research: Earth Surface*, 121: 1861–1875.

Leonardi, N., and Fagherazzi, S. 2014. How waves shape salt marshes. *Geology*, 42: 887–890.

Leonardi, N., Ganju, N. K., and Fagherazzi, S. 2016. A linear relationship between wave power and erosion determines salt-marsh resilience to violent storms and hurricanes. *Proceedings of the National Academy of Sciences of the USA*, 113: 64–68.

Loder, N. M., Irish, J. L., Cialone, M. A., and Wamsley, T. V. 2009. Sensitivity of hurricane surge to morphological parameters of coastal wetlands. *Estuarine, Coastal and Shelf Science*, 84: 625–636.

Long, A. J., Waller, M. P., and Stupples, P. 2006. Driving mechanisms of coastal change: Peat compaction and the destruction of late Holocene coastal wetlands. *Marine Geology*, 2251–4: 63–84.

Ludlum, D. M. 1963. *Early American Hurricanes*. Boston, MA: American Meterological Society.

Marsooli, R., Orton, P. M., Georgas, N., and Blumberg, A. F. 2016. Three-dimensional hydrodynamic modeling of coastal flood mitigation by wetlands. *Coastal Engineering*, 111: 83–94.

Mcowen, C. J., Weatherdon, L. V., Bochove, J.-W. V. Sullivan, E., Blyth, S., Zockler, C., Stanwell-Smith, D., *et al.* 2017. A global map of saltmarshes. *Biodiversity Data Journal*, 5: 10.3897/BDJ.5.e11764.

McKee, K. L., and Cherry, J. A. 2009. Hurricane Katrina sediment slowed elevation loss in subsiding brackish marshes of the Mississippi River delta. *Wetlands*, 29: 2–15.

McLoughlin, S. M., Wiberg, P. L., Safak, I., and McGlathery, K. J. 2015. Rates and forcing of marsh edge erosion in a shallow coastal bay. *Estuaries and Coasts*, 38: 620–638.

Mendelsohn, R., Emanuel, K., Chonabayashi, S., and Bakkensen, L. 2012. The impact of climate change on global tropical cyclone damage. *Nature Climate Change*, 2: 205–209.

Millennium Ecosystem Assessment 2005. *Ecosystems and human well-being: Wetlands and water synthesis. Millennium Ecosystem Assessment*. Washington, DC.

Miller, K. G., Sugarman, P. J., Browning, J. V., Horton, B. P., Stanley, A., Kahn, A., Uptegrove J., and Aucott, M. 2009. Sea-level rise in New Jersey over the past 5000 years: Implications to anthropogenic changes. *Global and Planetary Change*, 66: 10–18.

Möller, I. 2012. Bio-physical linkages in coastal wetlands – implications for coastal protection. *Crossing Borders in Coastal Research: Jubilee Conference Proceedings*. https://doi.org/10.3990/2.170

Möller, I., Kudella, M., Rupprecht, F., Spencer, T., Paul, M., van Wesenbeeck, B. K., Wolters, G., et al. 2014. Wave attenuation over coastal salt marshes under storm surge conditions. *Nature Geoscience*, 7: 727–731.

Morgan, P. A., Burdick, D. M., and Short, F. T. 2009. The functions and values of fringing salt marshes in northern New England, USA. *Estuaries and Coasts*, 32: 483–495.

Morton, R. A., and Barras, J. A. 2011. Hurricane impacts on coastal wetlands: A half-century record of storm-generated features from Southern Louisiana. *Journal of Coastal Research*, 27: 27–43.

Muller, J., Collins, J. M., Gibson, S. and Paxton, L. 2017. Recent advances in the emerging field of paleotempestology. In: *Hurricanes and Climate Change*. Springer, Cham, pp. 1–33.

Nikitina, D. L., Kemp, A. C., Horton, B. P., Vane, C. H., van de Plassche, O., and Engelhart, S. E. 2014. Storm erosion during the past 2000 years along the north shore of Delaware Bay, USA. *Geomorphology*, 208: 160–172.

Oliva, F., Viau, A. E., Peros, M. C., and Bouchard, M. 2018. Paleotempestology database for the western North Atlantic basin. *Holocene*, 28: 1664–1671.

Orton, P., Talke, S., Jay, D., Yin, L., Blumberg, A., Georgas, N., Zhao, H., Roberts, H. J., and MacManus, K. 2015. Channel shallowing as mitigation of coastal flooding. *Journal of Marine Science and Engineering*, 3: 654–673.

Pielke Jr, R., Gratz, J., and Landsea, C. 2008. Normalized hurricane damage in the United States: 1900–2005. *Natural Hazards Review*, 29–42. https://ascelibrary.org/doi/10 .1061/%28ASCE%291527-6988%282008%299%3A1%2829%29

van de Plassche, O., Erkens, G., van Vliet, F., Brandsma, J., van der Borg, K., and de Jong, A. F. M. 2006. Salt-marsh erosion associated with hurricane landfall in southern New England in the fifteenth and seventeenth centuries. *Geology*, 34: 829–832.

van de Plassche, O., van der Borg, K., and de Jong, A. F. M. 1999. Sea level – climate correlation during the past 1400 yr. *Geology*, 26: 319–322.

van de Plassche, O., Wright, A. J., van der Borg, K., and de Jong, A. F. M. 2004. On the erosive trail of a 14th and 15th century hurricane in Connecticut (USA) salt marshes. *Radiocarbon*, 46: 1111–1150.

Postma, H. 1961. Transport and accumulation of suspended matter in the Dutch Wadden Sea. *Netherlands Journal of Sea Reasearch*, 1: 148–190.

Priestas, A., Mariotti, G., Leonardi, N., and Fagherazzi, S. 2015. Coupled wave energy and erosion dynamics along a salt marsh boundary, Hog Island Bay, Virginia, USA. *Journal of Marine Science and Engineering*, 3: 1041–1065.

Redfield, A. 1972. Development of a New England Salt Marsh. *Ecological Monographs*, 42: 201–237.

Redfield, A. C. 1965. Ontogeny of a salt marsh estuary. *Science*, 147: 50–55.

Resio, D. T., and Westerink, J. J. 2008. Modeling the physics of storm surges. *Physics Today*, 61: 33–38.

Schuerch, M., Vafeidis, A., Slawig, T., and Temmerman, S. 2013. Modeling the influence of changing storm patterns on the ability of a salt marsh to keep pace with sea level rise. *Journal of Geophysical Research: Earth Surface*, 118: 84–96.

Schwimmer, R. A., and Pizzuto, J. E. 2000. A model for the evolution of marsh shorelines. *Journal of Sedimentary Research*, 70: 1026–1035.

Shennan, I., and Horton, B. 2002. Holocene land- and sea-level changes in Great Britain. *Journal of Quaternary Science*, 175–6: 511–526.

Shepard, C. C., Crain, C. M., and Beck, M. W. 2011. The protective role of coastal marshes: A systematic review and meta-analysis. *PLOS ONE*, 6: e27374. https://doi .org/10.1371/journal.pone.0027374

Snedden, G. A., Cretini, K., and Patton, B. 2014. Inundation and salinity impacts to above- and belowground productivity in *Spartina patens* and *Spartina alterniflora* in the

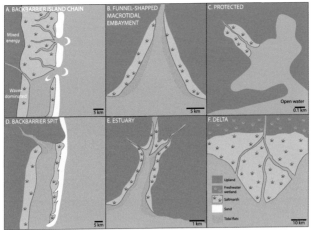

Figure 1.1 Broad types of salt marsh environments, including: (A) *Backbarrier Island Chains* (e.g., East and Gulf Coasts of USA; Algarve, Portugal; Frisian Islands, Germany); (B) *Funnel-shaped Macro-tidal Embayment* (e.g., The Wash, England; Mouth of Elbe River, Germany; Mont St. Michael Bay, France; Nushagak Bay, AK); (C) *Protected* (e.g., Sunborn Cove, Gouldsboro, ME; Etang de Toulvern, Bretagne, France); (D) *Backbarrier Spit* (e.g., Long Beach, WA; Cape Romain, SC; Hashirikotan barrier spit, Japan); (E) *Estuary* (e.g., Delaware and Chesapeake Bays; Rivers Esk and Eden, UK; Columbia River, USA; Lérez Estuary, Spain); (F) *Deltaic* (e.g., Mississippi River delta, LA; Yukon River delta, AK).

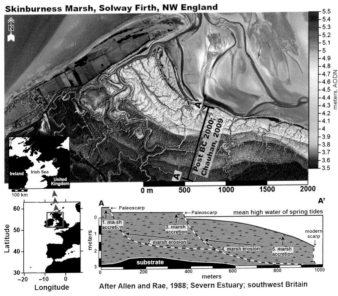

Figure 3.2 Digital elevation model of the Skinburness (Cumbria, UK) fringing salt marsh shows a series of terraces 0.5 m–0.1 m in relief. The delineated area accreted discontinuously over millennial time scales (Singh Chauhan, 2009). Cross section A–A' is from the margin of the Severn Estuary, another terraced marsh in the UK, and illustrates that the terraces are related to periods of marsh accretion and erosion (after Allen and Rae, 1987). AODN= Above Ordnance Datum Newlyn. Data used to create the digital elevation model was downloaded from data.gov.uk.

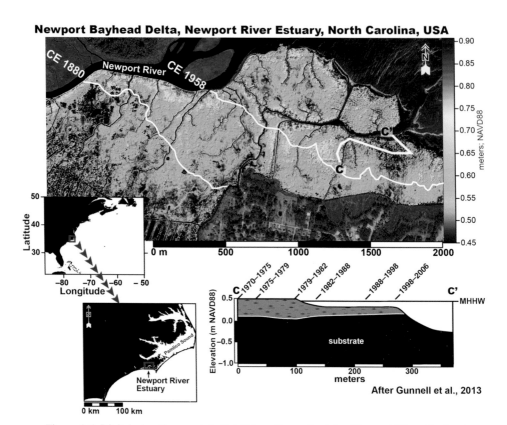

Figure 3.6 Digital elevation model of deltaic salt marsh at the Newport River Bayhead Delta, NC showing bayward accretion. Cross section C–C′ showing variations in deltaic salt marsh thickness and ages based on a time series of aerial photographs and cores (see Gunnell et al. 2013 for additional information) NAVD88=North American Vertical Datum of 1988.

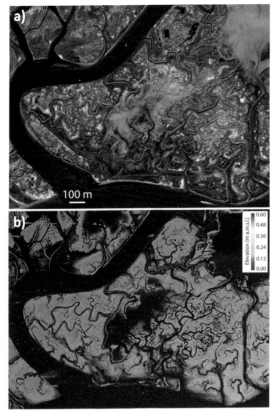

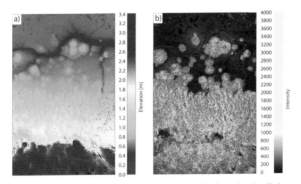

Figure 4.2 (a) Aerial photo and (b) color-coded elevation map derived from Lidar data (higher elevations in red, lower elevations in blue) of the San Felice salt marsh in the Venice lagoon, Italy (at approximately 45.79599°N, 1051253°E) showing tidal channels cutting through the marsh platform together with deposition patterns.

Figure 4.10 Morphological progression of a salt marsh platform in the Solway Firth, UK. In the top section, discontinuous vegetation patches have established on the intertidal mudflat and are anchoring the position of tidal creeks. In the early stages of marsh formation, vegetated patches have little influence on the mudflat topography. In the middle section, the low marsh has formed a continuous platform, although residual pools and channels are still visible. In the lower section, the high marsh sits approximately 1 m above the high marsh and is delineated by a subvertical scarp. At the foot of the scarp, erosion pools are visible. (a) Elevation map obtained by Terrestrial Laser Scanner (TLS) on 01/06/2017; (b) Intensity map obtained by TLS on 01/06/2017. High intensity values correspond to vegetated areas.

Figure 5.1 A: The orthopteran *Orchelimum fidicinium* eating *S. alterniflora* in a Georgia salt marsh. B: Horses, *Equus ferus*, grazing on *Spartina alterniflora* in a North Carolina salt marsh. C: Sheep *Ovis aries* grazing in a French salt marsh. D: The parasitic plant *Cuscuta* sp. parasitizing host plants in a salt marsh in Colombia. E: The decapod *Helice tientsinensis* feeding on *Suaeda salsa* in a Chinese salt marsh. F: The crane, *Grus japonensis*, searching for prey in a Chinese salt marsh. Photo credits: A, B, E, F: Qiang He; C, D: Steven Pennings.

Figure 5.2 A: Mussels, *Geukensia demissa*, facilitating the grass *S. alterniflora* in a US salt marsh. B: Zonation between the grass *S. alterniflora* on the left and the rush *Juncus roemerianus* on the right, in a Georgia salt marsh. C: The shrub *Tamarix chinensis* (center) facilitates the forb *Suaeda glauca* (arrow) beneath its canopy in a Chinese salt marsh. The salt-tolerant forb *S. salsa* occurs outside the shrub canopy. D: The arc-shaped head of an eroding creek in a Georgia salt marsh is both cleared of vegetation and excavated by the crab *Sesarma reticulatum*. E: Slumping of the creekbank in a Georgia salt marsh. F: Wrack disturbance along the creekbank of a Georgia salt marsh. Photo credits: A, B, C: Qiang He; D, E, F: Steven Pennings.

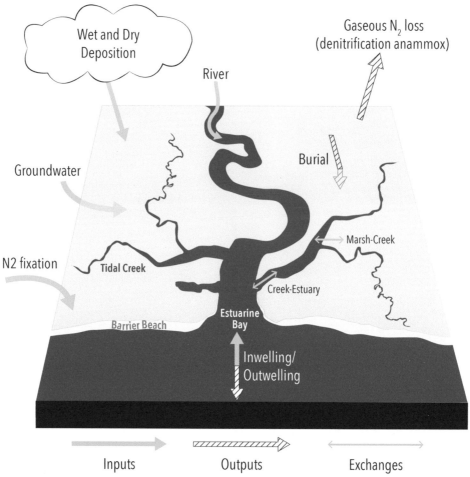

Wet and Dry
Deposition

Gaseous N$_2$ loss
(denitrification anammox)

River

Groundwater

Burial

N2 fixation

Marsh-Creek

Tidal Creek

Creek-Estuary

Estuarine
Bay

Barrier Beach

Inwelling/
Outwelling

Inputs   Outputs   Exchanges

Figure 6.1 The major sources of inputs and losses of N, P, and Si to a marsh estuarine complex. The sites of exchange are also illustrated.

Figure 7.2 Bioassay experiment (marsh organ) at North Inlet Estuary, SC with *S. alterniflora* growing in PVC pipes standing at different elevations, simulating differences in relative marsh elevation.

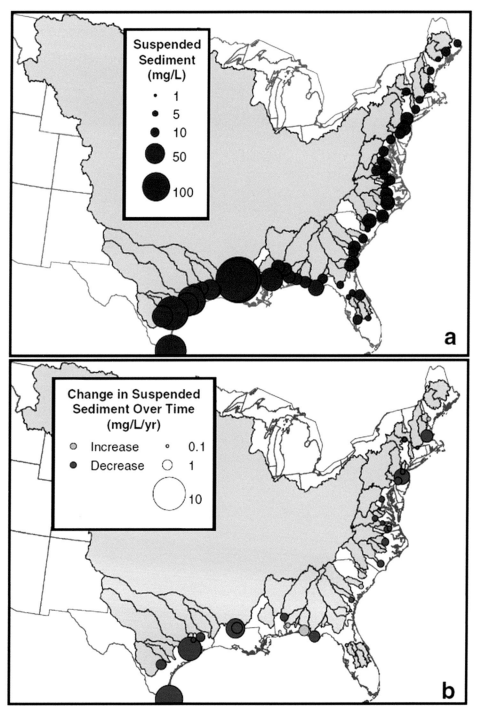

Figure 7.5 (a) Concentrations of suspended sediments in rivers, and (b) change in suspended sediment concentration over time in rivers draining to the US East and Gulf Coasts. From Weston (2014).

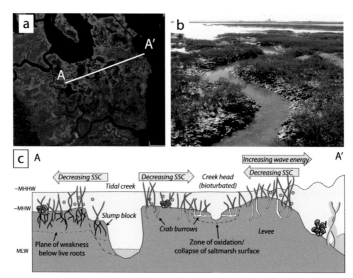

Figure 8.1 Salt marshes are comprised of vegetated marsh platforms dissected with tidal channels and smaller creeks, adjacent to extensive mudflats, shown here in aerial view for Great Bay, NJ (a) and on the ground in Bahia Blanca, Argentina (b). A few ecogeomorphic processes common in salt marshes are shown in the cross section of transect A–A', idealized in (c): natural levees form adjacent to larger bodies of water (channel or bay) where wave and tidal current energies are attenuated by vegetation and/or offshore oyster reefs, favoring sediment deposition; crab burrowing is prevalent across the platforms, but can be concentrated by creekbanks or interior creek heads; interior lowlands, defined as pans or ponds (see Perillo, 2019), are present where waterlogging is prevalent and vegetation is stunted or absent; live macrophyte rooting can extend up to 1 m depth, below which there can be a zone of weakness that can act as a failure plane along creekbanks and channel edges. SSC: suspended sediment concentration.

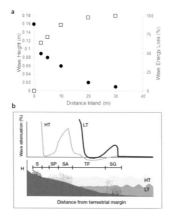

Figure 8.4 (a) Decreases in wave height (circles) and wave energy (squares) associated with friction along vegetated salt marsh platforms in Chesapeake Bay, Virginia, USA (from Knutson et al. 1982). (b) Idealized scheme of wave attenuation across the profile of a marsh ecosystem at high (HT) and low tide (LT). Waves attenuation change due to the kind of plants or bare sediments (TF). *SG = seagrass, SA = Spartina alterniflora, SP = Spartina patens, S = Sarcocornia.* (modified from Koch et al. 2009).

Figure 8.5 (a) An example of the extensive root network of belowground biomass of *Spartina alterniflora* (photo credit Amanda Davis). (b) The strength of the root network is measured using a hand-held shear vane, shown here in a *S. alterniflora* marsh in South Carolina. (c) Data from *S. alterniflora* marshes of Louisiana show a decrease in strength with depth, indicative of the limit of live rooting (Howes et al. 2010; Turner, 2011; Valentine and Mariotti, 2019). An increase in shear strength at deeper depths is from compaction and consolidation of sediment. (d) Zones of weakness at depth can be responsible for creekbank failures, shown here from a nutrient-enriched *S. patens* marsh in Plum Island (Deegan et al. 2012).

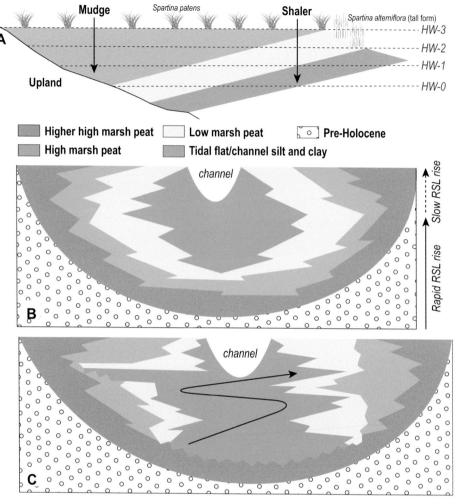

Figure 9.2 (a) Redfield's (1972) model of salt marsh evolution. The landward section is characterized by a transgressive sequence, according to the Mudge (1858) model. The Shaler (1886) model explains the regressive sequence in the seaward portion of the marsh. HW–0 is the position of high water at the beginning of marsh development. HW1–3 are various high water positions through time. Plant symbols are from the Integration and Application Network, University of Maryland Center for Environmental Science (ian. umces.edu/imagelibrary/). (b) Modification of the Redfield model showing how the rate of relative sea-level rise determines marsh evolution. Regressive sequences occur when the rate of relative sea-level rise exceeds the sedimentation rate. When relative sea-level rise slows down, the marsh progrades seaward over the tidal flat and a regressive sequence is formed. (c) Meandering tidal channels also play a role in salt marsh evolution. Only in sheltered places away from tidal channels and creeks are facies preserved that are controlled by relative sea-level rise and not disturbed by erosion.

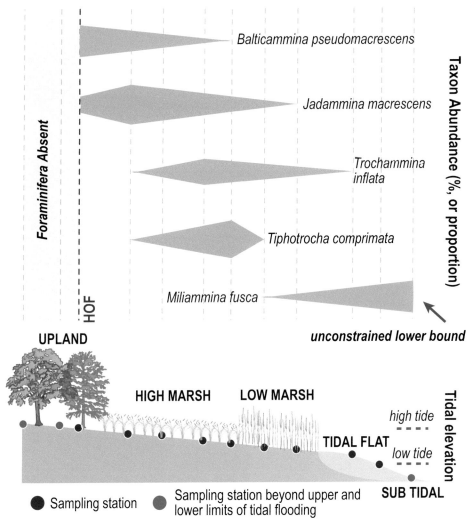

Figure 9.4 An example of intertidal zonation displayed by foraminifera in samples collected along the sloping surface of a salt marsh in eastern Maine. Plant zones are also shown. Redrawn and modified from Gehrels (1994).

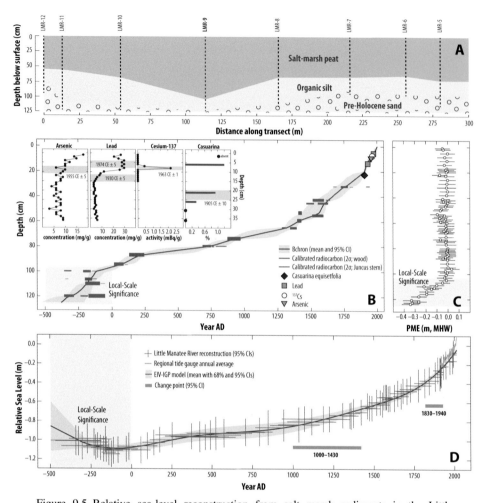

Figure 9.5 Relative sea-level reconstruction from salt marsh sediments in the Little Manatee River estuary, Gulf coast of Florida. (a) Stratigraphy underlying the site described from cores. Core LMR-9 was selected for detailed analysis. The lower part of the sequence represents a Shaler-type (colonizing) salt marsh when local-scale processes were inferred as the dominant cause of relative sea-level change. (b) Chronological data and age-depth model for LMR-9. Inset panels show downcore profiles of pollutants, radioisotopes, and pollen from the exotic genus *Casuarina*. Radiocarbon ages and chronological markers were combined using a Bchron age-depth model. (c) Reconstructed paleo-marsh elevations, or indicative meaning based on a regional-scale transfer function. (d) Relative sea-level reconstruction from LMR-9 combined with tide gauge measurements from Key West, Naples, St. Petersburg, and Fort Meyers. Modified from Gerlach et al. (2017).

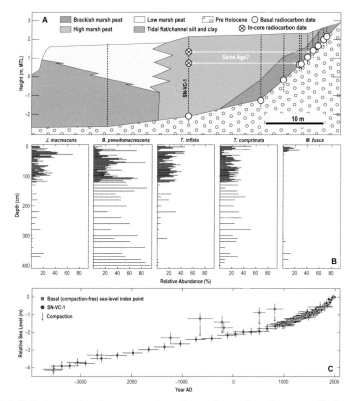

Figure 9.6 Relative sea-level reconstructions from salt marsh sediments at Sanborn Cove, eastern ME (Gehrels 1999). (a) Stratigraphy and radiocarbon dates. Core SN-VC-1 was used to produce a near-continuous reconstruction. (b) Foraminifera in core SN-VC-1. (c) Relative sea-level reconstructions from basal sea-level index points (blue squares) and from core SN-VC-1 (red circles). The offset between the two records is assumed to be the result of compaction (vertical arrows) of sediment in SN-VC-1.

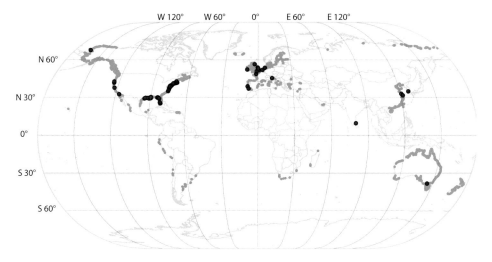

Figure 10.1 Locations of major research (white circles) on the effects of storms on salt marshes. Global marsh extent is indicated in black (from Mcowen et al. 2017). For a full list of research locations, please see http://arcg.is/1CCuXj.

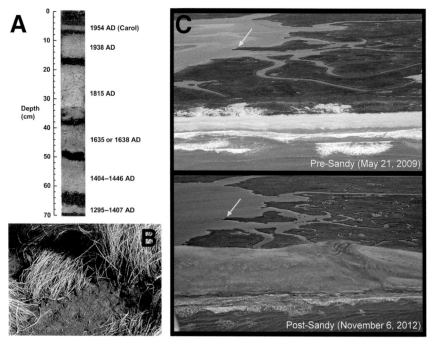

Figure 10.4 Field photos of various depositional impacts of storms on marshes. (a) Storm-induced overwash deposits from Succotash Marsh, East Matunuck, RI (modified from Cheung et al. 2007). (b) Example of sediment deposited on the marsh after Hurricanes Katrina and Rita (modified from Turner et al. 2006). (c) Aerial photos of overwash deposits left by Hurricane Sandy in Assateague Island, VA (modified from Sopkin et al. 2014).

Figure 10.5 Field photos of various erosional impacts of storms on marshes. (a) Eroding cliffs in the Westerschelde, the Netherlands (modified from van de Koppel et al. 2005). (b) Example of *Salicornia* spp. causing massive bank erosion in Tillingham, Essex, UK (modified from Möller 2012). c–e) Different kinds of marsh erosion (slumping, undercutting, and root scalping, respectively) in the Virginia Coast Reserve (modified from Fagherazzi et al. 2013). (f) Stratigraphic evidence of salt marsh peat overlain by tidal mud after a proposed erosional event in Sea Breeze Marsh, New Jersey (modified from Nikitina et al. 2014).

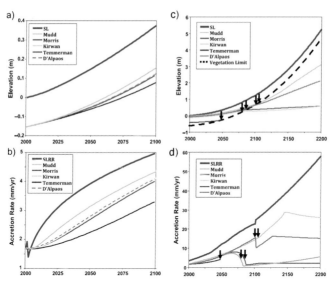

Figure 11.2 Response of marsh elevation (a,c) and accretion rate (b,d) to a conservative sea-level acceleration (IPCC A1B; a,b) and to the maximum scenario of Rahmstorf (2007) (c,d). The models shown are vertical models discussed in the text with the exception of the D'Alpaos et al. (2007) and the Kirwan amd Murray (2007) models, which are 2D planform models. Each model is parameterized for a different site. The response of each model is qualitatively similar despite differences in model form and parameters. From Kirwan et al. (2010).

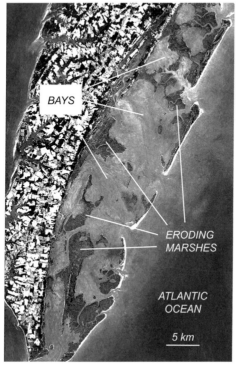

Figure 11.9 Bays and salt marshes in the Virginia Coast Reserve, USA (Landsat Image – Google Earth).

Figure 12.2 Mimic representations from literature (A) Non-specific rigid aquatic vegetation[58], (B) *Posidonia oceanica*[55], (C) *Spartina alterniflora*[73], (D) *Acorus camalus,*[51] (E) Non-specific rigid aquatic vegetation patch,[12] (F) *Enhalus acoroides,*[54] and (G) *Zostera marina*[65].

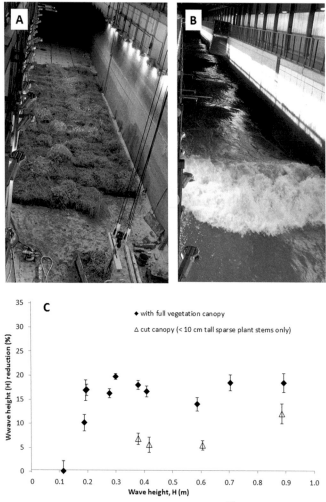

Figure 12.3 Large-scale flume experiment of Möller et al.[29] on wave dissipation over salt marshes (A) salt marsh re-assembly, (B) breaking waves over submerged marsh (2 m water depth), and (C) wave height reduction (%) of non-breaking waves over the 40-m long test section for monochromatic waves over intact and cut canopies.

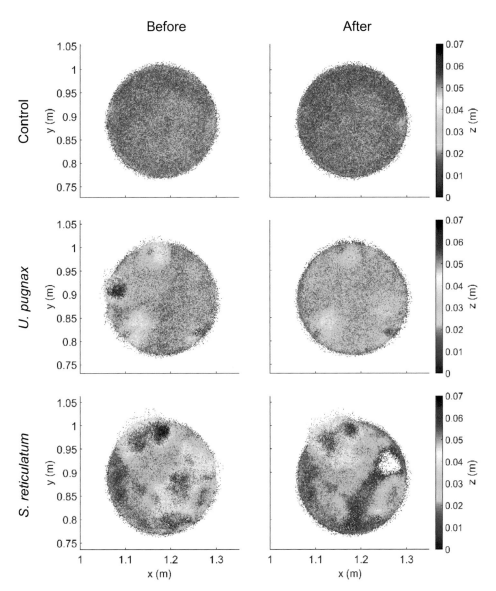

Figure 12.4 Laser scanning data showing bed morphological differences between controls, sediment burrowed by *U. pugnax*, and sediment burrowed by *S. reticulatum* before and after being eroded in the flume.

Observed change in surface temperature 1901–2012

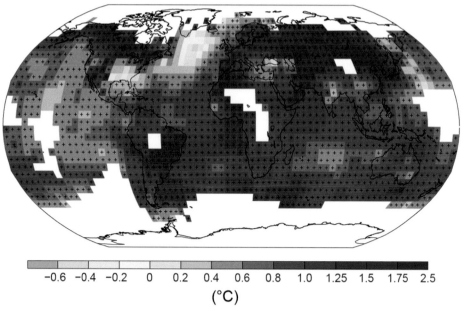

-0.6 -0.4 -0.2 0 0.2 0.4 0.6 0.8 1.0 1.25 1.5 1.75 2.5

(°C)

Figure 13.1 Map of observed temperature change from 1901 to 2012. Source: IPCC 2014.

Figure 13.4 Example of a marsh organ in the Apalachicola National Estuarine Research Reserve in Florida. Image courtesy of Jim Morris, University of South Carolina.

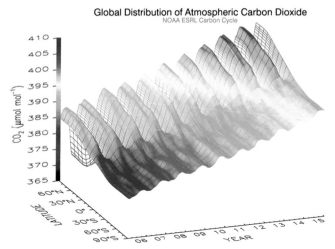

Figure 13.6 Atmospheric CO_2 concentrations from 2005 to 2014, showing both latitudinal and seasonal variations (Graph courtesy of Dr. Ed Dlugokencky and Dr. Pieter Tans, NOAA ESRL Carbon Cycle, www.esrl.noaa.gov/gmd/ccgg/).

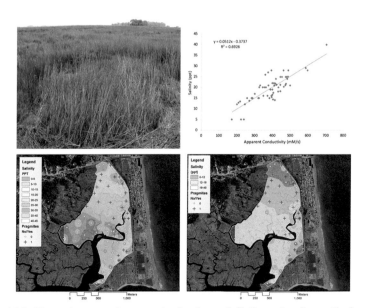

Figure 14.1 *Phragmites* control case study. 1a. (upper left) Example of a small, chemically treated *Phragmites* stand in Great Marsh, Newbury, MA following one growing season after treatment. Free from competition with *Phragmites*, the high marsh growth includes native *Spartina patens*, *P. odorata*, and *Salicornia depressa* (upper left); 1b. **(upper right)** Regression of salinity vs. apparent conductivity (upper right); 1c. **(lower left)** High resolution salinity gradients in a portion of Great Marsh located in Salisbury, MA (lower left); and 1d. **(lower right)** Reconfigured salinity map highlighting low salinity (green), moderate salinity (orange), and high salinity (pink) portions of the marsh. The moderate salinity area (orange) corresponds with published ranges of dense *Phragmites* proliferation (lower right).

Figure 14.2 Wave of *Littorina littorea* (common periwinkle) on eroded shoreline in York, ME (upper left). Planted cordgrass protected by wave baffle still damaged by snails (upper right). Snail association with planted cordgrass, now grazed and decomposing (lower left). Second year of growth of cordgrass at restoration site protected from snails by a fence of wire mesh (lower right).

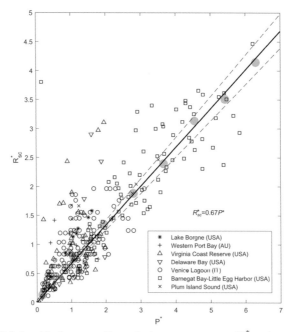

$R_{sc}^*=0.67P^*$

✳	Lake Borgne (USA)
+	Western Port Bay (AU)
△	Virginia Coast Reserve (USA)
▽	Delaware Bay (USA)
○	Venice Lagoon (IT)
□	Barnegat Bay-Little Egg Harbor (USA)
×	Plum Island Sound (USA)

Figure 15.6 Relationship between dimensionless wave power W^* and erosion rate R_{sc}^*. Gray circles indicate values obtained by averaging data points over regular bins (from Leonardi et al. 2015).

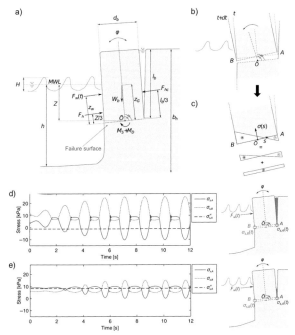

Figure 15.10 (a) Sketch of the cross-section of the system employed for the development of the dynamic model; (b) scheme of the failure surface and the stress distribution induced by (c) a small clockwise rotation of the block from equilibrium. (d) and (e) comparison between the behavior of the model with filled or empty tension crack. The blue continuous line represents the time evolution of the stress at the bottom of the tension crack in the inner point of the failure surface A. The green dashed line represents the stress at the external point of the failure surface B. Black dotted line represents the time evolution of the stress on point A in the static case (from Bendoni 2015).

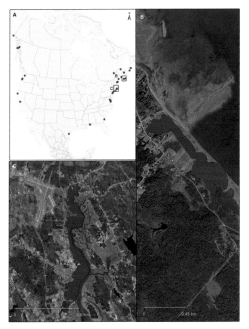

Figure 16.4 Location of salt marshes with the highest threat of coastal squeeze from steep slope (red, A). Boxes on the map show geographic location of an example at Quoddy Narrows, Lubec, Maine, USA (B) and Yarmouth, Nova Scotia, Canada (C).

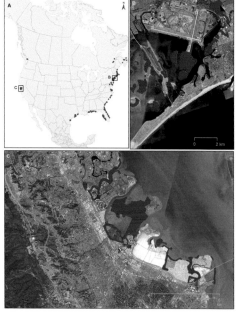

Figure 16.6 Salt marshes (red) under high threat of coastal squeeze from development (A), Jamaica Bay, New York (B) and San Francisco Bay, California (C).

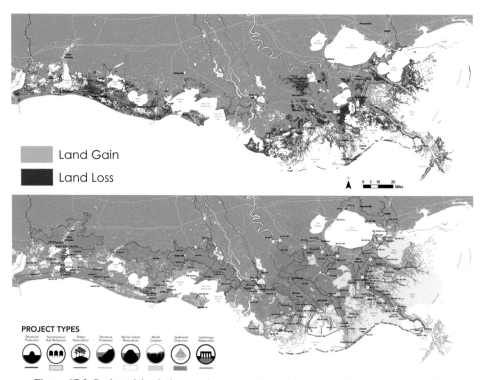

Figure 17.2 Projected land change along coastal Louisiana over the next 50 years if no action is taken (upper) and restoration projects implemented and planned under the 2017 Coastal Master Plan (lower). Projects include marsh creation (green), sediment diversion (white circles and brown area of influence), hydrological restoration (blue circles), barrier island restoration (red), and structural (pink) and non-structural risk reduction. (Modified from CPRA 2017)

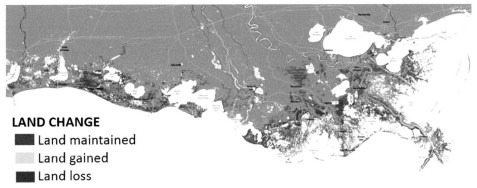

LAND CHANGE
■ Land maintained
☐ Land gained
■ Land loss

Figure 17.3 Projected 50 year outcome of fully enacted CMP under moderate environ-mental scenario (modified from CPRA 2017).

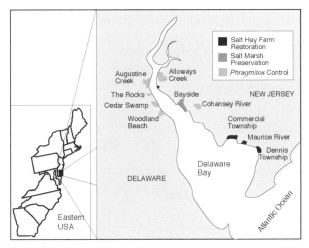

Figure 17.8 Map of Delaware Bay salt marsh restoration between New Jersey and Delaware from 1995 through present, showing locations of 5,800-ha salt marsh restoration and preservation that is being carried out to mitigate the loss of fin fish due to entrainment and impingement caused by once-through cooling at a Delaware Bay nuclear power plant. Wetlands are being preserved, restored from salt-hay farms by reintroducing flooding, and enhanced by removal of *P. australis* (reprinted from Mitsch and Gosselink 2015, with permission from John Wiley & Sons, Inc. Hoboken NJ.)

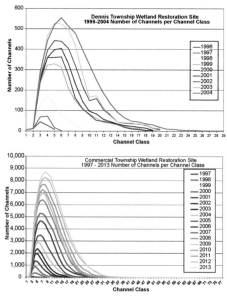

Figure 17.9 Total number of stream channels by channel class at Dennis Township (1996–2004) and Commercial Township (1997–2013) restored salt marshes on Delaware Bay (Provided with permission, Kenneth A. Strait, PSEG Service Corporation, Salem, NJ, and reprinted from Mitsch and Gosselink 2015, with permission from John Wiley & Sons, Inc. Hoboken NJ.)

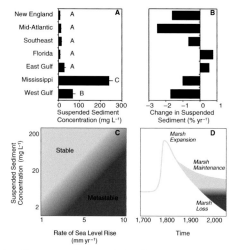

Figure 18.1 (A) Regionally averaged suspended sediment concentrations in rivers draining to the East and Gulf Coasts of the USA. Rivers that do not share the same letter have significantly different sediment concentrations and (B) proportional change in sediment concentration over time (in percent per year). Sediment supply is decreasing in nearly all regions of the USA This reduction will ultimately cross a threshold as marshes will not be able to create organic sediment or cannibalize from within at a sufficient rate to keep pace with sea level rise. (C) Conceptual model of tidal marsh stability as a function of suspended sediment concentration and rate of sea level rise (modified from Kirwan and Murray 2007) and D() conceptual diagram of changes in suspended sediment concentration in a U.S. river over time. (from Weston 2014).

Mississippi River deltaic plain: Implications for using river diversions as restoration tools. *Ecological Engineering*, 81: 133–139.

Stark, J., Van Oyen, T., Meire, P., and Temmerman, S. 2015. Observations of tidal and storm surge attenuation in a large tidal marsh. *Limnology and Oceanography*, 60: 1371–1381.

Stumpf, R. P. 1983. The process of sedimentation on the surface of a salt marsh. *Estuarine, Coastal and Shelf Science*, 17: 495–508.

Tate, A. S., and Battaglia, L. L. 2013. Community disassembly and reassembly following experimental storm surge and wrack application. *Journal of Vegetation Science*, 24: 46–57.

Temmerman, S., Bouma, T. J., Govers, G., Wang, Z. B., De Vries, M. B., and Herman, P. M. J. 2005. Impact of vegetation on flow routing and sedimentation patterns: Three-dimensional modeling for a tidal marsh. *Journal of Geophysical Research: Earth Surface*, 110: 1–18.

Temmerman, S., Meire, P., Bouma, T. J., Herman, P. M. J., Ysebaert, T., and De Vriend, H. J. 2013. Ecosystem-based coastal defence in the face of global change. *Nature*, 504: 79–83.

Turner, R. E., Baustian, J. J., Swenson, E. M., and Spicer, J. S. 2006. Wetland sedimentation from hurricanes Katrina and Rita. *Science*, 314: 449–452.

Tweel, A. W., and Turner, R. E. 2012. Landscape-scale analysis of wetland sediment deposition from four tropical cyclone events. *PLOS ONE*, 7(11). https://doi.org/10.1371/journal.pone.0050528

Walsh, K. J. E., McBride, J. L., Klotzbach, P. J., Balachndran, S., Camargo, S. J., Holland, G. J., ... Sugi, M. 2016. Tropical cyclones and climate change. *Wiley Interdisciplinary Reviews: Climate Change*, 7: 65–89.

Walters, D. C., and Kirwan, M. L. 2016. Optimal hurricane overwash thickness for maximizing marsh resilience to sea level rise. *Ecology and Evolution*, 6: 2948–2956.

Walters, D., Moore, L. J., Vinent, O. D., Fagherazzi, S., and Mariotti, G. 2014. Interactions between barrier islands and marshes affect island system response to sea level rise: Insights from a coupled model. *Journal of Geophysical Research: Earth Surface*, 119: 2013–2031.

Wamsley, T. V., Cialone, M. A., Smith, J. M., Ebersole, B. A., and Grzegorzewski, A. S. 2009. Influence of landscape restoration and degradation on storm surge and waves in southern Louisiana. *Natural Hazards*, 51: 207–224.

Watson, E. B., Wigand, C., Davey, E. W., Andrews, H. M., Bishop, J., and Raposa, K. B. 2017. Wetland Loss Patterns and Inundation-Productivity Relationships Prognosticate Widespread Salt Marsh Loss for Southern New England. *Estuaries and Coasts*, 40: 662–681.

Williams, H. F. L. 2012. Magnitude of Hurricane Ike storm surge sedimentation: Implications for coastal marsh aggradation. *Earth Surface Processes and Landforms*, 37: 901–906.

Wilson, K. R., Kelley, J. T., Croitoru, A., Dionne, M., Belknap, D. F., and Steneck, R. 2009. Stratigraphic and ecophysical characterizations of salt pools: Dynamic landforms of the webhannet salt marsh, wells, ME, USA. *Estuaries and Coasts*, 32: 855–870.

11

Understanding Marsh Dynamics

Modeling Approaches

SERGIO FAGHERAZZI, WILLIAM KEARNEY, GIULIO MARIOTTI,
NICOLETTA LEONARDI, AND WILLIAM NARDIN

11.1 Introduction

Salt marshes have received considerable scientific attention in recent years due to a combination of factors. Salt marshes host important ecosystems and store large quantities of carbon in their soils (Fagherazzi et al. 2004; Mudd et al. 2009). Currently salt marshes are endangered by accelerated sea-level rise triggered by global warming (Kirwan et al. 2010). A sharp reduction in sediment supply caused by the damming of rivers is also jeopardizing marsh survival along many coasts (Weston 2014). As a result, there is a need to determine the fate of marshlands in different settings in order to inform government and local communities and implement protection strategies. To this end, numerical models are playing an increasingly important role, because they can easily provide future scenarios of marsh conditions under different forcings. However, the evolution of salt marshes as a function of sea-level rise and sediment supply is relatively complex, because of feedbacks among hydrodynamics, sediment transport, and vegetation (Fagherazzi et al. 2012). As a result, marshes are continuously adjusting to a changing environment, in ways often difficult to predict. This intrinsic complexity has generated a flurry of numerical models, each emphasizing a different aspect of salt marsh evolution. It is thus becoming more and more accepted by the scientific community that a comprehensive model of salt marsh evolution is not feasible, given the number and variety of physical and biological processes at play. A detailed approach, based on the description of all possible processes acting at different spatial and temporal scales, has been slowly replaced by a more practical approach, in which separate models are built to address key important processes or to capture specific dynamics.

Salt marsh models are also affected by the different levels of understanding of distinct parts of the systems. Salt marsh hydrodynamics is captured in an excellent way by Navier–Stokes equations or shallow-water equations, given the predictability of the tides. Sediment dynamics are also tractable with equations based on physical principles, although with less precision compared to hydrodynamics. Vegetation dynamics and the effect of biota is in general so complex that often only empirical relationships are available. These relationships are frequently site-specific, and hard to apply in a different context. The coupling of these components creates unique hybrid models, with components that are physically based providing results at very high resolution and accuracy, and components that rely on empirical results, with a coarse representation of reality and low accuracy. This hybrid modelling approach strongly affects the overall quality of the results, with the parts represented at lower resolution and with limited physical insight controling the overall error.

Numerical models can be arrayed along an axis having at the two extremes "simulation models" and "exploratory models" (e.g., Murray 2013). Simulation models aim to reproduce the natural system as accurately as possible. On the contrary, highly simplified exploratory models aim to understand the general behavior of the system, by purposely avoiding the representation of as many processes as possible, and by following the "emergent property viewpoint" (e.g., Goldenfeld and Kadanoff 1999).

In this review chapter, we present an overview of current models for salt marsh dynamics, separately discussing vertical models, transect models, two-dimensional and three dimensional simplified models, and high resolution models. A particular emphasis will be placed on the distinction between simulation and exploratory models. A case study and a discussion paragraph on future modeling needs will close the chapter.

11.2 Vertical Models

Vertical models simulate the elevation of the marsh platform at a single point. In the literature, vertical models are sometimes called "zero-dimensional" models because the point is zero-dimensional. They can also be thought of as one-dimensional because they have a single degree of freedom (the platform elevation), or because they simulate the one-dimensional vertical stratigraphic profile. To avoid confusion, we will refer to these models as vertical models.

Such models are useful because the complex spatial behavior of hydrodynamics and sediment transport does not need to be explicitly simulated, there is a single degree of freedom (the platform elevation) whose qualitative dynamics can readily be studied with methods from dynamical systems theory (e.g., Marani et al. 2007, Fig. 11.1), and they form the basis of local process representations in higher dimensional models (e.g., Kirwan and Murray 2007; Mariotti and Fagherazzi 2010).

There are two different interpretations of the single point simulated within a vertical model. It can represent the dynamics of marsh elevation at a specific point within the marsh (Morris et al. 2002), or it can represent the vertical elevation spatially averaged throughout the marsh (Allen 1990; Temmerman et al. 2003; Marani et al. 2007). In practice, these two interpretations lead to identical models, and the relatively flat topography of marshes suggests that, on long time scales, spatial variability in the vertical dynamics is limited (Temmerman et al. 2003). Care must be taken when parameterizing the models that the selected parameters are representative of the scale at which the vertical model is being used.

Vertical models naturally fall on the exploratory end of the spectrum between exploratory models and simulation models (Murray 2013), and they are most often used to understand the qualitative behavior of marsh systems rather than to make quantitative predictions of marsh survival under particular scenarios of sea-level rise. Vertical models suggest that positive feedbacks between vegetation growth and marsh accretion allow

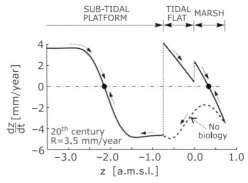

Figure 11.1 Alternative stable states in the model of Marani et al. (2007) under historical rates of sea-level rise. Two stable states are identified, one is an unvegetated subtidal platform (left circle) and one is a vegetated intertidal salt marsh (right circle). When no vegetation is present, the trajectory follows the dashed curve to the subtidal equilibrium, indicating the importance of the vegetation-mediated positive feedback in maintaining marsh elevation. From Marani et al. (2007)

marshes to survive moderate rates of sea-level rise because the accretion rate equilibrates with the rate of sea-level rise (Kirwan and Temmerman 2009; Kirwan et al. 2010; Fagherazzi et al. 2012). These predictions are supported by field observations, which suggest that marshes have persisted under historical rates of sea-level rise (Redfield 1965) and that marshes accrete at rates equal to or above that of relative sea-level rise (Cahoon et al. 2006). However, there is a threshold rate of sea-level rise beyond which the feedback is unable to maintain the system (Kirwan et al. 2010, Fig. 11.2). Under these rates of sea-level rise, marshes drown and are converted in other unvegetated intertidal or subtidal landforms.

While studying the qualitative dynamics of marshes is the main purpose of vertical models, many studies that develop vertical models also collect field data from specific field sites to parameterize the model (Allen 1990; French 1993; Morris et al. 2002; Temmerman et al. 2003; Marani et al. 2007; Mudd et al. 2009). The accuracy of these predictions remains an open question because of challenges in quantifying the uncertainty associated with model form in such simplified models (Murray et al. 2016).

11.3 A Generalized Vertical Model

Vertical models represent the evolution of a marsh platform relative to mean sea level, z, as a mass balance. The deposition of sediment and the accretion of organic matter raise the platform while decomposition, compaction, and relative sea-level rise lower it. Vertical models can therefore be represented by the following generalized equation:

$$\frac{dz}{dt} = M(z, B, S, H, C) + O(z, B, S, H, C) - P(z, B, S, H, C) - R \qquad (11.1)$$

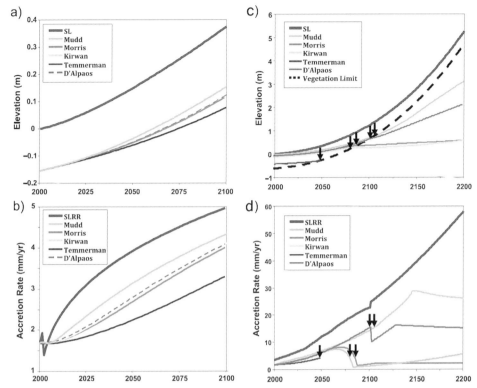

Figure 11.2 Response of marsh elevation (a,c) and accretion rate (b,d) to a conservative sea-level acceleration (IPCC A1B; a,b) and to the maximum scenario of Rahmstorf (2007) (c,d). The models shown are vertical models discussed in the text with the exception of the D'Alpaos et al. (2007) and the Kirwan amd Murray (2007) models, which are 2D planform models. Each model is parameterized for a different site. The response of each model is qualitatively similar despite differences in model form and parameters. From Kirwan et al. (2010). (A black and white version of this figure will appear in some formats. For the color version, please refer to the plate section.)

where M is the net deposition of mineral sediment on the surface, O is the net accretion of organic matter within the soil column, P is the compaction of the sediments, and R is the rate of sea-level rise. M, O, and P will be referred to below as *flux terms* since they represent a flow of sediment across a vertical point. The parameters B, C, and H represent collections of biological (B), sedimentary (S), hydrodynamic (H), and climatic (C) variables that are important to the marsh evolution. The biological component includes the above- and belowground macrophyte biomass as well as the biomass of microphytobenthos. The sedimentary component includes the suspended sediment concentration in the water column above the marsh and sediment properties such as grain size. The hydrodynamic component includes tidal range and the speed of tidal and wind-driven currents. The climatic component includes long-term climatic variables like temperature and precipitation, as well as a description of the storm regime. Exactly which parameters are chosen and how they are quantified is a modeling choice.

The flux terms, M, O, and P, are complex functions of the parameters B, S, H, and C, which themselves evolve over time and depend both on each other and on the elevation of the tidal platform. To close the model so that it can be analyzed or numerically simulated, the relationships between the flux terms and the parameters need to be specified. These closures can be either physical models, or empirical models fit to data.

A physical model is most often used to relate the water column suspended sediment concentration to biology and hydrodynamics (Temmerman et al. 2003; Marani et al. 2007; Mudd et al. 2009). The suspended sediment concentration represents a balance between sediment inputs entering the marsh system, deposition by settling, trapping of sediment by vegetation, and resuspension from tidal currents and wind waves. Even though this balance is specified as a mass conservation law, many of the terms in the balance, especially those representing the influence of vegetation on sedimentation, cannot be derived from first principles and must be specified empirically.

Empirical models are used, for example, to quantify the relationship between vegetation biomass and sediment trapping (Mudd et al. 2004; Marani et al. 2007; Mudd et al. 2009):

$$M_t(z, B, S, H) = \alpha B^\beta (H - z)S \tag{11.2}$$

where M_t is the component of inorganic sedimentation due to trapping, B is the vegetation biomass, $H - z$ is the flow depth, and S is the suspended sediment concentration. The parameters α and β characterize how the structure of vegetation causes sediment trapping and are usually chosen to fit independent data. Similarly, Morris et al. (2002) use a quadratic dependence of biomass on depth below mean high tide

$$B(z, H) = a(H - z)^2 + b(H - z) + c \tag{11.3}$$

where H is the elevation of mean high tide, and a, b, and c are parameters fit to biomass data collected from plots. Typically, vegetation is planted at different elevations in PVC pipes (the so-called marsh organ, Fig. 11.3) and harvested after a few years to measure biomass.

Regardless of the form the model takes, the key qualitative feature of vertical models is a positive feedback between the depth of the marsh below the water surface, mineral sediment deposition, and organic matter accretion. As the depth increases, this feedback causes deposition to increase, which in turn reduces the depth of the marsh. This feedback can arise when deposition depends on the length of inundation over a tidal cycle (Allen 1990; French 1993; Temmerman et al. 2003). A longer hydroperiod associated with marshes lower in the tidal frame causes more deposition and the necessary positive feedback. The feedback can also arise when deposition depends on the vegetation biomass (through either enhanced sediment trapping or the production of organic matter) and when the vegetation biomass increases with depth below the water surface (Morris et al. 2002). Both the physical and biological mechanisms can also be combined in a single model (Marani et al. 2007; Mudd et al. 2009, Fig. 11.1). Thresholds emerge when these feedbacks are limited by a maximum depth. This is usually represented as a depth beyond which vegetation drowns and ceases to enhance accretion.

Figure 11.3 Field apparatus to measure vegetation biomass as a function of marsh elevation. (Modified after Mudd et al. 2009)

Vertical models can be used to simulate the evolution of the platform elevation in time. For example, Figure 11.2 indicates the response of the marsh elevation to an increase in sea level. Depending on sediment supply and the rate of sea-level rise, the marsh elevation can track sea level (Fig. 11.2a,b) or fall beyond (Fig. 11.2c,d). In the second case the marsh drowns.

11.4 Two-Dimensional Transect Models

Moving forward from one-dimensional "point models", an additional level of realism in simulating marsh evolution is obtained by including a lateral spatial component. This is achieved with "transect models" often referred to as two-dimensional models (Temmerman et al. 2004; Marani et al. 2013; Ratliff et al. 2015). The transect generally starts at a distinct edge of the marsh, such as the bank of a channel or the beginning of a mudflat, and ends a few hundreds of meters in the marsh interior (Fig. 11.4). A simplified tidal flow is then calculated along the transect direction, a reference suspended sediment concentration (SSC) is imposed at the marsh edge, and the sediment is transported laterally. This transport is often calculated using the advection–diffusion equation (Marani et al. 2013):

$$\frac{\partial (hc)}{\partial t} = \frac{\partial \left(Kh \frac{\partial c}{\partial x} + Uhc \right)}{\partial x} + E - D \tag{11.4}$$

where h is the water depth, c is the suspended sediment concentration, K is the turbulent diffusion, U is the water velocity, D is the sediment deposition, and E is the sediment erosion, which is generally set equal to zero on the marsh platform. Alternatively, a simplified form of sediment transport, based on mass balance considerations, was

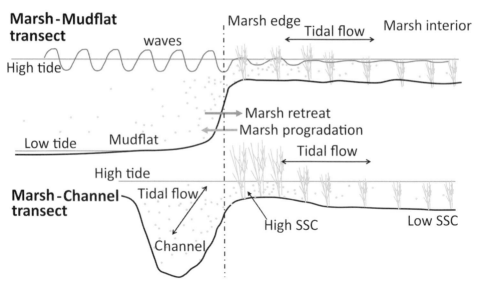

Figure 11.4 Cartoon of a marsh-mudflat and marsh-channel transects. Suspended sediment concentration (SSC) is highest near the marsh edge than in the marsh interior, a spatial pattern that can be obtained with a transect model by using the one-dimensional advection diffusion equation.

implemented by Temmerman et al. (2004). In both cases, the evolution of the marsh elevation stems from the competition between sea-level rise and the total (organic + inorganic) sediment deposition. Differently from one-dimensional models, transect models are able to simulate spatial variations in SSC and sediment deposition. Specifically, these models predict higher SSC and deposition near the marsh edge than in the marsh interior, in agreement with field observations (Cahoon et al. 2000; Christiansen et al. 2000). Consequently, these models are able to predict the formation of natural levees (Temmerman et al. 2004), which are often found on the sides of marsh channels. By including different types of marsh vegetation with different response to tidal inundation, transect models are also able to reproduce vegetation zonation (Marani et al. 2013). The models predict that the marsh interior – as it receives less sediment – will drown for smaller rates of sea-level rise than the marsh edge (Ratliff et al. 2015).

A further improvement to the marsh transect models was achieved by coupling the marsh with the mudflat (Mariotti and Fagherazzi 2010). Differently from previous models, wind waves were included in the framework and the marsh edge was not specified at priori, allowing it to migrate along the transect. This study revealed the inherently unstable nature of the marsh–mudflat edge, which either retreats or prograldes depending on the competition between sediment supply and wave-induced marsh edge erosion. Subsequently, Tambroni and Seminara (2012) developed a similar model with a more detailed hydrodynamics and with the inclusion of wind driven circulation. They found qualitative similar but quantitatively different results, highlighting how the delivery of sediment to the marsh edge depends on the delicate sediment balance in the mudflats and large tidal channels.

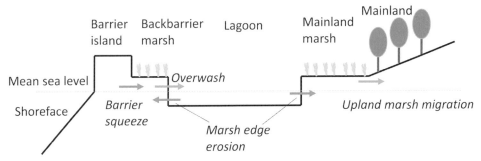

Figure 11.5 Cartoon of an idealized barrier-lagoon-mainland transect. Marshes are present both behind the barrier (backbarrier marshes) and on the landward side of the lagoon (mainland marshes).

A slightly different approach was undertaken in the model of Mariotti and Carr (2014), where the marsh–mudflat transect was simplified with only three variables: the elevation of the marsh, the elevation of the mudflat, and the position of the marsh edge. This study highlighted how the retreat of the marsh edge releases sediment, which in turn increases deposition on the marsh platform and thus increases marsh resilience to sea-level rise. Using the same approach, Kirwan et al. (2016) expanded the model by including upland marsh migration and introducing a spatially explicit representation of the marsh portion (similarly to Marani et al. 2013). They showed that, for a certain range of rates of sea-level rise and sediment supply, the total marsh area could expand even if the marsh edge retreats.

Transect models have also been used to study the coupled evolution of marshes and barrier islands. This coupling was first performed using GEOMBEST, a model that computes the evolution of a shoreface-barrier-bay profile using a kinematic approach (Stolper et al. 2005). Both a backbarrier marsh and a mainland marsh were added to the model (Walters et al. 2014) (Fig. 11.5). In addition to marsh edge erosion by waves, the effect of overwash in aiding the vertical accretion and progadation of backbarrier marshes was included. The authors found that the presence of a backbarrier marsh slows down barrier migration and that the overwash sediment flux from the shoreface aids the persistence of marshes "perched behind barrier islands. A similar study was performed by Lorenzo-Trueba and Mariotti (2017), who coupled the marsh-mudflat model of Mariotti and Carr (2014) with the barrier island model of Lorenzo-Trueba and Ashton (2014). Despite their formulation used a different schematization and different equations for the morphodynamics than GEOMBEST (Walters et al. 2014), a similar dynamics of the perched backbarrier marsh was found. Furthermore Lorenzo-Trueba and Mariotti (2017) emphasized how the supply of mud to the backbarrier can have a cascade effect on the evolution of barrier islands, and that the retreat of the barrier island can squeeze the backbarrier marsh.

11.5 Simplified Planar and Three-Dimensional Models

The models presented in this section are models belonging to the middle and second half of the "simulation–exploratory axis. These models include simplifications in the hydrodynamic and/or in the geomorphic response of the system, and are well suited when computational power is limited or when the state of knowledge is inadequate. These simplifications were adopted for various reasons including the exploration of emergent properties, or the simple focus on what were retained as the main external agents. The models explored in this section are representative of three main dynamics governing the salt marsh–tidal flat continuum, namely: tidal network development, tidal flats and internal ponds expansion/contraction, and the evolution of the marsh boundary due to wind waves and sea-level rise.

The first simplified model we explore here simulates the ontogeny of tidal networks (D'Alpaos et al. 2005; 2007). According to the hydrodynamic module of this model, the free surface water elevation, and the drainage directions across the marsh surface during flood periods, can be determined by solving the following Poisson-like equation (Rinaldo et al. 1999):

$$\nabla^2 \eta_1 = \frac{\Lambda}{(\eta_0 - z_0)^2} \frac{\partial \eta_0}{\partial t} \tag{11.5}$$

In Equation 11.5, η_1 is the local deviation of the water surface from the average tidal elevation, η_0, z_0 is the average bottom elevation, and Λ is a linearized friction coefficient. Equation 11.5 is valid under the following assumptions: (1) the tidal propagation in intertidal areas nearby the channels is dominated by friction; (2) the propagation of water within creek networks is instantaneous compared to that of shallow salt marshes or flats; (3) The spatial variations of the instantaneous water surface, η_1, and fluctuations in bed level are significantly smaller than the instantaneous average water depth, $\eta_0 - z_0$. These assumptions make the model more suitable for a short tidal embayment (Rinaldo et al. 1999).

The morphological module of this model starts from the following assumptions: (1) tidal channels develop through the incision of channels heads (landward side), where shear stress values are maximum; (2) Given a channel width-to-depth ratio, $\beta = B/D$, the cross-sectional area of the channel, B/β^2, is proportional to the drainage area A, following a Jarret's type of law, i.e. $B \propto \sqrt{\beta A^{\alpha_A}}$, where α_A is close to 1. Given all sites where the shear stress, τ, exceed a critical value, τ_c, the selection of the site to erode is done by considering a Boltzmann-like probability $P(S) \propto e^{-\frac{E(S)}{T}}$, which recalls the annealing procedure of metals. In this equation, $E(S)$ is represented by the value of $\eta_1(S)$ over the unchannelized portion of the basin, S, and T is a constant equivalent of the Gibb's temperature parameter.

More recent model updates (e.g., D'Alpaos et al. 2007) maintain the same simplified hydrodynamic assumptions, but the morphodynamic model is based on the calculation of erosive and depositional fluxes evaluated through the advection diffusion equation, and classic sediment transport formulations (e.g., Einstein and Krone 1962; Mehta 1984). The model also includes parameterizations for vegetation dynamics (Morris et al. 2002; Mudd et al. 2004).

Following similar hydrodynamic and sediment transport parameterizations for the evaluation of sediment fluxes, Kirwan and Murray (2007) explored the three-dimensional development of marsh platforms and tidal channels and found that for moderate sea-level

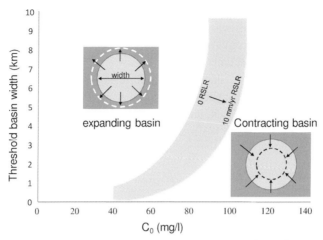

Figure 11.6 Critical basin width as a function of the external sediment input concentration and sea-level rise. Blue band comprises sea-level rise scenarios from 0 to 10 mm/yr. Above the critical width value, the salt marsh surrounding the basin erodes, and the basin collapse. (Adapted from Mariotti and Fagherazzi 2013)

rise values, accretion rates are maintained equal to the rate of sea-level rise. In contrast, high rates of sea-level rise lower the salt marsh surface, reduce biomass productivity, and widen the channel network.

Mariotti and Fagherazzi (2013) presented a simplified model for the expansion of tidal flats and marsh ponds. The model suggests that enclosed basins above a critical width are inevitably going to expand by eroding the surrounding salt marshes due to locally generated wind waves. As the tidal flat expands, a positive feedback between the size of the tidal flat basin and the magnitude of the waves generated within the basin causes a further deterioration and collapse of the surrounding marsh (Fig. 11.6). The model is three-dimensional and accounts for both lateral migration and vertical adjustment of tidal flat and salt marshes. In the model the tidal flat basin is approximated as a cylinder carved into a salt marsh, whose width, w, changes depending on the competition of eroding and accreting forces according to the following equation:

$$\frac{dw}{dt} = 2(B_e - B_a) \tag{11.6}$$

The erosive force, B_e, is considered proportional to the locally generated wave energy and to an erodibility coefficient. The marsh boundary progradation is set equal to:

$$B_a = k_a w_s C_r \rho^{-1} \tag{11.7}$$

where k_a is a non-dimensional constant related to the marsh boundary geometry, w_s is the settling velocity, C_r is the reference concentration, and ρ is the dry bed density. The basin depth is instead updated based on the redistribution of sediments eroded or accumulated from the marsh edge, sea-level rise, and exchange of sediment fluxes.

Leonardi and Fagherazzi (2014, 2015) used a cellular automata model to investigate the response of salt marsh boundaries to different wave climates and extreme events, and to test whether these environments could reach self-organized criticality. The model is based on the analogy between salt marsh erosion and the chemical etching (corrosion) of metals, and accounts for the fact that salt marshes are not uniform. A variability in erosional resistance is present across the marsh that might be caused by natural heterogeneities, including biological and ecological factors and different soil properties. The model consists of a 2-D square lattice where each cell, i, has a randomly distributed resistance value, r_i, and an erosion rate:

$$E_i = \alpha P^\beta \exp\left(\frac{r_i}{P}\right) \tag{11.8}$$

where P is wave power, and α and β are non-dimensional constants. The first part of Equation (11.8) follows theoretical and empirical investigations of salt marsh erosion, suggesting a proportionality between erosion rate and wave power (e.g., Schwimmer 2001; Priestas et al. 2015; Leonardi et al. 2016). The second part of the equation is meant to take into account the variability in erosional resistance induced by the variety of biological and ecological processes affecting each portion of the salt marsh. The erosion rate has two extreme limits, such that for low wave power the system is highly disordered (each element i has a different erosion rate), while for very high wave power, only a weak disorder is present (all elements have similar erosion rates). The equation is such that when the wave power is very low, variability in erosional resistance is more important. For example, during low wave energy conditions, areas with denser vegetation could erode significantly less. On the contrary, when the wave energy is very high, for example under the continuous action of very energetic storms, different salt marsh portions are eroded at the same rate because the variability in resistance is small when compared to the main external agent (term in parenthesis goes to zero, and the exponential goes to one). At every simulation time step, one cell at the boundary is eroded at random with a probability $p_i = \frac{E_i}{\Sigma E_i}$, where the sum refers to all the cells that can be potentially removed because located at the boundary (Leonardi and Fagherazzi 2014, 2015). Weaker cells (high erosion rate) are thus more likely to be removed. Results suggest that for high wave energy conditions, erosion proceeds uniformly with the generation of a smooth salt marsh boundary profile because each cell has similar resistance when compared to the main external driver (Fig. 11.7a). In contrast, when exposed to low wave energy, differential erosion rates are more relevant and affect the global system behavior with a marsh boundary which is rough and jagged (Fig. 11.7c). Statistically, high wave energy conditions correspond to a Gaussian distribution of erosion events (Fig. 11.7b), while for low wave-energy conditions, the frequency distribution of erosion events follows a long-tailed power law distribution with many low magnitude events, accompanied by occasional failures of large salt marsh blocks (Fig. 11.7d). Erosion events represent the number of cells (or meters of salt marsh) eroded within a time lapse Δt. In fact, for low wave energy, a long time period might be required to erode resistant cells, but once the very resistant cells are eroded new weak sites are uncovered and rapidly disintegrate. As a

HIGH WAVE ENERGY

smooth marsh boundary, gaussian frequency-magnitude distribution erosion events

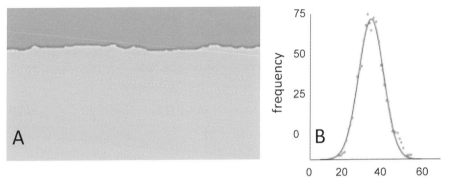

LOW WAVE ENERGY

rough marsh boundary, logarithmic frequency-magnitude distribution erosion events

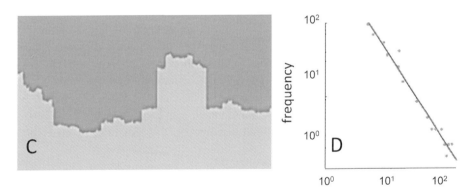

Figure 11.7 Salt marsh boundary geometry and frequency distribution of erosion events for high (A, B), and low (C, D) wave energy conditions. In the model, the magnitude of erosion events is the number of eroded cells per unit of time $\Delta t \Delta t$. Low wave energy conditions correspond to a rough marsh boundary (C), and long-tailed distribution of erosion events (D). The high energy case corresponds to a smooth boundary (A), and a Gaussian distribution of erosion events having a well-defined mean value (B). (Adapted from Leonardi and Fagherazzi 2014)

consequence, unpredictable large failures occur leading to a long-tailed distribution. Model results agree well with field measurements and suggest that under low wave-energy conditions the system might reach self-organized criticality (SOC) (Leonardi and Fagherazzi 2014).

Finally, the SLAMM (Sea Level Affecting Marsh Model) model is a simplified tool which can be used to simulate shoreline modification and wetlands conversion under the long-term action of sea-level rise. The model incorporates different wetland categories and the transition of a computational cell from a wetland type to another is governed by the minimum elevation of the cell according to the following equation:

$$ME_t = ME_{t-1} + \Delta t(AR) - SLR_t \qquad (11.9)$$

Where *ME* is the minimum cell elevation, *AR* is the site specific accretion and/or sedimentation rate, *SLR* is the rise in sea level, and *t* is time. If the *ME* of the cell is lower than the lower boundary of that wetland category, a fraction of that cell will be converted to another wetland category of lower elevation. A fraction of the cell could be also lost by erosion (Chu-Agor et al. 2011). The model requires the following input parameters: wetlands habitat with elevation ranges, a precise digital elevation model, tidal range, accretion rates of the marsh, and tidal flat, erosion rates of both marsh and tidal flat, regional subsidence level and /or uplift, sea-level rise predictions (e.g., Craft et al. 2009).

11.6 High Resolution Three-Dimensional Models

The aim of this section is to review the most recent high resolution models of hydrodynamics coupled to marsh evolution. The two-dimensional and three-dimensional high resolution models studied in this section focus on three different ways to represent marsh evolution and its interaction with water flows, specifically: (1) feedbacks between flow routing and vegetation in the model Delft-3D (Lesser et al. 2004); (2) wetland-estuarine-shelf interactions studied with FVCOM (Chen et al. 2003); (3) feedbacks between sea-level rise and salt marsh productivity analysed by the Hydro-MEM model (Alizad et al. 2016).

The first high resolution marsh model was set up by Temmerman et al. (2005, 2007) who studied the impact of vegetation on flow routing and sedimentation patterns with the 3D hydrodynamic and morphodynamic model Delft-3D (Lesser et al. 2004).

Delft-3D flow module computes flow characteristics (such as water depth, turbulence characteristics, flow velocities, and directions) dynamically in time over a three-dimensional computational grid. The flow computations are based on a finite-difference solution of the three-dimensional shallow water equations with a k-ε turbulence closure model (Rodi 1980).

Temmerman et al. (2005) used Delft-3D to study the relative impact of vegetation, high resolution topography, and water level fluctuations on sedimentation patterns in a tidal marsh landscape. The model incorporates the three-dimensional effect of vegetation on flow (drag and turbulence). The influence of vegetation on drag gives rise to an extra source term of friction force, $F(z)$, in the momentum equations caused by cylindrical plant structures:

$$F(z) = \frac{1}{2}\rho_0\varphi(z)n(z)|u(z)|u(z) \qquad (11.10)$$

where ρ_0 is the fluid density, $\varphi(z)$ is the diameter of cylindrical plant structures at height z above the bottom, $n(z)$ is the number of plant structures per unit area, $u(z)$ is the horizontal flow velocity.

As long as the water level is below the top of the vegetation canopy, differences in flow resistance between vegetated and non-vegetated areas result in faster flow routing over non-vegetated areas, so that vegetated zones are flooded with water coming from non-vegetated parts, with flow directions relatively perpendicular to the vegetation boundary. At the vegetation edge, flow velocities are reduced and sediments are rapidly trapped. In contrast, in bare areas between vegetation patches, flow velocities are enhanced, resulting in reduced sedimentation or erosion. As the water level overtops the vegetation, the flow paths described above change to large-scale sheet flow crossing both vegetated and non-vegetated areas.

Nardin and Edmonds (2014) applied the Delft-3D model and its vegetation module based on Baptist et al. (2007) on deltaic marshes (Fig. 11.8). The model suggests that intermediate vegetation heights and densities are optimal for enhancing sediment deposition on deltaic marshes (Fig. 11.8). Other analyses conducted by Nardin et al. (2016) on river deltas reveal the key role of the spatial distribution of vegetation in trapping sediment

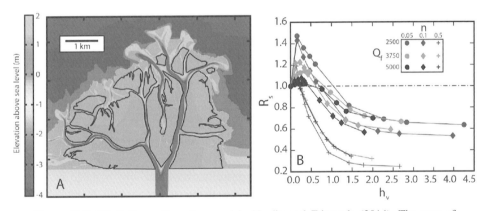

Figure 11.8 (A) Delta configuration used in Nardin and Edmonds (2014). The map of bathymetry shows river delta morphology generated by Delft-3D. Solid black lines surround marsh platforms. (B) Modelled sediment fluxes for marsh platforms under vegetated and non-vegetated conditions. Non-dimensional sedimentation RS (ratio between the total amount of sediments trapped in vegetated deltaic islands with respect to the unvegetated case) increases as a function of vegetation height, hv and then decreases. The inset plot indicates the flood discharge (Qf) and vegetation density (n = mD, where m is the number of vegetation stems per square meter and D is the stem diameter) in each simulation. (Adapted from Nardin and Edmonds 2014)

in marsh platforms at the head of the delta. These dynamics control the overall slope of the delta and therefore its resilience with respect to sea-level rise.

Chen et al. (2008) simulated salt marsh flooding and drying in an estuarine environment with the unstructured-grid Finite-Volume Coastal Ocean Model (FVCOM) (Chen et al. 2003). FVCOM reproduced the three-dimensional structure of tidal currents, the amplitudes and phases of the tidal wave, and salinity gradients in an intricate estuarine environment formed by tidal creeks and intertidal salt marshes. The study shed light on the effect of vegetation roughness on water flow. These results also suggested that a model can generate unrealistic flow fields and water fluxes if it does not resolve the complex geometry of tidal creeks. A mass-conservative unstructured-grid model is required to exactly and efficiently reproduce tidal currents in a complex geometrically controlled estuarine environment. Another high resolution modeling application in a marsh system was performed by Zhao et al. (2010), who applied FVCOM in Plum Island Sound, a bay along the Massachusetts coast, USA. The model computational domain developed with a high horizontal resolution of 10–200 m includes the entire intertidal region. FVCOM predicted the salinity field in the bay and marshes, and how it changed as a function of tidal and riverine forcing.

Hydro-MEM is another example of spatially explicit model that couples tidal propagation to salt marsh vegetation (Alizad et al. 2016). The model accounts for the growth of *Spartina* species common in temperate salt marshes, and the impact of sea-level rise on vegetation production. The model uses the two-dimensional depth-integrated Advanced CIRCulation finite element model (ADCIRC; Luettich et al. 1992) to simulate tidal hydrodynamics. The model is able to simulate the highly variable tidal hydroperiod across the salt marsh platform, which is one of the main parameters controlling vegetation biomass. Plant biomass and marsh accretion as a function of sediment supply and vegetation density are captured by the Marsh Evolution Model (MEM) (Morris et al. 2002). The Hydro-MEM model provides a powerful technique to measure the complex spatial dynamics of salt marsh vegetation and to predict landscape evolution under sea-level rise projections.

11.7 Case Study: The Virginia Coast Reserve, USA

We further point out the lack of a single model able to capture all the processes controlling salt marsh dynamics. The development of such model is also not recommended, because it would require a large number of parameters, possibly over fitting any available dataset and hindering a correct validation. Here we advocate the use of a suite of models, each addressing important dynamics of the system. A suite of models better captures the large range of spatial and temporal scales, by addressing each scale separately. As a case study for this approach we briefly describe our model activities at the Virginia Coast Reserve, one of the Long-Term Ecological Research sites in the USA (Fig. 11.9). Early vertical models of marsh accretion supported by field data showed that the marshes in these lagoons are keeping pace with sea-level rise in the vertical direction (Kirwan et al. 2010). The focus of the modeling research then shifted to possible horizontal variations in marsh extension, and in particular to lateral movements of the boundary between marshes and bays. A transect

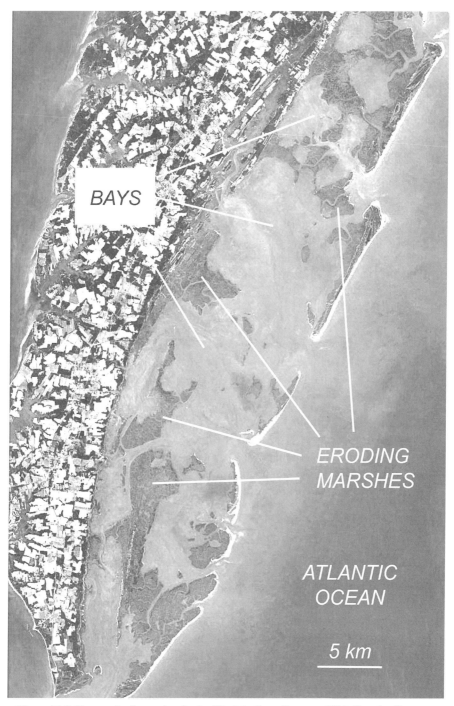

Figure 11.9 Bays and salt marshes in the Virginia Coast Reserve, USA (Landsat Image – Google Earth). (A black and white version of this figure will appear in some formats. For the color version, please refer to the plate section.)

model was developed, accounting for both progradation by sedimentation and erosion by waves (Mariotti and Fagherazzi 2010) the model indicated that the boundary is not stable, and the marsh is either expanding or contracting. To understand the consequences of these boundary dynamics on the entire bay system we then developed a two-dimensional simplified model for the evolution of entire bays bordered by salt marshes (Mariotti and Fagherazzi 2013). This model showed that indeed a stable marsh extension is never reached, and tidal bays larger than a critical dimension will always expand whereas small tidal bays will always shrink. We then focused our attention on the rate of marsh erosion, the role of hurricanes and violent storms, and whether it is possible to forecast marsh erosion from the geometry of the boundary. A statistical model of wave erosion during storms indicated that erosion rates of marsh boundaries is relatively constant and site specific, with extreme events as hurricanes barely affecting this rate (Leonardi et al. 2016). A cellular automata model further showed that it is possible to predict the rate of marsh erosion from the planar geometry of the boundary, with corrugated boundaries indicating slow erosion and linear boundaries suggesting fast deterioration (Leonardi and Fagherazzi 2015). A puzzling aspect of the Virginia Coast Reserve marshes is the absence of riverine inputs, so researchers always wondered how these marshes formed and accreted in time. A three-dimensional high-resolution model of sediment transport showed that the sediment necessary to build the marshes is imported in the bays during storms, with larger storms importing more sediment (Castagno et al. 2018). The same high resolution model indicated that vegetated marshes are important for the stability of the entire bay, increasing the overall sediment trapping by 10% (Nardin et al. 2018). The take home messages from our suite of models are therefore the following:

(1) Salt marshes in the Virginia Coast Reserve keep pace with sea-level rise and acquire sediment during storms.
(2) The marsh boundary is eroding away at a relatively constant rate, irrespectively of hurricanes; this erosion will never stop because equilibrium does not exist.
(3) The presence of marshes is critical for the stability of the entire barrier island system.

Only a suite of models can achieve these important results.

11.8 Future Research Needs

Recent years have seen a considerable improvement in our ability to forecast salt marsh evolution with numerical models. However, much still needs to be done. The representation of vegetation dynamics in numerical models is still in its infancy. Some of the relationships commonly used for the interactions between vegetation and hydrodynamics are based on laboratory experiments (e.g., Nepf 1999, Mendez and Losada 2004, Zhong and Nepf 2010) and there is a need to retool them in real settings. Moreover, vegetation is often schematized in simple geometric forms (as for instance cylinders) in order to compute friction and turbulence parameters. These simple representations hardy capture the complexity of marsh vegetation. Marsh species display a variety of canopies, with wide differences in density, height, roots, number and size of leaves, or stiffness of the stems (Neumeier 2005).

Indicative of this complexity is the difference between smooth cordgrass (*Spartina alterniflora*), that stands vertical on the marsh platform, and saltmarsh hay (*Spartina patens*), the weak stems of which are bent by flow forming dense cowlicks (Bertness 1991). We are clearly far from capturing this morphological diversity of vegetation – even within the same plant genus – in numerical models. More field investigations are needed to determine the hydrodynamic behavior of different vegetation canopies.

Models of marsh evolution must also include how plants respond to changes in hydroperiod, elevation, and flow dynamics. Relationships linking vegetation biomass to local marsh morphology have been proposed (e.g., Morris et al. 2002). However, very often these relationships were derived for specific geographic locations, and lack generality. More comparative studies looking at dynamics of vegetation canopies across large geographic gradients are needed to develop general relationships.

More research in needed to better quantify the organic accretion term. Salinity plays a strong role in temperate plant biomass, reducing the amount of organic material entering the soil. Changes in salinity have not been adequately addressed within marsh accretion models.

Distributed models, capturing the complex spatial evolution of marshes, need spatially distributed datasets. However, vegetation, sedimentological, and hydrodynamic components of marsh models often rely on data collected at few locations (point measurements), which do not capture the spatial distribution of physical attributes. Recent advances in remote sensing can provide valuable information at the landscape scale. Remote sensing data can inform models, or be utilized to validate results. Among others, Light Detection and Ranging (Lidar) data can provide the high resolution topography of a marsh; hyperspectral images enable vegetation classification in great detail, while Surface Water Ocean Topography (SWOT) can be used to quantify water levels in marsh channels. The integration of remote sensing data within marsh models will be the focus of much research in future years.

Acknowledgments

S. F. was supported by NSF awards DEB-1237733 (VCR-LTER program), and OCE-1637630 (PIE-LTER program).

References

Alizad, K., Hagen S. C., Morris, J. T., Bacopoulos, P., Bilskie, M. V., Weishampel, J. F., and Medeiros, S. C. 2016. A coupled, two-dimensional hydrodynamic-marsh model with biological feedback. *Ecological Modelling*, 327: 29–43.

Allen, J. R. L. 1990. Salt-marsh growth and stratification: A numerical model with special reference to the Severn Estuary, southwest Britain. *Marine Geology* 95: 77–96.

Baptist, M. J., Babovic, V., Uthurburu, J. R., Keijzer, M., Uittenbogaard, R. E., Mynett, A., and Verwey, A. 2007. On inducing equations for vegetation resistance. *Journal of Hydraulic Research*, 45: 435–450.

Bertness, M. D. 1991. Zonation of *Spartina patens* and *Spartina alterniflora* in New England salt marsh. *Ecology*, 72: 138–148.

Cahoon, D. R., French, J. R., Spencer, T., Reed, D., and Möller, I. 2000. Vertical accretion versus elevational adjustment in UK saltmarshes: An evaluation of alternative methodologies. *Geological Society of London Special Publication*, 175: 223–238.

Cahoon, D. R., Hensel, P. F., Spencer, T., Reed, D. J., McKee, K. L., and Saintilan, N. 2006. Coastal wetland vulnerability to relative sea-level rise: Wetland elevation trends and process controls. In *Wetlands and Natural Resource Management*. In: J. T. A. Verhoeven, . Beltman, R. Bobbink, and D. F. Whigham, eds, Berlin, Heidelberg: Springer Berlin Heidelberg. pp. 271–292.

Castagno, K. A., Jiménez-Robles, A. M., Donnelly, J. P., Wiberg, P. L., Fenster, M. S. and Fagherazzi, S., 2018. Intense storms increase the stability of tidal bays. *Geophysical Research Letters*, 45: 5491–5500.

Chen, C., Liu, H., and Beardsley, R. 2003. An unstructured grid, finitevolume, three-dimensional, primitive equations ocean model: Application to coastal ocean and estuaries, *Journal of Atmospheric Oceanic Technology*, 20: 159–186.

Chen, C., Qi, J., Li, C., Beardsley, R. C., Lin, H., Walker, R., and Gates, K., 2008. Complexity of the flooding/drying process in an estuarine tidal-creek salt-marsh system: An application of FVCOM. *Journal of Geophysical Research: Oceans*, 113(C7).

Christiansen, T., Wiberg, P. L., and Milligan, T. G. 2000. Flow and sediment transport on a tidal salt marsh surface. *Estuarine and Coastal Shelf Science*, 50: 315–331.

Chu-Agor, M. L., Muñoz-Carpena, R., Kiker, G., Emanuelsson, A. and Linkov, I., 2011. Exploring vulnerability of coastal habitats to sea level rise through global sensitivity and uncertainty analyses. *Environmental Modelling & Software*, 26: 593–604.

Craft, C., Clough, J., Ehman, J., Joye, S., Park, R., Pennings, S., Guo, H. and Machmuller, M., 2009. Forecasting the effects of accelerated sea-level rise on tidal marsh ecosystem services. *Frontiers in Ecology and the Environment*, 7: 73–78.

D'Alpaos, A., Lanzoni, S., Marani, M., Fagherazzi, S. and Rinaldo, A., 2005. Tidal network ontogeny: Channel initiation and early development. *Journal of Geophysical Research: Earth Surface*, 110(F2).

D'Alpaos A., Lanzoni, S., Marani, M., and Rinaldo, A. 2007. Landscape evolution in tidal embayments: Modeling the interplay of erosion, sedimentation, and vegetation dynamics. *Journal of Geophysical Research: Earth Surface* 112 (F1). Wiley Online Library.

D'Alpaos, A., and Marani, M. 2016. Reading the signatures of biologic–geomorphic feedbacks in salt-marsh landscapes. *Advances in Water Resources*, 93: 265–275.

Einstein, H. A., and Krone, R. B. 1962. Experiments to determine modes of cohesive sediment transport in salt water, *Journal of Geophysical Research*, 67: 1451–1461.

Fagherazzi, S., Kirwan, M. L., Mudd, S. M., Guntenspergen, G. R., Temmerman, S., D'Alpaos, A., Koppel, J., et al. 2012. Numerical models of salt marsh evolution: Ecological, geomorphic, and climatic factors. *Reviews of Geophysics*, 50(1): 1–28.

Fagherazzi S., Marani M., and Blum L.K. (eds) 2004. *The Ecogeomorphology of Tidal Marshes*, Vol. 59, Coastal and Estuarine Studies. American Geophysical Union, Washington, D.C.

French, J. R. 1993. Numerical simulation of vertical marsh growth and adjustment to accelerated sea-level rise, North Norfolk, U.K. *Earth Surface Processes and Landforms* 18: : 63–81.

Goldenfeld, N., and Kadanoff, L. P. 1999. Simple lessons from complexity. *Science*, 284: 87–89.

Kirwan, M. L., Guntenspergen, G. R., D'Alpaos, A., Morris, J. T., Mudd, S. M., and Temmerman, S. 2010. Limits on the adaptability of coastal marshes to rising sea level. *Geophysical Research Letters*, 37: 1–5.

Kirwan, M. L., and Murray, A. B. 2007. A coupled geomorphic and ecological model of tidal marsh evolution. *Proceedings of the National Academy of Sciences of the United States of America*, 104: 6118–22.

Kirwan, M. L., and Temmerman, S. 2009. Coastal marsh response to historical and future sea-level acceleration. *Quaternary Science Reviews*, 28: 1801–8.

Kirwan, M. L., Walters, D. C., Reay, W. G., and Carr, J. A. 2016. Sea level driven marsh expansion in a coupled model of marsh erosion and migration. *Geophysical Research Letters*, 43: 4366–4373.

Leonardi, N., and Fagherazzi, S. 2014. How waves shape salt marshes. *Geology*, 42: 887–890.

Leonardi, N., and Fagherazzi, S. 2015. Effect of local variability in erosional resistance on large-scale morphodynamic response of salt marshes to wind waves and extreme events. *Geophysical Research Letters*, 42: 5872–5879.

Leonardi, N., Ganju, N. K., and Fagherazzi, S. 2016. A linear relationship between wave power and erosion determines salt-marsh resilience to violent storms and hurricanes. *Proceedings of the National Academy of Sciences of the United States of America*, 113: 64–68.

Lesser, G., Roelvink, J., Van Kester, J., and Stelling, G. 2004. Development and validation of a three-dimensional morphological model. *Coastal Engineering*, 51: 883–915.

Lorenzo-Trueba, J., and Ashton, A. D. 2014. Rollover, drowning, and discontinuous retreat: Distinct modes of barrier response to sea-level rise arising from a simple morphodynamic model. *Journal of Geophysical Research: Earth Surface*, 119: 779–801.

Lorenzo-Trueba, J., and Mariotti, G. 2017. Chasing boundaries and cascade effects in a coupled barrier-marsh-lagoon system. *Geomorphology*, 290: 153–163.

Luettich, R. A., Westerink, J. J., and Scheffner, N. W. 1992. ADCIRC: An advanced three-dimensional circulation model for shelves, coasts, and estuaries. I: Theory and methodology of ADCIRC-2DD1 and ADCIRC-3DL. In: *Technical Rep. No. DRP-92-6*. U.S. Army Engineer Waterways Experiment Station, Vicksburg, MS.

Marani, M., D'Alpaos, A., Lanzoni, S., Carniello, L., and Rinaldo, A. 2007. Biologically-controlled multiple equilibria of tidal landforms and the fate of the Venice Lagoon. *Geophysical Research Letters*, 34: L11402.

Marani, M., Lio, C. D., and D'Alpaos, A. 2013. Vegetation engineers marsh morphology through multiple competing stable states. *Proceedings of the National Academy of Sciences of the United States of America*, 110: 3259–3263.

Mariotti, G., and Carr, J. 2014. Dual role of salt marsh retreat: Long-term loss and short-term resilience. *Water Resources Research*, 50: 2963–2974.

Mariotti, G., and Fagherazzi, S. 2010. A numerical model for the coupled long-term evolution of salt marshes and tidal flats. *Journal of Geophysical Research: Earth Surface,*. 115: F01004.

Mariotti, G. and Fagherazzi, S. 2013. Critical width of tidal flats triggers marsh collapse in the absence of sea-level rise. *Proceedings of the National Academy of Sciences of the United States of America*, 110: 5353–5356.

Mehta, A. J. 1984. Characterization of cohesive sediment properties and transport processes in estuaries. In: A. J. Mehta, ed., *Estuarine Cohesive Sediment Dynamics, Lecture Notes on Coastal and Estuarine Studies*, vol. 14, Springer, New York, pp. 290–315.

Mendez, F. J. and Losada, I. J. 2004. An empirical model to estimate the propagation of random breaking and nonbreaking waves over vegetation fields. *Coastal Engineering*, 51: 103–118.

Morris, J. T., P. V. Sundareshwar, C. T. Nietch, B. Kjerfve, and D. R. Cahoon 2002. Responses of coastal wetlands to rising sea level, *Ecology*, 83: 2869–2877.

Mudd, S. M., Fagherazzi, S., Morris, J. T., and Furbish, D. J. 2004. Flow, sedimentation, and biomass production on a vegetated salt marsh in South Carolina: Toward a predictive model of marsh morphologic and ecologic evolution. In: S. Fagherazzi, M. Marani, and L. K. Blum, eds., *The Ecogeomorphology of Salt Marshes, Coastal Estuarine Studies*, vol. 59, American Geophysical Union, Washington, DC, pp. 165–188.

Mudd, S. M., Howell, S. M. and Morris, J. T. 2009. Impact of dynamic feedbacks between sedimentation, sea-level rise, and biomass production on near-surface marsh stratigraphy and carbon accumulation. *Estuarine, Coastal and Shelf Science*, 82: 377–389.

Murray, A. B. 2003. Contrasting the goals, strategies, and predictions associated with simplified numerical models and detailed simulations. In: P. R. Wilcock and R. M. Iverson, eds, *Prediction in Geomorphology*, American Geophysical Union, Washington DC, pp. 151–165.

Murray, A. B., Gasparini, N. M., Goldstein, E. B., and van der Wegen, M. 2016. Uncertainty quantification in modeling earth surface processes: More applicable for some types of models than for others. *Computers & Geosciences*, 90: 6–16.

Nardin, W. and Edmonds, D. A. 2014. Optimum vegetation height and density for inorganic sedimentation in deltaic marshes. *Nature Geoscience*, 7: 722–726.

Nardin, W., Edmonds, D. A., and Fagherazzi, S. 2016. Influence of vegetation on spatial patterns of sediment deposition in deltaic islands during flood. *Advances in Water Resources*, 93: 236–248.

Nardin, W., Larsen, L., Fagherazzi, S., and Wiberg, P., 2018. Tradeoffs among hydro-dynamics, sediment fluxes and vegetation community in the Virginia Coast Reserve, USA. *Estuarine, Coastal and Shelf Science*, 210: 98–108.

Nepf, H. M., 1999. Drag, turbulence, and diffusion in flow through emergent vegetation. *Water Resources Research*, 35: 479–489.

Neumeier, U., 2005. Quantification of vertical density variations of salt-marsh vegetation. *Estuarine, Coastal and Shelf Science*, 63: 489–496.

Priestas, A. M., Mariotti, G., Leonardi, N., and Fagherazzi, S., 2015. Coupled wave energy and erosion dynamics along a salt marsh boundary, Hog Island Bay, Virginia, USA. *Journal of Marine Science and Engineering*, 3: 1041–1065.

Rahmstorf, S. 2007. A semi-empirical approach to projecting future sea-level rise. *Science*, 315: 368–370.

Ratliff, K. M., Braswell, A. E., and Marani, M. 2015. Spatial response of coastal marshes to increased atmospheric CO2. *Proceedings of the National Academy of Sciences of the United States of America*, 112, 15580–15584.

Redfield, A. C. 1965. Ontogeny of a salt marsh estuary. *Science* 147: 50–55.

Rinaldo, A., Fagherazzi, S., Lanzoni, S., Marani, M., and Dietrich, W. E. 1999. Tidal networks: 2. Watershed delineation and comparative network morphology. *Water Resources Research*, 35: 3905–3917.

Rodi, W., 1980. *Turbulence Models and Their Application in Hydraulics*. International Association for Hydro-Environment Engineering and Research, Delft.

Schwimmer, R. A., 2001. Rates and processes of marsh shoreline erosion in Rehoboth Bay, Delaware, USA. *Journal of Coastal Research*, 17: 672–683.

Stolper, D., List, J. H., and Thieler, E. R. 2005. Simulating the evolution of coastal morphology and stratigraphy with a new morphological-behaviour model (GEOMBEST). *Marine Geology*, 218: 17–36.

Tambroni, N., and Seminara, G. 2012. A one-dimensional eco-geomorphic model of marsh response to sea level rise: Wind effects, dynamics of the marsh border and equilibrium. *Journal of Geophysical Research: Earth Surface*, 117: F03026.

Temmerman, S., Bouma, T. J., Govers, G., Wang, Z. B., De Vries, M. B., and Herman, P. M. J. 2005. Impact of vegetation on flow routing and sedimentation patterns: Three-dimensional modeling for a tidal marsh. *Journal of Geophysical Research: Earth Surface*, 110: F04019.

Temmerman, S., Bouma, T. J., Van de Koppel, J., Van der Wal, D., De Vries, M. B. and Herman, P. M. J. 2007. Vegetation causes channel erosion in a tidal landscape. *Geology*, 35: 631–634.

Temmerman, S., Govers, G., Meire, P., and Wartel, S. 2003. Modelling long-term tidal marsh growth under changing tidal conditions and suspended sediment concentrations, Scheldt estuary, Belgium. *Marine Geology* 193: 151–169.

Temmerman, S., Govers, G., Meire, P., and Wartel, S. 2004. Simulating the long-term development of levee–basin topography on tidal marshes. *Geomorphology*, 63: 39–55.

Walters, D., Moore, L. J., Duran Vinent, O., Fagherazzi, S., and Mariotti, G. 2014. Interactions between barrier islands and backbarrier marshes affect island system response to sea level rise: Insights from a coupled model. *Journal of Geophysical Research: Earth Surface*, 119: F003091.

Weston, N. B. 2014. Declining sediments and rising seas: An unfortunate convergence for tidal wetlands. *Estuaries and Coasts*, 37: 1–23.

Zhao, L., Chen, C., Vallino, J., Hopkinson, C., Beardsley, R. C., Lin, H., and Lerczak, J., 2010. Wetland-estuarine-shelf interactions in the Plum Island Sound and Merrimack River in the Massachusetts coast. *Journal of Geophysical Research: Oceans*, 115: C10039.

Zong, L. and Nepf, H., 2010. Flow and deposition in and around a finite patch of vegetation. *Geomorphology*, 116: 363–372.

12

Understanding Marsh Dynamics

Laboratory Approaches

CHARLIE E. L. THOMPSON, SARAH FARRON, JAMES TEMPEST, IRIS MÖLLER,
MARTIN SOLAN, AND JASMIN GODBOLD

Salt marshes are valuable but complex biophysical systems with associated ecosystems. This presents numerous challenges when trying to understand and predict their behaviour and evolution, which is essential to facilitate their continued and sustainable use, conservation and management[1]. Detailed understanding of the hydrodynamics, sediment dynamics, and ecology that control the system is required, as well as their numerous interactions[2,3], but is complicated by spatial and temporal heterogeneity at a range of scales[4,5]. These complex interactions and feedbacks between the physical, biological, and chemical processes can be investigated in situ following natural, unintentional, or intentional manipulation[6], but the mechanistic basis of any observations are confounded by the presence of collinear variables. Hence, laboratory investigations can be beneficial, as they provide the opportunity for systematic testing of subsets of coastal processes, mechanisms, or conditions typical of salt marsh systems, in the absence of confounding variables. With appropriate scaling, this allows a better understanding of the overall function of the salt marsh, and better predictions of their evolution.

There are many challenges related to laboratory investigations of such complex environments. Being able to accurately scale-up any findings to marsh, system, or regional levels is essential; assessing the appropriate simplification of multiple interdependent processes, while not overlooking key interactions is challenging; and achieving sufficient replication to suitably capture the heterogeneity of the system logistically challenging. In short, one must determine how best to simply represent the complexity of the salt marsh system, to balance an understanding of mechanistic processes and applicability at a system scale.

This chapter provides an overview of good practice when performing laboratory investigations of salt marsh dynamics and processes, divided into three sections, each investigating a different component of a vertical profile through the salt marsh. Section 12.1 Above the Bed considers the benthic boundary layer and interactions between biogenic roughness, currents and waves. Section 12.2 At the Bed examines bed stability and exchange across the sediment-water interface. Section 12.3 Within the Bed looks at some of the hidden, often overlooked, processes that occur below the salt marsh surface. The interactions between these interlinked sections, and appropriate scaling of the findings will then be considered in the discussion.

12.1 Above the Bed: Biogenic Roughness and Fluid Flows

Aboveground biomass and roughness elements interact with flow properties to affect the structure of the benthic boundary layer[7], flow magnitudes and directions[8,9], wave attenuation[10], turbulence generation[11], applied bed shear stress[12], erosion,[13] and deposition[14]. These interactions are complex, often exhibiting both positive and negative feedbacks[15]. Because large-scale experiments investigating interactions between deformable bodies such as vegetation and flows are difficult[16], the best approach is often to analyse small-scale processes and upscale the results to infer their macroscale implications[17], with an awareness that the subsequent scaling is difficult in inherently non-linear systems[18].

The use of laboratory flume studies is a practical way to model these small-scale processes, and isolate individual mechanisms and characteristics of flow–roughness interactions. Despite the complexity and associated difficulties in isolating individual flow effects, laboratory studies have identified general trends apparent when flows interact with salt marsh roughness elements, in particular vegetation. Flow behaviour appears closely correlated to submergence ratios (water depth relative to canopy height), especially when considering wave dissipation[19,20]. Where water depth exceeds canopy height, flow restriction by vegetation[21,22] deflects above canopy flows into areas of higher flow velocities and turbulence, known as skimming flow[11]. These exhibit characteristic logarithmic vertical flow profiles[21–23], coinciding with the height where cumulative vertical canopy biomass reaches 85%–90%[7,11,24–26]. Within the canopy the flow typically decelerates, although local accelerations can occur due to vertical variations in canopy stem density or biomass[11,21,27], alongside changes in turbulence production[22,23]. In emergent conditions flows are constrained by the vegetation, and canopy gaps create complex localized areas of higher velocities and turbulence[11].

These laboratory investigations are undertaken at ranges of scales utilizing real (section 12.1.1) or mimic (section 12.1.2) individual plants or animals, patches, or full marsh sections (section 12.1.3), to elucidate the controlling mechanisms, individually and in combination. The majority use laboratory flumes, ranging from small desk-top studies[28], to prototype (1:1) scales[29], which allow close control over experimental variables, and simplify three-dimensional systems into two-dimensional problems.

12.1.1 Live Vegetation in Laboratory Investigation

12.1.1.1 Flow Structure Effects

Laboratory flume experiments utilizing live salt marsh vegetation typically consist of monospecific canopies with grass-like architectures such as *Spartina* or *Puccinellia* spp.[7,11,21,28]. The canopies are subjected to steady[21] or oscillatory currents[30] and probes positioned before, within, and behind the canopy are used to record water level, velocity and turbulence characteristics (Fig. 12.1).

Vertical velocity profiles are constructed and depth averaged velocity ($\bar{U}$), applied bed shear stress (τ_0), or roughness length (z_0)[21] quantified as:

Side View

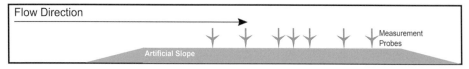

Plan View

Figure 12.1 A typical experimental design for a vegetation patch in a unidirectional recirculating flume. Example shows a submerged patch, narrower than the flume width. A slope raises the flume bed to level of the saltmarsh bed surface or aboveground biomass. Probe measurement positions in red, consider free-stream, before, within, and post patch flow changes. For narrow patches, measurements are taken to assess acceleration around patch edges. Full flume calibration is used to determine the optimum patch positioning within the fully developed benthic boundary layer.

$$U = \frac{1}{h} \int_0^h U_z . dz$$

Eq.12.1

where $U_{(z)}$ is velocity at height (z) above the bed, and h is water depth.

$$\tau_0 = \rho u_*^2$$

Eq.12.2

where ρ is fluid density, u_* the friction velocity

$$u_* = \frac{U_g \kappa}{\ln\left(\frac{z}{z_0}\right)}$$

Eq.12.3

and κ is von Karman's constant (0.41).

Sometimes flow complexity within the canopy makes it difficult to identify coherent velocity profiles[11,31,32]. Under steady flow conditions, the friction slope ($\frac{\partial h}{\partial x}$: change in water depth [h] across canopy length [x]) can be used to quantify changes in momentum and drag (C_d) through the canopy[33]

$$C_d a = \frac{-2g\delta h}{U^2 \delta x}$$

Eq.12.4

Where a is canopy density, and g acceleration due to gravity. However, this application can complicate separation of momentum losses caused by the bed versus the canopy, and the relative contributions of momentum losses depends on both vegetation structure and water depth[34].

Similarly, wave energy losses may be quantified from reductions in wave height as it travels through the canopy[35]

$$\frac{H}{H_0} = exp\left(-k_i x\right) \qquad \text{Eq.12.5}$$

where H is local wave height, H_0 is incident wave height, and k_i the wave decay coefficient.

Alternatively, force transducers can be attached directly to the vegetation[30].

Within canopy turbulence is usually represented by turbulent kinetic energy density (TKE)[36], based on variance of turbulent fluctuations (') of downstream (u), cross-stream (v) and vertical (w) velocity components.

$$\text{TKE} = \frac{1}{2}\rho\left(\overline{u'^2} + \overline{v'^2} + \overline{w'^2}\right) \qquad \text{Eq.12.6}$$

12.1.1.2 Plant Morphology

Changes to flow and boundary layer structure are subsequently related to vegetation morphology. Natural salt marsh contains many plant elements (stems, leaves, shoots/branches, and flowers) contributing to complex plant morphologies[37]. *Spartina alterniflora*, for example, has a grass-like architecture (generating relatively predictable vertical flow profiles[23,38,39]) whereas *Atriplex portulacoides*[24] is more shrubby. Variations in morphology and vegetation density lead to variable vertical biomass within the canopy, affecting local flow velocities, momentum losses[31], and the development of skimming flow[11]. There is therefore a requirement to accurately measure plant morphology. This can be done manually[40] which is time consuming, but recent advances in technologies such as laser scanning may help data availability at larger scales[41].

This is further complicated by variability in mechanical features such as buoyancy and flexibility. Buoyancy may redistribute canopy biomass toward the water's surface[31]; however, little is known about its effect on flow modification[43] or its variability amongst species or plant elements. Flexibility modifies canopy shape and can influence flow momentum losses, skimming flow development, and vertical velocity profiles. This appears particularly important for wave energy dissipation as wave period and frequency can often force the canopy to flex in or out of phase with the incoming wave[43,44]. Flexibility[30,45] varies by species[46], and measurements focus on stem flexibility, despite other plant features exhibiting varying flexibility thresholds, which can significantly affect momentum losses[44]. This limits our understanding of the importance of plant flexibility to flow modification and requires further study. This complexity has, in part led to the development of simplified representations of salt marsh species for laboratory experimentation in the form of mimic plants.

12.1.2 Mimic Representations of Vegetation in Flume Experiments

While many laboratory experiments make use of natural vegetation, there are a number of arguments which might lead to the use of artificial or mimic plants. These include:

(1) *Availability*. Restrictions may relate to required quantities, transportation logistics, seasonal availability or "at risk" status (e.g., *Zostera*, www.OSPAR.org).
(2) *Maintenance*: Challenges keeping plants healthy during acclimatization to avoid shock responses[47] can require significant investments in space and environmental control.

Figure 12.2 Mimic representations from literature (A) Non-specific rigid aquatic vegeta-tion[58], (B) *Posidonia oceanica*[55], (C) *Spartina alterniflora*[73], (D) *Acorus camalus,*[51] (E) Non-specific rigid aquatic vegetation patch,[12] (F) *Enhalus acoroides,*[54] and (G) *Zostera marina*[65]. (A black and white version of this figure will appear in some formats. For the color version, please refer to the plate section.)

(3) *Prevention:* Minimizing environment damage, or perturbation of the natural system[48]; or,

(4) *Specificity:* The requirement for explicit traits to be isolated (e.g., stiffness, buoyancy, leaf length, stem length)[49].

Mimic roughness elements have been successfully utilized for studies of hydrodynamic interaction (unidirectional currents[50,51], waves[35,52–55], combined waves and currents,[49,56] and plant motions[57]), submergence ratios, patchiness[7,17,58–63], roughness[64,65], sediment dynamics[66,67], and ecosystem engineering[68]. Many studies address multiple topics, and mimic use is not restricted to vegetation[69,70]. The types of mimic used range from plastic imitations of aquatic vegetation[66] to simple nails in a board[28] (Fig. 12.2). While mimics may not fully represent the structural or mechanical properties of their natural counter-parts[2,26,71,72], they have proven a useful tool when investigating roughness–flow interactions.

12.1.2.1 Designing Mimics for Laboratory Investigations

The experimental methods and measurements are typically the same as for natural vegeta-tion (section 12.1.1), but the requirement of the mimic to replicate specific behaviors or movements of the plants must be considered. These requirements depend on the scientific questions being asked, and the species being investigated. For example, it is typical for vegetation with rigid stems to be represented by rigid cylinders of equivalent height and depth, whereas, when considering flexible vegetation, bending, and "monami"[57] behaviors must be considered These approaches are combined when flexible vegetation includes stiffened areas near the base to represent stems or sheaths[55].

Common traits considered when designing mimics can be split into two categories: subjective, matching general bending characteristics, movement types, or leaf geometry; or quantitative, using scaling or engineering arguments to match or scale natural characteris-tics. Table 12.1 provides a summary of these approaches from literature. The former tends

Table 12.1 *Mimic use in published literature, with justifications of material choices, canopy densities, and plant arrangements where available*

Type	Mimiced Species	Material and Justification	Canopy Density and Arrangement	Reference
Non-Specific Vegetation	Rigid vegetation	Finish nails (rigid cylinders). Chosen due to availability		Tollner et al 1982[28] (from Lopez and Garcia, 1998[34])
	Bushes and shrubs	Semirigid rubberized horsehair: stiffness and density not measured		Wu et al, 1999[59]
	Reed stems; bladed grasses bent by flow; watercress	Rigid vertical cylinders; rigid angled metal strips; commercial aquarium plants		Elliott, 2000[66]
	Seagrass	Vinyl plastic. Flexural rigidity matched to natural seagrass. Ratio of bending resistance to drag. Leaf geometry.		Nepf & Vivoni, 2000[33]
	Non-specific	Flexible glass fiber. Ratio of fluid kinetic energy to elastic potential energy (fiber length to bending length)		Alben et al, 2002[16]
	Vegetation stems	Rigid cylinders		Stone & Shen, 2002[60]
	Non-specific	Rigid cylinders		Hu et al, 2014
	Aquatic vegetation	Rigid cylinders	Representative of field conditions	Murphy et al, 2007[58]
	Aquatic vegetation; matching to existing literature	Flexible plastic strips. Flexural rigidity same order of magnitude to aquatic vegetation. Undeflected height matched.	Total frontal area per vegetation volume	Okamoto & Nezu, 2009[57]
	Aquatic vegetation	Rigid cylinders. Height matched to natural emergent vegetation ranges	Non-dimensional blockage number; solid volume fraction. Chosen as representative of natural aquatic vegetation.	Shi et al, 2016[15]

Table 12.1 (*cont.*)

Type	Mimiced Species	Material and Justification	Canopy Density and Arrangement	Reference
Posidonia	*P. oceanica*; representative of any highly flexible submerged aquatic plants	Polyethylene. Leaf geometry typical of natural observations		Folkard, 2011[63]
	P. oceanica	Polyethylene material density; modulus of elasticity; coefficient of kinetic friction. Matched to upper limits of natural plants. Leaf geometry.	Typical of natural densities	Folkard, 2005[17]
	P. oceanica	PVC foam combined with PVC pipe. Modulus of elasticity; material density; leaf geometry, Cauchy number. Matched to natural leaves	Average vs. sparse. Matched to natural densities.	Stratigaki et al, 2011[55]
Zostera	*Z. marina*	Polypropylene ribbon. Leaf geometry; positive buoyancy; flexural stiffness compared but not matched	Mean max. densities in field.	Fonesca & Koehl, 2006[62]
	Z. marina	Rigid cylinders for stems, polyethylene film for leaves, modulus of elasticity: material density: leaf geometry; ratio of blade buoyancy to blade rigidity (to ensure waving motion).	Representative of dense aquatic meadows	Ghisalberti & Nepf, 2006[65]
	Salt marsh vegetation; *Z. noltii*	Cable ties; polyribbon. Movement under waves; measured bed similarity: leaf geometry, upper limit of natural plants.	Covers range of natural densities for *noltii*	Paul et al, 2012[49]
	Zostera-like; *Spartina*-like	Cable ties; unspecified flexible plastic. Leaf geometry; bending behavior; ease of availability for high densities		Bouma, 2005[68]

Table 12.1 (*cont.*)

Type	Mimiced Species	Material and Justification	Canopy Density and Arrangement	Reference
Spartina	*S. alterniflora*	Rigid cylinders; flexible tubing of similar rigidity and density to natural plants		Augustin et al, 2009[61]
	S. alterniflora	Polyolefin tubing. Reproduces swaying motion; buoyancy in emergent conditions; stem geometry; material density; modulus of elasticity	Representative of natural meadows	Anderson & Smith, 2014[73]
Enhalus	*E. acoroides*	Rigid cylinders, polyethylene plastic; flexible vegetation; 1:30 scaling; Young's modulus; stiffness property; Froude scaling of stiffness; leaf geometry	Typical of intertidal marshes	John et al, 2015[74]
Acorus	*A. camalus*	Rigid cylinders; geometric		Panigrahi & Khatua, 2015[51]
Laminariales	Kelp	Polypropylene. Specific gravity of material		Asano et al, 1988 (from Kobayashi et al, 1993[52])
	Kelp	Flexible plastic, manufactured to specific density; stiffness; flexural rigidity	Length of canopy sufficient for establishment of uniform flows	Wilson et al, 2003[64]
Animal Models	Non-specific epibenthic structures (animal/plant roughness)	Rigid cylinders		Eckman and Nowell, 1984[69]
	Animal tubes	Rigid single cylinder		Eckman, 1985[70]
	Animal tubes	Bamboo stakes	Comparable to field measurements	Bouma et al 2007[50]

to be used where the key parameters matched are geometrical (height or diameter). The quantitative measures include:

The *modulus of elasticity* (Young's modulus, E, Nm^{-2}) representing stiffness (e.g., [55,65,74])

$$E = \frac{FL_0}{A\Delta L}$$
Eq.12.7

Where F is force exerted on the mimic, L_o its original length, A the cross-sectional area the force is applied to, and ΔL the resulting length change.

The *flexural rigidity* (J, $Pa.m^3$) representing resistance to bending (e.g.[31,57,62,64])

$$J = EI$$
Eq.12.8

Where I is the inertial moment of the vegetation element.

The *Cauchy number* (Ca), a dimensionless ratio of inertial and elastic forces (e.g., [55])

$$C_a = \rho_f \frac{U^2 S^2}{E}$$
Eq.12.9

Where U is characteristic velocity (ms^{-1}), ρ_f fluid density, and S the Slenderness number ($S = L'/l$; L' and l are the maximum and minimum cross-sectional dimensions).
and, the material density (ρ_s; kgm^{-3}) (e.g., [52,55,65]).

Mimics are typically designed at prototype scale (1:1), but where larger species are downscaled, Froude scaling may be used[74] as stiffness parameters can be difficult to scale.

While many researchers expend considerable effort to match as many traits as possible and produce the most accurate possible mimics, some mimic choice justifications happen post-hoc: often more informed by availability and ease of construction than by mathematical or physical scaling arguments. This is common where isolated properties are modelled, generalized across vegetation types (i.e. flexibility) rather than specific species.

12.1.2.2 Canopy Construction

As well as considering the properties of individual plants, one must consider that canopy densities and geometries also have a significant influence on flow, as outlined in section 12.1.1. Plant distribution is an important consideration when designing mimic experiments, and approaches taken range from matching typical field densities[17,35,49,50,58,62,65,67], to investigating generalized density ranges[52,56,66,68]. Canopies can occupy the whole width of the flume, replicating flows deep within canopies and avoiding problems associated with acceleration around patches (ensuring two-dimensional conditions), while others may be narrower than the flume width, to avoid edge effects[67]. This decision is often dictated by the width and geometry of the flume used, although the former approach is preferred by the authors, as it prevents flow deflection around the canopy. Where patchiness is investigated, this tends to be represented by one or two isolated patches, with completely bare regions between[63], which while not fully representing natural patchiness is often a necessary constraint for flume work; or for assessing fundamental processes such as wake formation and extent[17]. Current research aims to assess the importance of small-scale heterogeneity in canopies, to aid with parameterization of canopy roughness. There is currently no research which has used mimics to represent entire community structures.

12.1.3 Representing Complete Communities or Ecosystems

A growing body of experimental and field evidence[10,29] quantifying flow dissipation across salt marsh vegetation suggests that salt marshes could be an integral part of coastal flood and erosion protection. However, physical evidence is typically limited to relatively shallow water depths and low wave energy conditions. Surge reduction measurements over marshes appear to match modelled scenarios, but models are highly sensitive to morphological configuration[75], and flume investigations often require significant down-scaling. Dissipation may be limited under extreme storm conditions, as suggested by observations of vegetation debris accumulation after major storms[76], but field measurements under storm conditions are difficult, and limitations to flume size or flow generation capabilities meant that dissipation during energetic conditions (high water depths, current speeds, and wave heights) remained untested until recently[29].

Approximately 200 m^2 of natural salt marsh was relocated into a 300 m long, 5 m wide and 7 m deep flume where storm surge conditions could be simulated at prototype scale (Fig. 12.3). The study concluded that:

(a) the influence of vegetation on waves was significant even under relatively high water depths and small waves (30 cm waves reduced by $>15\%$ over 40 m of a 70 cm high canopy in 2 m water depth);
(b) vegetation was damaged under sustained impact, although this is likely species specific[46];
(c) the marsh bed (vegetation stems and roots only) has significant stabilizing effects (although less than the full vegetation canopy); and
(d) the marsh bed is remarkably stable[77] confirming the importance of roots on bed stability (see section 12.3.2), and highlighting possible species-specific effects.

These findings have been supported in subsequent large-scale flume experiments[78] which showed wave dissipation over salt marshes submerged at lower (but still relatively high) water depths (1 m).

While providing valuable insight, large-scale flume experiments are still limited by availability of facilities, prohibitive costs, logistical challenges in experimental setup, and associated methodological problems. As outlined in section 12.1.2, growing[78] or relocating[29] the quantity of salt marsh vegetation required for large-scale flume experiments is challenging. Constructing the salt marsh at sufficient spatial scales (200 m^2) is time-consuming and difficult to replace or modify without significant effort. This may make large-scale experiments more susceptible to hysteresis effects whereby previous experimental conditions influence subsequent system behavior (e.g., gradual removal of vegetation throughout an experiment[29]). It is also difficult to isolate the hydrodynamic resistance of the vegetation from that of the marsh bed. Smaller-scale plant-flow interactions within the salt marsh may also be difficult to determine, although comparing single species (section 12.1.1) and community experiments may help. Despite such limitations, true-to-scale experiments are necessary to

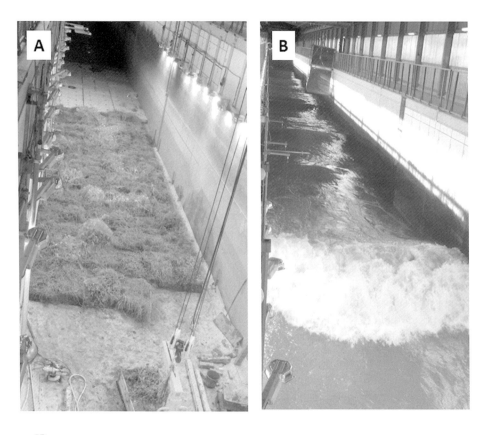

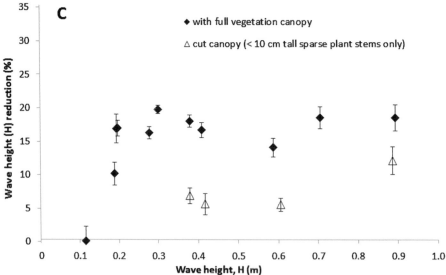

Figure 12.3 Large-scale flume experiment of Möller et al.[29] on wave dissipation over salt marshes (A) salt marsh re-assembly, (B) breaking waves over submerged marsh (2 m water depth), and (C) wave height reduction (%) of non-breaking waves over the 40-m long test section for monochromatic waves over intact and cut canopies. (A black and white version of this figure will appear in some formats. For the color version, please refer to the plate section.)

determine how coastal wetlands may be incorporated into future flood management and sea-defense solutions.

12.1.4 Non-Vegetation Biogenic Roughness

Non-vegetation biogenic elements can also significantly influence flow structure, bed roughness and turbulence generation. For example, bottom boundary layers can be modified by mussel beds[79], both in terms of physical roughness and the momentum input of the exhalent jets of the mussels, and increases in vertical accelerations have been observed around polychaete tubes[80]. Laboratory experiments of such effects are not common, and typically focus on single species, single individuals, or mimics[69,70]. The methods however remain similar to those used to assess vegetation effects, with extra consideration toward the sediments used (see section 12.2 for more details) and acclimatization of individuals.

12.2 At the Bed: Bed Stability, Sediment Transport and Sediment-Water Interface Exchanges

The interface between the sediment and water column is key to the stability of the bed, erosion, deposition, and the morphology of the salt marsh. It is a biologically active zone in which the direct action of the flow on the sediment is mitigated by the action of microbes, flora, and fauna in ways which act to either increase[81,82] or decrease[83] stability: sometimes simultaneously[84,85]. The spatial heterogeneity of many of these processes makes them difficult to capture accurately in the field, and laboratory investigations which isolate specific species, behaviours, or mechanisms are extremely useful in deconstructing the relative importance of the processes occurring at the bed.

In this region of the salt marsh it becomes important to carefully consider the sediment used for experiments. It is often necessary to match the sediment type and structure to field conditions, to replicate habitats and behaviours, or prevent stress responses. Sediments can be collected from the field; however, transporting them can result in structural changes through vibration, compaction/consolidation, and water loss. The combined effects of these disturbances and changes in organism behavior during transport and storage can lead to significantly different shear strengths[86] and erosion thresholds[87] from those observed in situ. Transportation and storage times should be therefore be minimized, and changes to sediment properties documented. Tollhurst et al.[87] also found that user bias in site selection can be significant in areas with significant small-scale spatial variation. Minimizing this requires randomized sediment sampling. Alternatively, created, abiotic sediments can provide a replicable substrate allowing direct comparisons to be made between experimental treatments, although later relating this to field conditions may be more challenging.

12.2.1 Biostabilization

Biostabilization often refers to organism growth and extracellular polymeric substance (EPS) production adhering particles, or providing protection from direct fluid forces[88].

We can expand this definition to include any biologically mediated increase in bed stability, including flow modification and wave attenuation provided by biogenic roughnesses, as discussed in section 12.2.1. However, here we focus on microbial biostabilization.

Biofilms are "complex consortia"[89] of prokaryotes, microbial eukaryotes (diatoms, protozoa, and fungi[90]), organic material (including EPS) and inorganic solids. In salt marshes, they undergo chronobiological rhythms (including UV exposure, wetting and drying cycles, temperature and salinity changes, and hydrodynamic forces[90]) leading to vertical migration that generates EPS excretion[91], which stabilizes the bed surface. Algal biofilm presence can be related to bioturbator presence, bed stability, and sediment availability[92], but change in microbial populations affects sediment stability successionally[91,89], is temporally variable[91], and site specific[93]. Biofilm growth duration is also important, and successive resuspensions events prevents growth[93].

Heterogeneity and factors such as macrofaunal bioturbation make it difficult to assess the effect of biofilms in the field, so their effect on bed stability is often investigated using flumes, where consideration must be given to how to sample, grow, and sustain biofilms, and how to assess their nature, structure, composition, and activity[89]. Experimental design usually involves incremental increases in horizontally applied bed shear stresses until bed failure[94], commonly utilizing annular flumes. Alternatively, these increases in shear stress can be achieved through instruments such as the Cohesive Strength Meter (CSM)[95] or Eromes[94], which are harder to compare to naturally occurring shear stresses, but are rapid and useful for intercomparison. Biofilm influence on bed stability is represented using a biostabilization index (BI) by comparing the critical erosion thresholds of biostabilized (B) and abiotic control (C) beds[89].

$$BI = \left(\frac{\tau_0 B}{\tau_0 C}\right) \times 100 \qquad\qquad \text{Eq.12.10}$$

12.2.1.1 Establishing Laboratory Biofilms

Biofilms for laboratory experiments can be created from settled beds of artificial (e.g., kaolinite[88,89]) or naturally collected sediments[88], and combining with collected in situ water[93] or enrichment with collected epipelic biofilms. These rely on naturally occurring microbes to provide the biofilm building blocks . Where natural sediments are used, these are typically sieved (from 0.5[96] to 2 mm[97]) to remove macrofauna, and are stored or frozen to destroy existing meiofauna[97] or biofilms. Growth lights and varying photoperiod are used to control microalgae growth rates[97,98]. Lags of several days (3[99] to 8[89]) are expected before sustained growth over the next 6[89] to 13[99] days, if nutrient limitation[91] is avoided. It is possible to isolate specific microbes, using antibiotics to remove bacteria[93], light restriction to prevent photosynthesis[93], or both. Alternatively, isolated microbial polymer exudates[95] from the bacterium *Xanthom onascan pestris* (Xantham gum) are often used to represent EPS in the absence of a biofilm. Abiotic control sediment can be created by addition of sodium hypochlorite[91,96], light restriction, induced anoxia, freezing, or purchasing abiotic sediments (e.g., kaolinite clay).

12.2.1.2 Biofilm Health and Function

Early research into biostability rarely directly addressed the health or function of the biofilm. However, as analysis techniques have advanced, it is possible to determine the structural nature of biofilms, the relative proportion of the microbe populations or EPS production levels. To avoid disturbing sediment beds used in experiments, removable slides or petri dishes are often placed within the experimental chambers during biofilm growth and removed for assessment before the sediments are exposed to flows. Microscopy is useful in assessing biofilm structure. This includes light microscopy assessments of eroded floc morphology[100], or environmental scanning electron microscopy and scanning confocal laser microscopy to assess biofilm structure and composition[89] Holographic cameras[95] can be used to assess the structure and types of resuspended particle, including proportions of clean, EPS-associated, and EPS-aggregated particles. Carbohydrate analysis[101] can assess biofilm mass and growth rates; the colloidal EPS carbohydrate fraction[102], diatom production[103], or proteic fractions[104]. Chlorophyll a analysis allows estimates of algal and cyanobacterial populations[105]. Phospholipid fatty acid analysis[106] is used to assess microbial responses to environmental conditions[89] and community structure[100], and denaturing gradient gel electrophoresis measures DNA sequences; combined these are used to assess viable biomass, community structure, and metabolic activity[89]. Pulse amplitude modulation fluorometry can be used to assess microphytobenthos physiological states, and photosynthetic active biomass[91]. It is even possible to isolate microbes likely to be responsible for EPS production[107], shown to correlate well with sediment stability[98].

12.2.1.3 Research Trajectory

Insights into the complex interactions and feedbacks between microbial biofilms, sediments, and flow have required the development of new technologies[90], as well as an integrative approach[81]. While great strides have been made in the past two decades, there are still a number of questions yet to be fully addressed[81], including those relevant to small-scale heterogeneity's effects on sediment stability (e.g.,[96]), linking descriptive and structural parameters to functionality, and the impact of anthropogenic forcing and climate change. In particular there is a need for laboratory approaches to try and mimic natural conditions as closely as possible to gain valid insight[81], and assess the potential effect of disturbance, through comparisons of cores collected in the field with sieved, or lab grown biofilms[98].

12.2.2 Bioturbation

Bioturbation can have significant effects on sediment stability, with implications for long-term accretion and erosion patterns. Bioturbation tends to increase surface roughness, reduce critical erosion thresholds, and increase erosion rates through changes in water content, bulk density, permeability, rheology, or the removal of biostabilizers (grazing). Laboratory flumes can be useful when studying the effects of bioturbation on marsh erosion, allowing close control of confounding factors, precise measurements, and higher replication than is feasible in field settings while including dynamic physical forcing to determine the response on infaunal invertebrate behaviour. However, as yet only a few key salt marsh species have been investigated, largely because stabilization by vegetation has generally been considered the more significant factor in salt marsh environments.

12.2.2.1 Effects on Sediment Geotechnical Properties

Bioturbation in general, and burrowing in particular, tends to increase sediment water content and decrease sediment density, reducing shear strength and destabilizing the sediment[108–111]. IAdditionally, the pellets and burrow formation repackages formerly compacted sediment, changing its texture and effective grain size by creating larger aggregates of grains[112–114]. This repackaging can increase the potential for sediment erosion, as the softer, less dense mud has a lower threshold shear stress[111].

To ensure comparability with field measurements, the bulk density, water content, and shear strength of sediments used in laboratory experiments should be carefully matched. This ensures comparable levels of sediment stability, but also helps preserve natural bioturbator behavior. This often necessitates collection of in situ sediment, while considering necessary removal of "unwanted" biological components[115] through freezing[116], anoxia[117,118], or sieving[119,120], and the resulting effects this may have on sediment structure. Changes in bulk density and water content measurements can be used to assess the effects of bioturbation. However, since changes may be limited to the surface or pelletized sediment, standard methods may lack the resolution needed to capture changes. Shear strength may be measured with a shear vane, penetrometer, or similar, but similarly integrates shear strength over depth, typically at centimeter scales[86]. Integration of geotechnical properties over these scales is therefore a key limitation in methodological design.

12.2.2.2 Bed Roughness Changes

Increased surface roughness increases turbulence, in turn, increasing shear stress and erosion[111,121]. Biogenic structures, such as tubes, tracks, pellets, mounds, and burrow structures, create greater obstruction to flow than flat sediment surfaces, facilitating greater mobilization and transport of bioturbated sediment compared to undisturbed sediment. Despite its importance to sediment stability, roughness created by bioturbation can be difficult to quantify. Roughness length (z_0) can be determined from the logarithmic part of the flow profile (Eq. 1.3)[122]; although flume geometry is a consideration, as annular flumes for example, compress the boundary layer, preventing this method[123]. Alternatively roughness can be inferred from turbulence based methods (e.g., TKE [Eq. 1.6] or Reynolds stress)[124].

Where roughness length calculation is not possible, other means of assessing surface roughness include visual height and spacing assessments, stereophotogrammetry[125,126], or terrestrial laser scanning, and reported as root mean squared (rms) roughness height, height distribution histogram, average peak spacing (or average roughness), average peak width, roughness power spectra, or similar[125,127,128].

12.2.2.3 Bed Stability

Sediment erodibility is defined relative to measured current velocities and estimated bed shear stresses, expressed in terms of a critical erosion threshold, erosion rate, and/or mass of sediment eroded per unit surface area[129]. The combined effects of bioturbation, outlined above, often reduces the critical threshold for erosion[113]. For some species, this is dependent on population density, for example, low densities of the bivalve *Macoma balthica* associated with a well-developed microphytobenthos, are related to low erosion rates and high critical erosion velocities, whereas, at high densities with reduced microphytobenthos, critical erosion thresholds are reduced by ~60% and erosion rates increased 10 to 100-fold[92,130]. In some species, population density affects erosion rates rather than critical erosion threshold[86]. A study of the burrowing crab species *Neohelice granulate* revealed that densely spaced burrows with funnel-shaped openings trap sediment reducing erosion rates, whereas sparse burrows with aboveground structures, closed openings, or in high velocity flow environments tend to increase erosion rates[131].

A common method for calculating critical erosion thresholds uses optical backscatter sensor (OBS) time-series, calibrated against suspended sediment concentrations (SSC). Velocities required to erode a specific mass of sediment can be calculated from linear regression of log (SSC) or log(maximum erosion rate) and current velocity[123,132]. The critical shear stress is calculated as described in section 12.1.1, or measured directly with a stress sensor[133,134]. Erosion rate (volume or mass eroded per unit area or time) may be calculated using the OBS calibration curve[123]. However, the appropriateness of OBS depends on laboratory setup and the type of sediment and bioturbation being evaluated. For example, the return pump in a recirculating flume breaks up pelletized sediment, changing the effective grain size and affecting OBS calibration, as can variable ambient light levels. In cases where OBS use is not appropriate, a simple visual assessment may be used[121]. Imaging techniques (section 12.2.2.6) can also be used to quantify mixing depths and bioturbation rates.

It is important to keep in mind the differences between the erosion threshold and rate. The erosion threshold is a measure of surface stability, but the erosion rate indicates the amount of sediment resuspension and removal[135]. A decoupling of threshold and rate[136] has been observed for some species. The amphipod *Corophium volutator* inhibits diatom films, lowering erosion thresholds, but also binds sediment into its burrows, reducing sediment transport rates[86]. As single species may have opposing effects on these two measures of erodibility, both should be measured to fully elucidate the impacts of bioturbation on sediment stability.

12.2.2.4 Surface Assessment of Salt Marsh Bioturbation

A recent study by Farron et al.[214] of the effects of burrowing marsh crabs on sediment erosion illustrates some of the factors which must be considered when studying bioturbation in the laboratory. A non-recirculating racetrack flume was used to compare bioturbation by two crab species: *Uca pugnax*, a deposit feeder which constructs simple, chimney-like burrows, and *Sesarma reticulatum*, an herbivorous crab which builds extensive, interconnected burrows. Bioturbation by dense populations of *S. reticulatum* has been found to facilitate rapid expansion of tidal creeks and extensive creek bank erosion.

The study contextualizes the effects of the two species, and, determines whether *S. reticulatum* increased sediment erodibility significantly more than other common burrowing species. Gender ratios were kept constant, since females tend to burrow more than males, and population densities per unit area were equivalent to field observations. Use of unvegetated marsh sediment collected from the edges of pools and channel banks, removed the confounding effects of vegetation and rooting on sediment erodibility, and favoured preferential burrowing habitat[137]. This simplification isolates the effects of the studied species, but the absence of vegetation may have facilitated greater erosion than typical under vegetated conditions and encouraged burrowing activity. In addition, both species were observed to preferentially dig along the sides of their experimental containers, likely due to the reduced obstruction to burrowing and increased stability provided by the container walls.

Critical thresholds for erosion of different particle types (individual grains, flocs, or pellets), and aboveground burrow structures were evaluated by visual assessment while gradually increasing flow velocities as the flume tank drained. Shear strength was calculated from acoustic Doppler velocimeter measurements using the TKE method[36].

Terrestrial laser scanning (Z + F Laser Imager 5003) quantified the impact of burrowing on surface roughness (as standard deviation of surface elevation) and surface erosion volumes (as before and after elevation change) (Fig. 12.4) which were converted to eroded mass using measured bulk density. Experimental flow velocities exceeded typical day-to-day conditions but were comparable to storm-induced velocities. Therefore, while critical threshold velocities may be considered representative of in situ erosion patterns, the erosion rates measured were only applicable for storm conditions.

Burrowing and feeding by *S. reticulatum*, and to a lesser extent *U. pugnax*, increases surface roughness (by 124% and 34% respectively), decreased the erosion threshold (by up to 50% in both cases), and increased the eroded sediment volume at high velocities (1132% and 60% respectively). The observed burrowing-facilitated erosion was of a similar magnitude to average rates of marsh surface accumulation, providing a mechanism for sediment loss in areas that have been heavily burrowed by *S. reticulatum*, and potentially leading to large-scale morphological changes.

12.2.2.5 Within Bed Assessments of Bioturbation

The impact of bioturbation can extend several centimetres into the sediment, dependent on species and habitat[138]. In the field, short-term variations in species behaviour at and below the sediment–water interface can be observed using time-lapse sediment profile imaging camera systems (t-SPI[139]), whilst a combination of fluorescent luminophores (particulate tracers) and an optically modified sediment profile camera (f-SPI) can allow quantification

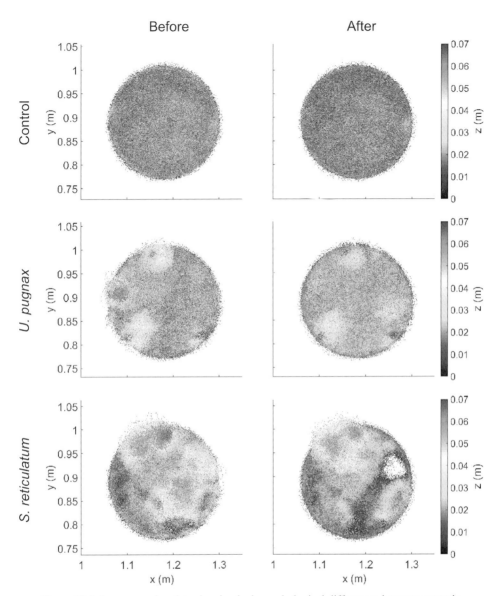

Figure 12.4 Laser scanning data showing bed morphological differences between controls, sediment burrowed by *U. pugnax*, and sediment burrowed by *S. reticulatum* before and after being eroded in the flume. (A black and white version of this figure will appear in some formats. For the color version, please refer to the plate section.)

of particle reworking depths and rates[138,140], and assessments of potential changes to community structure and ecosystem function[141].

SPI has a long history of use in subtidal sediments[142], but no record of use for in situ saltmarsh observations in the peer-reviewed literature. Laboratory f-SPI, however, has the potential to allow quantification of belowground bioturbation in these systems[143]. To assess bioturbation by mudflat species in the laboratory, field-collected sediments are sieved to remove macrofauna[141], and placed in transparent, square-sided aquaria to a sufficient depth to allow natural burrow formation. Individuals of the macrofaunal species of interest are added and allowed to acclimatize, and maintained in an environmental chamber that controls for confounding factors such as temperature. Luminophores are added as a thin surface layer, and the individual allowed to establish natural bioturbation behaviours. Daily f-SPI images are taken under UV illumination, allowing quantification of the mean, median, and maximum vertical luminophore distributions and reworking rates. While there are several considerations to be made when interpreting this data, including spatial constraints (from the aquaria sides and depth of sediment) on bioturbation behaviour; the effect of isolation of individuals; differences between luminophores and substrate grain size; temporal resolutions; and the two-dimensional nature of the images; f-SPI provides a rapid, non-invasive means to visualize and quantify bioturbation.

For salt marsh environments, where sieving would result in unwanted changes to the sediment structure, intact cores are collected and monitored (Fig. 12.5). Salt marsh macrofaunal invertebrate communities are rarely studied, but suggest a low level of species richness (typically <20–25 species)[144] with biomass dominated by surficial modifiers (e.g., bivalves *M. balthica* and *Tapes decussatus*) and, numerically, by epifaunal/surficial species (e.g. gastropod, *Hydrobia ulvae;* amphipod *C. volutator*) or species inhabiting the interstitial spaces within the root-sediment matrix (e.g., Oligochaetes). Collectively, these species have low bioturbation potential, and the extent of vertical particle mixing should be minimal. Preliminary f-SPI images suggest that this is true (Fig. 12.5), but that they can significantly modify the sediment surface and may be very important for the relocation of organic matter. Current work on UK saltmarshes (Solan, pers. comms.) aims to confirm these findings.

12.2.3 Solid-Transmitted Stress

Erosion on tidal flats and salt marshes can be accelerated by the transport of granular material[145–148], and areas of rapid retreat and erosion have been linked to discarded or

Figure 12.5 Longitudinal section of a saltmarsh core showing limited vertical particle mixing of particulate tracers (green in color) illuminated under (A) white light and (B) UV light. Cores are 12 cm wide and obtained from Tillingham, Essex, UK.

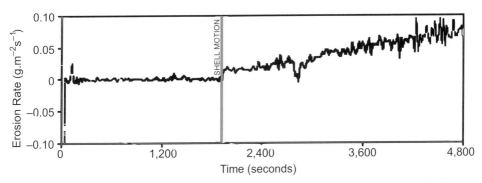

Figure 12.6 Erosion rate fo,r a remolded cohesiv,e sediment bed with a, single *Cerastoderma, edulis* shell. Note erosion coincides with the threshold of shell motion. After Thompson et al, 2002

disarticulated shells[149]. These impart a solid-transmitted stress to the bed[150,151] (the Ballistic Momentum Flux [BMF]). Laboratory flumes, in particular annular flumes, have proved useful in quantifying BMF[149,152,153], partly because they contain eroded sediment, making accurate determination of erosion rates and thresholds (described in section 12.2.1) possible. Experiments show that BMF can induce erosion[153] beyond that possible under fluid forces alone (Fig. 12.6), with shells and shell fragments inducing erosion and lowering bed stability[149,152,154]. The magnitude of the effect is related to shell morphology[149] and densities below those which afford bed armouring. These often overlooked processes can significantly impact sediment fluxes from the bed, and accelerate salt marsh bank retreat through oversteepening[149]. Initial work in this area is limited to small-scale experiments, and often individual shells, but there is scope to expand research to larger scales and in particular the effects of waves on solid transmitted stresses.

12.3 Within the Bed: the Hidden Processes

The sediment surface is traditionally considered as a solid boundary, experiencing elevation changes through erosion, deposition, and biological modifications as discussed in sections 12.1 and 12.2. The reality is that the upper layers of the sediment are dynamic regions of interaction, exchange, and transport; processes often hidden from view, but having significant effects on sediment stability and the exchange of solutes and particles between the bed and the water column. Many of the processes discussed in sections 12.1 and 12.2 including root stabilization[155], bioturbation[138], and EPS biostabilization[156] extend for several centimeters below the sediment surface, and depending on the porosity of the bed sediment, advective flows can flush the interstitial spaces between the sediment grains[157]. The depth at which the sediment is influenced by these processes is key to sediment stability and resistance to erosion, but also to benthic biogeochemical cycling, including oxygen penetration depths[158,159], redux potential depth[160], carbon[161,162], and nutrient cycling[163,164], all of which influence the suitability of the sediment as a habitat through complex feedback mechanisms. Many of the difficulties in understanding these processes relate to observing processes that occur below the sediment surface, and measuring at sufficient vertical resolution, which are easier to overcome in a laboratory setting.

12.3.1 Burrow Structures and Complexity

Burrowing species can significantly affect benthic processes, including carbon cycling[165], solute exchange[166], microbial community structure[166], sediment stability[167], and marsh-edge erosion[168]. To relate this biological mediation to specific ecosystem properties, it is essential to understand the effects of their biogenic structures[169]; however, insights about organism–sediment relationships below the sediment surface have largely been two-dimensional[170]. While burrow structures have been described[171–173], most visualization techniques (e.g., SPI[174,175]) are limited to two dimensions and do not examine vertical and horizontal co-location of biogenic structures, invertebrates, and marsh features such as roots. Three-dimensional visualization techniques are often destructive (e.g., sieving, burrow castings[176–178], or core slicing[179]), use artificial sediments[180], or do not capture dynamic processes.

Developments in the use of high-resolution computed tomography (CT[181,182]) and positron emission tomography[183], have made it possible to image organisms and quantitatively examine their burrow structures[169,170,181,184–187], and particle movements induced by their behaviour[182]. CT scanning utilizes both medical (e.g.,[182]) or micro-focus (μ-CT) engineering scanners (e.g.,[169,170]), allowing a balance between scan speed and image resolution. The resulting X-ray attenuation images can be related to sediment bulk density to identify burrow presence through simple, threshold based determinations[185–187], and new techniques allow the estimation of internal volumes[182], complexity[170], connectivity, and spacing (Fig. 12.7).

12.3.2 Root Stabilization

The effects of aboveground vegetated biomass is recognized (see section 12.1), but the stabilization potential of roots is less well understood[188] for salt marsh species. While usually positive, increasing sediment shear strength (by $>500\%$[189]), and reducing flow induced erosion rates (by $>60\%$[190]) in terrestrial settings, the effect is species dependent[191], and more work is needed to determine the effects of fine root structure, and root morphology[155] on stabilization.

Laboratory investigations of these processes usually require the extraction of entire plants, or plant communities, complete with sediments, to a depth sufficient to retain the root system. This requires working at a prototype scale, although samples can be relatively small in diameter[155,190]. Replication and validation against in situ measurements can quanitfy potential sampling artifacts[188] (section 12.2). Depending on the context, fluid forces are either directed at the exposed root sections (e.g.,[155,188]) simulating wave and current attack on exposed cliffs and banks or allowed to flow over the salt marsh sections[192] as a tidal flow. Jet erosion tests[193] indicate that while roots increased soil stabilization, there

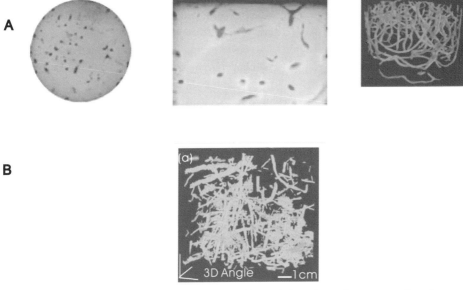

Figure 12.7 A. Transverse core slices taken 50 mm below the sediment-water interface (left), coronal core slices through the rotational centre of the core (center), and reconstructed three-dimensional burrow models (right) for *Hediste diversicolor*. Adapted from Hale et al. (2014). B. CT images of the roots of *Atriplex portulacoides* in the surface 10cm of a salt marsh system. Image courtesy of Dr. Yining Chen

were no significant correlations between critical shear stresses and root parameters. However, jet erosion (similar to the CSM) does not induce typical tidal or wave induced stress patterns[94] and results should be interpreted with caution. Accurate assessments of root parameters are also difficult, and when extracted through sieving and hand measurement, the smallest root structures may be underreported. CT scanning technology (Section 12.3.1, Fig. 12.7) now provides visualization of the roots allowing stability effects to be related to root morphology.

This field is relatively underdeveloped, particularly for salt-marsh species, where land-based analogues do not fully represent the variable forces at work, but there is scope to address many aspects of root stability in the laboratory.

12.4 Discussion: Integrating a Multi-layer Problem

The three sections described within this chapter are clearly interlinked, and often overlapping. However, the structure provides a convenient way to consider a highly complex environment, and reflects the way in which many marsh processes are considered in the laboratory.

It is possible to integrate processes across the entire vertical profile, and there are examples where the complete vertical profile of the marsh is considered, either through series of separate but related experiments[191], or by considering the marsh section as a whole[29,92,191]. These tend to be full-scale experiments, in which entire sections of the marsh are collected and exposed to experimental testing. Here, all the potential mechanisms of stabilization and destabilization that have been discussed are inherently included although not necessarily measured. The challenge then becomes decoupling the relative effects of each of process on the stability of the marsh. Here, comparison against process-specific, mechanistic testing provides the necessary insight for interpreting the results. As such, it is the combined use of whole-system and process/mechanism/trait-specific experimentation which provides the best approach to understanding the processes controlling the morphology, function, and evolution of the salt marsh environment.

12.4.1 Scaling-up and Projecting Forward: Utilizing Laboratory Results to Interpret Marsh Environments

The ultimate purpose of research designed to better understand the salt marsh environment is to understand, manage and predict their evolution. Laboratory-based, targeted experimental research requires careful design to allow for subsequent upscaling to successful landform-, local- or regional- scale application; in particular considering the heterogeneity and range of temporal and spatial scales involved and the potential scale dependence of saltmarsh processes[5]. Neither observations nor models alone are likely sufficient for regional assessment; interdisciplinary approaches integrated with experimental testing may be more successful[196,197], and improvements in process understanding and scaling techniques[197,198] require combining experimental, observational, and predictive research.

Careful design of laboratory experiments is needed to identify appropriate temporal and spatial scales to represent field conditions[199], considering that different scales may be important for different variables. Detailed individual studies of representative sediment types, habitats, species, and communities spanning an appropriate range of seasonal and hydrodynamic conditions may offer the best mechanistic insights for improved understanding and prediction[200]. Logistical limitations to experimental design are unavoidable, but scaling concerns can be addressed through careful planning (e.g., [196,200,201]). Some key considerations are: the representativeness of data to the desired application or outputs[198]; numbers of observations needed to address uncertainty; identification of processes not currently considered[198]; and the benefits of interdisciplinary, holistic approaches to parameterization[197].

One successful method for efficient scaling is to use assessments of regional variability, to step-down to the local, site and then lab scales, providing a clear trajectory for the subsequent upscaling required for marsh-scale applications, in contrast to experimental designs based on isolated variables alone[201]. This approach is common in salt marsh research, where most experimental work is performed at full scale (i.e., using intact plants or fauna), or based on simplifications of field observations. There is a valuable place for such stand-alone, mechanistic experiments which test specific parameters, and consideration of typical field conditions (e.g., plant densities, or ranges of leaf flexibility) allow the application of these mechanistic insights to the larger field environment.

An additional consideration when scaling experimental findings is that salt marsh environments are dynamic, and change temporally in response to sedimentological, hydrodynamic, and climatic drivers at different scales (from seasonal to millennial). Longer-scale temporal changes may have significant effects on any future scenario prediction, for example: we lack sufficient record lengths to capture long-term changes through field observation, but the manipulative nature of experimental laboratory work is well placed to address these problems.

12.4.2 Future Scenarios

Much of this chapter has focused on good practice in investigating sediment–fluid–biological interactions in the laboratory, and the latest methodologies offering improved resolution, accuracy, or new capabilities in quantifying processes, interactions, and mechanisms. Next generation models will also require understanding about the uncertainty imposed by environmental change and anthropogenic activities. Very little is known about the sensitivity of coastal ecosystems to changing conditions over time, yet current practice generally assumes that sediment–fluid–biological interactions will remain unchanged under future contexts.

Multiple aspects of community dynamics (e.g., species diversity, community composition, behavior, growth, and plant/faunal traits) and physical processes (e.g., water density, salinity, temperature, flow regimes) are likely to be modified under future scenarios. Salt marsh ecosystems are likely to have significant, but varied responses to the multiple drivers of climate change. Increased atmospheric CO_2, warming temperatures and moderate changes in sea level may lead to increases in primary productivity and marsh accretion[202–209], leading toward promising strategies for climate change mitigation and adaption[210]. However, climate change can also lead to rapid changes of marsh zonation[211], marshes unable to keep pace with larger sea-level rises[212] and significant erosional losses[213] from increasing and strengthening coastal storms.

A key challenge going forward will be to address disconnects between the representation of coastal systems in experimental systems and the context in which plant–sediment–fluid–biological interactions are moderated in natural systems under environmental and anthropogenic forcing. For example, experimental incorporation of novel conditions, such as the combined effects of elevated temperature, ocean acidification, and flow regimes, are likely to reveal differences in the way sediment is transported and how organisms interact with each other and their environment, which may lead to outcomes not previously considered. Such a multiple-stressor approach is necessary to parameterize models that explore the consequences of near-future scenarios, establish levels of uncertainty, and in providing evidence to support management decisions that relate to the governance of coastal processes and morphology that are important for maintaining ecosystem integrity and are of value in the provision of societal needs.

References

1. JNCC. An Overview of Coastal Saltmarshes, Their Dynamic and Sensitivity Characteristics for Conservation and Management. UK: JNCC, 2003.
2. Tempest JA, Möller I, Spencer T. A review of plant-flow interactions on salt marshes: The importance of vegetation structure and plant mechanical characteristics: Salt marsh plant-flow interactions. *Wiley Interdiscip Rev Water* 2015;2: 669–681.
3. Townend I, Fletcher C, Knappen M et al. A review of salt marsh dynamics: A review of salt marsh dynamics. *Water Environ J* 2011;25: 477–488.
4. van Wesenbeeck BK, van de Koppel J, Herman PMJ et al. Does scale-dependent feedback explain spatial complexity in salt-marsh ecosystems? *Oikos* 2008;117: 152–159.
5. Wang H, van der Wal D, Li X et al. Zooming in and out: Scale dependence of extrinsic and intrinsic factors affecting salt marsh erosion: Factors on salt marsh edge erosion. *J Geophys Res Earth Surf* 2017;122: 1455–1470.
6. Reise K. Sediment mediated species interactions in coastal waters. *J Sea Res* 2002;48: 127–141.
7. Shi Z, Pethick JS, Pye K. Flow structure in and above the various heights of a saltmarsh canopy: a laboratory flume study. *J Coast Res* 1995: 1204–1209.
8. Temmerman S, Bouma TJ, Govers G et al. Impact of vegetation on flow routing and sedimentation patterns: Three-dimensional modeling for a tidal marsh: Vegetation impact on flow routing. *J Geophys Res Earth Surf* 2005;110, DOI: 10.1029/2005JF000301.
9. Chen Y, Li Y, Cai T et al. A comparison of biohydrodynamic interaction within mangrove and saltmarsh boundaries: Bio-hydrodynamics within mangrove and salt-marsh boundaries. *Earth Surf Process Landforms* 2016;41: 1967–1979.
10. Möller I, Spencer T, French JR et al. Wave transformation over salt marshes: A field and numerical modelling study from North Norfolk, England. *Estuarine, Coast Shelf Sci* 1999;49: 411–426.
11. Neumeier U. Velocity and turbulence variations at the edge of saltmarshes. *Continental Shelf Res* 2007;27: 1046–1059.
12. Shi B, Wang YP, Yang Y et al. Determination of critical shear stresses for erosion and deposition based on in situ measurements of currents and waves over an intertidal mudflat. *J Coast Res* 2015;316: 1344–1356.
13. Lo V, Bouma T, van Colen C, Airoldi L. Interactive effects of vegetation and grain size on erosion rates in salt marshes of the Northern Adriatic Sea. *ESCA Local Meeting report.* 2016.
14. Reed DJ. Sediment dynamics and deposition in a retreating coastal salt marsh. *Estuarine, Coast Shelf Sci* 1988;26: 67–79.
15. Shi Y, Jiang B, Nepf HM. Influence of particle size and density, and channel velocity on the deposition patterns around a circular patch of model emergent vegetation patch on deposition patterns. *Water Resour Res* 2016;52: 1044–1055.
16. Alben S, Shelley M, Zhang J. Drag reduction through self-similar bending of a flexible body. *Nature* 2002;420: 479–481.
17. Folkard AM. Hydrodynamics of model *Posidonia oceanica* patches in shallow water. *Limnol Oceanogr* 2005;50: 1592–1600.
18. Cowell PJ, Thom BG. Morphodynamics of coastal evolution. In: Carter, W, Woodroffe, CD. (eds.). *Coastal Evolution : Late Quaternary Shoreline Morphodynamics* Cambridge University Press, Cambridge, 1994, 33–86.
19. Yang SL, Shi BW, Bouma TJ et al. Wave attenuation at a salt marsh margin: A Case study of an exposed coast on the Yangtze Estuary. *Estuaries Coasts* 2012;35: 169–182.

20. Ysebaert T, Yang S-L, Zhang L et al. Wave attenuation by two contrasting ecosystem engineering salt marsh macrophytes in the intertidal pioneer zone. *Wetlands* 2011;31: 1043–1054.
21. Shi Z, Pethick JS, Burd F et al. Velocity profiles in a salt marsh canopy. *Geo-Marine Lett* 1996;16: 319–323.
22. Neumeier U, Amos CL. The influence of vegetation on turbulence and flow velocities in European salt-marshes. *Sedimentology* 2006;53: 259–277.
23. Leonard LA, Croft AL. The effect of standing biomass on flow velocity and turbulence in *Spartina alterniflora* canopies. *Estuarine, Coast Shelf Sci* 2006;69: 325–336.
24. Leonard LA and Reed DJ. Hydrodynamics and sediment transport through tidal marsh canopies. *Journal of Coastal Research* 2017; SI36: 459–469.
25. Neumeier U, Ciavola P. Flow resistance and associated sedimentary processes in a *Spartina maritima* salt-marsh. *J Coast Res* 2004;202: 435–447.
26. Neumeier U. Quantification of vertical density variations of salt-marsh vegetation. *Estuarine, Coast Shelf Sci* 2005;63: 489–496.
27. Pethick J, Leggett D, Husain L. Boundary layers under salt marsh vegetation developed in tidal currents. In: Thornes J. (ed.). *Vegetation and Erosion*. John Wiley and Sons, Chichester, 1990, 113–124.
28. Tollner E., Barfield B., Hayes J. Sedimentology of erect vegetal filters. *J. Hydraulics Division* 1982;108: 1518–1531.
29. Möller I, Kudella M, Rupprecht F et al. Wave attenuation over coastal salt marshes under storm surge conditions. *Nat Geosci* 2014;7: 727–731.
30. Bouma TJ, De Vries MB, Herman PMJ. Comparing ecosystem engineering efficiency of two plant species with contrasting growth strategies. *Ecology* 2010;91: 2696–2704.
31. Tempest J. *Hydrodynamic effects of salt marsh canopies and their prediction using remote sensing techniques*. 2017. PhD Thesis. Cambridge University.
32. Horn R, Richards K. Flow–vegetation interactions in restored floodplain environments. *Hydroecology and Ecohydrology*. John Wiley & Sons, Ltd, Chichester, 2008, 269–294.
33. Nepf HM, Vivoni ER. Flow structure in depth-limited, vegetated flow. *J Geophys Res Ocean* 2000;105: 28547–28557.
34. Lopez F, Garcia M. Open-channel flow through simulated vegetation: Suspended sediment transport modeling. *Water Resour Res* 1998;34: 2341–2352.
35. Anderson ME, Smith JM. Wave attenuation by flexible, idealized salt marsh vegetation. *Coast Eng* 2014;83: 82–92.
36. Stapleton KR, Huntley DA. Seabed stress determinations using the inertial dissipation method and the turbulent kinetic energy method. *Earth Surf Process Landforms*, 20: 807–815.
37. Wilson CAME, Stoesser T, Bates PD. Modelling of open channel flow through vegetation. *Computational Fluid Dynamics: Applications in Environmental Hydraulics*. John Wiley & Sons, Chichester, 2005, 395–428.
38. Leonard LA, Luther ME. Flow hydrodynamics in tidal marsh canopies. *Limnol Oceanogr* 1995;40: 1474–1484.
39. Christiansen T, Wiberg PL, Milligan TG. Flow and sediment transport on a tidal salt marsh surface. *Estuarine, Coast Shelf Sci* 2000;50: 315–331.
40. Lightbody AF, Nepf HM. Prediction of velocity profiles and longitudinal dispersion in salt marsh vegetation. *Limnol Oceanogr* 2006;51: 218–228.
41. Paulus S, Schumann H, Kuhlmann H, Léon, J. High-precision laser scanning system for capturing 3D plant architecture and analysing growth of cereal plants. *Biosyst Eng* 2014;121: 1–11.

42. Luhar M, Nepf HM. Flow-induced reconfiguration of buoyant and flexible aquatic vegetation. *Limnol Oceanogr* 2011;56: 2003–2017.

43. Bradley K, Houser C. Relative velocity of seagrass blades: Implications for wave attenuation in low-energy environments. *J Geophys Res* 2009;114, DOI: 10.1029/2007JF000951.

44. Albayrak I, Nikora V, Miler O, O'Hare, M. Flow-plant interactions at a leaf scale: Effects of leaf shape, serration, roughness and flexural rigidity. *Aquat Sci* 2012;74: 267–286.

45. Feagin RA, Irish JL, Möller I et al. Short communication: Engineering properties of wetland plants with application to wave attenuation. *Coast Eng* 2011;58: 251–255.

46. Rupprecht F, Möller I, Evans B et al. Biophysical properties of salt marsh canopies — Quantifying plant stem flexibility and above ground biomass. *Coast Eng* 2015;100: 48–57.

47. Lagerspetz KY. Thermal acclimation without heat shock, and motor responses to a sudden temperature change in Asellus aquaticus. *J Therm Biol* 2003;28: 421–427.

48. Díaz S, Symstad AJ, Stuart Chapin F et al. Functional diversity revealed by removal experiments. *Trends Ecol & Evol* 2003;18: 140–146.

49. Paul M, Bouma T, Amos C. Wave attenuation by submerged vegetation: Combining the effect of organism traits and tidal current. *Mar Ecol Prog Ser* 2012;444: 31–41.

50. Bouma TJ, van Duren LA, Temmerman S et al. Spatial flow and sedimentation patterns within patches of epibenthic structures: Combining field, flume and modelling experiments. *Cont Shelf Res* 2007;27: 1020–1045.

51. Panigrahi K, Khatua KK. Prediction of velocity distribution in straight channel with rigid vegetation. *Aquat Procedia* 2015;4: 819–825.

52. Kobayashi N, Raichle AW, Asano T. Wave attenuation by vegetation. *J Waterw Port, Coastal, Ocean Eng* 1993;119: 30–48.

53. John BM, Shirlal KG, Rao S. Effect of artificial sea grass on wave attenuation- an experimental investigation. *Aquat Procedia* 2015;4: 221–226.

54. John BM, Shirlal KG, Rao S et al. Effect of artificial seagrass on wave attenuation and wave run-up. *Int J Ocean Clim Syst* 2016;7, DOI: 10.1177/1759313115623163.

55. Stratigaki V, Manca E, Prinos P et al. Large-scale experiments on wave propagation over *Posidonia oceanica*. *J Hydraul Res* 2011;49: 31–43.

56. Hu Z, Suzuki T, Zitman T et al. Laboratory study on wave dissipation by vegetation in combined current–wave flow. *Coast Eng* 2014;88: 131–142.

57. Okamoto T-A, Nezu I. Turbulence structure and "Monami" phenomena in flexible vegetated open-channel flows. *J Hydraul Res* 2009;47: 798–810.

58. Murphy E, Ghisalberti M, Nepf H. Model and laboratory study of dispersion in flows with submerged vegetation. *Water Resour Res* 2007;43. https://doi.org/10.1029/2006WR005229

59. Wu F-C, Shen HW, Chou Y-J. Variation of roughness coefficients for unsubmerged and submerged vegetation. *J Hydraul Eng* 1999;125: 934–42.

60. Stone BM, Shen HT. Hydraulic resistance of flow in channels with cylindrical roughness. *J Hydraul Eng* 2002;128: 500–506.

61. Augustin LN, Irish JL, Lynett P. Laboratory and numerical studies of wave damping by emergent and near-emergent wetland vegetation. *Coast Eng* 2009;56: 332–340.

62. Fonseca MS, Koehl MAR. Flow in seagrass canopies: The influence of patch width. *Estuarine, Coast Shelf Sci* 2006;67: 1–9.

63. Folkard AM. Flow regimes in gaps within stands of flexible vegetation:Laboratory flume simulations. *Environ Fluid Mech* 2011;11: 289–306.

64. Wilson CAME, Stoesser T, Bates PD, Stoesser, T. Open channel flow through different forms of submerged flexible vegetation. *J Hydraul Eng* 2003;129: 847–853.

65. Ghisalberti M, Nepf H. The structure of the shear layer in flows over rigid and flexible canopies. *Environ Fluid Mech* 2006;6: 277–301.

66. Elliott AH. Settling of fine sediment in a channel with emergent vegetation. *J Hydraul Eng* 2000;126: 570–577.

67. Shi Y, Jiang B, Nepf HM. Influence of particle size and density, and channel velocity on the deposition patterns around a circular patch of model emergent vegetation. *Water Resour Res* 2016; 52: 1044–1055.

68. Bouma TJ, De Vries MB, Low E et al. Trade-offs related to ecosystem engineering: A case study on stiffness of emerging macrophytes. *Ecology* 2005;86: 2187–2199.

69. Eckman JE, Nowell ARM. Boundary skin friction and sediment transport about an animal-tube mimic. *Sedimentology* 1984;31: 851–862.

70. Eckman JE. Flow disruption by an animal-tube mimic affects sediment bacterial colonization. *J Mar Res* 1985;43: 419–435.

71. Wilson CAME, Yagci O, Rauch H -P, Stoesser, T. Application of the drag force approach to model the flow-interaction of natural vegetation. *Int J River Basin Manag* 2006;4: 137–46.

72. Paul M, Henry P-YT, Thomas RE. Geometrical and mechanical properties of four species of northern European brown macroalgae. *Coast Eng* 2014;84: 73–80.

73. Anderson ME, Smith JM. Wave attenuation by flexible, idealized salt marsh vegetation. *Coast Eng* 2014;83: 82–92.

74. John BM, Shirlal KG, Rao S. Effect of artificial vegetation on wave attenuation – an experimental investigation. *Procedia Eng* 2015;116: 600–606.

75. Loder NM, Irish JL, Cialone MA, Wamsley, TV. Sensitivity of hurricane surge to morphological parameters of coastal wetlands. *Estuarine, Coast Shelf Sci* 2009;84: 625–636.

76. Spencer T, Brooks SM, Evans BR et al. Southern North Sea storm surge event of 5 December 2013: Water levels, waves and coastal impacts. *Earth-Science Rev* 2015;146: 120–145.

77. Spencer T, Möller I, Rupprecht F et al. Salt marsh surface survives true-to-scale simulated storm surges. *Earth Surf Process Landforms* 2016;41: 543–552.

78. Maza M, Lara JL, Losada IJ et al. Large-scale 3-D experiments of wave and current interaction with real vegetation. Part 2: Experimental analysis. *Coast Eng* 2015;106: 73–86.

79. van Duren LA, Herman PMJ, Sandee AJJ, Heip, CHR. Effects of mussel filtering activity on boundary layer structure. *J Sea Res* 2006;55: 3–14.

80. Carey DA. Particle resuspension in the benthic boundary layer induced by flow around polychaete tubes. *Can J Fish Aquat Sci* 1983;40: s301–s308.

81. Gerbersdorf SU, Wieprecht S. Biostabilization of cohesive sediments: Revisiting the role of abiotic conditions, physiology and diversity of microbes, polymeric secretion, and biofilm architecture. *Geobiology* 2015;13: 68–97.

82. Decho A. Microbial exopolymer secretions in ocean environments: Their role(s) in food webs and marine processes. In: Barnes M (ed.). *Oceanography and Marine Biology: An Annual Review*. Aberdeen University Press, 1990.

83. Andersen TJ. Seasonal variation in erodibility of two temperate, microtidal mudflats. *Estuarine, Coast Shelf Sci* 2001;53: 1–12.

84. Graf G, Rosenberg R. Bioresuspension and biodeposition: A review. *J Mar Syst* 1997;11: 269–278.

85. Passarelli C, Olivier F, Paterson DM et al. Organisms as cooperative ecosystem engineers in intertidal flats. *J Sea Res* 2014;92: 92–101.
86. Grant J, Daborn G. The effects of bioturbation on sediment transport on an intertidal mudflat. *Neth J Sea Res* 1994;32: 63–72.
87. Tolhurst T., Riethmüller R, Paterson D. In situ versus laboratory analysis of sediment stability from intertidal mudflats. *Cont Shelf Res* 2000;20: 1317–1334.
88. Droppo IG, Lau YL, Mitchell C. The effect of depositional history on contaminated bed sediment stability. *Sci Total Environ* 2001;266: 7–13.
89. Droppo IG, Ross N, Skafel M, Liss, SN. Biostabilization of cohesive sediment beds in a freshwater wave-dominated environment. *Limnol Oceanogr* 2007;52: 577–589.
90. Van Colen C, Underwood GJC, Serôdio J, Paterson, DM. Ecology of intertidal microbial biofilms: Mechanisms, patterns and future research needs. *J Sea Res* 2014;92: 2–5.
91. Ubertini M, Lefebvre S, Rakotomalala C, Orvain, F. Impact of sediment grain-size and biofilm age on epipelic microphytobenthos resuspension. *J Exp Mar Biol Ecol* 2015;467: 52–64.
92. Widdows J, Brown S, Brinsley M. et al. Temporal changes in intertidal sediment erodability: influence of biological and climatic factors. *Cont Shelf Res* 2000;20: 1275–1289.
93. Lundkvist M, Grue M, Friend PL, Flindt, MR. The relative contributions of physical and microbiological factors to cohesive sediment stability. *Cont Shelf Res* 2007;27: 1143–1152.
94. Widdows J, Friend PL, Bale AJ et al. Inter-comparison between five devices for determining erodability of intertidal sediments. *Cont Shelf Res* 2007;27: 1174–1189.
95. Black KS, Sun H, Craig G et al. Incipient erosion of biostabilized sediments examined using particle-field optical holography. *Environ Sci & Technol* 2001;35: 2275–2281.
96. Neumeier U, Lucas CH, Collins M. Erodibility and erosion patterns of mudflat sediments investigated using an annular flume. *Aquat Ecol* 2006;40: 543–554.
97. da S. Quaresma V, Amos CL, Flindt M. The influences of biological activity and consolidation time on laboratory cohesive beds. *J Sediment Res* 2004;74: 1527–1404.
98. Madsen KN, Nilsson P, Sundbäck K. The influence of benthic microalgae on the stability of a subtidal sediment. *J Exp Mar Biol Ecol* 1993;170: 159–177.
99. Wolf G, Picioreanu C, van Loosdrecht MCM. Kinetic modeling of phototrophic biofilms: the PHOBIA model. *Biotechnol Bioeng* 2007;97: 1064–1079.
100. Stone M, Emelko MB, Droppo IG, Silins, U. Biostabilization and erodibility of cohesive sediment deposits in wildfire-affected streams. *Water Res* 2011;45: 521–534.
101. Pacepavicius G, Lau YL, Liu D et al. A rapid biochemical method for estimating biofilm mass. *Environ Toxicol Water Qual* 1997;12: 97–100.
102. DuBois M, Gilles KA, Hamilton JK et al. Colorimetric method for determination of sugars and related substances. *Anal Chem* 1956;28, DOI: 10.1021/ac60111a017.
103. Kochert AG. Carbohydrate determination by the phenol–sulfuric acid method. In: Hellebust JA, Craigie JS (eds.), *Handbook of Phycological Methods: Physiological and Biochemical Methods*. Cambridge: Cambridge University Press, 1978, 95–97.
104. Bradford MM. A rapid and sensitive method for the quantitation of microgram quantities of protein utilizing the principle of protein-dye binding. *Anal Biochem* 1975;72: 248–54.
105. Lorenzen CJ. Determination of chlorophyll and pheo-pigments: Spectrophotometric equations. *Limnol Oceanogr* 1967;12: 343–346.

106. White DC, Davis WM, Nickels JS et al. Determination of the sedimentary microbial biomass by extractible lipid phosphate. *Oecologia* 1979;40: 51–62.

107. Eaton J W., Moss B. The estimation of numbers and pigment content in epipelic algal populations: Estimation of epipelic algal populations. *Limnol Oceanogr* 1966;11: 584–595.

108. Nowell A. Flow environments of aquatic benthos. *Annu Rev Ecol Syst* 1984;15: 303–328.

109. Gerdol V, Hughes R. Effect of *Corophium volutator* on the abundance of benthic diatoms, bacteria and sediment stability in two estuaries in southeastern England. *Mar Ecol Prog Ser* 1994;114: 109–115.

110. Mazik K, Elliott M. The effects of chemical pollution on the bioturbation potential of estuarine intertidal mudflats. *Helgol Mar Res* 2000;54: 99–109.

111. Le Hir P, Monbet Y, Orvain F. Sediment erodability in sediment transport modelling: Can we account for biota effects? *Cont Shelf Res* 2007;27: 1116–1142.

112. Fernandes S, Sobral P, Costa MH. *Nereis diversicolor* effect on the stability of cohesive intertidal sediments. *Aquat Ecol* 2006;40: 567–579.

113. Widdows J, Brinsley M, Pope N. Effect of *Nereis diversicolor* density on the erodability of estuarine sediment. *Mar Ecol Prog Ser* 2009;378: 135–143.

114. Grabowski RC, Droppo IG, Wharton G. Erodibility of cohesive sediment: The importance of sediment properties. *Earth-Science Rev* 2011;105, 101–120.

115. Tolhurst TJ, Chapman MG, Underwood AJ et al. Technical Note: The effects of five different defaunation methods on biogeochemical properties of intertidal sediment. *Biogeosciences* 2012;9: 3647–3661.

116. Hale R, Jacques RO, Tolhurst TJ. Cryogenic defaunation of sediments in the field. *J Coast Res* 2015;316: 1537–1540.

117. Gamenick I, Jahn A, Vopel K et al. Hypoxia and sulphide as structuring factors in a macrozoobenthic community on the Baltic Sea shore: colonisation studies and tolerance experiments. *Mar Ecol Prog Ser* 1996;144, 73–85.

118. Thrush SF, Whitlatch RB, Pridmore RD et al. Scale-dependent recolonization: The role of sediment stability in a dynamic sandflat habitat. *Ecology* 1996;77: 2472–2487.

119. Kristensen E, Blackburn TH. The fate of organic carbon and nitrogen in experimental marine sediment systems: Influence of bioturbation and anoxia. *J Mar Res* 1987;45: 231–257.

120. Gilbert F, Rivet L, Bertrand J-C. The in vitro influence of the burrowing polychaete *Nereis diversicolor* on the fate of petroleum hydrocarbons in marine sediments. *Chemosphere* 1994;29: 1–12.

121. Nowell AR., Jumars PA, Eckman JE. Effects of biological activity on the entrainment of marine sediments. *Mar Geol* 1981;42: 133–153.

122. Orvain F, Sauriau P, Sygut A et al. Interacting effects of *Hydrobia ulvae* bioturbation and microphytobenthos on the erodibility of mudflat sediments. *Mar Ecol Prog Ser* 2004;278: 205–223.

123. Needham HR, Pilditch CA, Lohrer AM et al. Density and habitat dependent effects of crab burrows on sediment erodibility. *J Sea Res* 2013;76: 94–104.

124. Thompson CEL, Williams JJ, Metje N et al. Turbulence based measurements of wave friction factors under irregular waves on a gravel bed. *Coast Eng* 2012;63: 39–47.

125. Wheatcroft RA. Temporal variation in bed configuration and one-dimensional bottom roughness at the mid-shelf STRESS site. *Cont Shelf Res* 1994;14: 1167–1190.

126. Lyons AP, Fox WLJ, Hasiotis T et al. Characterization of the two-dimensional roughness of wave-rippled sea floors using digital photogrammetry. *IEEE J Ocean Eng* 2002;27: 515–524.

127. Briggs KB. Microtopographical roughness of shallow-water continental shelves. *IEEE J Ocean Eng* 1989;14: 360–367.

128. Jackson DR, Briggs KB. High-frequency bottom backscattering: Roughness versus sediment volume scattering. *J Acoust Soc Am* 1992;92, DOI: 10.1121/1.403966.

129. Widdows J, Brinsley M. Impact of biotic and abiotic processes on sediment dynamics and the consequences to the structure and functioning of the intertidal zone. *J Sea Res* 2002;48: 143–156.

130. Widdows J, Brinsley M, Salkeld P, Lucas, CH. Influence of biota on spatial and temporal variation in sediment erodability and material flux on a tidal flat (Westerschelde, The Netherlands). *Mar Ecol Prog Ser* 2000;194: 23–37.

131. Escapa M, Perillo GME, Iribarne O. Sediment dynamics modulated by burrowing crab activities in contrasting SW Atlantic intertidal habitats. *Estuarine, Coast Shelf Sci* 2008;80: 777–780.

132. Widdows J, Brinsley MD, Bowley N, Barrett, C. A benthic annular flume for in situ measurement of suspension feeding/biodeposition rates and erosion potential of intertidal cohesive sediments. *Estuarine, Coast Shelf Sci* 1998;46: 27–38.

133. Grant J, Gust G. Prediction of coastal sediment stability from photopigment content of mats of purple sulphur bacteria. *Nature* 1987;330: 244–246.

134. Gust G. Skin friction probes for field applications. *J Geophys Res* 1988;93: 14121–14132.

135. Blanchard G, Sauriau P, Cariou-Le Gall V et al. Kinetics of tidal resuspension of microbiota: Testing the effects of sediment cohesiveness and bioturbation using flume experiments. *Mar Ecol Prog Ser* 1997;151: 17–25.

136. Thompson CEL, Couceiro F, Fones GR et al. In situ flume measurements of resuspension in the North Sea. *Estuarine, Coast Shelf Sci* 2011;94: 77–88.

137. Ringold P. Burrowing, root mat density, and the distribution of fiddler crabs in the eastern United States. *J Exp Mar Biol Ecol* 1979;36: 11–21.

138. Teal L, Bulling M, Parker E, Solan, M. Global patterns of bioturbation intensity and mixed depth of marine soft sediments. *Aquat Biol* 2010;2: 207–218.

139. Solan M, Kennedy R. Observation and quantification of in situ animal-sediment relations using time-lapse sediment profile imagery (t-SPI). *Mar Ecol Prog Ser* 2002;228: 179–191.

140. Solan M, Wigham B, Hudson I et al. In situ quantification of bioturbation using time-lapse fluorescent sediment profile imaging (f-SPI), luminophore tracers and model simulation. *Mar Ecol Prog Ser* 2004;271: 1–12.

141. Murray F, Solan M, Douglas A. Effects of algal enrichment and salinity on sediment particle reworking activity and associated nutrient generation mediated by the intertidal polychaete Hediste diversicolor. *J Exp Mar Biol Ecol* 2017;495: 75–82.

142. Germano J, Rhoads D, Valente R et al. The use of sediment profile imaging (SPI) for environmental impact assessments and monitoring studies: Lessons learned from the past four decades. *Ocn. and Mar. Biol.: An Annual Rev.* 2011;49: 235–298.

143. Schiffers K, Teal LR, Travis JMJ Solan, M. An open source simulation model for soil and sediment bioturbation. *PloS one* 2011;6: e28028.

144. Wood C., Hawkins S., Godbold J. et al. Coastal Biodiversity and Ecosystem Service Sustainability (CBESS) Macrofaunal Community Metrics - Total Abundance (TA), Total Biomass (TB), Species Richness (SR), Evenness (J) and Community Bioturbation Potential (BPc) in Mudflat and Saltmarsh Habitats. NERC Environmental Information Data Centre, 2015.

145. Bishop C, Skafel M, Nairn R. Cohesive profile erosion by waves. Coastal Engineering 1992. American Society of Civil Engineers, 1993, https://doi.org/10.1061/9780872629332.227.

146. Skafel MG. Laboratory measurement of nearshore velocities and erosion of cohesive sediment (till) shorelines. *Coast Eng* 1995;24: 343–349.

147. Skafel MG, Bishop CT. Flume experiments on the erosion of till shores by waves. *Coast Eng* 1994;23: 329–348.

148. J. R. L. Allen. Mixing at turbidity current heads, and its geological implications: ERRATUM. *SEPM J Sediment Res* 1971;41: 889.

149. Thompson CEL, Amos CL. The impact of mobile disarticulated shells of *Cerastoderma edulis* on the abrasion of a cohesive substrate. *Estuaries* 2002;25: 204–214.

150. Amos CL, Daborn G., Christian H. et al. In situ erosion measurements on fine-grained sediments from the Bay of Fundy. *Mar Geol* 1992;108: 175–196.

151. Bagnold RA. *The Physics of Blown Sand and Desert Dunes.* Methuen, New York.

152. Amos C., Sutherland T, Cloutier D, Patterson, S. Corrasion of a remoulded cohesive bed by saltating littorinid shells. *Cont Shelf Res* 2000;20: 1291–1315.

153. Thompson CEL, Amos CL. Effect of sand movement on a cohesive substrate. *J Hydraul Eng* 2004;130: 1123–1125.

154. Quaresma V da S, Amos CL, Bastos AC. The influence of articulated and disarticulated cockle shells on the erosion of a cohesive bed. *J Coast Res* 2007;236: 1443–1451.

155. Ford H, Garbutt A, Ladd C et al. Soil stabilization linked to plant diversity and environmental context in coastal wetlands. *J Veg Sci : Off Organ Int Assoc Veg Sci* 2016;27: 259–268.

156. Chen XD, Zhang CK, Paterson DM et al. Hindered erosion: The biological mediation of noncohesive sediment behavior: eps mediating sediment erosion. *Water Resour Res* 2017;53: 4787–4801.

157. Huettel M, Rusch A. Advective particle transport into permeable sediments-evidence from experiments in an intertidal sandflat. *Limnol Oceanogr* 2000;45: 525–533.

158. Howes BL, Howarth RW, Teal JM Valiela, I. Oxidation-reduction potentials in a salt marsh: Spatial patterns and interactions with primary production1: Salt marsh redox potentials. *Limnol Oceanogr* 1981;26: 350–360.

159. Kristensen E. Organic matter diagenesis at the oxic/anoxic interface in coastal marine sediments, with emphasis on the role of burrowing animals. *Hydrobiologia* 2000; 426: 1–24.

160. Gerwing TG, Gerwing AM, Hamilton DJ et al. Apparent redox potential discontinuity (aRPD) depth as a relative measure of sediment oxygen content and habitat quality. *Int J Sediment Res* 2015;30: 74–80.

161. Fanjul E, Escapa M, Montemayor D et al. Effect of crab bioturbation on organic matter processing in South West Atlantic intertidal sediments. *J Sea Res* 2015;95: 206–216.

162. Kostka JE, Gribsholt B, Petrie E et al. The rates and pathways of carbon oxidation in bioturbated saltmarsh sediments. *Limnol Oceanogr* 2002;47: 230–240.

163. Negrin VL, Spetter CV, Asteasuain RO et al. Influence of flooding and vegetation on carbon, nitrogen, and phosphorus dynamics in the pore water of a Spartina alterniflora salt marsh. *J Environ Sci* 2011;23: 212–221.

164. Patrick WH, Delaune RD. Nitrogen and phosphorus utilization by *Spartina alterniflora* in a salt marsh in Barataria Bay, Louisiana. *Estuar Coast Mar Sci* 1976;4: 59–64.

165. Gutiérrez JL, Jones CG, Groffman PM et al. The contribution of crab burrow excavation to carbon availability in surficial salt-marsh sediments. *Ecosystems* 2006;9: 647–658.
166. Kristensen E, Kostka JE. Macrofaunal burrows and irrigation in marine sediment: Microbiological and biogeochemical interactions. Interactions between Macro- and Microorganisms in Marine Sediments. American Geophysical Union, 0, 125–157.
167. Luckenbach MW. Sediment stability around animal tubes: The roles of hydrodynamic processes and biotic activity1: Stability around tubes. *Limnol Oceanogr* 1986;31: 779–787.
168. Talley TS, Crooks JA, Levin LA. Habitat utilization and alteration by the invasive burrowing isopod, *Sphaeroma quoyanum*, in California salt marshes. *Mar Biol* 2001;138: 561–573.
169. Hale R, Mavrogordato MN, Tolhurst TJ et al. Characterizations of how species mediate ecosystem properties require more comprehensive functional effect descriptors. *Sci Reports* 2014;4: 6463.
170. Hale R, Boardman R, Mavrogordato MN et al. High-resolution computed tomography reconstructions of invertebrate burrow systems. *Sci Data* 2015;2: 150052.
171. Coelho V, Cooper R, de Almeida Rodrigues S. Burrow morphology and behavior of the mud shrimp *Upogebia omissa* (Decapoda: Thalassinidea: Upogebiidae). *Mar Ecol Prog Ser* 2000;200: 229–240.
172. Yunusa IAM, Braun M, Lawrie R. Amendment of soil with coal fly ash modified the burrowing habits of two earthworm species. *Appl Soil Ecol* 2009;42: 63–68.
173. Salvo F, Dufour SC, Archambault P et al. Spatial distribution of *Alitta virens* burrows in intertidal sediments studied by axial tomodensitometry. *J Mar Biol Assoc United Kingd* 2013;93: 1543–1552.
174. Rhoads DC, Cande S. Sediment profile camera for in situstudy of organism-sediment relations. *Limnol Oceanogr* 1971;16: 110–114.
175. Rhoads D, Germano J. Characterization of organism-sediment relations using sediment profile imaging: An efficient method of remote ecological monitoring of the seafloor (Remots™ System). *Mar Ecol Prog Ser* 1982;8: 115–128.
176. Nickell L, Atkinson R. Functional morphology of burrows and trophic modes of three thalassinidean shrimp species, and a new approach to the classification of thalassinidean burrow morphology. *Mar Ecol Prog Ser* 1995;128: 181–197.
177. Koo BJ, Kwon KK, Hyun J-H. The sediment-water interface increment due to the complex burrows of macrofauna in a tidal flat. *Ocean Sci J* 2005;40, DOI: 10.1007/BF03023522.
178. Seike K, Goto R. Combining in situ burrow casting and computed tomography scanning reveals burrow morphology and symbiotic associations in a burrow. *Mar Biol* 2017;164, DOI: 10.1007/s00227-017-3096-y.
179. Oug E, Høisœter T. Soft-bottom macrofauna in the high-latitude ecosystem of Balsfjord, northern Norway: Species composition, community structure and temporal variability. *Sarsia* 2000;85: 1–13.
180. Downie H, Holden N, Otten W et al. Transparent soil for imaging the rhizosphere. *PloS one* 2012;7: 44276.
181. Rosenberg R, Davey E, Gunnarsson J et al. Application of computer-aided tomography to visualize and quantify biogenic structures in marine sediments. *Mar Ecol Prog Ser* 2007;331: 23–34.
182. Rosenberg R, Grémare A, Duchêne J et al. 3D visualization and quantification of marine benthic biogenic structures and particle transport utilizing computer-aided tomography. *Mar Ecol Prog Ser* 2008;363: 171–182.

183. Delefosse M, Kristensen E, Crunelle D et al. Seeing the unseen–bioturbation in 4D: Tracing bioirrigation in marine sediment using positron emission tomography and computed tomography. *PloS One* 2015;10: e0122201.
184. Perez KT, Davey EW, Moore RH et al. Application of computer-aided tomography (CT) to the study of estuarine benthic communities. *Ecol Appl* 1999;9: 1050–1058.
185. Mermillod-Blondin F, Marie S, Desrosiers G et al. Assessment of the spatial variability of intertidal benthic communities by axial tomodensitometry: Importance of fine-scale heterogeneity. *J Exp Mar Biol Ecol* 2003;287: 193–208.
186. Michaud E, Desrosiers G, Long B et al. Use of axial tomography to follow temporal changes of benthic communities in an unstable sedimentary environment (Baie des Ha! Ha!, Saguenay Fjord). *J Exp Mar Biol Ecol* 2003;285–286: 265–282.
187. Dufour SC, Desrosiers G, Long B et al. A new method for three-dimensional visualization and quantification of biogenic structures in aquatic sediments using axial tomodensitometry: CT scan of biogenic structures. *Limnol Oceanogr Methods* 2005;3: 372–380.
188. Feagin RA, Lozada-Bernard SM, Ravens TM et al. Does vegetation prevent wave erosion of salt marsh edges? *Proc Natl Acad Sci United States Am* 2009;106: 10109–10113.
189. Tengbeh GT. The effect of grass roots on shear strength variations with moisture content. *Soil Technol* 1993;6: 287–295.
190. M. Mamo, G. D. Bubenzer. Detachment rate, soil erodibility, and soil strength as influenced by living plant roots part i: Laboratory study. *Trans ASAE* 2001;44: 1167–1174.
191. Chen Y, Thompson CEL, Collins MB. Saltmarsh creek bank stability: Biostabilisation and consolidation with depth. *Cont Shelf Res* 2012;35: 64–74.
192. De Baets S, Poesen J, Gyssels G Knapen, A. Effects of grass roots on the erodibility of topsoils during concentrated flow. *Geomorphology* 2006;76: 54–67.
193. Khanal A, Fox GA. Detachment characteristics of root-permeated soils from laboratory jet erosion tests. *Ecol Eng* 2017;100: 335–343.
194. Katuwal S, Vermang J, Cornelis WM et al. Effect of root density on erosion and erodibility of a loamy soil under simulated rain. *Soil Sci* 2013;178: 29–36.
195. F. Ghidey, E. E. Alberts. Plant root effects on soil erodibility, splash detachment, soil strength, and aggregate stability. *Trans ASAE* 1997;40: 129–135.
196. Painting SJ, van der Molen J, Parker ER et al. Development of indicators of ecosystem functioning in a temperate shelf sea: A combined fieldwork and modelling approach. *Biogeochemistry* 2013;113: 237–257.
197. Queirós AM, Birchenough SNR, Bremner J et al. A bioturbation classification of European marine infaunal invertebrates. *Ecol Evol* 2013;3: 3958–3985.
198. Steiner N, Deal C, Lannuzel D et al. What sea-ice biogeochemical modellers need from observers. *Elem Sci Anthr* 2016;4, p.000084. DOI: 10.12952/journal. elementa.000084.
199. Morrisey D, Howitt L, Underwood A et al. Spatial variation in soft-sediment benthos. *Mar Ecol Prog Ser* 1992;81: 197–204.
200. Savchuk OP. Nutrient biogeochemical cycles in the Gulf of Riga: Scaling up field studies with a mathematical model. *J Mar Syst* 2002;32: 253–280.
201. Thompson CEL, Silburn B, Williams ME et al. An approach for the identification of exemplar sites for scaling up targeted field observations of benthic biogeochemistry in heterogeneous environments. *Biogeochemistry* 2017;135: 1–34.
202. Kirwan ML, Mudd SM. Response of salt-marsh carbon accumulation to climate change. *Nature* 2012;489: 550–553.

203. Morris JT, Sundareshwar PV, Nietch CT et al. Responses of coastal wetlands to rising sea level. *Ecology* 2002;83: 2869–2877.
204. Langley JA, McKee KL, Cahoon DR et al. Elevated CO2 stimulates marsh elevation gain, counterbalancing sea-level rise. *Proc Natl Acad Sci United States Am* 2009;106: 6182–6186.
205. Cherry JA, McKee KL, Grace JB. Elevated CO2 enhances biological contributions to elevation change in coastal wetlands by offsetting stressors associated with sea-level rise. *J Ecol* 2009;97: 67–77.
206. Kirwan ML, Guntenspergen GR, Morris JT. Latitudinal trends in *Spartina alterniflora* productivity and the response of coastal marshes to global change. *Glob Chang Biol* 2009;15: 1982–1989.
207. Charles H, Dukes JS. Effects of warming and altered precipitation on plant and nutrient dynamics of a New England salt marsh. *Ecol Appl : Publ Ecol Soc Am* 2009;19: 1758–1773.
208. Mudd SM, Howell SM, Morris JT. Impact of dynamic feedbacks between sedimentation, sea-level rise, and biomass production on near-surface marsh stratigraphy and carbon accumulation. *Estuarine, Coast Shelf Sci* 2009;82: 377–389.
209. Kirwan ML, Guntenspergen GR. Feedbacks between inundation, root production, and shoot growth in a rapidly submerging brackish marsh: Marsh root growth under sea level rise. *J Ecol* 2012;100: 764–770.
210. Duarte CM, Losada IJ, Hendriks IE et al. The role of coastal plant communities for climate change mitigation and adaptation. *Nat Clim Chang* 2013;3: 961–968.
211. Donnelly JP, Bertness MD. Rapid shoreward encroachment of salt marsh cordgrass in response to accelerated sea-level rise. *Proc Natl Acad Sci United States Am* 2001;98: 14218–14223.
212. Hartig EK, Gornitz V, Kolker A et al. Anthropogenic and climate-change impacts on salt marshes of Jamaica Bay, New York City. *Wetlands* 2002;22: 71–89.
213. Scavia D, Field JC, Boesch DF et al. Climate change impacts on U.S. Coastal and marine ecosystems. *Estuaries* 2002;25: 149–164.
214. Farron, SJ, Hughes, Z, FitzGerald, DM, and Storm, KB. The impacts of bioturbation by common marsh crabs on sediment erodibility: A laboratory flume investigation. *Estuarine, Coastal and Shelf Science* 2020;238: 1–11.
215. Asano, T, Tsutsui, S, and Sakai, T. 1988. Wave damping characteristics due to seaweed. Proceedings of the 35th Coastal Engineering Conference in Japan. JSCE. 138–142.

Part III

Marsh Response to Stress

13

Climatic Impacts on Salt Marsh Vegetation

KATRINA L. POPPE AND JOHN M. RYBCZYK

The salt marsh response to a changing climate may be more complex than that of either terrestrial or marine ecosystems because salt marshes exist at the interface of land and sea and both bring changes to the marsh. Climate change may exacerbate anthropogenic-related stresses that salt marsh plants are already experiencing, limiting their resilience (Keddy 2011). In this chapter we discuss major climate change impacts likely to affect salt marshes including temperature, sea level rise (SLR), salinity, CO_2, freshwater flow, sediment, and nutrients, and consider how salt marsh plants respond to these impacts and potential interactions of these impacts. Specifically, we explore changes in plant productivity and decomposition rates, aboveground and belowground biomass, and stem density as they are central to understanding marsh responses on a larger scale, with implications for species composition, elevation change, nutrient cycling, carbon sequestration, food webs, and ultimately marsh survival. Although this chapter is focused on salt marshes, examples from tidal fresh and brackish marshes are also included to a limited extent where relevant.

13.1 Temperature

Temperature change is an obvious and important climate change factor, probably the first that comes to mind for many, and its effects on plant production and decomposition have been fairly well studied. However, few have examined temperature effects specifically on salt marsh plants (Charles and Dukes 2009), and most of this work has taken place on the Atlantic coast of North America. Increasing temperatures worldwide are a legitimate concern; globally averaged land and ocean surface temperatures have increased by 0.85°C from 1880 to 2012 (Fig. 13.1), and projections for the end of the twenty-first century (2081–2100) exceed 2°C relative to the 1850–1900 period, for the majority of IPCC greenhouse gas emission scenarios (IPCC 2014).

Aside from extreme low and high temperatures which can do immediate damage to plants, even gradual, moderate temperature changes such as those predicted from climate change can directly affect day-to-day plant processes. Temperature directly impacts plant photosynthesis, respiration, transpiration, and absorption of water and nutrients, and optimal growth requires a fine balance between these processes. Most of these processes become faster with a rise in temperature, because enzyme activity and most chemical

Observed change in surface temperature 1901–2012

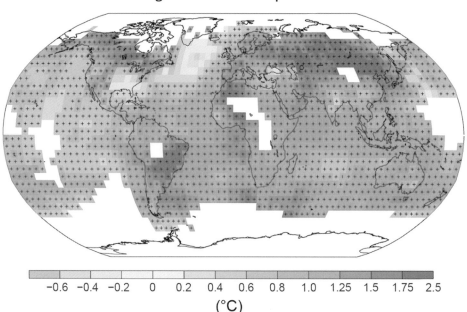

Figure 13.1 Map of observed temperature change from 1901 to 2012. Source: IPCC 2014. (A black and white version of this figure will appear in some formats. For the color version, please refer to the plate section.)

reaction rates are temperature-sensitive. However, at excessively high temperatures (that a species is not adapted to), enzymes and proteins become denatured, and water loss becomes overwhelming. All plant species have their own minimum, maximum, and optimum temperature for growth, so a global increase in temperature may affect some species more than others. In fact, temperature can be the primary factor in defining a species' range (Gedan et al. 2011). C_4 plants (including many grasses such as *Spartina*) may fare better under a warming climate than C_3 plants (including sedges), because C_4 plants achieve maximum growth at higher temperatures (Dunn et al. 1987).

Studies of temperature effects on marsh plants have typically taken one of three approaches: (1) controlled laboratory experiments, (2) field-based measurements, and (3) comparative studies of field measurements across a latitudinal gradient. Each approach has its own strengths and weaknesses. Laboratory studies have the ability to illuminate cause and effect relationships, but do not often represent *in situ* conditions. Field-based measurements allow for a more accurate quantification of the net effect of multiple factors, but can be plagued by confounding factors. Similarly, comparative studies can be complicated by confounding factors but can identify overarching patterns across multiple studies. Although it is not often feasible to include more than two or three factors in one study, the combination of experiments, field studies, gradient studies, and modeling should allow for an improved understanding of temperature effects and interactions with other factors.

13.1.1 Productivity/Biomass

In general, warmer temperatures have been shown to enhance marsh plant productivity, but not all plants respond equally. In some cases, aboveground biomass is affected while in other cases only belowground biomass or stem height responds. For example, Charles and Dukes (2009) measured a 24% increase in *Spartina alterniflora* shoot biomass and stem height with experimental daytime warming over one growing season in a New England salt marsh, while biomass decreased in their *Spartina patens-Distichlis spicata* community (Fig. 13.2). They suggest this selective temperature effect may be due to increased allocation to leaves in the *S. alterniflora* community. However, other studies have observed increased *D. spicata* biomass with warming (Gedan and Bertness 2009, Coldren et al. 2016), and it is unclear why certain species would increase in some cases and decrease in others. In Northern Europe, Gray and Mogg (2001) examined the temperature effect on both a C_3 species (*Puccinellia maritima*) and a C_4 species (*Spartina anglica*). Both species responded with increased biomass, but the *S. anglica* response was seen most in belowground biomass. In addition to increasing the rate of photosynthesis, warming may also enhance plant productivity on an annual scale by delaying senescence, effectively lengthening the growing season (Shaver et al. 2000, Cleland et al. 2007).

Of course, the temperature effect may be beneficial for growth only up to a point, as mentioned earlier. Field experiments measuring gas exchange show net photosynthesis and productivity being greatest at summer high temperatures, beyond which rates start to decline (Shea 1977, Giurgevich and Dunn 1979). Kirwan et al. (2009) observed a latitudinal gradient in *S. alterniflora* productivity with biomass measurements, along the Gulf and East coasts of North America, from Texas to Nova Scotia, predicting a productivity increase of 27 g m^{-2} yr^{-1} with every 1°C increase but only in northern marshes (Fig. 13.3). If southern marshes are currently closer to their optimum temperature (Giurgevich and Dunn 1979), additional warming beyond an approximately 30–35°C threshold would decrease southern marsh productivity (Kirwan et al. 2009).

13.1.2 Effect on Decomposition

Several studies have shown that experimental warming causes an increase in decomposition rates, which is supported by the observation that decomposition varies with seasonal temperature differences (Montagna and Ruber 1980). For example Valiela et al. (1985) reported a loss of biomass as a function of increased water temperature, likely due to increased decomposition. Carey et al. (2017) then used that relationship to predict a 9.4% increase in low marsh decomposition rates after 1960. Charles and Dukes (2009) also reported an increase in decomposition in their *S. alterniflora* community. Some observe the decomposition increase in terms of reduced belowground biomass (e.g., Coldren et al. 2016, Crosby et al. 2017) while others measure decomposition directly (e.g., Kirwan et al. 2014). By measuring decomposition directly, Kirwan et al. (2014) performed a gradient study, comparing organic matter decay rates across a seasonal and latitudinal gradient over

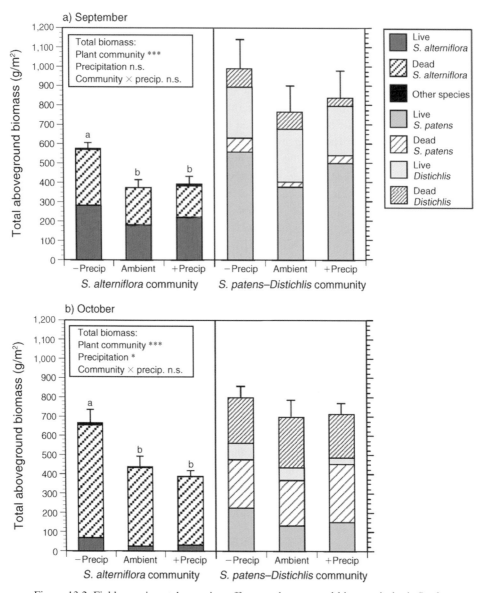

Figure 13.2 Field experimental warming effect on aboveground biomass in both *S. alterniflora* and *S. patens-D. spicata* communities sampled in (a) September and (b) October. (Charles and Dukes 2009).

a 3-year period in the northeastern USA. They found a 3–6% increase in decomposition rates for every degree of warming, reporting Q_{10} rates of 1.3–1.5 (see Box 13.1). However, because this study used buried cotton strips as a proxy for native soil organic material, actual decomposition rates may differ slightly from those reported, but the seasonal and latitudinal trends remain revealing.

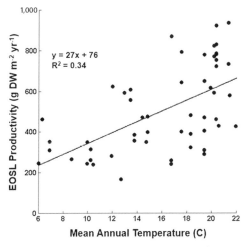

Figure 13.3 Observed end of season live (EOSL) *Sp. alterniflora* productivity compared with mean annual temperature, as measured across a latitudinal gradient along the Gulf and East coasts of North America. Figure courtesy of Matthew Kirwan, Virginia Institute of Marine Science.

Box 13.1: Q_{10} Temperature Coefficient:

The Q_{10} temperature coefficient is the factor by which a given reaction rate increases with a 10°C rise in temperature. The Q_{10} is often reported in plant studies in reference to the rates of photosynthesis, respiration, or decomposition, and it allows for a comparison of temperature sensitivity among physiological processes (e.g., photosynthesis vs. decomposition) or among species. Biomass changes are sometimes used as a proxy for a change in production or decomposition rates. Generally, most biological reaction rates tend to double or triple with every 10°C rise ($Q_{10} \approx 2$–3), but this is based on evidence rather than on first principles (Davidson and Janssens 2006). The Q_{10} temperature coefficient is defined as:

$$Q_{10} = \left(\frac{R_2}{R_1} \right)^{\left(\frac{10}{T_2 - T_1} \right)}$$

where Q_{10} is the factor by which the reaction rate increases with a 10°C rise in temperature; R_1 is the measured reaction rate at T_1; R_2 is the measured reaction rate at T_2; T_1 and T_2 are two different temperatures at which the reaction is measured. No specific temperatures are required but T_1 and T_2 must be reported in Celsius or Kelvin.

13.1.3 Net Effect on Productivity vs Decomposition and Interactions

Not many studies have examined both productivity and decomposition responses to temperature increases, though one might wonder how these two rates balance out, if both are generally enhanced by warming. Kirwan and Mudd (2012) modeled both productivity and decomposition rates in a hypothetical *S. alterniflora* marsh in response to a 4°C increase, and their predicted decomposition rates exceeded production rates. However, the balance was shifted when other factors such as SLR were included. When SLR was increased, productivity overwhelmed decomposition. Kirwan et al. (2009) concluded that although some marsh will be lost due to SLR, warming will increase productivity to compensate for what is lost. Salinity may also moderate the temperature effect, in such a way that salinity stress lowers the optimum temperature and effectively reduces the potential positive effects of warming on productivity (Kirwan et al. 2009). In addition, warming may increase salinity in some environments, resulting in a negative feedback on productivity. In other words, although warming itself may directly enhance productivity, it may also indirectly reduce productivity through changes in salinity.

13.1.4 Change in Species Composition, Diversity

Temperature increases are expected to alter not only individual plant growth and decay but also marsh species richness and community composition. Species richness is expected to decline, as stress-tolerant species expand and outcompete others. For example, when diverse salt marsh forb pannes in New England were experimentally warmed, *S. patens* outcompeted other species (Gedan and Bertness 2009). McKee et al. (2012) compared species richness in Australian marshes and observed lower species richness in lower latitude (warmer) marshes. This exemplifies the centrifugal model of community organization, with stressful conditions reducing diversity, and the "peripheral habitat" expanding to leave only the most tolerant species to persist (Keddy 1990).

13.1.5 Mangrove Expansion

Some marshes may even become mangrove forests as the freezing events that limit mangrove occurrence become less frequent (Osland et al. 2013, Short et al. 2016). At the edge of the current mangrove range extent, mangroves are already expanding poleward into salt marshes on at least five continents (Saintilan et al. 2014), and they are expected to continue expanding, particularly into areas where freeze events occur no more than once every 12 years (Henry and Twilley 2013). However, a warming experiment on co-occurring marsh and mangrove plants found the marsh plant (*D. spicata*) to respond better to summer warming than the mangrove (*Avicennia germinans*) (Coldren et al. 2016), suggesting that marshes may outcompete mangroves during warm periods between freeze events. Therefore, although mangrove propagules are expected to expand into marshes in many areas and cause a shift in the dominant community, chronic warming between freeze events may slow their progression.

13.1.6 Drought

Droughts may increase in some areas due to increased evapotranspiration, potentially in combination with reduced precipitation. Drought intensity and/or duration is likely to increase by the end of the twenty-first century on a regional to global scale, but particularly in dry regions (IPCC 2014). Drought can affect marsh plants in two major ways: (1) increased porewater salinity (the direct effects of which are discussed further in a later section), and (2) soil redox changes leading to porewater acidification and mobilization of metals such as iron, aluminum, and magnesium (Hughes et al. 2012). Some have linked drought to acute marsh dieback (AMD) events, either because of salinity (Hughes et al. 2012) or soil chemistry changes (McKee et al. 2004). However, the association between drought and dieback is still tenuous. Alber et al. (2008) found evidence of a linkage between drought and AMD in the southeast and Gulf coasts of theUSA, but not along the mid-Atlantic or northeast US coasts. If drought does not cause complete dieback, it can still result in more subtle changes to marsh plant growth. Drought stress may reduce root and shoot biomass as well as nutrient uptake (Brown et al. 2006). C_4 plants are generally better adapted to drought conditions than C_3 plants, although even within C_4 plants, some species such as *S. alterniflora* appear less tolerant than others (Mendelssohn et al. 2006). For example, within C_4 plants, high marsh species tend to me more heat-tolerant, with their narrower leaves allowing for less latent heat loss and less need for water for transpirational cooling, and indeed high marshes face drier and saltier conditions than low and middle marshes (Maricle et al. 2007). Therefore, in areas experiencing increased drought intensity or duration, we may expect to see the expansion of more drought- and salt-tolerant high marsh species.

13.1.7 Effect of Temperature on Accretion

With temperature potentially altering productivity and decomposition rates and perhaps even other structural characteristics such as stem height, warming temperatures may have an indirect effect on sediment accretion rates. Will accretion rates increase due to enhanced productivity, or decrease due to more rapid decomposition? The existing literature addressing this question is inconclusive, suggesting that the effect may be species- or site-specific. Carey et al. (2017) reported no difference in accretion rates in Rhode Island marshes compared to a cooler period 30 years ago. In the more recent, warmer period, they found that organic matter contributed less to accreted material, likely due to higher rates of sediment organic matter decomposition, despite enhanced productivity rates. The accretion response likely differs in marshes with organic-rich versus mineral-rich sediment. In mineral-rich marshes such as those on the west and southeast coasts of North America, accretion rates may be less negatively affected by warming-induced decomposition, and more positively affected by the sediment trapping benefit of increased aboveground biomass or stem height. In organic-rich marshes, changes to accretion rates will depend more on the balance between production and decomposition. Turner et al. (2004) suggested that belowground biomass is more important than aboveground biomass in determining accretion rates in Louisiana coastal marshes. Therefore marsh elevation there will depend on how root biomass in particular responds to warming. Kirwan and Mudd (2012) predicted increased accretion rates due to enhanced aboveground production, but only in the short term. In the longer term, increased accretion led to lower productivity rates which fed back to lower accretion rates. Over time, marshes will stabilize at an elevation dependent on rates of plant growth and decay, SLR, and sediment supply.

13.1.8 Summary

Although the direct effects of rising temperatures have been well studied, the indirect ecosystem impacts are still somewhat equivocal (Shaver et al. 2000). In general, both photosynthesis and respiration are enhanced by warming, though this effect appears to be more pronounced in cooler climates. And though many studies report enhanced productivity in the short- and medium-term, long-term measurements are less clear (Teal and Howes 1996), suggesting that the relationship between temperature and productivity may be short-lived or less important than other factors in the long term.

13.2 Sea-Level Rise

As salt marshes typically exist within a relatively narrow elevation band, marsh plants might seem vulnerable to changes in sea level associated with anthropogenic climate change. Too much tidal flooding may deprive the plants of light or oxygen, while not enough flooding can lead to stressful soil salinity levels. Reports of past marsh loss have increased concerns over the possibility of salt marshes drowning out of existence due to SLR, with the loss of all the ecosystem goods and services that they provide. Yet, through feedbacks between plants and their environment, those seemingly vulnerable marsh plants have the ability to maintain a stable elevation relative to sea level, facilitating their own continued existence amidst a changing environment.

Hydrology is perhaps the primary factor controlling the structure and function of salt marshes (Keddy 2011). The extent and duration of flooding regulates nearly every component of these ecosystems, including productivity, decomposition, species composition, elevation, and carbon sequestration. Thus, changes in flooding regimes due to SLR are likely to have dramatic impacts on salt marsh plants. Moderate SLR increases can sometimes be beneficial, flushing salts and providing nutrients and sediment, but at some point flooding starts to become a stressor, limiting growth due to lowered oxygen levels, potentially higher salinity, and sulfide buildup (Lamers et al. 2013). Models predict 0.30–0.80 m SLR by 2100, corresponding to rates of 2–17 mm/yr in 2,100 (IPCC 2014), but others have used semi-empirical models that predict a less conservative range of 0.40–1.60 m by 2,100 (Rahmstorf 2007). Relative SLR will be geographically variable, with some areas such as the Mississippi delta experiencing relative SLR rates much higher than the global average, and other areas less (Day et al. 2008). Unless marshes are able to accrete mineral and organic material at a rate that keeps pace with rising sea levels, SLR is very likely to modify plant production.

13.2.1 Increased Biomass

Perhaps one of the most widely used methods for studying the effect of SLR on marsh plant production is the use of "marsh organs." Marsh organs are PVC tubes filled with marsh plants and sediment, allowing for measurements of plant growth over a range of experimentally manipulated elevations to test the effects of different inundation frequencies on a small scale (Fig. 13.4). They typically extend both lower and higher than the adjacent marsh, so the optimal elevation for plant productivity can be identified. Some plant species demonstrate a parabolic productivity–inundation curve with a peak biomass at their optimal elevation, as seen with *S. alterniflora* (Morris et al. 2002) (Fig. 13.5), *Schoenoplectus americanus* and *S. patens* (Kirwan and Guntenspergen 2013), *S. americanus* (Langley et al. 2013), *Bolboschoenus maritimus*, *Spartina foliosa*, and *Carex lyngbei* (Janousek et al. 2016). However, other studies have observed productivity merely decreasing with increased flooding when studying different species,

Figure 13.4 Example of a marsh organ in the Apalachicola National Estuarine Research Reserve in Florida. Image courtesy of Jim Morris, University of South Carolina. (A black and white version of this figure will appear in some formats. For the color version, please refer to the plate section.)

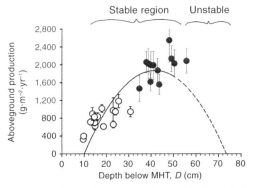

Figure 13.5 Observed productivity of *S. alterniflora* with depth below mean high tide (MHT) from both high marsh (open circles) and low marsh (solid circles) sites. Stable (solid line) or unstable (dashed line) combinations of equilibrium productivity, biomass, and depth are shown. (Morris et al. 2002)

sometimes within the same study, such as with *Salicornia pacifica* and *Juncus balticus* (Janousek et al. 2016), *S. patens* (Langley et al. 2013, Tobias and Nyman 2017), *S. americanus* and *Schoenoplectus acutus* (Schile et al. 2017). Even the same species may show a different response in different locations (e.g., Morris et al. 2002 versus Watson et al. 2017). Perhaps other factors such as salinity, nutrients, or soil drainage play a role in modifying the plant's ability to tolerate flooding stress (Langley et al. 2013). Some species, particularly high marsh species, may simply exhibit optimal growth near zero inundation levels. This variation among species and locations points to a need for more research examining productivity–inundation relationships for a variety of species. Since marsh plants show either a temporary increase in production with increased flooding at a higher initial elevation, or a monotonic decrease at a lower elevation, as SLR accelerates over time the effect of SLR itself will eventually be entirely negative, which does not bode well for organic-rich marshes that depend on root production. Of course, some marshes will be able to accrete enough to keep pace with SLR, as discussed in Chapter 7, but it appears that increased flooding in general will be detrimental to marsh productivity in the long term.

13.2.2 Decomposition

Very little is known about the SLR effect on decomposition. It has often been assumed that increasingly anaerobic soils will lead to a reduction in decomposition of soil organic matter (Nyman and DeLaune 1991, Reed 1995, Miller et al. 2001), supported by the observation that decomposition rates decline with depth in the soil column due to shifts in electron acceptors (e.g., Mendelssohn et al. 1999, Pozo and Colino 1992). Yet, actual measurements of decomposition under varying flooding regimes suggest otherwise. For example, Kirwan et al. (2013) measured increased soil carbon loss with increased flooding, although the overall flooding influence was minor, suggesting that decomposition is relatively insensitive to SLR. But given the paucity of research in this area, the ability to predict the effects of SLR on decomposition is limited.

13.2.3 Species Composition

SLR is likely to cause a change in plant species composition, with more flood-tolerant species replacing less tolerant species. For example, in northeast North America, the low marsh plant *Spartina alterniflora* is expected to expand into high marsh *S. patens* populations (Donnelly and Bertness 2001, Smith 2009, Langley et al. 2013). In addition, since SLR is expected to increase salinity, species composition will also tend toward salt-tolerant species, perhaps decreasing species richness (Sharpe and Baldwin 2012). van Dobben and Slim (2011) monitored vegetation for 15 years and modeled species composition change over the next century due to SLR, predicting a change in species composition but no significant change in species diversity. Their model projected a decrease in common species and an increase in rare species adapted to extreme habitats.

13.2.4 Elevation Change

Flooding effects on aboveground growth are relatively well studied compared to below-ground growth. However belowground production may be far more important in determin-ing marsh accretion and resilience to SLR, particularly in organogenic marshes. Where marsh survival in minerogenic marshes primarily depends on sediment supply, survival in organogenic marshes, such as those on the east coast of North America, depends more on root production (Kirwan et al. 2010, Kirwan and Guntenspergen 2012). Turner et al. (2000) estimated that organic accumulation, particularly belowground, is five times more import-ant than inorganic accumulation for east coast salt marshes with low sediment supply. Even on the southeast coast of North America, Stagg et al. (2016) examined the major processes contributing to elevation change (and consequently SLR resilience) and identified root zone expansion as the primary determinant of elevation change.

13.2.5 Habitat Resilience

When the rate of SLR becomes high enough that plant growth and hence vegetation-induced accretion rates can no longer keep up, vegetation die-off and shallow water ponds can develop. Just as biogeomorphic feedbacks can allow marsh accretion to keep pace with SLR, eventually SLR may reach a tipping point that leads to an alternate stable state, stabilized with a feedback between plant mortality and elevation loss (Fagherazzi et al. 2006). Plant die-off and pool formation typically begin in the marsh interior due to excessive flooding, and potentially other factors such as salt stress or herbivory (Alber et al. 2008). These unvegetated pools can make the marsh more prone to erosion and subsidence, which then allows for additional flooding, with the pools continuing to expand. Pool enlargement is considered an early indicator of a submerging marsh that is not keeping up with SLR, exemplified by marshes of the Mississippi delta (Wells and Coleman 1987, Nyman et al. 1993), Venice Lagoon (Carniello et al. 2009), and Chesapeake Bay (Stevenson et al. 1985, Kearney et al. 1988).

Fortunately, however, salt marshes that succumb to SLR through die-off may be in the minority. Most salt marshes are gaining elevation at rates that equal or exceed historical rates of SLR (French 2006) and those that do not often have other anthropogenic disturb-ances to blame (Kirwan and Megonigal 2013). But even in situations where suspended sediment concentrations are low or the rate of SLR becomes too high, a marsh that cannot keep pace in the vertical dimension could still maintain, or even expand, total marsh area in the horizontal dimension if it has the opportunity to migrate inland (Kirwan et al. 2016). Of course, marsh expansion through transgression is only possible where uplands are undevel-oped and gently sloping. Marshes in areas that are developed with defensive structures lack this opportunity, as is the case in many northwest European (Wolters et al. 2005) and U.S. Pacific coast estuaries (Stralberg et al. 2011).

13.2.6 Summary

The SLR effect is variable across species, and is dependent on the starting point elevation relative to a species' optimal elevation for growth. If a plant exists above its optimal elevation relative to sea level, SLR is expected to result in increased shoot and root growth in the short term, but below its peak elevation (or where no optimal elevation is apparent), SLR is expected to cause a decrease in shoot and root growth. Fortunately SLR appears to have little to no effect on decomposition, though very few have studied this relationship.

13.3 Salinity

Salinity is an important factor in determining tidal marsh community structure due to its influence on biogeochemical pathways. A major feature of temperate tidal marshes is plant zonation along a salinity gradient (Odum 1988), therefore tidal marsh ecologists have long been interested in how plant species and communites respond to salinity changes (Howard and Mendelssohn 1999). The dominant drivers of increased salinity include: SLR, reduction of riverine freshwater inputs, channel excavation for navigation/freshwater diversions, and storm surge (Herbert et al. 2015), though drought and increased evapotranspiration can affect salinity as well. Many of these are either caused or could be exacerbated by climate change. Alhough there are many mechanisms by which climate change can increase marsh salinity, it may also decrease salinity in some areas, if freshwater inputs are increased.

Although tidal marsh plants are adapted to some level of salinity, the species-specific tolerance range can be narrow, and sometimes just a small increase in salinity can push a plant out of its optimal zone. Even if a plant persists in spite of a change in salinity, it may still be stressed, indicated by lower growth rates, shorter plants or lower stem densities, with mortality only at the extreme end of its tolerance.

Research on salinity effects appears to be concentrated in the northern Gulf of Mexico, perhaps because that region has been experiencing increased droughts, storm events, freshwater diversion, and particularly high rates of relative SLR for decades (DeLaune and Pezeshki 1994), whereas others are only beginning to worry about it. Consequently, the majority of plant species studied for salinity tolerance are specific to that region, but the mechanisms involved may be applicable to other regions. In addition, marsh salinity studies appear to be dominated by greenhouse and manipulative experiments each focused on one or two plant species, although there are a few studies documenting landscape-level response, either using interannual salinity variability or spatial salinity gradients to determine the response.

13.3.1 Plant Adaptations to Salinity

Salt-tolerant plants (halophytes) have ways of keeping salt out of their cells, either by preventing salt entry into the plant, or by minimizing salt concentrations in the cytoplasm (Munns 2002). Some plants utilize salt exclusion mechanisms to reduce salt transport to their leaves, while others compartmentalize salts in cell vacuoles, effectively preventing it from harming the cells themselves. Salt-tolerant plants also exhibit osmoregulation which allows them to aquire water in the face of low external water potential (Cronk and Fennessy 2001).

13.3.2 Decreased Production

Increased porewater salinities tend to reduce marsh plant productivity, although some species are more sensitive than others. The reduction in growth is primarily attributed to stomatal closure, as well as a decrease in leaf photosynthetic capacity from excessive tissue ion concentrations (DeLaune and Pezeshki 1994). Studies often use biomass as an indicator of productivity changes, and many studies have reported a decrease in plant biomass as a result of increased salinity (Linthurst and Seneca 1981, De Leeuw et al. 1990, Broome et al. 1995, Howard and Mendelssohn 1999). For example, Howard and Mendelssohn (1999) tested the effects of increased salinity on four common coastal Louisiana oligohaline marsh plant species (*Eleocharis palustris*, *Panicum hemitomon*, *Sagittaria lancifolia*, and *Scirpus americanus*) with a greenhouse experiment. Their study included the effect of not only salt concentration, but also the rate of salinity increase and duration of exposure over one growing season. They found that both aboveground and belowground biomass and stem height were significantly reduced for all species under the elevated salinity level (12 g L^{-1} compared to 0 g L^{-1}), although some species showed a response sooner than others. Stagg et al. (2017) tested whether the landscape-scale response would reflect patterns observed by others in greenhouse and manipulative field experiments, using a natural spatial salinity gradient from fresh to saline to determine the salinity effect on aboveground and belowground production. They found that higher salinities negatively affected rates of primary production, but the magnitude of the effect differed by wetland type. Although the fresh marsh was most sensitive, the intermediate, brackish, and saline marshes also showed declines. Others have also documented salt-induced declines in productivity even in the most salt-tolerant species (Naidoo et al. 1992), indicating that, although salt marsh plants may be adapted to higher salinity levels, they still respond to it as a stress when it exceeds its normal range.

13.3.3 Root Response

There is still some question as to how salinity affects marsh plant roots. Some say that roots are more robust than shoots with regard to salt stress (Munns 2002). However, salinity has been shown to disrupt nutrient uptake (Criddle et al. 1989) and also cause water stress and toxicity when it accumulates in the roots (Greenway and Munns 1980, Poljakoff-Mayber 1988). Stagg et al. (2017) reported finding 50–80% fewer studies in their literature review included belowground production, and only a handful included both aboveground and belowground production, so more research focusing on the salinity effect on roots may be helpful.

13.3.4 Decreased Stem Height and Density

Several studies have reported salinity-induced reductions in plant stem heights and stem densities (Linthurst and Seneca 1981, Zedler et al. 1986, Broome et al. 1995, Howard and Mendelssohn 1999). Even plants growing in the hypersaline Tijuana River estuary in southern California, adapted to high salinity conditions, responded with greater stem heights when salinity was reduced by winter flooding. Interestingly, while winter flooding increased stem heights, higher summer flows increased stem densities instead. It is thought that reduced salinity early in the growing season (from winter flooding) stimulates height growth, whereas reduced salinity in the summer is too late to stimulate height growth but is still able to stimulate growth as vegetative reproduction (Zedler et al. 1986).

13.3.5 Decreased Stem Elongation

Leaf elongation is another measure of plant growth that can be affected by salinity. Mendelssohn and McKee (1992) observed a decrease in *S. patens* leaf elongation when grown in salinities greater than 12 ppt. Similarly, Naidoo et al. (1992) reported higher salinities causing decreased leaf elongation for both *S. alterniflora* and *S. patens*, but only under drained conditions. In other words, both species are adapted to flooding and high salinities, but they are more tolerant of high salinities when flooded instead of drained.

13.3.6 Increased Decomposition/Soil Organic Matter

It appears evident that elevated salinity tends to negatively impact the growth of tidal marsh plants, but how does it affect decomposition? Craft (2007) studied the relationship between salinity (as a proxy for freshwater input) and soil properties in Georgia tidal marshes and compared results with similar published data from marshes across the USA. Though he found no relationship between salinity and aboveground or belowground production, elevated salinity did appear to cause an increase in root decomposition rates, which in turn lowered soil organic content and vertical accretion rates, leading to the conclusion that soil organic matter and accretion rates are mediated by freshwater inputs through the salinity effect on decomposition. However, other studies have documented a decrease in soil organic content correlated with higher salinities (Linthurst and Seneca 1981, Morrissey et al. 2014).

13.3.7 Interactions

Some studies include an analysis of potential interactions between salinity and other environmental factors, most commonly considering flooding but also including aeration, nutrients, depth, soil type, and metals. Most find no interaction between salinity and other factors (Linthurst and Seneca 1981, Broome et al. 1995, Naidoo et al. 1992, Stagg et al. 2017). However, Willis and Hester (2004) observed an interaction between salinity and soil organic content, with production being greater in mineral soils over organic soils when grown under a stressful salinity regime.

13.3.8 Summary

As sea levels rise, tidal marshes will likely experience gradually more saline conditions, at least in the summer, while more frequent storms may increase marsh salinity more dramatically and quickly. At a community level, we expect to see more vulnerable species succumbing to salinity stress while more tolerant species expand (Zedler 1983, Howard and Mendelssohn 1999, McKee et al. 2012, Herbert et al. 2015). Anthropogenic limits on marsh transgression will result in an overall shift to more saline wetlands (Williams et al. 1999, Visser et al. 2013). This shift unfortunately may come at the cost of plant production (Grime 1988), although this will depend on the degree of salinity change and the salinity tolerance of the species involved.

13.4 Carbon Dioxide

Although CO_2 is the main driver of climate change, which results in increased global mean temperatures, land ice melting, SLR, ocean acidification, precipitation, and salinity changes, each of which has its own particular impact on marsh plant production and decomposition, in this section we focus on the direct effect of elevated atmospheric CO_2 levels. Because plants need CO_2 to photosynthesize, it is often assumed that higher atmospheric CO_2 levels will have a positive effect on plant productivity and we may actually be better off; crops will thrive, forestry will benefit, and marshes will become even more productive. Generally, this may be true, but some plants, or even parts of plants, may be more affected than others, and the CO_2 response is complicated when additional environmental factors are considered.

Atmospheric concentrations have increased by over 40% since the Industrial Revolution (IPCC 2014), now exceeding 400 ppm annually since 2013, and currently increasing by approximately 2 ppm per year. And just as temperature increases vary by latitude, so too does atmospheric CO_2. Global annual average CO_2 concentrations are generally greater in the northern hemisphere and at higher latitudes because the majority of global emissions take place in the northern hemisphere (Fig. 13.6).

13.4.1 Productivity

Elevated atmospheric levels of CO_2 are generally expected to increase photosynthetic rates in C_3 plants by increasing CO_2 concentrations in plant leaves, which favors the use of CO_2 by Rubisco, over its normal affinity for O_2. Thus, an increase in atmospheric CO_2 should theoretically increase photosynthesis because of its effect on Rubisco (Farquhar et al. 1980, Woodrow and Berry 1988, Bowes 1993). C_4 plants, in contrast, are not as sensitive to CO_2 enhancements because for them, the photosynthetic pathway is already essentially saturated at current CO_2 levels, therefore CO_2 is not limiting (Woodward et al. 1991). Elevated CO_2 levels can also modify plant hormones and cell division (Urban 2003), further contributing to increases in not only growth rates but also plant biomass and stem density, with aboveground and belowground responses differing in some cases. For example, Erickson et al. (2007) observed a sustained CO_2 enhancement of C_3 plant biomass over an 18-year study period, with shoot biomass increasing more than root biomass. Wolf et al. (2007), however, observed a larger increase in belowground biomass than aboveground biomass, and others have noted a primarily belowground response (Curtis et al. 1990, Langley et al. 2009). Shoot density has also been shown to increase with CO_2 (Drake 1992, Rasse et al. 2005, Wolf et al. 2007). We

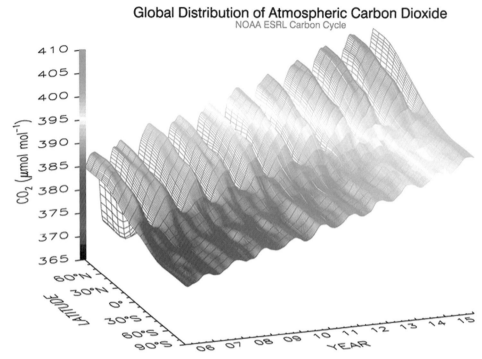

Figure 13.6 Atmospheric CO_2 concentrations from 2005 to 2014, showing both latitudinal and seasonal variations (Graph courtesy of Dr. Ed Dlugokencky and Dr. Pieter Tans, NOAA ESRL Carbon Cycle, www.esrl.noaa.gov/gmd/ccgg/). (A black and white version of this figure will appear in some formats. For the color version, please refer to the plate section.)

can thus conclude that elevated CO_2 generally has a positive effect on plant growth, but the specific location and the expression of the response is not yet predictable. CO_2 may also modify the species composition of marsh communities, because increased CO_2 is expected to give C_3 plants a competitive advantage over C_4 plants (Polley et al. 2003, Cherry et al. 2009), although some have suggested that plants in mixed communities may not respond the same as those in isolation (Gray and Mogg 2001, McKee et al. 2012). This question has not yet been well-studied, but Erickson et al. (2007) did in fact observe a shift from C_4 to C_3 plants in a mixed salt marsh community, over an 18-year period.

13.4.2 Long-Term Effect

Some studies have shown that the positive effect of CO_2 on photosynthesis is not maintained in the long-term as the photosynthetic apparatus responds to a buildup of carbohydrates, and acclimates by producing less Rubisco, but study results are variable. For example, Jacob et al. (1995) found *Scirpus olneyi* plants in Chesapeake Bay to sustain a high photosynthetic rate in the field under elevated CO_2 conditions relative to those grown under ambient conditions, but Rubisco content declined by 43% during this period, indicating a decline in carboxylation efficiency. In contrast, Rasse et al. (2005) observed short-term acclimations over the first nine years, measured as declining net ecosystem exchange after an initial increase but the longer-term trend over their 18-year study period was that of increasing productivity due to elevated CO_2 levels. In other words, if studies are conducted over a longer period than the typical two to five year period, we may see plants overcome this initial acclimation phase of unchanged or decreasing productivity, to demonstrate sustained increased productivity.

13.4.3 Decomposition

If it appears that CO_2 is all about stimulating plant growth, decomposition complicates the situation. Studies focused on the decomposition response are not very common, but existing studies indicate that an increase in soil organic matter decomposition may actually reduce the positive growth effects of CO_2 to some degree (Ball and Drake 1998, Megonigal et al. 2004). For example Wolf et al. (2007) found that, although CO_2 enhancement increased belowground biomass, it also increased soil organic matter decomposition rates, perhaps due to the oxygen released by roots elevating soil redox potential. This suggests a positive (reinforcing) feedback, where high CO_2 levels increase root biomass, boosting the remineralization of buried soil carbon and pumping more CO_2 back into the atmosphere.

13.4.4 Elevation Change

Since CO_2 impacts plant growth, and plant growth can influence surface elevation change (see Chapter 7), it follows that CO_2 may have an indirect impact on marsh elevation change. CO_2 increases elevation gain primarily through belowground shoot expansion, particularly with C_3 plants (Fig. 13.7; Cherry et al. 2009, Langley et al. 2009). Ratliff et al.

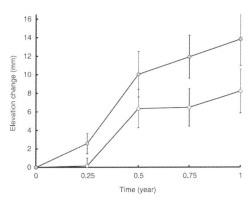

Figure 13.7 Elevation change in ambient CO_2 (380 ppm; open circles) and elevated CO_2 (720 ppm; gray circles) treatments, in a greenhouse experiment including a mix of *Spartina patens* and *S. americanus*. Total elevation gain was more than 50% greater at elevated CO_2 levels. Surface elevations were measured with a miniature surface elevation table. Error bars represent 1 SE (n = 30). (Cherry et al. 2009)

(2015) modeled elevation change and included a CO_2-induced increase in aboveground and belowground biomass, which resulted in an increased resilience of the marsh to SLR and reduced sediment inputs. Since their model did not incorporate a CO_2 effect on decomposition, they suggested a negative (stabilizing) effect wherein CO_2 enhances plant biomass and sequesters more CO_2, contrasting with the positive feedback suggested by Wolf et al. (2007) and described above.

13.4.5 Interactions

Of course, in nature no stressor acts in isolation, and recent studies have examined the interactions between CO_2 and other stressors, primarily salinity and flooding. Salinity and flooding are both known to negatively impact the growth of certain marsh plants, as discussed earlier. Interestingly, the presence of these stressors may actually enhance the CO_2 effect on plant growth, although the reasons for this have not yet been well studied (Fig. 13.8; Rozema et al. 1991, Lenssen et al. 1993, Idso and Idso 1994, Cherry et al. 2009). For example, over the course of a 17-year study, Rasse et al. (2005) observed an increase in the CO_2 effect on C_3 shoot density during higher salinity years. But again, results are varied. Erickson et al. (2007) reported an increased CO_2 effect on C_3 aboveground biomass during higher salinity years but a lower CO_2 effect on belowground biomass. Additionally, their C_4 and mixed C_3-C_4 communities showed a decreased CO_2 effect on aboveground biomass in higher salinity years. Still others report no interaction between CO_2 and salinity (Yeo 1999, Centritto 2002, Centritto et al. 2002). CO_2 effects also appear to be modified by flooding conditions, with a more consistent CO_2 response under constant flooding conditions as opposed to intermittent flooding (Langley et al. 2009).

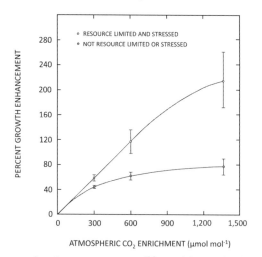

Figure 13.8 Plant growth enhancement versus CO_2 enrichment level under both stressful (open circles) and non-stressful conditions (solid circles), as the mean response from a review of 342 papers (Idso and Idso 1994). Stresses included suboptimal light, water, nutrients, and high levels of salinity, air pollution, and temperature.

13.4.6 Summary

As atmospheric CO_2 concentrations continue to increase, productivity is expected to increase, which may give C_3 plants a competitive advantage over C_4 plants. An increase in aboveground productivity and stem density may help to raise accretion rates and increase marsh resilience to SLR, while an increase in belowground productivity would do the same as long as productivity is not outweighed by higher decomposition rates. The presence of other stressors may enhance the positive effect of CO_2 on plant growth, thus it is important to consider multiple factors and their interactions.

13.5 Changes in Freshwater, Nutrient, and Sediment Inputs

Freshwater, sediment, and nutrient inputs are difficult to predict in a generalized way because they are expected to increase in some areas and decrease in others due to climate change. They may also increase in some seasons and decrease in others. This can have a substantial effect on estuarine salt marsh plants, however. Freshwater inputs to salt marshes are expected to increase where watersheds experience more precipitation, resulting in a concomitant increase in sediment and nutrient delivery to the marsh. Conversely, a decrease in precipitation will bring reduced freshwater, sediment, and nutrient inputs. Global climate models show highly variable predicted precipitation patterns from one place to the next, but in general, increases are expected in the tropics and at high latitudes, while decreases are expected in the subtropics and at mid latitudes (IPCC 2014). However, in some locations, a change in the amount of precipitation may be less important than a shift in the timing of major rainfall events on an annual scale. In the Pacific Northwest USA, for instance, precipitation is predicted to decrease by 30% in the summer, and increase in every other season for a total annual increase of 3–5% (Mote and Salathé 2010), worsening both winter floods and summer droughts. Some of that precipitation will also shift from snow to rain, further worsening those seasonal extremes in streamflow (Lee et al. 2016). And despite an annual increase in precipitation, the lower summer flows may increase salt marsh salinity during the growing season, when plants are likely more sensitive to salinity changes.

Increased freshwater inputs are generally expected to be beneficial to marshes by both lowering salinity and bringing additional sediment and nutrients. Recent Mississippi River freshwater diversions have provided an interesting natural experiment in increased freshwater inputs, with the intent to mimic past flood events and increase marsh resilience, but also inadvertently providing a glimpse of expected future conditions due to climate change. A greenhouse study tested the effect of salinity and nutrient conditions associated with the Caernarvon Mississippi River diversion on *S. patens* plants, to understand the effect freshwater reintroduction to a brackish marsh (DeLaune et al. 2005). Both aboveground biomass and nutrient uptake were enhanced in the fertilized treatments (fertilized with nitrogen (N), phosphorus (P), and potassium (K)), particularly at low (0 ppt) salinity levels compared to ambient (8 ppt) levels, demonstrating that freshwater reintroduction is beneficial to *S. patens* growth, due to the combination of lowered salinity and increased nutrient supply. Interestingly, lowered salinity levels alone had no significant effect on plant growth, but increased nutrients alone did increase plant growth, while the combination of the two provided the greatest benefit. Marsh plant production is typically limited by nitrogen (N) availability (Mendelssohn and Morris 2000), thus marsh plants typically respond to N enrichment with an increase in aboveground productivity (e.g., DeLaune et al. 2005, Anisfeld and Hill 2011, Davis et al. 2017), although there is some variability, with sometimes an observed decrease or no change (e.g., Darby and Turner 2008, Deegan et al. 2012, Hanson et al. 2016, Johnson et al. 2016). The variability in response is likely due to interactions between N loading and other factors such as species composition, salinity, flooding, ambient N concentration, and other nutrients (Ratliff et al. 2015). The belowground response, however, is even more variable and does not lend itself to general-izations (Fig. 13.9; Morris et al. 2013, Ratliff et al. 2015).

Although N is typically considered the major limiting nutrient for salt marsh growth, P may also play a role in limiting plant growth (DeLaune and Pezeshki 1994, Lissner et al. 2003). For example, the addition of P to a South Carolina salt marsh did not produce any growth response, but the addition of N and P together resulted in a greater growth response than that from N alone, suggesting that P may become limiting when N inputs reach a threshold (Morris 1988). In addition, P may be the primary limiting factor for soil bacterial activity, affecting rates of soil respiration and carbon turnover (Sundareshwar et al. 2003).

Greater freshwater inputs are also expected to deliver more sediment to tidal marshes, which is beneficial in raising marsh elevations, particularly in sediment-limited marshes vulnerable to SLR. These sediment inputs can be crucial to marsh survival, because salt marshes cannot persist over time without substantial mineral inputs (Giosan et al. 2014), both to maintain relative elevation and maintain soil strength. Using the Mississippi River diversion example again, DeLaune et al. (2003) examined accretion rates in a *S. patens* marsh at varying distances from the Caernarvon

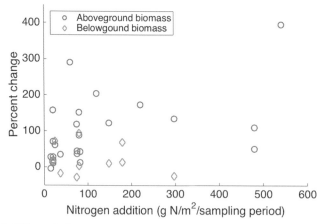

Figure 13.9 Effect of nitrogen addition on aboveground and belowground marsh plant biomass. Data shown are from literature review sources. (Ratliff et al. 2015)

diversion, finding mineral accumulation rates an order of magnitude higher at sites closest to the freshwater source and thus more resilient to SLR compared to the farthest sites. Another Mississippi River delta study by Day et al. (2011) demonstrated that a salt marsh receiving a larger freshwater input from the Mississippi River had soil strength about 10 times greater than a marsh without this freshwater input. In addition, mineral sediments provide additional nutrients and metals that can be beneficial such as iron, which neutralizes sulfide, a compound toxic to marsh plants.

It should be noted that although this chapter has focused on the effects of climate change alone, human modification of freshwater, nutrient, and sediment inputs contribute substantially, if not more, to the changes in these conditions. Dams, diversions, and levees modify freshwater flow and prevent sediments from reaching the salt marsh. An estimated 20% of the global sediment supply is held back by dams and reservoirs and prevented from reaching the coast (Syvitski et al. 2005). Of the sediments that do reach the coast, some of that supply bypasses the marsh as it is directed straight to the open ocean by diversions and levees. Agricultural fertilizers and municipal waste also modify nutrient inputs, potentially leading to eutrophication. Although nutrient inputs may seem generally beneficial to marsh plant growth, an excessive nutrient supply can be detrimental (Kirwan and Megonigal 2013), possibly reducing soil strength (Turner 2011, Deegan et al. 2012) and increasing macroalgal growth that outcompetes marsh macrophytes (Wasson et al. 2017).

13.5.1 Summary

Increased freshwater input, and the increased sediment and nutrient loading that typically accompany it, are generally expected to enhance plant growth and marsh survivability in marshes cut off from their historical freshwater supply. Sediments directly contribute to marsh accretion, while freshwater and nutrients may indirectly contribute, through the increase in above- and belowground production. Belowground growth raises marsh elevation, while aboveground growth can provide more organic litter, and more effectively trap suspended sediments. However, in marshes with higher ambient nutrient levels, increased nutrient inputs can lead to harmful eutrophication. The nutrient loading effect on salt marsh resilience thus remains a controversial topic. Anthropogenic activities related to watershed management are important to consider with their ability to increase or decrease these factors independently of climate change.

13.6 Net Effect of MultiFactored Climate Change

Considering multiple factors in studies of climate change effects on marsh plants can be difficult, but researchers are increasingly considering them, in both experimental and modeling studies. For example, Kirwan and Mudd (2012) concluded from a modeling study that the temperature effect on organic matter accumulation in a *Spartina* marsh was more subtle than SLR, although the temperature effect was greater at higher sea levels. The N enrichment effect reportedly increases with added CO_2 and SLR (Langley et al. 2013), while the CO_2 enrichment effect decreases with added N over the long term due to community shifts to more N-responsive C_4 species that happen to be less responsive to CO_2 (Langley and Megonigal 2010). This leads N addition to have a much more substantial impact on the plant community than CO_2 (Reich 2009, White et al. 2012), although SLR and salinity stress may enhance the CO_2 effect (Langley et al. 2009).

In consideration of potential interactions, some claim that the salt marsh response to multifactored change is generally less than the sum of its parts, in other words the magnitude of the response to each driver tends to decrease with an increasing number of drivers (Leuzinger et al. 2011, Langley and Hungate 2014). However, others hypothesize that plants may be able to tolerate moderate changes in single stressors, but their adaptability to multiple combined stressors may be much lower, heightening the response (Mendelssohn and Morris 2000). For instance, salinity and temperature are two factors expected to increase decomposition rates, and these two factors acting simultaneously may raise decomposition rates to a greater degree than the sum of the two alone (Wu et al. 2017). Schile et al. (2017) reported that salinity had an amplifying effect on the SLR-induced biomass reduction. Cherry et al. (2009) reported that CO_2 reduced the negative effects of both SLR and salinity on plant productivity. Perhaps the cumulative effect depends on whether the factors considered are stressors or aids to growth. Some stressors appear to be synergistic, while beneficial factors may ameliorate some amount of that stress. Interactions are also likely to vary geographically due to differences in physical conditions (Pennings et al. 2005), therefore more research may help lead us to a better understanding of the variability in these interactions, and hence the full range of possible responses.

References

Alber, M., Swenson, E. M., Adamowicz, S. C. and Mendelssohn, I. A. 2008. Salt marsh dieback: an overview of recent events in the US. *Estuarine, Coastal and Shelf Science*, 80:1–11.

Anisfeld, S. C., and Hill, T. D. 2011. Fertilization effects on elevation change and below-ground carbon balance in a Long Island Sound tidal marsh. *Estuaries and Coasts*, doi:10.1007/s12237-011-9440-4.

Ball, A. S., and Drake, B. G. 1998. Stimulation of soil respiration by carbon dioxide enrichment of marsh vegetation. *Soil Biology and Biochemistry*, 30:1203–1205.

Baustian, M. M., Stagg, C. L., Perry, C. L., Moss, L. C., Carruthers, T. J. B., and Allison, M. 2017. Relationships between salinity and short-term soil carbon accumulations rates from marsh types across a landscape in the Mississippi River Delta. *Wetlands*, 37:313–324.

Bowes, G. 1993. Facing the inevitable: plants and increasing atmospheric CO_2. *Annual Review of Plant Physiology & Plant Molecular Biology*, 44:309–332.

Broome, S. W., Mendelssohn, I. A., and McKee, K. L. 1995. Relative growth of *Spartina patens* (Ait.) Muhl. and *Scirpus olneyi* Gray occurring in a mixed stand as affected by salinity and flooding depth. *Wetlands*, 15(1):20–30.

Brown, C. E., Pezeshki, S. R., and DeLaune, R. D. 2006. The effects of salinity and soil drying on nutrient uptake and growth of *Spartina alterniflora* in a simulated tidal system. *Environmental and Experimental Botany*, 58:140–148.

Carey, J. C., Moran, S. B., Kelly, R. P., Kolker, A. S., and Fulweiler, R. W. 2017. The declining role of organic matter in New England salt marshes. *Estuaries and Coasts*, 40:626–639.

Carniello, L., Defina, A. and D'Alpaos, L. 2009. Morphological evolution of the Venice lagoon: evidence from the past and trend for the future. *Journal of Geophysical Research*, 114:F04002.

Centritto, M. 2002. The effects of elevated [CO_2] and water availability on growth and physiology of peach (*Prunus persica*) plants. *Plant Biosystems*, 136:177–188.

Centritto, M., Lucas, M. E., and Jarvis, P. G. 2002. Gas exchange, biomass, whole-plant water-use efficiency and water uptake of peach (*Prunus persica*) seedlings in response to elevated carbon dioxide concentration and water availability. *Tree Physiology*, 22:699–706.

Charles, H., and Dukes, J. S. 2009. Effects of warming and altered precipitation on plant and nutrient dynamics of a New England salt marsh. *Ecological Applications*, 19(7):1758–1773.

Cherry, J. A., McKee, K. L., and Grace, M. B. 2009. Elevated CO_2 enhances biological contributions to elevation change in coastal wetlands by offsetting stressors associated with sea-level rise. *Journal of Ecology*, 97:67–77.

Cleland, E. E., Chuine, I., Menzel, A., Mooney, H. A., and Schwartz, M. D. 2007. Shifting plant phenology in response to global change. *Trends in Ecology and Evolution*, 22:357–365.

Coldren, G. A., Barreto, C. R., Wykoff, D. D., Morrissey, E. M., Langley, J. A., Feller, I. C., and Chapman, S. K. 2016. Chronic warming stimulates growth of marsh grasses more than mangroves in a coastal wetland ecotone. *Ecology*, 97(11):3167–3175.

Craft, C. 2007. Freshwater input structures soil properties, vertical accretion, and nutrient accumulation of Georgia and U.S. tidal marshes. *Limnology and Oceanography*, 52 (3):1220–1230.

Criddle, R. S., L. D. Hansen, R. W. Breidenbach M. R. Ward, and R. C. Huffaker. 1989. Effects of NaCl on metabolic heat evolution rates by barley roots. *Plant Physiology* 90(1):53–58.

Cronk, J. K., and M. S. Fennessy. 2001. *Wetland Plants: Biology and Ecology*. CRC Press, Boca Raton, Florida.

Crosby, S. C., Angermeyer, A., Adler, J. M., Bertness, M. D., Deegan, L. A., Sibinga, N. and Leslie, H. M. 2017. *Spartina alterniflora* biomass allocation and temperature: implications for salt marsh persistence with sea-level rise. *Estuaries and Coasts*, 40:213–223.

Curtis, P.S., Balduman, L. M., Drake, B. G., and Whigman, D. F. 1990. Elevated atmospheric CO_2 effects on belowground processes in C_3 and C_4 estuarine marsh communities. *Ecology*, 71(5):2001–2006.

Darby, F. A., and Turner, R. E. 2008. Effects of eutrophication on salt marsh root and rhizome biomass accumulation. *Marine Ecology Progress Series*, 363:63–70.

Davidson, E. A., and Janssens, I. A. 2006. Temperature sensitivity of soil carbon decomposition and feedbacks to climate change. *Nature*, 440:165–173.

Davis, J., Currin, C., and Morris, J. T. 2017. Impacts of fertilization and tidal inundation on elevation change in microtidal, low relief salt marshes. *Estuaries and Coasts*, 40:1677–1687.

Day, J. W., Christian, R. R., Boesch, D. M., Yáñez-Arancibia, A., Morris, J., Twilley, R. R., Naylor, L. et al. 2008. Consequences of climate change on the ecogeomorphology of coastal wetlands. *Estuaries and Coasts*, 31:477–491.

Day, J., Kemp, P., Reed, D., Cahoon, D., Boumans, R., Suhayda, J., and Gambrell, R. 2011. Vegetation death and rapid loss of surface elevation in two contrasting

Mississippi delta salt marshes: the role of sedimentation, autocompaction and sea-level rise. *Ecological Engineering*, 37:229–240.

Deegan, L., Johnson, D. S., Warren, R. S., Peterson, B. J., Fleeger, J. W., Fagherazzi, S., and Wollheim, W. M. 2012. Coastal eutrophication as a driver of salt marsh loss. *Nature*, 490:388–392.

DeLaune, R. D., Jugsujinda, A., Peterson, G., and Patrick, W. 2003. Impact of Mississippi River freshwater reintroduction on enhancing marsh accretionary processes in a Louisiana estuary. *Estuarine, Coastal and Shelf Science*, 58:653–662.

DeLaune, R. D., and Pezeshki, S.R. 1994. The influence of subsidence and saltwater intrusion on coastal marsh stability: Louisiana Gulf Coast, U.S.A. *Journal of Coastal Research*, 12:77–89.

DeLaune, R. D., Pezeshki, S., and Jugsujinda, A. 2005. Impact of Mississippi River freshwater reintroduction on *Spartina patens* marshes: responses to nutrient input and lowering of salinity. *Wetlands*, 25:155–161.

De Leeuw, J., Olff, H., and Bakker, J. P. 1990. Year-to-year variation in peak above-ground biomass of six salt-marsh angiosperm communities as related to rainfall deficit and inundation frequency. *Aquatic Botany*, 36:139–151.

Donnelly, J. P., and Bertness, M. D. 2001. Rapid shoreward encroachment of salt marsh cordgrass in response to accelerated sea-level rise. *Proceedings of the National Academy of Sciences of the USA* 98:14218–14223.

Drake, B. G. 1992. A field-study of the effects of elevated CO_2 on ecosystem processes in a Chesapeake Bay wetland. *Australian Journal of Botany*, 40:579–595.

Dunn, R., Thomas, S. M., Keys, A. J., and Long, S. P. 1987. A comparison of the growth of C_4 grass *Spartina anglica* with the C_3 grass *Lolium perenne* at different temperatures. *Journal of Experimental Botany*, 38(188):433–441.

Erickson, J. E., Megonigal, J. P., Peresta, G., and Drake, B. G. 2007. Salinity and sea level mediate elevated CO_2 effects on C_3-C_4 plant interactions and tissue nitrogen in a Chesapeake Bay tidal wetland. *Global Change Biology*, 13:202–215.

Fagherazzi, S., Carniello, L. D'Alpaos, L., and Defina, A. 2006. Critical bifurcation of shallow microtidal landforms in tidal flats and salt marshes. *Proceedings of the National Academy of Sciences of the USA*, 103:8337–8341.

Farquhar, G. D., von Caemmerer, S., and Berry, J. A. 1980. A biochemical model of photosynthetic CO_2 assimilation in leaves of C_3 species. *Planta*, 149:78–90.

French, J. 2006. Tidal marsh sedimentation and resilience to environmental change: exploratory modelling of tidal, sea-level and sediment supply forcing in predomin-antly allochthonous systems. *Marine Geology*, 235:119–136.

Gedan, K. B., Altieri, A. H., and Bertness, M. D. 2011. Uncertain future of New England salt marshes. *Marine Ecology Progress Series*, 434:229–237.

Gedan, K. B., and Bertness, M.D. 2009. Experimental warming causes rapid loss of plant diversity in New England salt marshes. *Ecology Letters*, 12:842–848.

Giosan, L., Syvitski, J., Constantinescu, S. D., and Day, J. 2014. Protect the world's deltas. *Nature*, 516:31–33.

Giurgevich, J. R, and Dunn, E.L. 1979. Seasonal patterns of CO_2 and water vapor exchange of the tall and short height forms of *Spartina alterniflora* Loisel in a Georgia salt marsh. *Oecologia*, 43:139–156.

Gray, A. J., and Mogg, R. J. 2001. Climate impacts on pioneer saltmarsh plants. *Climate Research*, 18:105–112.

Greenway, H., and Munns, R. 1980. Mechanisms of salt tolerance in nonhalophytes. *Annual Review of Plant Physiology*, 31(1):149–190.

Grime, J. P. 1988. The C-S-R model of primary plant strategies–origins, implications and tests. pp. 371–393. In: *Plant Evolutionary Biology*, S. K. Jain (ed.), Chapman & Hall, London.

Hanson, A., Johnson, R., Wigand, C., Oczkowski, A., Davey, E., and Markham, E. 2016. Responses of *Spartina alterniflora* to multiple stressors: changing precipitation patterns, accelerated sea level rise, and nutrient enrichment. *Estuaries and Coasts*, 39:1376–1385.

Henry, M. K., and Twilley, R. R. 2013. Soil development in a coastal Louisiana wetland during a climate-induced vegetation shift from salt marsh to mangrove. *Journal of Coastal Research*, 29:1273–1283.

Herbert, E. R., Boon, P., Burgin, A. J., Neubauer, S. C., Franklin, R. B., Ardón, M., Hopfensperger, K. N., Lamers, L. P. M., and Gell, P. 2015. A global perspective on wetland salinization: ecological consequences of a growing threat to freshwater wetlands. *Ecosphere*, 6(10):206.

Howard, R. J., and Mendelssohn, I. A. 1999. Salinity as a constraint on growth of oligohaline marsh macrophytes. I. Species variation in stress tolerance. *American Journal of Botany*, 86(6):785–794.

Hughes, A. L. H., Wilson, A. M., and Morris, J. T. 2012. Hydrologic variability in a salt marsh: assessing the links between drought and acute marsh dieback. *Estuarine, Coastal and Shelf Science*, 111:95–106.

Idso, K. E., and Idso, S. B. 1994. Plant responses to atmospheric CO_2 enrichment in the face of environmental constraints: a review of the past 10 years' research. *Agricultural and Forest Meteorology*, 69:153–203.

IPCC, 2014: *Climate Change 2014: Synthesis Report. Contribution of Working Groups I, II and III to the Fifth Assessment Report of the Intergovernmental Panel on Climate Change* [Core Writing Team, R.K. Pachauri and L.A. Meyer (eds.)]. IPCC, Geneva, Switzerland.

Jacob, J., Greitner, C. and Drake, B. G. 1995. Acclimation of photosynthesis in relation to Rubisco and non-structural carbohydrate contents and *in situ* carboxylase activity in Scirpus olneyi grown at elevated CO_2 in the field. *Plant, Cell and Environment*, 18:875–884.

Janousek, C. N., Buffington, K. J., Thorne, K. M., Guntenspergen, G. R., Takekawa, J. Y., and Dugger, B. D. 2016. Potential effects of sea-level rise on plant productivity: species-specific responses in northeast Pacific tidal marshes. *Marine Ecology Progress Series*, 548:111–125.

Johnson, D. S., Warren, R. S., Deegan, L. A., and Mozdzer, T. J. 2016. Saltmarsh plant responses to eutrophication. *Ecological Applications*, 26(8):2649–2661.

Kearney, M. S., Grace, R. E., and Stevenson, J. C. 1988. Marsh loss in Nanticoke Estuary, Chesapeake Bay. *Geographical Review*, 78:205–220.

Keddy, P. A. 1990. Competitive hierarchies and centrifugal organization in plant communities. pp. 265–290. In: *Perspectives on Plant Competition*, Grace, J.B. and D. Tilman (eds.), Academic Press, Inc., San Diego, USA.

Keddy, P.A. 2011. *Wetland Ecology: Principles and Conservation*. Vol. 2. Cambridge University Press, Cambridge, U.K.

Kirwan, M. L., and Guntenspergen, G. R. 2013. Feedbacks between inundation, root production, and shoot growth in a rapidly submerging brackish marsh. *Journal of Ecology*, 100:764–770.

Kirwan, M. L., Guntenspergen, G. R., D'Alpaos, A., Morris, J. T., Mudd, S. M., and Temmerman, S. 2010. Limits on the adaptability of coastal marshes to rising sea level. *Geophysical Research Letters*, 37: L23401.

Kirwan, M. L., Guntenspergen, G. R., and Langley, J. A. 2014. Temperature sensitivity of organic-matter decay in tidal marshes. *Biogeosciences*, 11:4801–4808.

Kirwan, M. L., Guntenspergen, G. R., and Morris, J. T. 2009. Latitudinal trends in *Spartina alterniflora* productivity and the response of coastal marshes to global change. *Global Change Biology*, 15:1982–1989.

Kirwan, M. L., Langley, J. A., Guntenspergen, G. R. and Megonigal, J. P. 2013. The impact of sea-level rise on organic matter decay rates in Chesapeake Bay brackish tidal marshes. *Biogeosciences*, 10:1869–1876.

Kirwan, M. L., and Megonigal, J. P. 2013. Tidal wetland stability in the face of human impacts and sea-level rise. *Nature*, 504:53–60.

Kirwan, M. L., and Mudd, S. M. 2012. Response of salt-marsh carbon accumulation to climate change. *Nature*, 489:550–553.

Kirwan, M. L., Temmerman, S., Skeehan, E. E., Guntenspergen, G. R., and Fagherazzi, S. 2016. Overestimation of marsh vulnerability to sea level rise. *Nature Climate Change*, 6:253–260.

Lamers, L., Govers, L., Janssen, I., Geurts,J. Van der Welle, M., Van Katwijk, M., Van der Heide, T., Roelofs, J., and Smolders, A. 2013. Sulfide as a soil phytotoxin–a review. *Frontiers in Plant Science*, 4:1–4.

Langley, J. A., and Hungate, B. A. 2014. Plant community feedbacks and long-term ecosystem responses to multi-factored global change. *AoB Plants*, 6:plu035. doi:10.1093/aobpla/plu035

Langley, J. A., McKee, K. L., Cahoon, D. R., Cherry, J. A., and Megonigal, J. P. 2009. Elevated CO_2 stimulates marsh elevation gain, counterbalancing sea-level rise. *Proceedings of the National Academy of Sciences of the USA*, 106(15):6182–6186.

Langley, J. A., and Megonigal, J. P. 2010. Ecosystem response to elevated CO_2 levels limited by nitrogen-induced plant species shift. *Nature*, 466:96–99.

Langley, J. A., Mozdzer, T. J., Shepard, K. A., Hagerty, S. B., and Megonigal, J. P. 2013. Tidal marsh plant responses to elevated CO_2, nitrogen fertilization, and sea level rise. *Global Change Biology*, 19: 1495–1503.

Lee, S. -Y., Hamlet, A. F., and Grossman, E. E. 2016. Impacts of climate change on regulated streamflow, hydrologic extremes, hydropower production, and sediment discharge in the Skagit River Basin. *Northwest Science*, 90(1):23–43.

Lenssen, G. M., Lamers, J., Stroetenga, M., and Rozema, J. 1993. Interactive effects of atmospheric CO_2 enrichment, salinity and flooding on growth of C_3 (*Elymus athericus*) and C_4 (*Spartina anglica*) salt marsh species. *Vegetatio*, 104/105:379–388.

Leuzinger, S., Luo, Y., Beier, C., Dieleman, W., Vicca, S., and Korner, C. 2011. Do global change experiments overestimate impacts on terrestrial ecosystems? *Trends in Ecology & Evolution*, 26:236–241.

Linthurst, R. A., and Seneca, E. D. 1981. Aeration, nitrogen and salinity as determinants of *Spartina alterniflora* Loisel. Growth response. *Estuaries*, 4(1):53–63.

Lissner, J., Mendelssohn, I. A., Lorenzen, B., Brix, H., Mckee, K. L., and Miao, S. 2003. Interactive effects of redox intensity and phosphate availability on growth and nutrient relations of *Cladium jamaicense* (Cyperaceae), *American Journal of Botany* 90: 736–748.

Maricle, B. R., Cobos, D. R., and Campbell, C. S. 2007. Biophysical and morphological leaf adaptations to drought and salinity in salt marsh grasses. *Environmental and Experimental Botany*, 60:458–467.

McKee, K. L., Mendelssohn, I. A., and Materne, M. D. 2004. Acute salt marsh dieback in the Mississippi River deltaic plain: a drought-induced phenomenon? *Global Ecology and Biogeography*, 13:65–73.

McKee, K., Rogers, K., and Saintilan, N. 2012. Response of salt marsh and mangrove wetlands to changes in atmospheric CO_2, climate, and sea level. *Global Change Ecology and Wetlands*, 1:63–96.

Megonigal, J. P., Hines, M. E., and Visscher, P. T. 2004. Anaerobic metabolism: linkages to trace gases and aerobic processes. pp. 317–424. In: *Biogeochemistry*, W.H. Schlesinger (ed.), Elsevier-Pergamon, Oxford, UK.

Mendelssohn, I. A., and McKee, K. L. 1992. Indicators of environmental stress in wetland plants. pp. 603–624. In: *Ecological Indicators*, McKenzie, D. H., D. E. Hyatt, and V. J. MacDonald (eds.), Elsevier Applied Science, New York, NY, USA.

Mendelssohn, I. A., McKee, K. L., Hester, M. W., Lin, Q., McGinnis, T., and Willis, J. 2006. *Brown marsh task II.1: integrative approach to understanding the causes of salt marsh dieback – determination of salt marsh species tolerance limits to potential environmental stressors*. Report submitted to the Louisiana Department of Natural Resources, Baton Rouge, LA.

Mendelssohn, I. A., and Morris, J. T. 2000. Ecophysiological controls on the growth of Spartina alterniflora. pp. 59–80. In: *Concepts and Controversies in Tidal Marsh Ecology*. N. P. Weinstein,, and D. A. Kreeger (eds.). Kluwer Academic Publishers, New York.

Mendelssohn, I. A., Sorrell, B. K., Brix, H., Schierup, H. H., Lorenzen, B., and Maltby, E. 1999. Controls on soil cellulose decomposition along a salinity gradient in a Phragmites australis wetland in Denmark. *Aquatic Botany*, 64:381–398.

Miller, W. D., Neubauer, S. C., and Anderson, I. C. 2001. Effects of sea level induced disturbance on high salt marsh metabolism. *Estuaries*, 24:357–367.

Montagna, P. A., and Ruber, E. 1980. Decomposition of *Spartina alterniflora* in different seasons and habitats of a northern Massachusetts salt marsh, and a comparison with other Atlantic regions. *Estuaries*, 3:61–64.

Morris, J. T. 1988. Pathways and controls of the carbon cycle in salt marshes. pp. 497–510. In: *The Ecology and Management of Wetlands, Volume 1, Ecology of Wetlands*, D. D. Hook, (ed.), Croom Helm Ltd., Beckenham, UK.

Morris, J. T., Shaffer, G. P., and Nyman, J. A. 2013. Brinson Review: perspectives on the influence of nutrients on the sustainability of coastal wetlands. *Wetlands*, 33:975–988.

Morris, J. T., Sundareshwar, P. V., Nietch, C. T., Kjerfve, B., and Cahoon, D. R. 2002. Responses of coastal wetlands to rising sea level. *Ecology*, 83(10):2869–2877.

Morrissey, E. M., Gillespie, J. L., Morina, J. C., and Franklin, R. B. 2014. Salinity affects microbial activity and soil organic matter content in tidal wetlands. *Global Change Biology*, 20:1351–1362.

Mote, P. W., and Salathé Jr, E. P. 2010. Future climate in the Pacific Northwest. *Climatic Change*, 102:29–50.

Munns, R. 2002. Comparative physiology of salt and water stress. *Plant, Cell and Environment*, 25:239–250.

Naidoo, G., McKee, K. L., and Mendelssohn, I. A. 1992. Anatomical and metabolic responses to waterlogging and salinity in *Spartina alterniflora* and *S. patens* (Poaceae). *American Journal of Botany*, 79(7):765–770.

Nyman, J. A., and DeLaune, R. D. 1991. CO_2 emission and soil Eh responses to different hydrological regimes in fresh, brackish and saline marsh soils. *Limnology and Oceanography*, 36:1406–1414.

Nyman, J. A., DeLaune, R. D., Roberts, H. H.,and Patrick, W. H. 1993. Relationship between vegetation and soil formation in a rapidly submerging coastal marsh. *Marine Ecology Progress Series*, 96:269–279.

Odum, W. E. 1988. Comparative ecology of tidal freshwater and salt marshes. *Annual Review of Ecology and Systematics*, 19:147–176.

Osland, M. J., Enwright, N., Day, R. H., and Doyle, T. W. 2013. Winter climate change and coastal wetland foundation species: salt marshes vs. mangrove forests in the southeastern United States. *Global Change Biology*, 19:1482–1494.

Pennings, S. C., Grant, M. -B., and Bertness, M. D. 2005. Plant zonation in low-latitude salt marshes: disentangling the roles of flooding, salinity and competition. *Journal of Ecology*, 93:159–167.

Poljakoff-Mayber, A. 1988. Ecological-physiological studies on the responses of higher plants to salinity and drought. *Science Review of Arid Zone Research*, 6:163–183.

Polley, W. H., Johnson, H. B.,and Derner, J. D. 2003. Increasing CO_2 from subambient to superambient concentrations alters species composition and increases above-ground biomass in a C_3/C_4 grassland. *New Phytologist*, 160:319–327.

Pozo, J., and Colino, R. 1992. Decomposition processes of *Spartina maritime* in a salt marsh of the Basque Country. *Hydrobiologia*, 231:165–175.

Rahmstorf, A. 2007. A semi-empirical approach to projecting future sea-level rise. *Science*, 315:368–370.

Rasse, D. P., Peresta, G., and Drake, B. G. 2005. Seventeen years of elevated CO_2 exposure in a Chesapeake Bay wetland: sustained but contrasting responses of plant growth and CO_2 uptake. *Global Change Biology*, 11:369–377.

Ratliff, K. M., Braswell, A. E., and Marani, M. 2015. Spatial response of coastal marshes to increased atmospheric CO_2. *Proceedings of the National Academy of Sciences of the USA*, 112(51):15580–15584.

Reed, D. J. 1995. The response of coastal marshes to sea-level rise: survival or submergence? *Earth Surface Processes*, 20:39–48.

Reich, P. B. 2009. Elevated CO_2 reduces losses of plant diversity caused by nitrogen deposition. *Science*, 326:1399–1402.

Rozema, J., Dorel, F., Janissen, R., Lenssen, G. M., Broekman, R. A., Ar, W. J., and Drake, B. G. 1991. Effect of elevated CO_2 on growth, photosynthesis and water relations of salt marsh grass species. *Aquatic Botany*, 39:45–55.

Saintilan, N., Wilson, N., Rogers, K., Rajkaran, A., and Krauss, W. K. 2014. Mangrove expansion and salt marsh decline at mangrove poleward limits. *Global Change Biology*, 20:147–157.

Schile, L. M., Callaway, J. C., Suding, K. N., and Kelly, N. M. 2017. Can community structure track sea-level rise? Stress and competitive controls in tidal wetlands. *Ecology and Evolution*, 7:1276–1285.

Sharpe, J. P., and Baldwin, H. A. 2012. Tidal marsh plant community response to sea-level rise: a mesocosm study. *Aquatic Botany*, 101:34–40.

Shaver, G. R., Canadell, J., Chapin, F. S., Gurevitch, J., Harte, J., Henry, G., Ineson, P., et al. 2000. Global warming and terrestrial ecosystems: a conceptual framework for analysis. *Bioscience*, 50:871–882.

Shea, M. L. 1977. *Photosynthesis and photorespiration in relation to phenotypic forms of Spartina alterniflora*. PhD thesis, Yale University, New Haven, Connecticut.

Short, F. T., Kosten, S., Morgan, P. A., Malone, S., and Moore, G. E. 2016. Impacts of climate change on submerged and emergent wetland plants. *Aquatic Botany*, 135:3–17.

Smith, S. M. 2009. Multi-decadal changes in salt marshes of Cape Cod, MA: photographic analyses of vegetation loss, species shifts and geomorphic change. *Northeastern Naturalist*, 16:183–208.

Stagg, C. L., Schoolmaster Jr., D. R., Piazza, S. C., Snedden, G., Steyer, G. D., Fischenich, C. J., and McComas, R. W. 2017. A landscape-scale assessment of above- and belowground primary production in coastal wetlands: implications for climte change-induced community shifts. *Estuaries and Coasts*, 40:856–879.

Stevenson, J. C., Kearney, M. S., and Pendleton, E. C. 1985. Sedimentation and erosion in a Chesapeake Bay brackish marsh system. *Marine Geology*, 67:213–235.

Stralberg, D., Brennan, M., Callaway, J. C., Wood, J. K., Schile, L. M., Jongsomjit, D., Kelly, M., Parker, V. T., and Crooks, S. 2011. *Evaluating tidal marsh sustainability in the face of sea-level rise: a hybrid modeling approach applied to San Francisco Bay. PLoS ONE*, 6(11): e27388.

Sundareshwar, P. V., Morris, J. T., Koepfler, E. K., and Fornwalt, B. 2003. Phosphorus limitation of coastal ecosystem processes. *Science*, 299:563–565.

Syvitski, J. P. M., Vörösmarty, C. J., Kettner, A. J., and Green, P. 2005. Impact of humans on the flux of terrestrial sediment to the global coastal ocean. *Science*, 308:376–380.

Teal, J. M., and Howes, B. L. 1996. Interannual variability of a salt-marsh ecosystem. *Limnology and Oceanography*, 41:802–809.

Tobias, V. D., and Nyman, J. A. 2017. Leaf tissue indicators of flooding stress in the above- and belowground biomass of *Spartina patens*. *Journal of Coastal Research*, 33(2):309–320.

Turner, R. E., Swenson, E. M., and Milan, C. S. 2000. Organic and inorganic contributions to vertical accretion in salt marsh sediments. pp. 583–595. In: *Concepts and Controversies in Tidal Marsh Ecology*, Weinstein, M.P., and D.A. Kreeger (eds.), Springer, Dordrecht, The Netherlands,.

Turner, R. E., Swenson, E. M., Milan, C. S., Lee, J. M. and Oswald, T. A. 2004. Below-ground biomass in healthy and impaired salt marshes. *Ecological Research*, 19:29–35.

Turner, R. E. 2011. Beneath the salt marsh canopy: loss of soil strength with increasing nutrient loads. *Estuaries and Coasts*, 34:1084–1093.

Urban, O. 2003. Physiological impacts of elevated CO_2 concentration ranging from molecular to whole plant responses. *Photosynthetica*, 41:9–20.

Valiela, I., Teal, J. M., Allen, S. D., Van Etten, R., Goehringer, D., and Volkmann, S. 1985. Decomposition in salt marsh ecosystems: the phases and major factors affecting disappearance of above-ground organic matter. *Journal of Experimental Marine Biology and Ecology*, 89:29–54.

van Dobben, H. F., and Slim, P. A. 2011. Past and future plant diversity of a coastal wetland driven by soil subsidence and climate change. *Climatic Change*, 110: 597–618.

Visser, J. M., Duke-Sylvester, S. M., Carter, J., and Broussard III, W. P. 2013. A computer model to forecast wetland vegetation changes resulting from restoration and protection in coastal Louisiana. *Journal of Coastal Research*, 67(4):51–59.

Wasson, K., Endris, R. C., Perry, D. C., Woolfolk, A., Beheshti, K., Rodriguez, M., et al. 2017. Eutrophication decreases salt marsh resilience through proliferation of algal mats. *Biological Conservation* 212:1–11.

Watson, E. B., Wigand, C., Davey, E. W., Andrews, H. M., Bishop, J., and Raposa, K. B. 2017. Wetland loss patterns and inundation-productivity relationships prognosticate widespread salt marsh loss for southern New England. *Estuaries and Coasts*, 40:662–681.

Wells, J. T., and Coleman, J. M. 1987. Wetland loss and the subdelta life cycle. *Estuarine, Coastal and Shelf Science*, 25:111–125.

White, K. P., Langley, J. A., Cahoon, D. R., and Megonigal, J. P. 2012. C_3 and C_4 biomass allocation responses to elevated CO_2 and nitrogen: contrasting resource capture strategies. *Estuaries and Coasts*, 35:1028–1035.

Williams, K., Pinzon, Z. S., Stumpf, R. P., and Raabe, E. A. 1999. *Sea-level rise and coastal forests on the Gulf of Mexico*. Open-file report 99-441, United States Geological Survey, St. Petersburg, FL.

Willis, J. M., and Hester, M. W. 2004. Interactive effects of salinity, flooding, and soil type on Panicum hemitomon. *Wetlands*, 24(1):43–50.

Wolf, A. A., Drake, B. G., Erickson, J. E., and Megonigal, J. P. 2007. An oxygen-mediated positive feedback between elevated carbon dioxide and soil organic matter decomposition in a simulated anaerobic wetland. *Global Change Biology*, 13:2036–2044.

Wolters, M., Garbutt, A., and Bakker, J. P. 2005. Salt-marsh restoration: evaluating the success of de-embankments in north-west Europe. *Biological Conservation*, 123:249–268.

Woodrow, I. E., and Berry, J. A. 1988. Enzymatic regulation of photosynthetic CO_2 fixation in C_3 plants. *Annual Review of Plant Physiology and Plant Molecular Biology*, 39:533–594.

Woodward, F. I., Thompson, G. B., and McKee, I. F. 1991. The effects of elevated concentrations of carbon dioxide on individual plants, populations, communities, and ecosystems. *Annals of Botany*, 67:23–38.

Wu, W., Huang, H., Biber, P., and Bethel, M. 2017. Litter decomposition of *Spartina alterniflora* and *Juncus roemerianus*: implications of climate change in salt marshes. *Journal of Coastal Research*, 33(2):372–384.

Yeo, A. 1999. Predicting the interaction between the effects of salinity and climate change on crop plants. *Scientia-Horticulturae-Amsterdam*, 78:159–174.

Zedler, J. B. 1983. Freshwater impacts in normally hypersaline marshes. *Estuaries*, 6:346–355.

Zedler, J. B., Williams, P., and Boland, J. 1986. Catastrophic events reveal the dynamic nature of salt-marsh vegetation in southern California. *Estuaries*, 9(1):75–80.

14

Impacts of Exotic and Native Species Invading Tidal Marshes

DAVID M. BURDICK, GREGG E. MOORE, AND KATHARYN E. BOYER

14.1 Introduction

14.1.1 Problems Invaders Can Cause in Tidal Marshes

As humans have spread across the globe, travel and trade have deliberately or inadvertently carried and released animals and plants as well as microbes into new geographies. With human populations concentrated along rivers and coasts, it is not surprising that many exotic species have been released in coastal areas and a few can survive and thrive, especially in habitats similar to those where they evolved. In tidal marshes, organisms experience some of the most extreme physical conditions on earth: temperatures from -20 to $40°C$, flooding twice a day but only a few times a month at higher elevations, sediments ranging from oxidized to severely reduced (Eh of $+700$ to -300 mV), soil salinity from hypersaline (40–90 ppt) to fresh depending on floodwater source and precipitation, and erosive forces from waves, currents, and ice at higher latitudes. Despite these harsh and variable conditions, there are many organisms adapted to tidal marshes, and new introductions and hybrids that can thrive given the opportunity.

Invading organisms can maintain small populations that appear to be benign or even beneficial, but those that attract our attention develop large populations and interfere with marsh structure or function and the dynamic balance of processes that maintain tidal marshes (see Chapters 7 and 8). In addition to exotic species introduced from other regions, there are native species of plants and animals that are poised to reproduce and spread rapidly in these marshes with changing physical or biological conditions (e.g., release from predation). We use some examples to show there is a continuum of invader impacts to tidal marshes, from slight to dramatic.

In New England, an introduced plant commonly found along shorelines where upper marsh edges become upland is *Elymus repens* (quackgrass). This plant does not seem to alter marsh ecology, though it is classified as a noxious weed in ten US states (this and other species, including photographs, can be found on USDA Plants: https://plants.sc.egov .usda.gov/java/; ITIS: www.itis.gov/; or Encyclopedia of Life: http://eol.org/. *E. repens* is mildly salt tolerant and is believed to have strong competitive abilities that may be aided by allelopathic exudates (Hierro and Callaway 2003).

Another well-known exotic in North America, *Lythrum salicaria* (Purple loosestrife) was introduced in the nineteenth century and has been found invading oligohaline (0.5–5 ppt) marshes as well as salt marshes where human interference in hydrology has reduced tidal exchange, allowing fresher species to thrive. Although invasions of this species are visually dramatic, research specifically looking for impacts to freshwater wetland habitats has reported mixed results (Anderson 1995, Blossey et al. 2001). Further investigations have showed reduced production by native plants (Farnsworth and Ellis 2001), but minor effects on uncommon plants (Denoth and Myers 2007, Flanagan et al. 2010) and mixed effects on specific bird species (Tavernia and Reed 2012), suggesting many of the effects are subtle and without dramatic impacts to habitat quality.

At the other extreme are species that infest and completely displace native marsh plants (Chambers et al. 1999) and can even cause loss of marshes (Silliman et al. 2005, Bertness et al. 2014). In the USA, a native variety of *Phragmites australis* was an uncommon plant of coastal marshes (Orson 1999, Saltonstall et al. 2004), but a Eurasian genotype has invaded US wetlands, particularly tidal marshes. It has been found to displace all marsh plants and form a monoculture (Chambers et al. 1999), with a variety of impacts to marsh structure and function that range from loss of native plants and faunal habitat to visual aesthetics and increased fire hazard (Burdick and Konisky 2003, Dibble and Meyerson 2013, Dibble et al. 2013). However, some changes can be viewed as beneficial (Guntenspergen et al. 2003, Kiviat 2013), especially when emergent marshes are faced with existential threats (e.g., sea level rise [SLR], Hauber et al. 2011). Marsh destruction can also result from overgrazing by animals, with geese, nutria, crabs and snails implicated. Other plant invaders create marsh by displacing mudflat habitat (*Spartina alterniflora* in China and US Pacific coast). This chapter will review some of the most notable exotic, native, and hybrid invaders of tidal marshes.

14.1.2 Marsh Systems and Invader Effects

Perennial plants occupy a central role in tidal marsh dynamics (Morris et al. 2002). They are able to slow waters, capture sediments, and infuse these sediments with 200–2,000 g dry weight/m^2 of root and rhizomes annually, allowing marshes to build (Nyman et al. 2006) at rates at or above historic SLR until an elevation approximating high tide is reached (Morris et al. this volume). Even though marshes come in many shapes and sizes and some may have different mechanisms for development and persistence (e.g., bluff toe marshes, Kelley et al. 1988), the general structure and continued development of most tidal marshes follows this pattern (Morris et al. 2002, Fagherazzi et al. 2012).

Invaders that initiate marsh development on mudflats create marsh where none existed and may displace important shellfish and shorebird habitat (Neira et al. 2007, Li et al. 2009). On the other hand, those that eat or otherwise destroy perennial plants disrupt the marsh feedback system (Burdick and Roman 2012) and marsh loss may occur. Thus, we can see that exotic invaders can result in minor, subtle changes but also dramatic changes to tidal marshes. Changes can range from a few to almost all conceivable functions, affect the balance of processes that maintain marshes (thereby resulting in marsh creation or loss), and interfere with marsh response to climate change. Of course, the loss of marsh can open the area to other habitats that have different functions and values (e.g., mudflat, reed swamp) and may even become novel habitat (Morse et al. 2014). Since resource managers were tasked to preserve and protect the coastal resources as they are (or were), and the plants and animals adapted to these systems have coevolved for millennia, we will focus on systemic changes brought about by invaders and consider them "bad" or unwanted by coastal managers.

14.2 Plant Invaders

14.2.1 *Phragmites australis*, Common Reed

On the coasts and throughout much of the USA, *Phragmites australis* is the quintessential invasive plant, with a cryptic origin and obscure differences with native varieties. Differences within the species were elucidated through comparisons of maternally inherited chloroplast genes (Saltonstall 2002) which set apart native East and Gulf Coast varieties (e. g., *Phragmites australis* subsp. *americanus*) that inhabit tidal marshes (Orson 1999) from a Eurasian invader (Saltonstall et al. 2004). The Eurasian variety likely arrived over 200 years ago with European settlers and, like other invasives, took decades to become established and begin to spread (Saltonstall 2002, Lelong et al. 2007). However, spread it has, from Canada to Florida and across the entire North American continent (Saltonstall 2002, Jodoin et al. 2008). *Phragmites* is a large perennial grass (5 m in height) that spreads by rhizomes. Much of the spread has been enabled by human activities within salt marshes (e. g., tidal restriction, filling, ditching) and other wetlands, and by connecting wetlands (e. g., transportation corridors; Bart et al. 2006, Lelong et al. 2007, Jodoin et al. 2008).

Many stands of invasive *Phragmites* are small and do not appear to impact marsh structure and function, but tidal marshes are dynamic environments and seasonal and annual differences in climate can allow rapid expansion and development of a dense monoculture (Minchinton 2002) that seems to affect almost every aspect of marsh structure and ecosystem service. Other stands have established in less stressful marshes and rapidly grew to monocultures (Chambers et al. 1999, Chambers et al. 2012), displacing native plants through a variety of mechanisms that begins with competition for light and nutrients (Konisky and Burdick 2004, Minchinton et al. 2006) but also include enhanced access to resources (Amsberry et al. 2000, Moore et al. 2012), wrack burial of competitors (Minchinton and Bertness 2003) and perhaps allelopathy (Rudrappa et al. 2007, Uddin et al. 2017).

Impacts on salt marshes from *Phragmites* invasion are numerous and often dramatic (reviewed by Burdick and Konisky 2003), but there are some benefits of *Phragmites* (Hershner and Havens 2008, Kiviat 2013). Alarming resource managers, *Phragmites* invasions result in significant large-scale displacement of native plants mostly at upper marsh edges where diversity is typically greatest, but also across entire marshes (see Box 14.1, Meyerson et al. 2009). Numerous large rhizomes along with roots tend to flatten and elevate the soil surface (Windham and Lathrop 1999, Rooth and Stevenson 2000), which eliminates microtopography and access of small forage fish to important feeding and spawning areas (Able et al. 2003, Balouskus and Targett 2012, Dibble and Meyerson 2012). Researchers have found that the abundance and diversity of invertebrates are reduced in *Phragmites* monocultures (Raichel et al. 2003). Bird use also appears to be reduced, especially for marsh dependent species (Benoit and Askins 1999, Trocki and Paton 2006). *Phragmites* marshes can reduce storm surge and flooding and may allow marshes to better adapt to SLR through rapid accretion (Rooth and Stevenson 2000), but also has been identified at slowing marsh migration over adjacent uplands (Smith 2013), a process critical for climate adaptation. The rhizomes and roots oxidize the typically anaerobic soils of salt marshes, detoxifying sulfides but also oxidizing stored organic matter in the peat, reducing carbon storage (Windham and Meyerson 2003), a more recent ecosystem service recognized of marshes. So, while it fixes more CO_2 due primarily to its greater size and areal biomass, it enhances oxidation of reduced carbon in the peat. Soil oxidation by *Phragmites* also affects nitrogen cycling (Windham and Meyerson 2003). Finally, *Phragmites*-dominated marshes block scenic vistas prized by residents and tourists.

Efforts to eradicate or control this plant have been reviewed recently (Hazelton et al. 2014). Although most efforts have not been successful, two approaches seem to work. If the invasion has occurred where the hydrology of a salt marsh has been altered (e.g., tidal restriction) restoring full tides to increase salinity can limit the invader to upland edges (Buchsbaum et al. 2006, Smith et al. 2009, Day et al. this volume), and negative impacts to fauna reversed (Dibble et al. 2013). Where natural conditions support rapid *Phragmites* expansion, herbicide use as part of a long-term landscape-scale management effort is recommended (Hazelton et al. 2014) as described in Box 14.1, the case study for the Great Marsh.

14.2.2 *Spartina* Species, Cordgrass

Spartina species have evolved to occupy a key role in salt marshes; they are typically the overwhelmingly dominant species and drivers of salt marsh development and maintenance. Often growing lower in elevation than other plants in salt marshes, *Spartina* species inhabit some of the most stressful environments and have a variety of adaptations to tidal flooding, seasonal freezing, variable salinity, and soil toxins, and are likely to develop more adaptations in the future through genetic variation and natural selection (Anderson and Treshow 1980). Infertile hybrids are well known from native sympatric species (*S. patens* x *S. pectinata* = *S. caespitosa*) as well as introduced crosses with native species (*S. alterniflora* x *S. maritima* = *S. townsendii*). However, polyploidy is common in this genus and *S. townsendii* became a new fertile hybrid, *S. anglica*, over a century ago, leading to invasive problems in Europe and, subsequently other continents (Strong and Ayres 2013).

Spartina introductions have often been purposeful – by people who either did not recognize the value of native habitats or understand the threats posed by these species. Both accidental and intentional introductions of a dozen species have resulted in several invasive hybrids that backcross with parents to form hybrid swarms that are fertile (as in the polyploidy case of *S. anglica*) and very difficult to eliminate (Strong and Ayers 2013). For example, in San Francisco Bay, *S. alterniflora* was intentionally introduced at least twice in the 1970s, subsequently hybridized with native *S. foliosa*, and spread to many locations throughout the central and southern portion of the bay (Daehler and Strong 1997, Ayres et al. 2004). *S. alterniflora*, which is native to western Atlantic coasts of North and South America, is perhaps the most widely distributed species that has led to invasive problems across Atlantic and Pacific coasts (e.g., England: Marchant 1967, USA: Daehler and Strong 1996, and China: Zhang et al. 2004). The most expansive *Spartina* invasion to date was by *S. alterniflora* across more than 100,000 ha of tidal salt marsh in eastern China (Wan et al. 2009). The largest invasion in the USA was by *S. alterniflora* in Willapa Bay, Washington, across more than 27,000 ha of intertidal lands (Strong and Ayres 2013).

Although some *Spartina* introductions do not advance in acreage or cause detrimental effects (reviewed by Strong and Ayres 2013), strong impacts have been documented for many *Spartina* infestations. These can include loss of native plant species; e.g., the hybrid *S. alterniflora* x *foliosa* invasion of San Francisco Bay led to local extirpation of *S. foliosa* across much of the invaded region (Ayres et al. 2003). In addition, invading *Spartina* can often grow seaward of the most flood-adapted native vascular plants, appropriating mudflats and altering functions through conversion to marsh. This habitat conversion has been most vivid when there was no native *Spartina* present in the lower marsh, as was the case when *S. alterniflora* invaded in Washington state (Simenstad and Thom 1995) or *S. anglica* covered open mudflats at Mont Saint Michel in France (Kahn 1973). In San Francisco Bay, hybrid *Spartina* grows lower than the native, and invaded mudflats show distinct declines in infaunal invertebrate numbers and diversity when compared to non-invaded mudflats (Neira et al. 2007). Similar declines in invertebrate numbers were observed following invasion of mudflats by *S. anglica* in the Dutch Wadden Sea; such forfeiture of mudflat may have important implications for higher trophic levels, such as waterbirds, that previously foraged in invaded areas (Tang and Kristensen 2010). Concordantly, *S. alterniflora* invasion in the Yangtze River Estuary in China, in addition to competitive exclusion of native plants, shifted the invertebrate community composition in mudflats to the detriment of shorebirds (Li et al. 2009). Trophic effects of *Spartina* invasion are sometimes viewed as positive; *S. alterniflora* invasion of mangroves in the Zhangjiang Estuary in China led to increased nutrition for fishes (Feng et al. 2015) and supplemented seagrass-dominated fish diets in *S. anglica*-invaded areas of southeastern Australia (Hindell and Warry 2010). However, invasive *S. alterniflora* x *S. foliosa* may be less palatable to Canada geese than the native *S. foliosa* in San Francisco Bay, and this might encourage persistence and spread of the invader (Grosholz 2010). *Spartina* invasion may be positive for certain species or functional groups that react to increased habitat structure, organic matter, or chemical changes to the sediments resulting from plant growth and senescence (e.g., shifting from algal-based to detritus-based foods: Levin et al. 2006). This can be cause for concern when some species facilitated are themselves non-native, but facilitation of native and even rare species also occurs when *Spartina* invades (e.g., the federally endangered Ridgway's rail in San Francisco Bay benefited from *Spartina* invasion: Overton et al. 2014, Casazza et al. 2016).

Attempts to control *Spartina* invasions have met with a variety of outcomes and, in some cases, difficult constraints of competing management goals. In Willapa Bay, Washington, initial efforts to control *S. alterniflora* invasion of mudflats failed during the first decade, but by 2004 new herbicide formulations and large-scale application were resulting in effective control (Patten 2002), with the invader now nearly eliminated (Strong and Ayres 2013). In San Francisco Bay, an extensive program of surveys and control efforts (largely with herbicides) reduced the invasion of *S. alterniflora* x *S. foliosa* hybrids by more than 95% between 2005 and 2012 (Rohmer et al. 2014). Phenotypic similarity between invasive and native *Spartina* complicates control work and genetic tests necessarily guide identification of the hybrid. This control work has been further complicated by the favorable habitat provided by hybrid *Spartina* for endangered rail nesting and thus there are limits to where eradication can occur (Casazza et al. 2016). However, efforts to date have produced intended outcomes, and native cordgrass has recruited to some areas in the most northerly and southerly extents of the invaded area where native *S. foliosa* seed was still present, while active revegetation through nursery propagation and outplanting is underway in the central portion of the invaded area where natural *S. foliosa* recruitment is unlikely (Kerr et al. 2016). Site-specific measures to encourage survivorship (e.g., caging out Canada Geese at some sites, protecting plantings in higher energy areas) are contributing to establishment and spread in some sites (Thornton 2018). In contrast, in China, invasive *S. alterniflora* continues to expand its acreage and transform mudflats and mangroves into salt marsh. Although this may assist in stabilization of shorelines, habitat conversions and losses of biodiversity are a great concern and probably an intractable problem at this point, considering the vast acreages affected (Strong and Ayres 2013).

14.2.3 *Lepidium latifolium*, Perennial Pepperweed

Perennial pepperweed (*Lepidium latifolium*), an herbaceous perennial native to Europe, Asia, and the Mediterranean (Lyeik 1989), is now widely distributed in the USA. It is believed to have been introduced in the 1930s (Robbins et al. 1951) and has been documented in each of the continental western states (Young et al. 1997) and parts of the midwest where it has been well established since the late 1990s. Once considered rare in New England, its rapid spread through northeastern states including New York, Connecticut, Massachusetts, and New Hampshire was only recently documented (Orth et al. 2006), while Connolly and Hale (2016) provide evidence of its proliferation in Rhode Island within the last two years. Pepperweed commonly occurs in freshwater wetlands including riparian areas, moist mountain meadows, and seasonal wetlands (Weber 1989). Once considered a nuisance only in those habitats, its significance as an invader of salt marshes has taken more time to realize (Reynolds and Boyer 2010).

Like many invasive plants, pepperweed gains a competitive advantage over native plants through rapid, robust growth and shading plants of lower stature at maturity. Like other species invading tidal marshes, it uses early season freshets from snow melt or heavy spring rains to germinate in more hospitable conditions, starts growth earlier in the season than its native competitors, and recovers from control efforts more quickly in fresher marshes (Blank and Young 1997, Boyer and Burdick 2010, Reynolds and Boyer 2010). Its combination of high stem density, high initial leaf area, and the formation of secondary inflorescences can create a dense canopy that persists throughout much of the spring and summer, allowing it to out-compete native plants (Renz and Blank 2004). Following aboveground senescence in fall, a dense litter layer can rapidly form, blocking light and further inhibiting emergence of other species in the seed bank (Renz and DiTomaso 1998). While its germination rate is not strong in truly saline conditions, its tolerance of high salinity (Blank and Young 1997) and extended submergence (Chen et al. 2002) once established, have helped make this plant a very successful invader.

Like *Phragmites*, pepperweed is capable of affecting the structure and function of tidal marshes. Reynolds and Boyer (2010) found that important factors including soil moisture, salinity, pH, canopy architecture light penetration and possibly insect/spider assemblages differed between pepperweed stands and nearby native *Salicornia pacifica* stands at similar elevations in San Francisco Bay. Their work suggests that pepperweed monocultures influence tidal marsh functions such as biogeochemical cycling and food web support. Reminiscent of *Phragmites*' role as an "ecosystem engineer" bringing about changes in the marsh from which it benefits, it is likely that pepperweed may initiate and maintain the differences observed. While not in tidal wetlands, Renz and Blank (2004) found pepperweed influenced plant-soil relationships. In their study of pepperweed in sodic soils, increases in N, Mg, and Ca were observed in invaded versus non-invaded areas, showing another way that this plant can alter soil conditions. Ironically, these characteristics may have made perennial pepperweed a suitable plant for restoration or containment of sodic soils – but the plant's aggressiveness as an invader of other, more sensitive habitats strongly discourages against consideration as a restoration tool.

Managers throughout the USA have found that perennial pepperweed is neither easy nor inexpensive to control. While management of upland/rangeland habitat can be accomplished efficiently and with measurable benefits (Eiswerth et al. 2005), control of pepperweed in tidal wetlands is considerably more expensive due to difficulty traversing salt marshes and locating scattered patches. Multiple studies have shown that non-chemical control (e.g., hand pulling, disking or mowing alone) is ineffective for large-scale control (Renz and DiTomaso 1999). However, if stands are small or newly established, hand-pulling and removal of plants can result in some localized success and has other benefits, such as involving the public in volunteer and outreach capacities that increase awareness of the threat (Liz Duff, Massachusetts Audubon, personal communication). For landscape level management, however, chemical control remains the most effective option, particularly when combined with non-chemical approaches.

Initially, the herbicide that has proven most effective for pepperweed was chlorsulfuron in upland habitats (Young et al. 1998), but it is not registered for use in wetlands. Alternatively, Renz and DiTomaso (2006) found that glyphosate could be used for control in wetland and tidal areas, but its efficacy was greatly enhanced if combined with mowing (>80% reduction in test plots after one year of application). Similarly, glyphosate efficacy increased when coupled with hand pulling, achieving 80% reduction after 2 years (Boyer and Burdick 2010). That study further noted that glyphosate application after clipping stems, or imazapyr treatment (a more recently approved herbicide for aquatic applications), with or without clipping stems, was more effective than applying glyphosate alone. Perhaps the most important consideration for effective control that simultaneously limits management costs is to develop resolve among managers and act immediately to limit establishment and prevent spread.

BOX 14.1 *Phragmites* Control in the Great Marsh: a Case Study

The list of management approaches to control common reed (*Phragmites australis*) are many and they have varied results (Hazelton et al. 2014). Managers who are limited by budgets and personnel often opt for the chemical control approach. Over the past two decades, a management team has grown from early efforts by the Parker River National Wildlife Refuge in Newbury Massachusetts (USA) to include town and county planners, NGOs and the University of New Hampshire across the entire Great Marsh. The Great Marsh Estuary and its associated watershed spans 29 municipalities and includes 17,000 acres of tidal marsh in northeastern Massachusetts. A landscape level control plan using adaptive management helps define critical control areas and prioritizes funding for control and monitoring actions.

The Refuge first noted *Phragmites* in the late 1990s and began to control small stands lining pools, creeks and upland edges that were subject to hydrologic manipulation or disturbance. Refuge managers had a host of options to manage these stands, from flooding to draw-down, cutting, burning, etc. but moreover, they had time to experiment with a number of management options. Over time, the occurrences of *Phragmites* grew, despite some management successes. What was unique about these new stands is that they did not appear to be associated with any particular disturbance event, nor were they limited to edges of tidal creeks, pools, or water ways as might be expected in vegetative spread of viable rhizomes from an off-site source. Managers noted that many of these new stands appeared in the middle of continuous high marsh platform, perhaps originating from seed, and spread to become monocultures relatively quickly. After numerous work sessions reviewing field data collections (mapping *Phragmites* stand expansion, tracking pore water salinity, redox potential and sulfide concentrations, assessing plant community changes), the Refuge began a carefully documented program of selective spraying the herbicide Imazapyr on the largest stands. With very few exceptions, results were immediate: rapid culm death followed by limited regrowth. Each growing season, original stand locations were revisited and assessed. Most were sprayed only every other year while a few persistent stands were selectively treated more frequently. Vegetation monitoring revealed that the reduction (or in most cases, elimination) of *Phragmites* was replaced by a diversity of desirable native halophytes, including one native species that is otherwise uncommon in the marsh as a whole, saltmarsh fleabane (*Pluchea odorata*; Fig. 14.1a).

Following the overall success battling large stands of common reed, questions remained regarding the occurrence of small, isolated stands that cropped up throughout Great Marsh. A study of the pore water conditions was initiated to examine factors known to contribute to *Phragmites* invasions, e. g., salinity. Pore water was sampled for salinity, pH, redox potential and sulfide concentration. Significant differences were not noted for pore water pH, redox potential, nor sulfide concentration between the *Phragmites* stands and adjacent non-invaded areas measured in the marsh at depths of 10cm and 50cm (unlike the clear differences found by Chambers et al. 2003). Likewise, these parameters were not statistically different over time (June, July, and September sampling dates) between invaded and non-invaded areas. The key difference for this marsh was linked to pore water salinity, but that variation was only noted in the early season measurements (June) – which were significantly lower for invaded sites. By midsummer, no significant salinity differences were evident. Realizing the importance of reduced salinity on competitive advantages for *Phragmites*, we sought to map salinity over large areas of marsh, using electromagnetic induction (EMI) following the approach detailed by Moore et al. (2011).

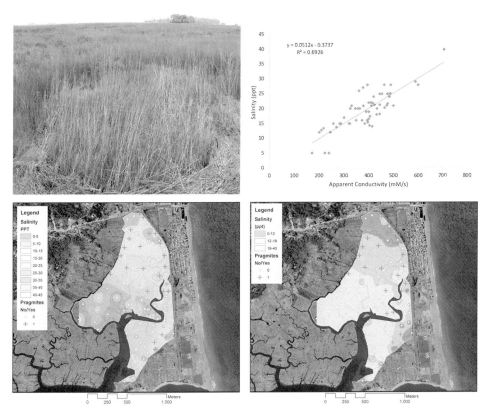

Figure 14.1 *Phragmites* control case study. 1a. (upper left) Example of a small, chemically treated *Phragmites* stand in Great Marsh, Newbury, MA following one growing season after treatment. Free from competition with *Phragmites*, the high marsh growth includes native *Spartina patens*, *P. odorata*, and *Salicornia depressa* (upper left); 1b. **(upper right)** Regression of salinity vs. apparent conductivity (upper right); 1c. **(lower left)** High resolution salinity gradients in a portion of Great Marsh located in Salisbury, MA (lower left); and 1d. **(lower right)** Reconfigured salinity map highlighting low salinity (green), moderate salinity (orange), and high salinity (pink) portions of the marsh. The moderate salinity area (orange) corresponds with published ranges of dense *Phragmites* proliferation (lower right). (A black and white version of this figure will appear in some formats. For the color version, please refer to the plate section.)

Working at the landscape scale, the invasive species management team now turned its attention to a large portion of the Great Marsh north of the Merrimac River which had not benefited from previous management efforts. In early summer, paired samples of pore water salinity using the sipper method (Portnoy and Valiela 1997) and apparent conductivity (EMI) were collected across the northern portion of the Great Marsh to create a standard curve. Measured salinity values ranged from 5 to 40 ppt, while corresponding conductivity values were 172mM/s and 704 mM/s. The regression illustrates a fairly strong relationship ($r^2 = 0.69$) given the high variability in the marsh. These data are consistent with site-specific standard curves generated for marshes in the past (Moore et al. 2011). The resulting conversion equation is illustrated in Fig. 14.1b. Following completion of the standard curve, numerous EMI measurements were combined to generate high-resolution salinity contour maps for the study area (Fig. 14.1c).

Examination of the distribution of salinity at this marsh show a pattern of salinity reduction from the open water southern bay to the upper, northern reaches of the marsh (Fig. 14.1c). Plant diversity was noted at each individual EMI sampling plot and the marsh habitat noted. In all cases, the presence of invasive *Phragmites australis* was noted. The results demonstrate that under present conditions, *Phragmites australis* is limited to the lower to moderate salinity portions of the Great Marsh site. Overall, this site had a mean salinity of 19.3±0.8 ppt, right at the upper limit of ideal pore water salinity conditions for proliferation of *Phragmites* mono-cultures in New England (12–19 ppt from Moore et al. 2011). When data were sorted by *Phragmites* presence/absence, *Phragmites*-free areas had a mean salinity of 19.9±0.9 ppt, while those areas currently inhabited by this invasive species were at 17.6±1.3 ppt – neatly within the ideal conditions. Figure 14.1c shows a high-resolution salinity map representing areas with and without *Phragmites* and in this case, suggests than there may be extensive areas that maintain suitable pore water salinity conditions to deter rapid invasion (southern and south-western portions of study area), while others may be perfectly suited for expansion of this plant (northern interior and eastern margins), and should be the focus of control efforts.

With extensive development of exotic *Phragmites* already present in this portion of the Great Marsh, there are real management challenges ahead. Fortunately, the success of *Phragmites* management at the adjacent Parker River National Wildlife Refuge, coupled with the high-resolution salinity mapping conducted in this study, suggest a prioritization scheme. Salinity mapping has identified core areas where *Phragmites* management is necessary and likely to be successful. To visualize priority management zones, Figure 14.1d recasts the original salinity map to show key ranges of salinity associated with *Phragmites* establishment. *Phragmites* present with the middle ranges of salinity (12–19 ppt, shown in orange in Fig. 14.1d) are the core target for management priority, given they will continue to expand. In contrast, higher salinity zones are areas of low *Phragmites* invasion likelihood, requiring less management/maintenance input over the long term to combat in the case of small-scale colonization. Finally, predominantly low salinity areas pose an interesting question for researchers. Do these fresher areas require more rigorous measures for control, or do competitors existing in these areas effectively control rapid spread of *Phragmites*?

The management team broadened its membership and approach to allow landscape scale analysis and actions to control exotic *Phragmites* in the Great Marsh. Where other methods were not appropriate or effective, herbicides have been used has been used to control *Phragmites* and these areas are quickly colonized by a diverse assemblage of native plants. While new tools can help prioritize control areas, they can also help managers and researchers develop new questions important for future control efforts.

14.3 Invading Fauna

Salt marsh distribution and productivity were once thought to be controlled by abiotic stressors (e.g., flooding, salinity) and competition between plants (Levine et al. 1998, Peter and Burdick 2010), but in the past 20 years several instances of top-down control have been demonstrated, often as a result of runaway grazer populations (Silliman et al. 2005, Bertness et al. 2014). Sometimes exotic animal introductions result in direct effects of over-grazing and marsh destruction, while in other cases populations of native animal grazers have been released by loss of predators, resulting in a trophic cascade. Salt marshes are presented with a variety of stressors, both abiotic and biotic, and catastrophic impacts attributed to invasive species may be limited to highly stressed systems in many cases.

14.3.1 Vertebrates

A variety of animals eat the aboveground (e.g., voles, deer) and belowground (e.g., geese, nutria) tissues of marsh grasses. Under most circumstances, the grazing pressure is within bounds of the ability of the plants to grow and reproduce and the system is considered to be in a dynamic balance. At other times the grazer population explodes, the plants are grazed to the extent they cannot recover, and the marsh is destroyed. This has been found for large areas of marsh surrounding Hudson Bay due to lesser snow geese (Jefferies et al. 2006), along the Texas marshes due to snow geese (Miller et al. 2005), and in Mid-Atlantic and Gulf Coast marshes due to nutria (Ford and Grace 1998). Dramatic losses by vertebrates are often associated with exhumation and consumption of rhizomes, which not only dislodges the plants but kills tissue that supports next year's growth. Once the plants are destroyed, the surface of the marsh collapses and the resulting abiotic conditions may not even support new marsh growth (e.g., Hudson Bay, Canada; Jefferies et al. 2006).

14.3.2 Invertebrates

Some invertebrates also have detrimental effects on marshes, and experimental exclosures (without grazers) and enclosures (with grazers) have demonstrated strong impacts. In New England, marsh loss, notably along creek banks, has been attributed to a nocturnal herbivorous crab, *Sesarma reticulatum* (Holdredge et al. 2009), which clips vegetation near the ground and consumes roots and rhizomes, leading to dead patches from several to tens of meters in length and width. Field experiments showed the presence of predators limits the damage done (Coverdale et al. 2012), but that declines in populations of predators may be due to overfishing (Altieri et al. 2012) or other reasons (e.g., declines in night herons). Loss of marsh in Rhode Island is thought to be the product of complex interactions between SLR and a variety of crab populations (Raposa et al. 2018). Findings there are consistent with observations further north (green crabs in Maine – Belknap and Wilson 2014) and south (crabs and SLR erode mid-Atlantic marshes – Hughes et al. 2009, Wilson et al. 2012). Similarly, on the US West coast, an exotic amphipod, *Sphaeroma quoianum,* burrows into the peat and may accelerate erosion in Oregon (Davidson and de Rivera 2010) and California salt marshes (Talley et al. 2001).

Salt marsh die-back along the Louisiana Gulf Coast and in the southeastern USA is a phenomenon observed by researchers at or following the turn of the century (McKee et al. 2004, Alber et al. 2008). In some cases, die-back appears to be exacerbated by the rough periwinkle, *Littoraria irrorata*, a native snail that grazes cordgrass leaves (see also Box 14.2). Further work has shown a complex relationship as the snail radulates the leaves, inoculates them with a fungus, and revisits the same sites to graze upon the fungus (rather than the leaf itself) (Silliman and Zieman 2001). When snail populations are so high that the marsh plants are killed, the snails move toward remaining plants as a wave. Loss of snail predators (blue crabs) and higher salinities arising from drought are two mechanisms thought to contribute to the snail wave (Silliman et al. 2005). In a South Carolina marsh, Kiehn and Morris (2009) found no top down control of cordgrass by this species of snail in marshes that averaged 81 snails/m^2; thus, another source of stress, such as drought may be required to initiate destructive snail waves.

In the lower Mississippi River Delta, introduced lineages of *Phragmites* are dominant, yet they are valued as one of the few emergent plants that can survive flooding and help protect landward marshes from erosion and loss (Hauber et al. 2011). Recently however, these stands have been found to be dying back. In an interesting twist, an exotic scale insect, *Nipponaclerda biwakoensis*, has been identified as a potential cause of *Phragmites* die-back in the Delta (Knight et al. 2018).

BOX 14.2 *Littorina littorea*, Common Periwinkle, Prevents Salt Marsh Re-Establishment

In New England, where the common periwinkle, *Littorina littorea*, was introduced from Europe over a century ago (Chapman et al. 2007), direct grazing impacts on the above- and belowground tissues of cordgrass appear to be minor (Bertness 1984, Tyrrell et al. 2008) but their presence was associated with enhanced erosion (Bertness 1984) or reduced sediment deposition (Tyrrell et al. 2008). Using a well-replicated field experiment, Tyrrell and colleagues found these snails had significant impacts, but only to flooding-stressed low marsh, and they pointed out their results were consistent with grazing of fungus-infected plants. Common periwinkles may increase infection by a pathogenic fungus, *Fusarium* (Elmer et al. 2013), and worsen marsh loss that has been associated with other stressors like SLR (Smith 2009).

In cases such as salt marsh restoration where rapid vegetative reproduction and expansion of cordgrass is relied upon, *L. littorea* can have devastating effects. Human overuse and shoreline erosion led to loss of sections of fringing marsh in York, Maine. A management plan was formulated to deter human use and restore the marsh by installing erosion control baffles (1.2 × 0.2 m rolls of wrack covered by jute fabric) and replanting the eroded shoreline with cordgrass (Burdick et al. 2013). Both commercially grown plugs and locally obtained bare-root shoots were planted on one-foot centers, but the plants grew poorly and died despite little-to-no human interference. These shoreline areas had high densities of common periwinkles that climbed and grazed on leaves, leaving radulations that browned and often killed the distal leaf blade. Snails would also break smaller plant stems. The snails often moved in waves and new shoots appeared to be particularly vulnerable (Fig. 14.2).

In response to these observations, a pilot study was conducted where new plants were grown in 0.5 m^2 exclusion cages constructed from stakes and erosion control fabric. Protected from snail activities, these plants were able to survive and grow. A new restoration effort was set up that used low fences of galvanized wire mesh (6 mm) to prevent waves of snails from traversing the planted area and this effort was successful (Fig. 14.2). Plant survival through two growing seasons was 92%. The sediment became covered with a film of green algae (*Vaucheria* spp.), which helped to retain finer grain sizes. The fencing did not exclude all snails from the planted area, however, and the plants still grew slowly; a subsample of plants showed snail damage on 80% of live shoots.

This case study shows that an invader with devastating effects on a shoreline restoration project is, under normal conditions, a factor that merely reduces marsh growth (as in Tyrrell et al. 2008), but under stressful conditions snails can limit the distribution of salt marsh, especially where they form a wave. As others have shown, there are many animal–plant interactions that do not pose a threat to salt marshes under normal conditions (e.g., Kiehn and Morris 2009), but where trophic cascades result in runaway populations (Coverdale et al. 2012) or other marsh stresses occur (Silliman et al. 2005), they can set back, alter or prevent marsh development or destroy existing marsh, resulting in an alternative stable state that limits salt marsh distribution (Jefferies et al. 2006).

14.4 Microbe Invaders

As we learn more about the microbes that live and interact with biological and physical processes in salt marshes, we may learn that populations of exotic microbes can become established or native species can explode in salt marsh locales, leading to significant impacts. As mentioned previously, rampant snail populations in southern marshes were shown to farm fungi on leaves of *S. alterniflora* and lead to waves of marsh die-off and loss (Silliman and Zieman 2001). In other investigations of marsh die-off along Atlantic and Gulf coasts of the USA, fungus (*Fusarium palustre*) and interactions with other stressors (e.g., drought) were suggested as potential causal agents (Elmer et al. 2013).

14.5 Approaches and Further Research to Control Current and Future Invasions

Large-scale invasions, like those seen from *Spartina*, *Lepidium*, or *Phragmites* are most successfully controlled when resource managers pool resources to use a landscape or regional scale approach. Control efforts should focus on the preferred habitat of the invader (*Phragmites* case study) to reduce its ability to spread. Treatments that allow recolonization by native plants quickly, as shown in the Great Marsh, will help prevent future invasions.

An outstanding question for coastal resources managers is whether the invader is contributing to a significant and permanent redistribution of coastal habitats that is "good" or "bad" for society. Invasions may indicate that other stressors operating in the marsh are leaving the marsh vulnerable to invasion. Alternatively, plant or animal invasions may become a stressor that leads to marshes becoming more vulnerable to other stressors, like SLR. If the invader can become 'naturalized', then perhaps affected tidal marshes can adapt

Figure 14.2 Wave of *Littorina littorea* (common periwinkle) on eroded shoreline in York, ME (upper left). Planted cordgrass protected by wave baffle still damaged by snails (upper right). Snail association with planted cordgrass, now grazed and decomposing (lower left). Second year of growth of cordgrass at restoration site protected from snails by a fence of wire mesh (lower right). (A black and white version of this figure will appear in some formats. For the color version, please refer to the plate section.)

and survive despite its presence. If not, then managers must decide if control efforts are worth the resources needed to be successful. Faunal invaders associated with cascading effects from loss or absence of a predator (common periwinkle case study) are difficult to address and research on how to replace predators is needed.

Many impacts (geese, snails, *Fusarium*) degrade salt marsh and lead to elevation loss (sometimes collapse), but some plant invasions also can build elevation (cordgrass, *Phragmites*) or interfere with marsh migration associated with SLR (*Phragmites* – Smith 2013). All are important considerations when managing marshes in a period of rapid SLR associated with climate change.

References

Able, K. W., Hagan, S. M., and Brown, S. A. 2003. Mechanisms of marsh habitat alteration due to Phragmites: Response of young-of-the-year mummichog (*Fundulus hetero-clitus*) to treatment for Phragmites removal. *Estuaries* 26:484–494.

Alber, M., Swenson, E. M., Adamowicz, S. C., and Mendelssohn, I. A. 2008. Salt marsh dieback: an overview of recent events in the US. *Estuarine, Coastal and Shelf Science* 80: 1–11.

Altieri, A. H., Bertness, M. D., Coverdale, T. C., Herrmann, N. C., and Angelini, C. 2012. A trophic cascade triggers collapse of a salt-marsh ecosystem with intensive recreational fishing. *Ecology*, 93: 1402–1410.

Amsberry, L., Baker, M. A., Ewanchuk, P. J., and Bertness, M. D. 2000. Clonal integration and the expansion of Phragmites australis. *Ecological Applications*, 10: 1110–1118.

Anderson, C. M., and Treshow, M. 1980. A review of environmental and genetic factors that affect height in *Spartina alterniflora* Loisel. (Salt marsh cord grass). *Estuaries*, 3:168–176.

Anderson M. G. 1995. Interaction between *Lythrum salicaria* and native organisms: a critical review. *Environmental Management*, 19: 225–231.

Ayres, D. R., Smith, D. L., Zaremba, K., Klohr, S., and Strong, D. R. 2004. Spread of exotic cordgrasses and hybrids (Spartina sp.) in the tidal marshes of San Francisco Bay, California, USA. *Biological Invasions*, 6:221–231.

Ayres, D. R., Strong, D. R., and Baye, P. 2003. Spartina foliosa (Poaceae)–a common species on the road to rarity. *Madrono*, 50:209–213.

Balouskus, R. G., and Targett, T. E. 2012. Egg deposition by Atlantic silverside, *Menidia menidia*: substrate utilization and comparison of natural and altered shoreline type. *Estuaries and Coasts*, 35:1100–1109.

Bart, D., Burdick, D., Chambers, R. and Hartman, J. M. 2006. Human facilitation of *Phragmites australis* invasions in tidal marshes: a review and synthesis. *Wetlands Ecology and Management*, 14:53–65.

Belknap, D. F. and Wilson, K. R. 2014. Invasive green crab impacts on salt marshes in Maine-sudden increase in erosion potential. Abstract. Northeast Section, Geological Society of America. 24 March 2014, Lancaster, PA.

Benoit, L. K., and Askins, R. A. 1999. Impact of the spread of *Phragmites* on the distribution of birds in Connecticut tidal marshes. *Wetlands*, 19: 194–208.

Bertness, M. D. 1984. Habitat and community modification by an introduced herbivorous snail. *Ecology*, 65: 370–381.

Bertness M. D., Brisson, C. P., Bevil, M. C., and Crotty, S. M. 2014. Herbivory drives the spread of salt marsh die-off. *PLoS ONE*, 9(3): e92916. doi:10.1371/journal.pone.0092916

Blank, R., and Young, J. 1997. Influence of invasion of perennial pepperweed on soil properties. pp. 11–13 In *Management of Perennial Pepperweed (Tall Whitetop)*. Special Report 972. Corvallis, OR: U.S. Department of Agriculture, Agricultural Research Service; Oregon State University, Agricultural Experiment Station.

Blossey, B., Skinner, L. C. and Taylor, J. 2001. Impact and management of purple loosestrife (*Lythrum salicaria*) in North America. *Biodiversity and Conservation*, 10:1787–1807.

Boyer, K. E. and Burdick, A. P. 2010. Control of *Lepidium latifolium* (perennial pepperweed) and recovery of native plants in tidal marshes of the San Francisco Estuary. *Wetlands Ecology and Management*, 18(6): 731–743.

Buchsbaum R. N., Catena, J., Hutchins, E., and Pirri, M. J. 2006. Changes in salt marsh vegetation, *Phragmites australis*, and nekton in response to increased tidal flushing in a New England salt marsh. *Wetlands*, 26:544–557.

Burdick, D. M., and Konisky, R. A. 2003. Determinants for expansion of *Phragmites australis*, common reed, in natural and impacted coastal marshes. *Estuaries*, 26: 407–416.

Burdick, D. M., and Roman, C. T. 2012. Salt marsh responses to tidal restriction and restoration. A summary of experiences. pp. 373–382 In: Roman, C.T. and D.M. Burdick (eds.) *Tidal Marsh Restoration: A Synthesis of Science and Practice*. Island Press. Washington.

Burdick, D., Peter, C., and Moore, G. E. 2013. Phase II of Tidal Marsh Restoration at Steedman Woods Reserve at York, Maine. Final report to Museums of Old York; accessed from UNH Scholars Repository: https://scholars.unh.edu.

Casazza, M. L., Overton, C. T., Bui, T. -V. D., Hull, J. M., Albertson, J. D., Bloom, V. K., Bobzien, S., et al. 2016. Endangered species management and ecosystem restoration: finding the common ground. *Ecology and Society*, 21(1):19.

Chambers R. M., Meyerson, L. A., and Dibble, K. L. 2012. Ecology of *Phragmites australis* and responses to tidal restoration. pp. 81–96. In: Roman, C.T. and D.M. Burdick (eds.) *Tidal Marsh Restoration*. Island Press. Washington, DC.

Chambers R. M., Meyerson L. A., and Saltonstall K. 1999. Expansion of *Phragmites australis* into tidal wetlands of North America. *Aquatic Botany*, 64:261–273.

Chambers, R. M., Osgood, D. T., Bart, D. J., and Montalto, F. 2003. *Phragmites australis* invasion and expansion in tidal wetlands: interactions among salinity, sulfide, and hydrology. *Estuaries*, 26:398–406.

Chapman, J. W., Carlton, J. T., Bellinger, M. R, and Blakeslee, A. M. H. 2007. Premature refutation of a human-mediated marine species introduction: the case history of the marine snail *Littorina littorea* in the Northwestern Atlantic. *Biological Invasions*, 9:995–1008.

Chen, H., Qualls, R. G., and Miller, M. C. 2002. Adaptive responses of *Lepidium latifolium* to soil flooding: biomass allocation, adventitious rooting, aerenchyma formation and ethylene production. *Environmental and Experimental Botany*, 48:119–128.

Connolly, B. A. and Hale, I. L. 2016. *Lepidium latifolium* (Brassicaceae): invasive perennial pepperweed observed in Rhode Island. *Rhodora*, 118(974):229–231.

Coverdale, T. C., Altieri, A. H., and Bertness, M. D. 2012. Belowground herbivory increases vulnerability of New England salt marshes to die-off. *Ecology*, 93:2085–2094.

Daehler, C. C., and Strong, D. R. 1996. Status, prediction and prevention of introduced cordgrass *Spartina* spp. invasions in Pacific estuaries, USA. *Biological Conservation*, 78:51–58.

Daehler, C., and Strong, D. 1997. Hybridization between introduced smooth cordgrass (*Spartina alterniflora*; Poaceae) and native California cordgrass (*S. foliosa*) in San Francisco Bay, California, USA. *American Journal of Botany*, 84:607–611.

Davidson, T. M., and de Rivera, C. E. 2010. Accelerated erosion of saltmarshes infested by the non-native burrowing crustacean *Sphaeroma quoianum*. *Marine Ecology Progress Series*, 419:129–136.

Denoth, M., and Myers, J. H. 2007. Competition between *Lythrum salicaria* and a rare species: combining evidence from experiments and long-term monitoring. *Plant Ecology*, 191:153–161.

Dibble, K. L., and Meyerson, L. A. 2012. Tidal flushing restores the physiological condition of fish residing in degraded salt marshes. *PLoS ONE* 7(9): e46161. doi:10.1371/journal.pone.0046161

Dibble, K. L., and Meyerson, L. A. 2013. The effects of plant invasion and ecosystem restoration on energy flow through salt marsh food webs. *Estuaries and Coasts*, 35

Dibble, K. L., Pooler, P. S, and Meyerson, L. A. 2013. Impacts of plant invasions can be reversed through restoration: a regional meta-analysis of faunal communities. *Biological Invasions*, 15:1725–1737.

Elmer, W. H., LaMondia, J. A., Useman, S., Mendelssohn, I. A., Schneider, R. W., Jimenez-Gasco, M. M., Marra, R. E., and Caruso, F. L. 2013. Sudden vegetation dieback in Atlantic and Gulf Coast salt marshes. *Plant Disease*, 97: 436–445.

Eiswerth, M., Singletary, L., Zimmerman, J., Johnson, W. 2005. Dynamic benefit–cost analysis for controlling perennial pepperweed (*Lepidium latifolium*): A Case Study. *Weed Technology* 19: 237–243.

Fagherazzi, S., Kirwan, M. L., Mudd, S. M., Guntenspergen, G. R., Temmerman, S., D'Alpaos, A., van de Koppel, J., et al. 2012. Numerical models of salt marsh evolution: Ecological, geomorphic, and climatic factors. *Reviews of Geophysics*, 50: RG1002.

Farnsworth, E. J., and Ellis, D. R. 2001. Is purple loosestrife (*Lythrum salicaria*) an invasive threat to freshwater wetlands? Conflicting evidence from several ecological metrics. *Wetlands*, 21:199–209.

Feng, J., Huang, Q., Qi, F., Guo, J., and Lin, G. 2015. Utilization of exotic *Spartina alterniflora* by fish community in the mangrove ecosystem of Zhangjiang Estuary: evidence from stable isotope analyses. *Biological Invasions*, 7: 2113–2121.

Flanagan, R. J., Mitchell, R. J., and Karron, J. D. 2010. Increased relative abundance of an invasive competitor for pollination, *Lythrum salicaria*, reduces seed number in *Mimulus ringens*. *Oecologia*, 164: 445–454.

Ford, M. A., and Grace, J. B. 1998. Effects of vertebrate herbivores on soil processes, plant biomass, litter accumulation and soil elevation changes in a coastal marsh. *Journal of Ecology*, 86: 974–982.

Grosholz, E. 2010. Avoidance by grazers facilitates spread of an invasive hybrid plant. *Ecology Letters*, 13:145–153.

Guntenspergen, G. R., Keough, J. R., and Weinstein, M. P. 2003. Phragmites technical forum and workshop: synthesis of scientific knowledge and management needs. *Estuaries*, 26:1–8.

Hauber, D. P., Saltonstall, K., White, D. A., and Hood, C. S. 2011. Genetic variation in the common reed, *Phragmites australis*, in the Mississippi River Delta marshes: Evidence for multiple introductions. *Estuaries and Coasts*, 34:851–862.

Hazelton, E. L. G., Mozdzer, T. J., Burdick, D. M., Kettenring, K. M., and Whigham, D. F. 2014. *Phragmites australis* management in the United States: 40 years of methods and outcomes. *AoB Plants* 6: plu001.

Hershner, C., and Havens, K. J. 2008. Managing invasive aquatic plants in a changing system: strategic consideration of ecosystem services. *Conservation Biology*, 22: 544–550.

Hierro, J. L., and Callaway, R. M. 2003. Allelopathy and exotic plant invasion. *Plant and Soil*, 256:29–39.

Hindell, J. S., and Warry F. Y. 2010. Nutritional support of estuary perch (*Macquaria colonorum*) in a temperate Australian inlet: Evaluating the relative importance of invasive *Spartina*. *Estuarine, Coastal and Shelf Science*, 90:159–167.

Holdredge, C., Bertness, M. D., and Altieri, A. H. 2009. Role of crab herbivory in die-off of New England salt marshes. *Conservation Biology*, 23: 672–679.

Hughes, Z. J., FitzGerald, D. M., Wilson, C. A., Pennings, S. C., Wieski, K., and A. Mahadevan. 2009. Rapid headward erosion of marsh creeks in response to relative sea level rise. *Geophysical Research Letters* 36: 5.

Jodoin, Y., Lavoie, C. L., Villeneuve, P., Theriault, M., Beaulieu, J., and Belzile, F. 2008. Highways as corridors and habitats for the invasive common reed *Phragmites australis* in Quebec, Canada. *Journal of Applied Ecology*, 45:459–466.

Jefferies, R. L., Jano, A. P., and Abraham, K. F. 2006. A biotic agent promotes large-scale catastrophic change in the coastal marshes of Hudson Bay. *Journal of Ecology*, 94:234–242.

Kahn, H. 1973. Paris Notebook, doldrums for French research. *New Scientist* 60;797–98.

Kelley, J. T., Belknap, D. F., Jacobson Jr., G. L., and Jacobson, H. A. 1988. The morphology and origin of salt marshes along the glaciated coastline of Maine, USA. *Journal of Coastal Research*, 4:649–665.

Kerr, D. W., Hogle, I. B., Ort, B. S., and Thornton, W. J. 2016. A review of 15 years of *Spartina* management in the San Francisco Estuary. *Biological Invasions*, 18:2247–2266.

Kiehn, W. M., and Morris, J. T. 2009. Relationships between *Spartina alterniflora* and *Littoraria irrorata* in a South Carolina salt marsh. *Wetlands*, 29:818–825.

Kiviat, E. 2013. Ecosystem services of Phragmites in North America with emphasis on habitat functions. *AoB PLANTS* 5: plt008. https://doi.org/10.1093/aobpla/plt008

Knight, I. A., Wilson, B. E., Gill, M., Aviles, L., Cronin, J. T., Nyman, J. A., Schneider, S. A., and Diaz, R. 2018. Invasion of *Nipponaclerda biwakoensis* (Hemiptera: Aclerdidae) and *Phragmites australis* die-back in southern Louisiana, USA. *Biological Invasions*, 20:2739–2744.

Konisky, R. A. and Burdick, D. M. 2004. Effects of stressors on invasive and halophytic plants. of New England salt marshes: a framework for predicting response to tidal restoration. *Wetlands*, 24: 434–447.

Lelong, B., Lavoie, C., Jodoin, Y., and Belzile. F. 2007. Expansion pathways of the exotic common reed (*Phragmites australis*): a historical and genetic analysis. *Diversity and Distributions*, 13:430–437.

Levin, L. A., Neira, C., and Grosholz, E. D. 2006. Invasive cordgrass modifies wetland trophic function. *Ecology*, 87:419–432.

Levine, J. M., Brewer, J. S. and Bertness, M. D. 1998. Nutrients, competition and plant zonation in a New England salt marsh. *Journal of Ecology*, 86: 285–292.

Li, B., Liao, C., Zhang, X., Chen, H., Wang, Q., Chen, Z., Gan, X. et al. 2009. *Spartina alterniflora* invasions in the Yangtze River Estuary, China: An overview of current status and ecosystem effects. *Ecological Engineering*, 35:511–520.

Lyeik, K. A. 1989. *Lepidium latifolium* L., a sea-shore species in Norway. *Blyttia*, 47:109–113.

Marchant. C. J. 1967. Evolution in *Spartina* (Gramineae): I. The history and morphology of the genus in Britain. *Botanical Journal*, 60:1–24.

McKee, K. L., Mendelssohn, I. A. and Materne, M. D. 2004. Acute salt marsh dieback in the Mississippi River deltaic plain: a drought induced phenomenon? *Global Ecology and Biogeography*, 13:65–73.

Meyerson, L. A., Saltonstall, K., and Chambers, R. M. 2009. *Phragmites australis* in coastal marshes of North America: A historical and ecological perspective. pp. 57–82. In: *Human Impacts on Salt Marshes: A Global Perspective*, ed. Silliman, B.R., Bertness, M. D., Grosholz, E. D. (eds.) University of California Press, Berkeley.

Miller, D. L., Smeins, F. E., Webb, J. W., and Yager, L. 2005. Mid-Texas, USA coastal marsh vegetation pattern and dynamics as influenced by environmental stress and snow goose herbivory. *Wetlands*, 25:648–658.

Minchinton, T. E. 2002. Precipitation during El Niño correlates with increasing spread of *Phragmites australis* in New England, USA, coastal marshes. *Marine Ecology Progress Series*, 242: 305–309.

Minchinton, T. E., and Bertness, M. D. 2003. Disturbance mediated competition and the spread of *Phragmites australis* in a coastal marsh. *Ecological Applications* 13: 1400–1416.

Minchinton, T. E., Simpson, J. C., and Bertness, M. D. 2006. Mechanisms of exclusion of native coastal marsh plants by an invasive grass. *Journal of Ecology*, 94: 342–354.

Moore, G. E., Burdick, D. M., Peter, C. R. and Keirstead, D. R. 2011. Mapping soil pore water salinity of tidal marsh habitats using electromagnetic induction in Great Bay Estuary, USA. *Wetlands*, 31:309–318.

Moore, G. E., Burdick, D. M., Peter, C. R. and Keirstead, D. R. 2012. Belowground biomass of *Phragmites australis* in coastal marshes. *Northeast Naturalist*, 19:611–626.

Morris, J. T., Sundareshwar, P. V., Nietch, C. T., Kjerfve, B. and Cahoon, D. R. 2002. Responses of coastal wetlands to rising sea level. *Ecology*, 83: 2869–2877.

Morse, N. B., Pellissier, P. A., Cianciola, E. N., Bereton, R. L., Sullivan, M. M., Shonka, N. K., Wheeler, T. B., and McDowell, W. H. 2014. Novel ecosystems in the Anthopocene: a revision of the novel ecosystem concept for pragmatic applications. *Ecology and Society*, 19:12.

Neira, C., Levin, L A., Grosholz, E. D. and Mendoza, G. 2007. Influence of invasive *Spartina* growth stages on associated macrofaunal communities. *Biological Invasions* 9:975–993.

Nyman, J. A., Walters, R. J., Delaune, R. D. and Patrick Jr., W. H. 2006. Marsh vertical accretion via vegetative growth. *Estuarine, Coastal and Shelf Science*, 69:370–380.

Orson, R. A. 1999. A paleoecological assessment of *Phragmites australis* in New England tidal marshes: changes in plant community structure during the last few millennia. *Biological Invasions*, 1:149–158.

Orth, J. F., Gammon, M., Abdul-Basir, F., Stevenson, R. D., Tsirelson, D., Ebersole, J., Speak, S. and Kesseli, R. 2006. Natural history, distribution, and management of *Lepidium latifolium* (Brassicaceae) in New England. *Rhodora*, 108: 103–118.

Overton, C. T., Casazza, M. L., Takekawa, J. Y., Strong, D. R., and Holyoak, M. 2014. Tidal and seasonal effects on survival rates of the endangered California clapper rail: Does invasive *Spartina* facilitate greater survival in a dynamic environment? *Biological Invasions*, 16:1897–1914.

Patten K. 2002. Smooth cordgrass (*Spartina alterniflora*) control with Imazapyr. *Weed Technology*, 16:826–32.

Peter, C. R., and Burdick, D. M. 2010. Can plant competition and diversity reduce the growth and survival of exotic *Phragmites australis* invading a tidal marsh? *Estuaries and Coasts*, 33: 1226–1236.

Portnoy, J. W., and Valiela, I. 1997. Short-term effects of salinity reduction and drainage on salt-marsh biogeochemical cycling and *Spartina* (cordgrass) production. *Estuaries*, 20:569–578.

Raichel, D. L., Able, K. W., and Hartman, J. M. 2003. The influence of *Phragmites* (common reed) on the distribution, abundance, and potential prey of a resident marsh fish in the Hackensack Meadowlands, New Jersey. *Estuaries*, 26: 511–521.

Raposa, K. B., McKinney, R. A., Wigand, C., Hollister, J. W., Lovall, C., Szura, K., Gurak Jr., J. A., et al. 2018. Top-down and bottom-up controls on southern New England salt marsh crab populations. *PeerJ*, 6, e4876. doi.org/10.7717/peerj.4876

Renz, M. J., and Blank, R. R. 2004. Influence of perennial pepperweed (*Lepidium latifolium*) biology and plant-soil relationships on management and restoration. *Weed Technology*, 18:1359–1363.

Renz, M. J., DiTomaso, and J. M. 1998. The effectiveness of mowing and herbicides to control perennial pepperweed in rangeland and roadside habitats. *Proceedings of the California Weed Science Society*, 50:178.

Renz, M. J., and DiTomaso, J. M. 1999. Biology and control of perennial pepperweed. *Proceedings of the California Weed Science Society*, 51: 13–16.

Renz, M. J., and DiTomaso, J. M. 2006. Early season mowing improves the effectiveness of chlorsulfuron and glyphosate for control of perennial pepperweed (*Lepidium latifolium*). *Weed Technology*, 20:32–36.

Reynolds, L. K., and Boyer, K. E. 2010. Perennial pepperweed (Lepidium latifolium): Properties of invaded tidal marshes. *Invasive Plant Science and Management*, 3(2): 130–138.

Robbins, W. W., Bellue, M. K., and Ball, W. S. 1951. *Weeds of California*. Sacramento, CA: California Department of Agriculture.

Rohmer, T., Kerr, D., and Hogle, I. 2014. San Francisco Estuary Invasive Spartina Project 2013 ISP monitoring and treatment report. *Prepared for the California State Coastal Conservancy*, Oakland, California, USA. www.spartina

Rooth, J. E., and Stevenson, J. C. 2000. Sediment deposition patterns in *Phragmites australis* communities: Implications for coastal areas threatened by rising sea-level. *Wetlands Ecology and Management*, 8:173–183.

Rudrappa, T., Bonsall, J., Gallagher, J. L., Seliskar, D. M., and Bais, H. P. 2007. Root-secreted allelochemical in the noxious weed *Phragmites australis* deploys a reactive oxygen species response and microtubule assembly disruption to execute rhizotoxicity. *Journal of Chemical Ecology*, 33: 1898–1918.

Saltonstall K. 2002. Cryptic invasion by a non-native genotype of the common reed, *Phragmites australis*, into North America. *Proceedings of the National Academy of Sciences of the USA*, 99:2445–2449

Saltonstall, K., Peterson, P. M., and Soreng, R. 2004. Recognition of *Phragmites australis* subsp. americanus (Poaceae: Arundinaceae) in North America: Evidence from morphological and genetic analyses. *Sida*, 21:683–692.

Silliman B. R., van de Koppel, J., Bertness, M. D., Stanton, L. E. and Mendelssohn, I. A. 2005. Drought, snails, and large-scale die-off of southern US salt marshes. *Science*, 310: 1803–1806.

Silliman B. R., and Zieman, J. C. 2001. Top-down control of *Spartina alterniflora* production by periwinkle grazing in a Virginia salt marsh. *Ecology*, 82: 2830–2845.

Simenstad, C. A., and Thom, R. M. 1995. *Spartina alterniflora* (smooth cordgrass) as an invasive halophyte in Pacific Northwest Estuaries. *Hortus Northwest*, 6:9–12; 38–40.

Smith, J. A. M. 2013. The role of *Phragmites australis* in mediating inland salt marsh migration in a mid-Atlantic estuary. *PLoS ONE* 8(5): e65091. doi.org/10.1371/journal.pone.0065091

Smith, S. M. 2009. Multi-decadal changes in salt marshes of Cape Cod, MA: photographic analyses of vegetation loss, species shifts, and geomorphic change. *Northeastern Naturalist*, 16:183–208.

Smith, S. M., Roman, C. T., James-Pirri, M. J., Chapman, Portnoy, K. J., and Gwilliam, E. 2009. Responses of plant communities to incremental hydrologic restoration of a tide-restricted salt marsh in southern New England (Massachusetts, U.S.A.). *Restoration Ecology*, 17:606–618.

Strong, D. R., and Ayres, D. R. 2013. Ecological and evolutionary misadventures of Spartina. *Annual Review of Ecology, Evolution, and Systematics*, 44:389–410.

Talley, T. S., Crooks, J. A., and Levin, L. A. 2001. Habitat utilization and alteration by the invasive burrowing isopod, *Sphaeroma quoyanum*, in California salt marshes. *Marine Biology*, 138:561–573.

Tang, M., and Kristensen, E. 2010. Associations between macrobenthos and invasive cordgrass, *Spartina anglica*, in the Danish Wadden Sea. *Helgoland Marine Research*, 64:321–329.

Tavernia, B. G., and Reed, J. M. 2012. The impact of exotic purple loosestrife (*Lythrum salicaria*) on wetland bird abundances. *The American Midland Naturalist*, 168: 352–363.

Thornton, W. 2018. *How do transplant source, restoration site, and herbivory influence Pacific cordgrass restoration?* Master's thesis, San Francisco State University.

Trocki, C. L., and Paton, P. W. C. 2006. Assessing habitat selection by foraging egrets in salt marshes at multiple spatial scales. *Wetlands*, 26:307–312.

Tyrrell, M., Dionne, M. and Edgerly, J. 2008. Physical factors mediate effects of grazing by a non-indigenous snail species on saltmarsh cordgrass (*Spartina alterniflora*) in New England marshes. *ICES Journal of Marine Science*, 65:746–752.

Uddin, N., Robinson, R. W., Buultjen, A., Al Haruna, A. U., and Shampa, S. H. 2017. Role of allelopathy of *Phragmites australis* in its invasion processes, *Journal of Experimental Marine Biology and Ecology*, 486: 237–244.

Wan, S., Qin, P., Liu, J., and Zhou, H. 2009. The positive and negative effects of exotic *Spartina alterniflora* in China. *Ecological Engineering*, 35:444–452.

Weber, W. A. 1989. Additions to the flora of Colorado. *Phytologia*, 67:429–437.

Wilson, C. A., Hughes, Z. J., and FitzGerald, D. M. 2012. The effects of crab bioturbation on Mid-Atlantic saltmarsh tidal creek extension: geotechnical and geochemical changes. *Estuarine, Coastal and Shelf Science*, 106: 33–44.

Windham, L., and Lathrop, V. 1999. Effects of *Phragmites australis* (common reed) invasion on aboveground biomass and soil properties in brackish tidal marsh of the Mullica River, New Jersey. *Estuaries*, 22:927–935.

Windham, L., and L. A. Meyerson. 2003. Effects of common reed (*Phragmites australis*) expansions on nitrogen dynamics of tidal marshes of the northeastern U.S. *Estuaries*, 26: 452–464.

Young, J. A., Palmquist, D. E., and Wotring, S. O. 1997. The invasive nature of *Lepidium latifolium*: A review, pp. 59–68. In: Brock J. A., Wade, M. Pysek, P., Green, D. (eds.). *Plant Invasions: Studies from North America and Europe*. Backhuys Publishers, Leiden, The Netherlands.

Young, J. A., Palmquist, D. E., and Blank, R. R. 1998. The ecology and control of perennial pepperweed. *Weed Technology*, 12(2): 402–405.

Zhang, R. S., Shen, Y. M., Lu, L. Y., Yan, S. G., Wang, Y. H., Li, J. L., and Zhang, Z. L. 2004. Formation of *Spartina alterniflora* salt marshes on the coast of Jiangsu Province, China. *Ecological Engineering*, 23:95–105.

15

Marsh Edge Erosion

MICHELE BENDONI, IOANNIS Y. GEORGIOU, AND ALYSSA B. NOVAK

15.1 Salt Marshes and Ecosystem Services

Salt marshes are coastal ecosystems located at the boundary between sea and land, generally in tidal environments, often covered by halophytic vegetation and periodically flooded by tide (Allen 2000).

The importance of salt marshes consists of providing valuable and useful ecosystem services to coastal communities (Costanza et al. 1997, Barbier et al. 2011, Gedan et al. 2011, Ojea et al. 2012). Many organisms, such as fishes, shrimps, crabs, mussels, and birds, find suitable habitats in salt marshes, which often become preferential areas for fish nurseries (Boesch and Turner, 1984). Salt marshes capture chemical compounds and act as sinks of carbon dioxide (up to 210 g CO_2 m^{-2}yr^{-1}; Chmura et al. 2003), nitrogen, and phosphorous, leading to a net reduction of nutrient inputs to estuaries (Gosselink and Pope 1974, Valiela and Teal 1979). Marshes and wetlands also play an important role in protecting shorelines through the dissipation of hydrodynamic energy from storm surges and waves due to the presence of halophytic vegetation (Möller and Spencer 2002, Riffe et al. 2011, Jadhav et al. 2013, Möller et al. 2014), and thus reducing the need for maintenance of sea walls or dikes protecting the hinterland, which represents an ecosystem-based coastal defense against floods, storm surges, and hurricanes (Costanza et al. 2008, Gedan et al. 2011; Shepard et al. 2011; Temmerman et al. 2013; Barbier et al. 2011). Similarly, shoreline erosion can be reduced due to the binding effect of plant roots on soil (King and Lester 1995, Coops et al. 1996, Chen et al. 2012), a paradigm that Feagin et al. (2009) challenged using flume studies, which highlights the need for continuing studies to better understand the role of plants on the morphodynamics of coastal systems.

Salt marshes provide some of the highest and most valuable ecosystem services among natural ecosystems (Bell 1997, Costanza et al. 1997, Chmura et al. 2003, Emerton and Kekulandala 2003, Costanza et al. 2008, Barbier et al. 2011). To present an order of magnitude example, Table 15.1 reports the estimated economic value in US$ of different ecosystem services provided by salt marshes.

Presently, salt marshes are threatened worldwide due to increasing anthropic pressure, subsidence, sea level rise, and the erosive processes brought about by storms which are expected to intensify with climate change (Webster et al. 2005, Knutson et al. 2010). In particular, those responsible for the retreat of the marsh edge are the main contributors to

Table 15.1 *Estimate of the economic value of several ecosystem services provided by salt marshes*

Ecosystem service	Value	Reference
Coastal protection (reduced hurricane damages in USA)	8236 US\$ ha^{-1}yr^{-1}	Costanza et al. (2008)
Flood attenuation in Colombo (Sri Lanka)	$5 \cdot 10^6$ US\$ yr^{-1}	Emerton and Kekulandala (2003)
Habitat for fisheries in east and west Florida coast (USA)	2980 US\$ ha^{-1}	Bell (1997)
Carbon sequestration worldwide	30.5 US\$ ha^{-1}yr^{-1}	Chmura et al. (2003)

Table 15.2 *Salt marsh edge reduction rate for several areas of the world*

Reduction rate (m/yr)	Environment	Reference
0.5–2.2	Venice Lagoon (Italy)	Day et al. (1998); Bendoni et al. (2016)
0.4–3.0	Westerschelde (Netherlands)	Van der Wal et al. (2008)
0.5–2.0	Virgina Coastal Reserve (Virgina, USA)	McLoughlin et al. (2014)
up to 2.3	Savannah River (Georgia, USA)	Houser (2010)
0.25–2.0	Barnegat Bay Little-Egg Harbor (New Jersey, USA)	Leonardi et al. (2016)
up to 2.5	Barataria Bay (Louisiana, USA)	Beland et al. (2017)

salt marsh area reduction (Schwimmer 2001, Gedan et al. 2009, Marani et al. 2011). Retreat rates of the order of decimeters to several meters per year can be seen in many areas of the world as reported in Table 15.2. As a consequence of salt marsh loss, essential ecosystem services are presently being lost at a high rate, and it is therefore of utmost importance to protect these environments to preserve the quality and quantity of the services they provide directly and indirectly to human communities.

15.2 Processes Promoting Salt Marsh Bank Erosion

In this section we describe the processes responsible for the lateral retreat of the salt marsh banks. A simple conceptual diagram is presented in Figure 15.1, showing all the interactions and feedback mechanisms among the processes and elements controlling bank retreat.

Tide controls the overall hydrodynamic framework producing currents and influencing wave generation and propagation, which are also affected by morphology and vegetation. Hydrodynamic forcing induces erosive processes and mass failures which are, in turn, conditioned by the characteristics of soil, local morphology and above- and belowground vegetation biomass. The morphodynamic evolution of the system exerts a feedback on the

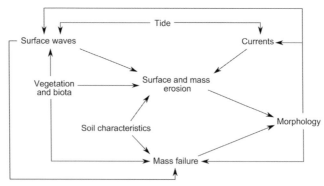

Figure 15.1 Interactions and links among between and elements leading to bank retreat. (From Bendoni 2015)

main hydrodynamic forcing linking all the processes involved. The conceptual schematic reported in Figure 15.1 does not account for salt marsh vertical evolution due to inorganic deposition, organic production, sediment compaction, and subsidence.

15.2.1 Mechanisms of Lateral Retreat

Lateral retreat is a complex mechanism by which salt marshes lose the majority of their spatial extent and can be identified as the landward migration of the marsh edge as a consequence of erosion; this is promoted by an ensemble of processes acting to remove or detach material from the marsh boundary through two main mechanisms: (1) surface and mass erosion and (2) mass failures (Winterwerp et. al 2012, Francalanci et al. 2013, Bendoni et al. 2016).

Surface erosion and mass erosion are defined as the continuous detachment of particles from a surface, both under drained and undrained conditions. This classification was introduced by Winterwerp et al. (2012) for environments with cohesive sediments, and includes the entrainment and suspension of flocs, as well as floc erosion processes. From a phenomenological point of view, here we extend the classification to estuarine environments where salt marshes are present. However, for simplicity, we make no distinctions between surface and mass erosion, and from hereon we consider them together and identify them by the term "surface erosion," which is envisaged as the quasi-continuous detachment of particles and aggregates of particles from the bank surface due to the shear stress induced by currents and waves or by the overpressure induced by wave impact. Length scales involved in surface erosion are much smaller compared to the marsh bank dimensions; surface erosion scale is order of centimeters, whereas bank retreat length scale is order of decimeters or meters.

Mass failure is a discontinuous process consisting of the detachment of much larger portions of material of different sizes from the bank scarp, characterized by length scale dimensions comparable to bank height. In more general terms, mass failures are episodic changes in the morphology of banks or cliffs in rivers, estuaries, or coasts

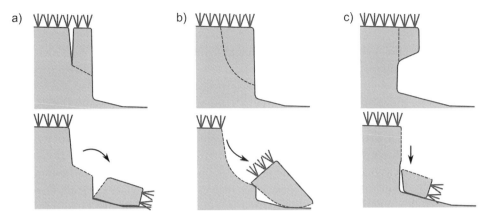

Figure 15.2 Mass failures modes related to salt marsh banks. (a) toppling failure; (b) rotational sliding; (c) cantilever failure (From Bendoni 2015)

which occur when destabilizing forces, such as gravity and hydrodynamic thrust, exceed resisting forces such as soil shear strength (Thorne and Tovey 1981, Van Eerdt 1985a, Allen 1989, Selby 1993).

Different typologies of mass failures can be identified including: cantilever failures, toppling, sliding, and others depending on the bank morphology. Characterizing features are cantilever profiles due to scour at the bank toe, vertical tension cracks produced by wetting and drying cycles, critical surfaces along which rotational sliding may occur. An illustrative conceptual sketch of the different mass failures typologies is provided in Figure 15.2.

15.2.2 Hydrodynamic Forcing

In tidal environments, where salt marshes are usually present, surface waves, tidal currents, and their mutual interaction are the main hydrodynamic factors affecting marsh retreat, together with boat wakes. Wave height in fetch-limited basins is controlled by water depth (Young and Verhagen 1996, Breugem and Holthuijsen 2007), which changes with tidal oscillations. Furthermore, the water level in front of the marsh edge influences whether waves strike the bank or overtop it, each of which produces different effects on erosion (Bendoni et al. 2016).

Several studies have focused on the hydrodynamic forcing, providing an analysis of flow characteristics and a quantitative measure of its potential effect on erosion: Callaghan et al. (2010) analyzed whether tidal currents or waves were more effective in causing erosion; Fagherazzi and Wiberg (2009) and Mariotti et al. (2010) focused on the effect of boundary conditions such as fetch length, water level, and predominant wind on wave height and wave power, and, as a consequence, on the fate of basins surrounded by salt marshes (Mariotti and Fagherazzi 2013). Tonelli et al. (2010) and Karimpour et al. (2016), through numerical modelling and field measurements, respectively, provided detailed analysis of the hydrodynamic features of the wave-induced flow at the edge of a salt marsh bank.

Callaghan et al. (2010) studied four different field sites located in the Westerschelde estuary, characterized by contrasting features in terms of lateral evolution (retreating or prograding) and wind exposure (sheltered or exposed). Waves and currents were measured at each site, and the wave model SWAN (Booij et al. 1999), forced by measured wind data, was validated and employed to support and improve the analysis. The authors concluded that bed erosion is more strongly affected by wind waves rather than tidal or wind-induced currents. Furthermore, spatial wave modelling revealed crucial information in assessing the vulnerability of a site to wave forcing: simulated waves in front of the marsh were higher for retreating salt marshes regardless of the marsh exposure to wind, suggesting that wave analysis based only on wind exposure is not always sufficient to predict the fate of marsh shoreline.

The growth and decay of wind waves depends on many factors including fetch length, water depth and water level, duration, and the predominant wind speed and direction. These dependencies were investigated by several authors using field data and models for wind wave generation and propagation (Fagherazzi and Wiberg 2009, Mariotti et al. 2010, Mariotti and Fagherazzi 2013, McLoughlin et al. 2014).

For example, Fagherazzi and Wiberg (2009) analyzed the response of the shallow basin of the Virginia Coast Reserve, Eastern Shore (Virginia, USA) to wind wave events, focusing on the induced bed shear stress. Wave heights were determined through the formulation proposed by Young and Verhagen (1996) for fetch-limited basins. Bed shear stress τ_b [N/m^2] was calculated as follows:

$$\tau_b = \frac{1}{2}f_w\rho u_b^2 \qquad \text{(Equation 15.1)}$$

where u_b [m/s] is the bottom orbital velocity determined by linear theory, ρ [kg/m^3] is the water density and f_w [-] is the wave friction factor (Fredsoe and Deigard, 1993) defined as:

$$f_w = 0.04\left(\frac{u_b T}{2\pi k_b}\right)^{-0.25} \qquad \text{(Equation 15.2)}$$

where T [s] is the wave period, and k_b [m] is the roughness length scale of the sediment bed and is equal to 2D$_{90}$. In order to quantify the response of the entire basin, they introduced an erosion factor EF [N] defined as follows:

$$EF = \sum_i A_i(\tau_{b,i} - \tau_{cr}) \qquad \text{(Equation 15.3)}$$

where the subscript i is related to the i-th location with area equal to A_i [m^2] and τ_{cr} [N/m^2] is the critical shear stress for bottom erosion, depending on the characteristics of the reference environment, which may vary from 0.2 to 1.0 N/m^2 or more for non-cohesive and cohesive sediment respectively (Amos et al. 2004). The authors determined the value of EF as a function of water depth and wind direction, showing the high sensitivity of the system susceptibility to erosion to variable boundary conditions.

An improvement of the analysis carried out by Fagherazzi and Wiberg (2009) was proposed by Mariotti et al. (2010), for the analysis of the system of shallow lagoons within the Virginia Coast Reserve, on the Atlantic side of the Delmarva Peninsula (USA). Since the wave impact on marsh edges is mainly responsible for marsh area reduction, the authors defined the Wave Factor at the marsh Boundary (*WFB*) [W/m] to quantify the erosive potential of wave impact as follows:

$$WFB = \frac{\sum_{t=1}^{T} \sum_{i=1}^{M} L_i(P_{i,t} - P_{cr})}{L_{mb}T} \qquad \text{(Equation 15.4)}$$

where $P_{i,t}$ [W/m] is the wave power at the marsh boundary element i, with length L_i [m], at time t [s], L_{mb} [m] is the total length of marsh boundary, and T [s] the total time on which *WFB* is averaged. Erosion occurs when the wave power is higher than a critical wave power P_{cr} [W/m] set by authors equal to 50 W/m. Wave power is calculated using linear theory through the following relation:

$$P = c_g \rho g \frac{H^2}{8} \qquad \text{(Equation 15.5)}$$

with g [m/s^2] the gravity acceleration, H [m] the wave height, and c_g [m/s] the wave group celerity given by linear theory.

Figure 15.3 shows how *WBF* is affected by the characteristics of the considered environment and subsequently, how the specific morphology of a site, as well as the wind field can be critical in determining the future evolution of the respective marsh shoreline.

Mariotti and Fagherazzi (2013) investigated the effect of the mutual interaction between hydrodynamics, wind dependent boundary conditions, and the morphological characteristics of a marsh shoreline using a dynamic model of a salt marsh basin. The model included marsh boundary progradation and retreat, and basin depth evolution in response to tidal and wind wave forcing. They showed that, all other things being equal, a critical width exists for each basin, beyond which a positive feedback arises between tidal flat widening and wave-induced marsh boundary erosion, leading to irreversible marsh loss.

Tonelli et al. (2010) employed a numerical model to analyze the effect of wave action on the marsh edge as a function of tidal elevation and wave height for different bank typologies. The numerical model solves the coupled set of Boussinesq (Madsen and Sorensen, 1992) and nonlinear shallow water equations depending on whether dispersion or nonlinear effects are predominant, using a finite volume method. The numerical experiments consider a bank edge subject to continuous wave forcing. Three different bank typologies are analyzed: vertical, sloping, and terraced, based on the morphology of the marsh boundaries of the Virginia Coast Reserve (USA). Wave thrust on the marsh surface is determined through the integral form of momentum conservation law.

They found that wave thrust is strongly dependent on tidal elevation. The thrust tends to increase with water level until the marsh surface is submerged, then experience a sharp reduction, both for the vertical and the sloping marsh scarp (Fig. 15.4). The terraced scarp shows a different behavior as waves break over the terrace, since in such a case it is the

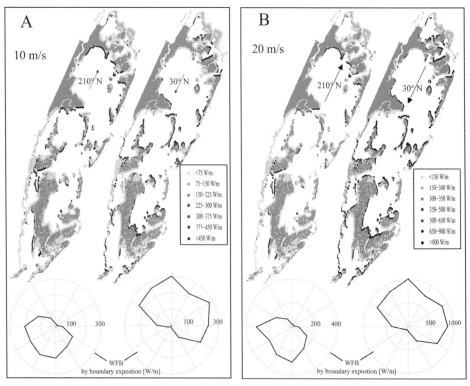

Figure 15.3 Wave power at the marsh boundary, averaged over 48 h of simulations, calculated for different wind speeds and directions. (A) wind speed of 10 m/s, (B) wind speed of 20 m/s. On the bottom *WFB*, for each wind condition, calculated as a function of marsh exposure. (Image from Mariotti et al. 2010)

lower terrace that absorbs most of the wave thrust. With reference to Figure 15.4 the following symbols are employed: ΔF_w [kN/m] is the difference between maximum and minimum wave thrust, $F_{w,max}$ [kN/m] is the maximum wave thrust, and $\sigma_{w,max}$ [kN/m^3] the maximum stress.

The above-mentioned studies, suggest that the way in which waves strike the marsh edge is fundamental in relating the resulting retreat rate or type of erosion. Water levels in front of the bank, which are generally controlled by tidal and subtidal motion, can alter the way waves are locally transformed, or more importantly, control how waves can differentially transfer part of their energy to the marsh surface leading to variable erosive potential.

A field measurement campaign carried out by Karimpour et al. (2016) in Terrebonne Bay, Louisiana (USA), during the passage of a storm surge, investigated the characteristics of the flow field at an inundated salt marsh edge. Though dimensional analysis of the field observations, they showed that a functional dependence exists between the wave orbital and current velocities, the incident wave height, the inundation depth, the ratio between water depth and stem vegetation height (submergence ratio), and the vegetation density. Furthermore, a functional relation to estimate current direction at the marsh edge from the

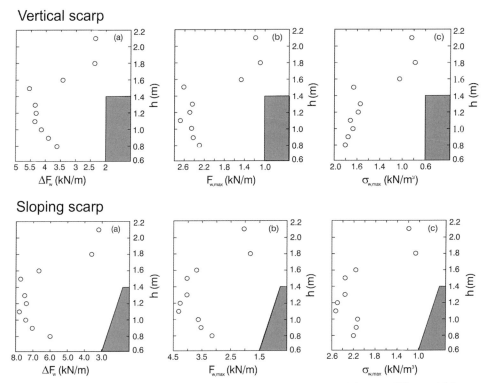

Figure 15.4 Wave thrust evolution for a given wave height ($H = 0.3$ m) at different tidal elevations for the vertical (upper pictures) and sloping (lower pictures) scarp. a) difference between maximum and minimum wave thrust; b) maximum wave thrust; c) maximum stress. (From Tonelli et al. 2010)

combined effects of local wave energy, inundation depth, submergence ratio, and vegetation density was found. The authors concluded that the flow field at the marsh edge is strongly influenced by a nonlinear combination of the hydrodynamic and vegetation characteristics at the site.

Another significant process which determines erosion in low energy environments such as small enclosed bays, tidal creeks, or sheltered marshes, is the wetting and drying cycle of marsh cliffs due to the tidal wave (Cola et al. 2008, Chen et al. 2011). Chen et al. (2011) found significant retreat rate on a sheltered marsh of the Beaulieu River Estuary (UK). Based on field measurements, they argued that a large amount of cycles of emergence and submergence may reduce the soil resistance to erosion, through the weakening of the bonds of soil particles. The toe of the saltmarsh bank was more frequently subject to alternation of emergence and submergence, resulting in a higher erodibility which promoted the formation of cantilever profiles and subsequent mass failures. Furthermore, through the modelling of the groundwater and soil saturation response of a salt marsh to tidal wave, Cola et al. (2008) showed that the cyclic oscillation of mean and effective stresses induced by the tide within the soil matrix, may facilitate the formation of tension cracks affecting the stability of marsh scarps.

In addition to wind waves, and depending on the coastal setting, salt marsh edges can be threatened by boat wakes. FitzGerald et al. (2011) reported that the passage of moving boats, particularly within confined waters and regions usually subject to low wave energy, can introduce an appreciable increase in wave energy that can have significant effects on the marsh shoreline.

The hydrodynamics of boat wakes depends on several factors: water depth, vessel length and speed, displacement (loading), hull shape, and the natural conditions of the water body, such as the presence of currents and wind waves (Maynord, 2001). To classify boat wakes (for displacement hull vessels) the depth-based and length-based Froude numbers (Fr_D [-] and Fr_L [-]) are employed (Sorensen, 1973):

$$Fr_D = \frac{V_b}{\sqrt{gh}} \qquad\qquad \text{(Equation 15.6a)}$$

$$Fr_L = \frac{V_b}{\sqrt{gL_b}} \qquad\qquad \text{(Equation 15.6b)}$$

where V_b [m/s] is the vessel speed, h [m] the water depth and L_b [m] the boat length. Wave height approaches a maximum as Fr_D [-] approaches 1 and Fr_L [-] approaches 0.5, with a consequent significant impact on the coastline.

Several authors investigated the effect of boat waves on marsh and beach shoreline (Hughes et al. 2007, Houser 2010, FitzGerald et al. 2011, Bilkovic et al. 2017) or on marsh vegetation (Silinski et al. 2015); these aspects will be addressed in section 15.3.2.

15.2.3 Effect of Vegetation

The evolution of the salt marshes is significantly affected by the presence (or absence) of vegetation. Aboveground and belowground biomass act differently on the morphodynamic evolution of salt marshes. On the one hand, aboveground vegetation modifies the flow field due to the presence of stems, which produce drag, dissipate energy, and create complex flow patterns (Kobayashi et al. 1993). As a consequence, vertical dynamics (or vertical accretion) of the marsh surface (or platform) are also influenced by plants through sediment trapping (Marani et al. 2010) as well as organic production (Morris et al. 2002). Belowground biomass on the other hand, affects the soil resistance due to the effect of the root mat (Chen et al. 2012).

The effectiveness of marsh vegetation in dissipating wave energy, acting as an obstacle to the free water movement, is widely recognized and accepted. Several researchers tested the effects of vegetation through field measurements and laboratory experiments, showing that wave height is considerably attenuated by vegetation even when the stems are completely submerged (Möller and Spencer 2002, Riffe et al. 2011, Jadhav et al. 2013, Möller et al. 2014). Furthermore, energy dissipation varies within the wave spectrum acting differently on sea waves and swell, where larger energy dissipation occurs for sea waves, particularly at the transition to the vegetated area, whereas swell wave energy is dissipated at a slower rate and is able to travel farther landward through the vegetation field (Jadhav et al. 2013).

To account for the presence of vegetation in phase-averaged wave propagation modelling, a dissipative term in the wave energy balance has been introduced by Mendez and Losada (2004). Such a dissipative term D_{veg} [J/m^2/s] is a function of the wave field through the root-mean-square wave height H_{rms}, [m], the wave number k [m^{-1}], wave frequency σ [s^{-1}], and water depth h [m], and the characteristics of the plant species through the stem width b_v [m], the vegetation unit density N_v [units/m^2], and the plant height αh [m]. Mathematically it reads:

$$D_{veg} = -\rho C_{Dv} b_v N_v f(k, \sigma, h, \alpha h) H_{rms}^3 \qquad \text{(Equation 15.7)}$$

where C_{Dv} [-] represents the bulk drag coefficient depending on the particular plant species and on the Keulegan–Carpenter number $KC = u_{rms} T_{rep}/b_v$ (Keulegan and Carpenter 1958), in which u_{rms} [m/s] is the root-mean-squared orbital velocity and T_{rep} [s] is a representative wave period. The KC number represents the ratio of drag forces to inertial forces for an object subject to an oscillatory flow field. A low value of KC number means the inertia force is dominant, whereas a high value means the drag force is the dominant one and the flows tends to behave as a uniform current.

Bulk drag coefficient C_{Dv} is determined empirically for marsh plants such as *Spartina alterniflora* (Jadhav et al. 2013, Ozeren et al. 2013), with values ranging from to 2 to 10. Values are obtained by $C_{Dv} = 70/KC^{0.86}$ (Jadhav et al. 2013) and $C_{Dv} = 0.036 + 50/KC^{0.93}$ (Ozeren et al. 2013), with KC number ranging from 10 to 120.

As stated at the beginning of the section, belowground biomass does not affect the flow field, but it acts on soil characteristics and erodibility. The strengthening effect of the root mat on soil, reducing erosion, is generally accepted in literature (Gray and Leiser 1982), and for salt marsh banks this effect was studied by King and Lester (1995) and Chen et al. (2012). The efficiency of plant species in reducing lateral retreat differs based on the characteristics of the vegetation (Van Eerdt 1985b), such as the density of the root mat and the tensile strength of the roots. Regardless, despite the damage potentially caused by waves on the aboveground vegetation, marsh substrates tend to remain stable and resistant to surface erosion (Möller et al. 2014).

Feagin et al. (2009) argued that salt marsh plants do not directly reduce erosion along the marsh edge, but the soil type is mainly responsible for the resistance against the erosive effect of waves. The authors found that the role of vegetation is to modify the soil parameters, thus indirectly affecting the lateral erosion rate. However, Feagin et al. (2009) refer to the process of particle removal from a volume of soil matrix where roots are present, and do not report on the role of vegetation in enhancing bank stability. Recently, Bendoni et al. (2016) developed a simple mathematical model based on the differential erosion of a bank profile due to the presence of vegetation, which accounts for mass failure events, to show that the presence of vegetation does not necessarily reduce the average retreat rate: a stronger layer at the bank top, due to the presence of vegetation roots, may induce the formation of cantilever profiles easily susceptible to mass failure events. Furthermore, Francalanci et al. (2013), using laboratory experiments of physical models of a salt marsh bank, found that the presence of vegetation promotes a delay in mass failures, but it does not have a significant effect in reducing the total eroded volume.

It is obvious from the results in the previous section that further efforts are required to deepen the study of the effects of vegetation on salt bank retreat. Moreover, it is important to stress that the edge of salt marshes are subject to both shear stress erosion and wave impact erosion and to date, in the literature, specific process-based or theoretical models describing the influence of vegetation on the erosion due to frontal wave impact are not available. The only exception is the simplified approach proposed by Mariotti and Fagherazzi (2010) where the presence of vegetation increases the critical wave power threshold P_{cr} [W/m] for the onset of wave impact erosion through the following formulation:

$$P_{cr,veg} = P_{cr}\left(1 + K_{veg}\frac{B}{B_{max}}\right) \qquad \text{(Equation 15.8)}$$

where K_{veg} [-] is a dimensionless parameter (Le Hir et al. 2007) which provides an estimate of the degree which the aboveground vegetation increases the resistance of soil to wave impact erosion, and B [g/m^2] and B_{max} [g/m^2] quantify respectively the aboveground biomass and its maximum value.

On the other hand, the effect of vegetation in reducing shear erosion via the root-mat has been modelled by Van der Meer et al. (2007) and Tuan and Oumeraci (2012) for an application that involved sea dikes. The effect of grass root reinforcement of the clay cover of sea dikes has been included into the formulation proposed by Mirtskhoulava (1991). The critical velocity (and subsequently the critical shear stress) for the onset of erosion of grass-permeated soil is considered to be a function of undrained shear strength of the soil c_u [N/m^2], tensile strength of the roots t_r [N/m^2] and root area ratio RAR [-] which is defined as roots area per unit area. Bendoni (2015) modified the proposed formulation to adapt it to soil mixtures characterized by the presence of sand and mud, obtaining:

$$\tau_{cr,veg} = \tau_{cr}C_{sv}^2\left[1 + \frac{1.2 \cdot RAR \cdot t_r}{f(\rho, \rho_s, d_a, p_m, c_u)}\right] \qquad \text{(Equation 15.9)}$$

where $C_{sv} = C_{cz}/C_{cz,veg}$ and C_{cz} [m$^{1/2}$/s] and $C_{cz,veg}$ [m$^{1/2}$/s] are respectively the Chezy coefficients for bare and vegetated soil, ρ_s [kg/m^3] is soil density, d_a [m] the average size of detaching aggregates (order of 0.003–0.005 m), and p_m [-] the mud fraction of the soil. In the proposed relation, the amount of root-mat and the mechanical characteristics are quantified by the RAR and the t_r respectively, which tend to increase the critical shear stress $\tau_{cr,veg}$. The effect of the interaction of the soil characteristics with the roots is quantified by the denominator $f(\rho, \rho_s, d_a, p_m, c_u)$.

Finally, Howes et al. (2010) showed that low salinity wetlands were preferentially eroded compared to higher salinity wetlands during Hurricane Katrina. They argued this was due to geotechnical differences related to different characteristics of the root mat of the local vegetation, particularly the depth of the roots (rooting depth). In high salinity systems, the deeper root-mat contributed to increase and homogenize shear strength with increasing depth (high rooting depth), while for low salinity wetlands, the shallower root-mat (low rooting depth) contributed to create a more superficial failure plane where an abrupt reduction in the shear resistance was observed. Such a reduction in shear resistance makes the low salinity wetlands more susceptible to erosion, as the critical shear resistance, proportional to the rooting depth, is more easily exceeded by wave-induced shear stress.

15.2.4 Human-Induced Processes

Human activities can directly or indirectly cause salt marsh loss. Researchers have hypothesized that recreational overfishing in New England has depleted large predators and led to an increase in *Sesarma reticulatum* (purple marsh crab) populations, expansion of communal *S. reticulatum*, and overgrazing of *Spartina alterniflora*. Altieri et al. (2012) showed that 80% of inter-site variation in the extent of salt marsh die off in Cape Cod (Massachusetts, USA) can be explained by inter-site variation in *S. reticulatum* herbivory, which is negatively correlated with predator density. Comparative data from salt marshes spanning the Atlantic coastline from Cape Cod to Long Island Sound (USA) further support the linkage between predator depletion and salt marsh die-off (Coverdale et al. 2013). Besides overgrazing on marsh plants, studies have shown that burrowing activity of crabs and other microfauna can further promote saltmarsh loss by weakening soil structure (Hughes et al. 2009, Davidson and Rivera 2010).

Historically, salt marshes were filled to form upland and/or drained or diked to create farmland in the USA. Filling has been identified as the major cause of the dramatic loss of salt marsh in the Boston Harbor region (Massachusetts, USA; Bromberg and Bertness 2005). Likewise, in Long Island salt marsh loss during the twentieth century is mostly due to dredging and fill projects. When salt marshes are dredged or filled, these activities can cause drowning of vegetation and increase the wave energy approaching the marsh boundary (Morris et al. 2002, D'Alpaos et al. 2007, Kirwan et al 2010, Mariotti and Fagherazzi 2010, Weston 2014, Smith 2015).

Nutrient enrichment, a global problem for coastal ecosystems can also be a driver of salt marsh loss. Deegan et al. (2012) showed in field experiments that nutrient levels commonly associated with coastal eutrophication increased aboveground leaf biomass, decreased the dense, belowground biomass of bank-stabilizing roots, and increased microbial decomposition of organic matter. Consequently, alterations in these key ecosystem properties reduced geomorphic stability and resulted in creek-bank collapse with significant areas of creek-bank marsh converted to unvegetated mudflat.

Man-made disasters such as oil spills have also resulted in marsh loss. Oil from the BP-Deepwater Horizon oil spill (April 2010) became concentrated on the marsh edge and enhanced the rate of decline of Louisiana salt marshes. Silliman et al. (2012) found the retreat rate of the marsh edges impacted by the oil spill was more than twice the retreat rate of those marshes not subject to contamination, during the period ranging from October 2010 and October 2011. They suggested the oil-induced death of stabilizing root mat promoted an acceleration of the erosive process. However, recovery of salt marsh vegetation was also found after roughly 1.5 year after the oil spill, observing a retreat rate levels comparable to those of non-oiled sites. Similarly, Beland et al. (2017) reported that oiled marshes in Louisiana eroded at rates more than 50% higher compared to non-oiled marshes when oiling of more than 20% occurred; however, they reported that the return to background erosion may take longer than 1.5 years as reported by Silliman et al. (2012) and anywhere between 3 and 6 years.

15.3 Modelling Strategies

In this section we present the different approaches employed by various researchers to study and analyze lateral bank retreat processes.

We start by introducing conceptual models which are employed to catch and schematize the principal features of the process of salt marsh edge erosion, in order to make it more easily intelligible by the reader and integrate it in the overall morphodynamic evolution of tidal environments. Then, we describe the proposed mathematical models used to interpret and explain measured data and observations and to make quantitative prediction regarding the future retreat rate of a marsh edge. Next, the models employed to describe, analyse, and quantify the mass failure mechanism are presented. Finally, we briefly introduce process-based models for long-term forecast of saltmarsh and tidal flat coupled evolution, which employ specific formulations to describe the retreat rate of the marsh edge.

15.3.1 Conceptual Models

Conceptual models are often used to qualitatively describe the lateral retreat mechanism and its implications on salt marsh dynamics, supporting the understanding of the processes driving lateral retreat or surface erosion, without providing a quantitative or mathematical interpretation.

Allen (1989), through field observations, analyzed and described various mass failures as a function of the characteristics of the physical environment and the soil type. He focused on different coastal areas of the UK including the Severn Estuary, which is a muddy environment, and Morecambe Bay and the Solway Firth characterized by the predominance of sand.

The author identified that muddy sediments with sparse root-mats yield tall cliffs subject to toppling failure and sliding (Fig. 15.2a, b) due to the presence of tension cracks. Such cracks can be ascribed to the shrinkage of the material during the dry season (Morris et al. 1992) or to the accumulation of shear and volumetric deformation induced by cyclic oscillation of effective stresses due to tidal excursion (Cola et al. 2008). Following crack formation, water can fill them as the tide rises, leading to an increase in pressure which in turn increases the thrust shoreward at low tide (Francalanci et al. 2013). Moreover, the slumping of the block identified by the presence of the crack can be accelerated by the effect of episodic wave forcing, which may promote a toppling-type failure (Bendoni et al. 2014).

Marshes with higher sand content on the other hand, covered by a dense root-mat which strengthens the upper portion of the soil, or marshes with stratigraphy that promotes a differential erosion between the toe and the top, tend to develop shorter cliffs, characterized by the presence of cantilever profiles due to the scour induced by waves and currents. Such

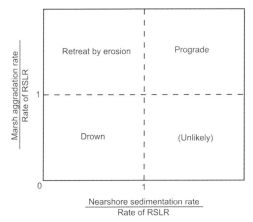

Figure 15.5 Three response of a marsh shoreline to local sea-level rise. (From Schwimmer and Pizzuto 2000)

morphology is generally subjected to cantilever failure once the scour exceeds a critical value (Fig. 15.2c).

Failure mechanisms are directly responsible for marsh retreat but the causes promoting them have to be facilitated by the hydrodynamic forcing at the marsh edge such as the interplay of waves, tide, and relative sea-level rise (RSLR).

The response of a marsh shoreline to local RSLR was first conceptualized by Schwimmer and Pizzutto (2000) where they identified three main states: retreat by erosion, progradation, or drowning. The first two states may occur when the rate at which the marsh surface elevation increases vertically, as a consequence of depositional processes, is equal or higher than the rate of the RSLR. Then, if the nearshore sedimentation is higher than the rate of RSLR, the marsh prograades, otherwise, it retreats. When the marsh aggradation rate is lower than the rate of RSLR, the marsh drowns (Fig. 15.5).

Wilson and Allison (2008) provided an interpretation of the evolution of the marsh profile, using coring and bathymetric surveys at various marsh shoreline environments in southeastern Louisiana (USA). They found an equilibrium profile exists among the shores of Baratraria Bay and Breton Sound and they argued it is mostly driven by wave activity instead of sediment characteristics due to the different soil compositions of the surveyed areas. They also developed a conceptual model describing the geomorphic evolution of a cross-shore transect of the marsh edge, accounting for the lateral retreat and the vertical translation of the equilibrium profile. This dynamic is produced by the interplay of wave erosion, RSLR, and deposition of estuarine sediments onto the marsh platform.

The model suggests that when RSLR is higher than marsh accretion or platform aggradation, wave forcing is able to laterally incise the marsh surface leading to the formation of a marsh scarp which is potentially more prone to erosion due to wave impact. Continuous RSLR, in combination with upward and landward translation of the equilibrium profile, induces the erosional bottom surface to reach a threshold where waves are no longer able to promote erosion due to a very low wave-induced bottom shear stress.

The effect of subsidence allows the accommodation depth to increase and receive sediments originating from the erosion of the marsh fringe or other sources (e.g., fluvial input from the Mississippi river), which deposit on the erosional surface maintaining the equilibrium profile.

Chauhan (2009) was inspired by field observations in Morecambe basin (UK) to derive a model characterizing the autocyclic marsh growth and erosion: the cyclical nature of erosional and progradational phases. When young marshes begin to accrete, scour tends to develop at the frontal edge of the shoreline at an elevation where water level is present most part of the time, meaning that at such an elevation wave forcing is more likely to attack the bank. The progression of the retreat, together with the erosion of the tidal flat driven by waves in front of the marsh, allows for the formation of accommodation space for sedimentation, leading to a new progradational phase. A subsequent erosional phase starts once the tidal flat is transformed into marsh surface. The author argued that the terraced morphology with the presence of several clifflets, observed in Morecambe Bay, is indeed the result of the autocyclic process characterized by distinct phases of marsh retreat.

Another conceptual model based on the interplay between continuous erosion and mass failures has been posited by Francalanci et al. (2013), starting from laboratory experiments conducted on vegetated and unvegetated marsh transects. The model describes continuous shifts among different states which may occur in the field, depending on the dominance of tide, wave action, and sediment availability. Starting from an equilibrium profile, the erosion of the basal bank leads to an unstable configuration prone to erosion; then, the interplay of waves, tide, and sediment fluxes may shift the system toward a new erosive phase or a renewed equilibrium profile.

The non-uniform, episodic shoreline erosion which can occur on salt marshes dissected by wave cut-gullies has been addressed by Priestas and Fagherazzi (2011) using field observations at the Rockefeller Wildlife Refuge, Louisiana (USA). They consider a cohesive scarp subject to wave forcing causing the removal of vegetation at specific locations due to small perturbation or irregularities present on the scarp edge; if the wave forcing is energetic enough (significant wave height at least 20 cm) the perturbation may enlarge and concentrate the wave load, leading to the development of the gullies which lengthen and widen as the headward erosion proceeds. If the wave forcing continues, the vegetation is also removed from the marsh platform between the gullies, leaving the lower layer without protection and thus more susceptible to erosion.

15.3.2 *Observational and Experimental-Based Prediction of Marsh Edge Evolution*

Field data of marsh shoreline evolution have been widely used to derive interpretative models or to forecast future trends. Generally, they were collected over medium to large temporal scales (order of years, decades) or they were obtained from aerial photographs covering wide tidal areas (Day et al. 1998, Schwimmer 2001, Callaghan et al. 2010, Howes et al. 2010, Marani et al. 2011, McLoughlin et al. 2014, Leonardi and Fagherazzi 2015). However, field data in delimited areas have also been collected by various groups (Feagin et al. 2009,

Houser 2010, Trosclair 2013, Bendoni et al. 2016) as well as field experiments that produced large and useful datasets used to interpret processes and mechanisms of retreat (Francalanci et al. 2013, Bendoni et al. 2014, Möller et al. 2014, Silinski et al. 2015).

Day et al. (1998) collected field data in the Venice Lagoon (Italy), reporting edge retreat ranging from 1.2 to 2.2 m/yr due to wave erosion, despite high sediment inputs; the eroded material from the marsh was subsequently deposited on the marsh surface, resulting in an average accumulation rate equal to 15.5 g/m²/day. Additionally, they observed the formation of new tidal channels dissecting the marsh surface, similar to the processes observed by Priestas and Fagherazzi (2011) in the Rockefeller Wildlife Refuge, Louisiana (USA). Nevertheless, neither study provided an estimate of the wave power impinging the marsh shoreline.

The first attempt to relate erosion rates and wave climate was proposed by Schwimmer (2001) which employed field data collected by several authors (Maurmeyer 1978, Swisher 1982, Phillips 1985, French 1990) in different areas, to determine a power relationship (nearly linear) between average or annualized wave power [W/m] and erosion rate [m/yr].

Later, Marani et al. (2011), carrying out a dimensional analysis of the involved quantities based on the Buckingam theorem, and using consecutive aerial photographs of the Venice lagoon (Italy), found a linear relation between average wave power P [W/m] and volumetric erosion rate per shoreline length R_{sc} [m²/yr $\equiv$ m³/m/yr]. The dimensional analysis output, lead to:

$$R_{sc} = a \cdot P \qquad \text{(Equation 15.10)}$$

providing a strong theoretical basis for the description of the retreat process. As reported by the authors, the relationship is clearly site-dependent and the characteristics of the analyzed environment, such as soil cohesion and vegetation typology affecting marsh "erodibility" (intended as the propensity to erosion), are implicitly accounted for into the coefficient a [m³/W/yr].

Another attempt to find a functional dependence between wave forcing and the rate at which the marsh edge retreats has been carried out by McLoughlin et al. (2014). They utilized shoreline retreat rates between 1957 and 2007 for four time periods for several salt marshes located in a coastal bay of the Virginia Coast Reserve (USA) to explore a relationship between the wave energy flux and the erosion rate, expressed as volumetric erosion rate per unit length [m²/yr] or linear erosion rate [m/yr]. To calculate the wave power impinging the marsh boundary, they compared the use of a parametric model (Young and Verhagen 1996) and the wave propagation model SWAN (Booij et al. 1999).

Results show that a strong correlation exists between the wave energy flux and the volumetric erosion rate per unit length expressed as [m²/yr] both at the local and global scales, in agreement with the results of Marani et al. (2011). A correlation between wave energy flux and linear erosion rate holds if data are averaged over the entire length of a marsh edge, suggesting that differential erosion is present.

In contrast to the results of Callaghan et al. (2010), they found that both the parametric model (Young and Verhagen 1996) and wave propagation model SWAN (Booij et al. 1999) are in agreement with respect to the determination of the wave power, even if the magnitude resulted in lower values compared to other studies employing similar methods in the same area (e.g., Mariotti et al. 2010). They argued the main reason for the differences might be due to the way by which the average wave power is calculated, especially when the marsh is submerged. McLoughlin et al. (2014) considered the wave energy flux equal to zero when the mean water level was above the marsh edge, whereas Mariotti et al. (2010) averaged the wave power over several tidal cycles regardless of whether the water level was above or below the marsh surface.

In an effort to compare several datasets from different environments, Leonardi et al. (2015) performed analysis on dimensionless quantities for the erosion rate R_{sc}^{*} [-] and the wave power P^{*} [-].

The authors showed that the response of the salt marsh edge to wind waves tends to remain linear (Fig. 15.6), without the presence of a critical threshold beyond which a dramatic acceleration in marsh retreat arises. They concluded that very intense meteorological events such as storms or hurricanes do not contribute significantly to marsh loss, whereas, moderate events with return period around 2.5 months are the chief drivers of such a process. While true, Trosclair (2013) demonstrated that even with a local erosion factor validated using local measurements of erosion rates and waves over the same period, forecasting short-term erosion trends due to storms can be quite difficult despite demonstrated success in predicting long-term trends.

A summary of the relationships proposed by various authors are reported in Table 15.3, maintaining, for each of them, the original dimensional quantities. It is obvious that there is a large variability among the relationships, likely due to the various environments analyzed and the approach employed for each study.

In order to better compare the results, some of the relationships are manipulated to obtain dimensionally consistent formulations with retreat rate R_{sc} expressed in [m^2/yr] and wave energy flux P in [W/m]. The formulation by Marani et al. (2011), McLoughlin et al. (2014), and Bendoni et al. (2016) remain unchanged. To convert the relation proposed by Schwimmer (2001), a regression analysis of the data used by the author applying the same relation is carried out. Since erosion rates are in [m/yr], and the height of the marsh bank lies between 30 and 90 cm, the rate is multiplied by 0.6 m to obtain [m^2/yr] and the wave power is expressed in [W/m]. To recover the volumetric erosion rate in [m^3/m/yr] from the relationship proposed by Mariotti and Fagherazzi (2013), the value of lateral retreat [m/yr] is multiplied by 1 m, assumed as reference bank height. It is important to note that, the multiplication of linear erosion rates by a specific bank height, even if not precise or a good representation of the local environment, does not affect significantly the order of magnitude of the proportionality coefficient a (Equation 15.10) between wave forcing end erosion rate.

In this way, the differences in the proportionality coefficients a are more easily recognizable (Table 15.4). The value of a is of the same order of magnitude for the studies proposed by Mariotti and Fagherazzi (2013), McLoughlin et al. (2014), and Bendoni et al. (2016), even if the latter found a significant variability for the a parameter in spatially close sites accounting for or neglecting the effect of mass failures in the retreat rate; a slightly lower value was proposed by Marani et al. (2011). This range of variability, despite being large, is likely due to the local site characteristics as well as the

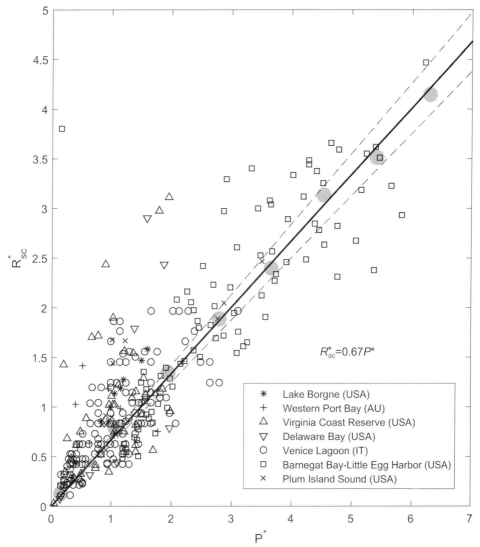

Figure 15.6 Relationship between dimensionless wave power W^* and erosion rate R_{sc}^*. Gray circles indicate values obtained by averaging data points over regular bins (from Leonardi et al. 2015). (A black and white version of this figure will appear in some formats. For the color version, please refer to the plate section.)

way in which wave forcing was calculated. The biggest departure among relationships resides with the results obtained by Schwimmer (2001). The value of a is three orders of magnitude lower. This might be due to the approach employed to determine the wave power impinging the marsh, since the value at different marsh locations considered by Schwimmer (2001) is on the order of kW/m, which is orders of magnitude higher than values reported by other authors. This results can be partially explained by the fact that

Table 15.3 *Relationships between the wave forcing and the salt marsh retreat proposed by several authors and quantified in different manners. The symbol "–" represents the negative number, whereas the symbol "–" represents the dash to indicate a range of values; h_b and h_m represent the top and the bottom of the marsh scarp*

Relation	R_{sc}	W	Environment	Reference
$R_{sc} = 0.35 \cdot P^{1.1}$	[m/yr]	[kW/m]	Rehobot Bay and others	Schwimmer (2001)
$R_{sc} = 0.0364 \cdot P$	[m²/yr]	[W/m]	Venice Lagoon	Marani et al. (2011)
$R_{sc} = 0.13 \cdot P + (0.05–0.08)$	[m/yr]	[W/m]	Hog Island Bay	McLoughlin et al. (2014)
$R_{sc} = 0.15 \cdot P – (0.02–0.05)$	[m²/yr]	[W/m]	Hog Island Bay	McLoughlin et al. (2014)
$R_{sc} \cdot (h_b – h_m) \cdot = 0.1 \cdot P$	[m/yr]	[W/m]	Mid Atlantic Coast	Mariotti and Fagherazzi (2013)
$R_{sc}^* = 0.67 \cdot P^*$	[-]	[-]	US, Australia, Italy	Leonardi et al. (2015)
$R_{sc} = 11.915 \cdot P$	[m/yr]	[kN/m]	Barnegat Bay Little Egg Harbor	Leonardi et al. (2016)
$R_{sc} = (0.098–0.648) \cdot P$	[m²/yr]	[W/m]	Venice Lagoon	Bendoni et al. (2016)

Table 15.4 *Comparison among relationships for several studies with consistent dimensional quantities*

Relation	Environment	Reference
$R_{sc} = 0.3320 \cdot 10^{-3} \cdot P$	Rehobot Bay and others	Schwimmer (2001)
$R_{sc} = 0.0364 \cdot P$	Venice Lagoon	Marani et al. (2011)
$R_{sc} = 0.10 \cdot P$	Mid Atlantic Coast	Mariotti and Fagherazzi (2013)
$R_{sc} = 0.15 \cdot P – (0.02–0.05)$	Hog Island Bay	McLoughlin et al. (2014)
$R_{sc} = (0.098–0.648) \cdot P$	Venice Lagoon	Bendoni et al. (2016)

Schwimmer (2001) computed the average wave power and considered sea state including times when water depth completely submerged the marsh surface. These conditions are generally associated with larger wave heights, potentially leading to an overestimation of the computed wave power (McLoughlin et al. 2014).

Field evidence relating the planimetric evolution of the salt marsh surface inspired the works of Leonardi and Fagherazzi (2014, 2015). By means of a cellular automata model and field data of Plum Island Sound, Massachusetts and Virginia Coast Reserve, Virginia (USA), they investigated the effect of different wave forcing on the response of the edge of a salt marsh, finding that the shape of the boundary is a geomorphic indicator of the evolutionary trend of the system. The model domain is cformed by a two-dimensional square lattice where cells have randomly distributed resistance to erosion r_i. At every time step a cell is eroded randomly with probability $p_i = E_i/\sum E_i$ where E_i is the erosion rate for cell i, the sum is extended to the erodible cell for such a time step, and only the neighbors of previously eroded

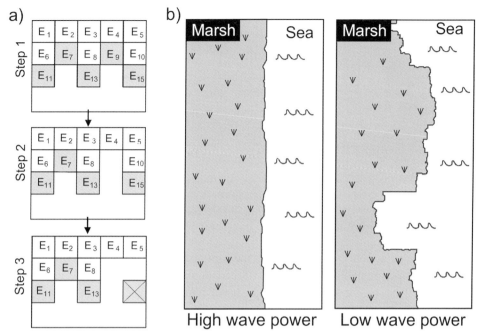

Figure 15.7 (a) Sketch of the cellular automata model mechanism; (b) results pattern from the cellular automata model. (Adapted from Leonardi and Fagherazzi 2014)

cells are susceptible to erosion (grey cells in Fig. 15.7a). Furthermore, a cell is removed from the lattice if it remains without neighbors (Step 3 in Fig. 15.7a). Wave power promoting erosion is calculated through the Young and Verhagen (1996) parametric model and the erosion rate E_i for a cell i is assigned by the relation:

$$E_i = \alpha P^\beta \exp\left(-\frac{r_i}{P}\right) \qquad \text{(Equation 15.11)}$$

with $\alpha = 0.35$ and $\beta = 1.1$.

Model results (Fig. 15.7b) show that a marsh edge subjected to erosion from high wave energy is uniform along the edge, while a low wave forcing leads to a jagged marsh boundary. This pattern is confirmed by the analysis of field data, indeed very exposed sites follow a Gaussian frequency distribution of the erosive events and uniform edge retreat rates, whereas less exposed marshes show a long tailed distribution of erosive events (lot of small events and few large events).

An improvement to the previous analysis was carried out by Leonardi and Fagherazzi (2015), where they run the cellular automata model with a uniform wave power and added randomly distributed high intensity events having magnitudes an order of magnitude higher than the background value. They found that the response of salt marshes exposed to a low wave forcing are sensitive to variations in the frequency of extreme events and mean wave energy forcing, whereas salt marshes generally exposed to high wave forcing are less affected by the variation in the frequency of extreme events.

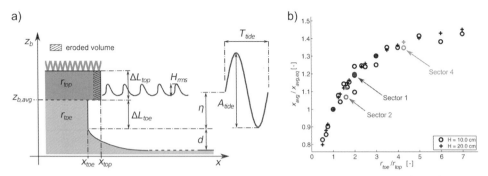

Figure 15.8 (a) Sketch of the model; (b) ratio $x_{avg}/x_{avg,eq}$ as a function of the ratio r_{toe}/r_{top} after 1,460 tidal cycles for different wave height (black circles: H_{rms} = 10 cm; black crosses: H_{rms} = 20 cm). Points associated to sectors 1, 2, and 4 are the results of the model runs with the ratio r_{toe}/r_{top} based on field data. Each sector represents a zone of the surveyed marsh.

Another type of field evidence, such as the presence of cantilever profiles observed in the field (Venice Lagoon, Italy) laid the foundation for the development of a simple mathematical model for the erosion of a marsh transect subject to continuous wave forcing, as a consequence of combined surface erosion and mass failures (Bendoni et al. 2016). The scarp, of height b_h [m] is divided into two layers of equal height ($\Delta L_{toe} = \Delta L_{top} = b_h/2$) with different erodibility, r_{toe}, r_{top} [m^3/W/yr] accounting for the presence of a vegetated layer on the top and a bottom layer with no vegetation or root-mat. Each layer is eroded when the mean water level falls within the corresponding range of influence as depicted in (Fig. 15.8a).

The cumulative retreat of each portion of the bank is computed by:

$$x_{top}(t + \Delta t) = x_{top}(t) + \frac{r_{top}W\Delta t}{\Delta L_{top}} \qquad \text{(Equation 15.12a)}$$

$$x_{toe}(t + \Delta t) = x_{toe}(t) + \frac{r_{top}W\Delta t}{\Delta L_{toe}} \qquad \text{(Equation 15.12b)}$$

where x_{top} [m] and x_{toe} [m] are respectively the position of the edge the upper and lower layer in horizontal direction and $x_{avg} = 0.5(x_{top} + x_{toe})$. Cantilever failure occurs when the scour exceeds a threshold based on field evidence, and such an event increase the cumulative retreat for values equal to the scour length. The model is run in two ways: with different values of r_{top} and r_{toe}, and with a uniform value of r equal to the average of r_{top} and r_{toe}. In the former case model output is x_{avg}, in the latter $x_{avg,eq}$. The model is also run with values of erodibility based on localized measurements of retreat rate performed on a salt marsh of Venice Lagoon. Sectors in Figure 15.8b refer to different areas of the monitored marsh.

Figure 15.8b shows that the increase in the difference in erodibility of the upper and lower layer tends to increase the cumulative lateral retreat with respect to the equivalent case with uniformly distributed erodibility. The former case is more susceptible to the formation of cantilever profiles, inducing more frequent failure events which, in turn, accelerate the overall erosion rate. This suggests that the presence of vegetation, increasing the resistance to erosion of the upper layer, does not necessarily guarantee, globally, a reduction of lateral erosion.

Feagin et al. (2009) employed both laboratory flume experiments and controlled field experiments to question the paradigm that salt marsh plants prevent the lateral erosion induced by wind waves. They collected several field samples of vegetated and unvegetated soil, and subjected them to continuous wave forcing in a laboratory wave flume. Their results suggested that the presence of plant does not significantly affect the erosion. They argued that the presence of salt marsh vegetation is not a pivotal factor for erosion; rather the soil typology is the primary element controlling erosion rates. Vegetation only indirectly affects erosion by modifying soil characteristics (such as organic matter content, bulk density, sediment size, and water content).

A similar behavior was found by the authors during a field campaign carried out at Jumbile Cove in Galveston Bay, Texas (USA) specifically located at the top of a marsh edge characterized by the initial formation of a small cliff. Several measurement stations were placed along the marsh edge using a rod sediment elevation table as an elevation benchmark. They treated half of their sites using herbicides to remove vegetation from the bank, allowing the comparison between a vegetated and unvegetated marsh approximately subjected to similar wave conditions. Their results suggest that there are no substantial differences in the erosive rates among the different stations studied. However, their results might be affected by the presence of live roots in the substrate which tend to maintain a low erosion rate regardless of the absence of the above ground vegetation stems, as shown by Möller et al. (2014) and Spencer et al. (2015).

Besides wind waves, vessel generated wakes can have an effect on marsh lateral retreat as shown by several authors (Price 2006, Houser 2010, Silinski et al. 2015, Bilkovic et al. 2017).

A field measurement campaign carried out by Houser (2010) at Fort Pulaski National Monument (Georgia, USA) from October 2007 to February 2008 showed that wind waves, rather than boat wakes, were responsible for the erosive trends of the marsh shoreline. This response is mainly due to the fact that during storms the water level is maintained at the marsh edge, allowing wind waves to reach the bank, whereas boat wakes are effective in damaging marsh surface only around the high tide water level, thus reducing the time span during which the marsh is potentially subject to erosion. Differently, through GIS-based analysis of aerial photographs, Price (2006) found that boat wakes are the primary cause of marsh retreat in the Guana Tolomato Matanzas National Estuarine Research Reserve (Florida). However, this was principally attributed to the location and morphology of the salt marsh monitored, which was sheltered from wind waves and only subjected to the passage of ships.

Silinski et al. (2015) studied the effect of different wave regimes on marsh vegetation with controlled experiments carried out in a wave flume. Two waves regimes characterized by long (vessel waves) and short (wind waves) wave period were tested and analyzed. Results show that seedlings and plants are able to withstand short-period waves, whereas long-period waves lead to plant tissue failure threatening plant survival.

15.3.3 Modelling Mass Failures

Mass failure is one of the main mechanisms responsible for salt marsh lateral retreat. In the literature, both cantilever failures and toppling-type failures have been addressed and mathematical models developed for each type.

For instance, van Eerdt (1985a) developed a cantilever failure model which accounts for the effect of soil properties and the tensile strength provided by roots. The author employed soil and vegetation properties and the geometry of the bank profile as input for the model using field data from the Eastern Scheldt (Netherlands).

A conceptual sketch of the model is reported in Figure 15.9a, where W_b [N/m] represents the weight of the overhanging block, $d_{s,cr}$ [m] is the maximum scour length, L_t [m] and L_c [m] measure respectively the extension of the fibers subject to tensile and compression stress, σ_{tmax} [m] is the tensile strength of the material, and σ_{cf} [m] the maximum compressive stress at failure.

The model is based on the following assumptions: failure is progressive from point A to point B starting from the top of the cliff; the initially vertical plane section remains plane. Based on the characteristics of soil, bank morphology, and the tensile strength of plant roots, one can identify the cliffs that are likely to fail through the following relation which identifies the maximum admissible scour $d_{s,cr}$:

$$d_{s,cr} = \frac{2}{3} \left(\frac{\sigma_{cf} L_{cf}^2 - \sigma_{tmax} L_t^2}{W_b} \right) \qquad \text{(Equation 15.13)}$$

Equation 15.13 prescribes that an increase of the tensile strength of a factor x produces an increase in the maximum scour of the order of $x^{1/2}$, all other things being equal (Fig. 15.9b).

The model described above was used and improved by Gabet (1998), including the effect of slumped blocks in protecting the bank toe from erosion, to investigate and predict the lateral dynamics of the tidal channels dissecting the surface of the salt marshes located in the San Francisco estuary, California, (USA).

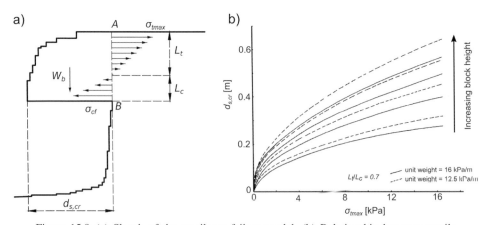

Figure 15.9 (a) Sketch of the cantilever failure model. (b) Relationship between tensile strength, block height and undermining width $d_{s,cr}$. (Modified from Ven Eerdt, 1985)

Toppling failure modelling was derived by Bendoni et al. (2014), based on flume experiments results. The model schematizes the behavior of a prismatic block of soil subjected to gravity, hydrostatic thrust, and hydrodynamic thrust caused by wave forcing. The block is a rigid body of mass m [kg] connected to the underlying homogeneous and isotropic soil by a system of spring and dashpot, and can rotate of an angle θ [rad] with respect to the midpoint of the failure surface. The soil is considered to have a viscoelastic behavior described by the Kelvin–Voigt model (Malkin and Isayev 2006). The failure of the block occurs when the tensile strength of the material is exceeded in at least one point of the failure surface due to block rotation induced by wave action. The governing equation of the system can be written as:

$$I\frac{d^2\theta}{dt^2} + c_d\frac{d\theta}{dt} + k_{sp}\theta = -F_h\frac{Z}{3} + F_{hc}\frac{l^*}{3} - F_wZ_w + W_b\frac{l}{2}(\theta_0 + \theta) \qquad \text{(Equation 15.14)}$$

where the rotational damping coefficient c_d [N·m·s] and the rotational spring coefficient k_{sp} [N·m] depend on the characteristics of the soil through the formulation proposed by Gazetas (1991), I [kg·m^2] is the moment of inertia of the block including the "added mass", F_h, F_{hc}, and F_w [N] are respectively the hydrostatic force of water in front of the bank and inside the crack and due to wave action, W_b [N] the block weight, θ_0 [rad] the initial rotation with respect to the vertical direction, and the other quantities are referred to in Figure 15.10. The stress on the failure surface is related to the rotation angle θ through the moment exerted by the block on the failure surface itself.

The model was applied to data collected from a salt marsh bank built in a wave flume and subjected to continuous wave forcing (Francalanci et al. 2013). Results show the model is able to identify the wave train responsible for the failure of the block. Furthermore, the application of the model to an ideal bank configuration with both presence and absence of water inside the crack (Fig. 15.10d, e respectively), allows one to observe that the dynamic behavior of the system due to the effect of waves is crucial in promoting the failure of the block. The stress acting on the point A in absence of waves (black dotted line) is indeed lower than the value in presence of wave forcing (blue continuous line), both with the crack filled or not with water.

15.3.4 Process-Based Models for Long-Term Forecast

The mechanism of bank retreat has been included in several mathematical models for the description of the long-term evolution of salt marshes. Such models account for the interplay between physical and biological processes, emphasizing the effect of local feedbacks in affecting the morphology of the system.

Van de Koppel et al. (2005) proposed a model characterized by the coupled dynamic of bed elevation and plant growth along a transect (cross shore direction) of a marsh bank. The erosion due to waves in the model is assumed proportional to bank slope, but not influenced by a specific wave climate. They showed that the salt marsh can reach a critical state at the edge between the vegetated area and the bare mud flat; at this position, the strong gradient in bed elevation makes the system vulnerable to wave attack and thus causing strong cliff erosion.

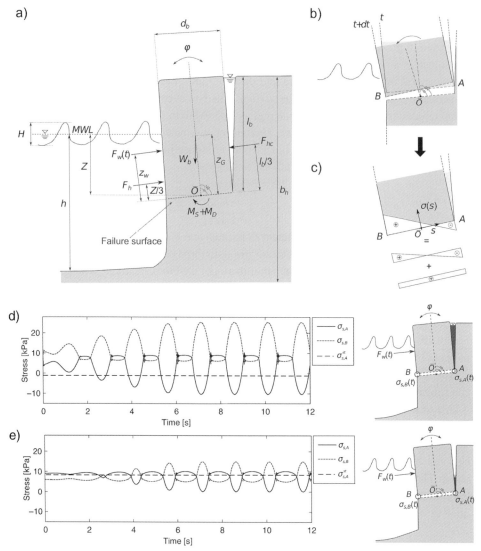

Figure 15.10 (a) Sketch of the cross-section of the system employed for the development of the dynamic model; (b) scheme of the failure surface and the stress distribution induced by (c) a small clockwise rotation of the block from equilibrium. (d) and
(e) comparison between the behavior of the model with filled or empty tension crack.
The blue continuous line represents the time evolution of the stress at the bottom of the
tension crack in the inner point of the failure surface A. The green dashed line
represents the stress at the external point of the failure surface B. Black dotted line
represents the time evolution of the stress on point A in the static case (from Bendoni
2015). (A black and white version of this figure will appear in some formats. For the color
version, please refer to the plate section.)

Mariotti and Fagherazzi (2010) improved the model of van de Koppel et al. (2005) to include the coupled evolution of tidal flat and salt marshes along a one-dimensional transect, adding the generation and propagation of wind waves from the bay toward the marsh shoreline. To simulate wave-induced edge erosion they employed a linear relation between wave power and volumetric erosion rate, and introduced a critical wave power for the onset of erosion. Through numerical experiments they found that high rate of sea level rise promotes deeper tidal flats which subsequently increases the transmission of wave energy arriving at the marsh edge, thus increasing marsh edge erosion.

A similar mathematical model was applied by Mariotti and Fagherazzi (2013) to salt marsh basins to show that irreversible marsh erosion can also occur in the absence of sea level rise, driven by the initial width of the basin surrounding the marsh. The model accounts for both horizontal marsh migration and vertical dynamics of vegetated areas and tidal flats, and considers inorganic sediment input via suspended sedimentin the bay. Similar to previous models, the retreat of the marsh cliff (scarp) is simulated through a linear relation between wave power and erosion rate (Table 15.3).

Leonardi and Fagherazzi (2014) employed a different modeling approach which explores the planimetric evolution of a marsh area subject to wave forcing by means of cellular automata simulations. In this approach, they propose a nonlinear relation between wave forcing and the erosion rate of a cell within the computational domain of the model, to describe the retreat of the marsh edge (Equation 15.11). They found that high energy wave climate produces a uniform erosion of marsh shoreline forming straight shorelines, whereas low wave energy climate produces jagged shorelines.

In summary, the process-based models used for long-term forecast of tidal environments characterized by the presence of salt marshes, employ a simple functional relationship to describe the retreat rate of the marsh edge based on the annualized or average wave power. This relation (Equation 15.10) is based on a single parameter incorporating the characteristics of each environment, and while it represents a strength that helps maintain the model simple, it may also present challenges or a limitation in the application of the model due to the loss of potential and still unknown nonlinear processes operating along marsh edge shorelines.

15.4 Outlook and Future Research Needs

This chapter presents a review of various studies that explore erosion of the edges of salt marshes, principally due to the effect of wind waves, as this is likely the chief process responsible for marsh surface reduction in most wetland systems worldwide. To date, the scientific advances in this field, while robust and informative, also raise new questions and provide insight into challenging issues, providing pathways for new research.

The retreat of marsh boundaries is indeed a very complex mechanism involving physical, chemical, and biological processes which act at different temporal and spatial scales with strong coupling and feedbacks. Mathematical and conceptual models are often used to manage coastal systems or help develop environmental policies and restoration strategies, or to implement protection measures for marsh shorelines; thus, knowing their limitations, applicability, and ways to improve predictability of retreat rates is imperative in future development of such models.

On the one hand, screening tools, which can be applied quickly and with relative ease, are necessary and likely attractive to coastal managers so this calls for the reduction of the number of variables involved in models; this approach suggest that accounting only for the significant variables could help simplify the problem and help create a consistent conceptual framework that would be simple to use. On the other hand, care needs to be exercised to avoid the loss of non-negligible feedbacks among variables initially not included, and whose effect may not be evident.

The result of our review identifies two main issues that deserve additional attention to aid in the predictability and widespread applicability of these models. The first issue is the characterization of the coefficient a (Equation 15.10) which attempts to link wave power to erosion rate in environments where the wave impact is the chief mechanism responsible for the erosion. The second issue is the modeling of the effect of vegetation on bank retreat and retreat, also in relation to mass failures.

The approach employed by many researchers to derive a relationship between wave forcing and volumetric erosion rate per unit width is quite robust, but limits the application of such a relationship to a specific site.

While the proportionality coefficient a implicitly accounts for most of the local characteristics of the site (e.g., soil properties, bank morphology, and the effect of site specific vegetation), it assumes a correlation of each of these characteristics to the erosion rate, and it ignores interactions among these characteristics. Furthermore, the coefficient is influenced by the procedure employed to determine it: the time span over which wave power is averaged; the individual methods by which the value of wave power is calculated at each site, weather empirical parametric wave models were used (Young and Verhagen 1996) or wave propagation models (e.g., SWAN, Booij et al. 1999); and the procedure adopted to determine the erosion rate at the site – for example, by measurements using erosion pins (Trosclair 2013, Bendoni et al. 2016) or by aerial photograph (Marani et al. 2011). All these aspects introduce individual and possibly cumulative uncertainties to the determination of a and my give rise to different values of the parameter even for the same site.

More effort is thus required to obtain functional relationships that directly account for the effects of soil properties, vegetation density and typology, scarp shape, and biological activity (micro- and macrofauna), to reduce the arbitrariness present in the approaches usually employed. This represents a rather complex task, since these variables act together with mutual influences. As a consequence, difficulties arise when the effects of single variables need to be quantified to establish a physical-based relationship. On the other hand, the advantage of having a formulation that provides an estimate of the tendency to erosion for any site based on the cumulative effect of selected site features, regardless of the need for site-specific surveys to determine those characteristics, does preclude the demand for long field surveys to establish erosion trends and associated wave fields.

Moreover, episodic mass failures cause punctuated erosion at a rate significantly higher than the instantaneous erosion rate, thus affecting the long-term average value. The treatment of the erosion parameter a as a time dependent random variable that is a function of the cumulative wave energy that reached the bank (e.g., similar to that proposed by Leonardi and Fagherazzi 2014, 2015), might be another useful approach for long-term forecast models. Including this discontinuous response of *a* in hydro-morphodynamic models would allow for more detailed estimation of bank retreat, whereby the interaction between erosion due to wave impact and mass failures can be simulated, even if, at first, this is only accomplished for short time scales.

The role of vegetation on marsh edge retreat requires additional studies. Obviously, the role of vegetation in mitigating erosion, directly or indirectly, is somewhat still debated in the literature, although there are obvious benefits. Rather than implicitly accounting for the presence or absence of vegetation in an attempt to establish a specific response on the retreat rate, it would be more useful and valuable to develop separate relationships which explain specific effects of vegetation. For instance, the effect of vegetation in attenuating waves when the bank is submerged (Möller et al. 2014) and the corresponding bottom shear stress as a function of submergence is the principal control on the erosion of the salt marsh horizontal surface (e.g., Neumeier and Amos 2006, Howes et al. 2010). Moreover, while the effect of vegetation in reducing/increasing the overall erosion rate when waves strike vertical edge of the bank is still debated (Feagin et al 2009, Mariotti and Fagherazzi 2010, Chen et al. 2012, Bendoni et al. 2016), relationships that correlate indirect effects of vegetation on soil properties can be developed and tested. Particular combinations of soil properties and bank morphology, along with the characteristics of wave forcing and the temporal distribution of water elevation in front the marsh bank, may lead to positive or negative effect of vegetation in reducing erosion.

For all the reasons above, additional field and laboratory experiments are necessary to better analyze these controls, identify dependencies and help develop more robust relationships at the spatial and temporal scales considered within the different modelling frameworks.

References

Allen, J. R. L. 1989. Evolution of salt-marsh cliffs in muddy and sandy systems: a qualitative comparison of British west-coast estuaries. *Earth Surface Processes and Landforms* 14(1): 85–92.

Allen, J. R. L. 2000. Morphodynamics of Holocene salt marshes: a review sketch from the Atlantic and Southern North Sea coasts of Europe. *Quaternary Science Reviews* 19(12):1155–1231.

Altieri, A. H., M. D. Bertness, T. C. Coverdale, N. C. Herrmann, and C. Angelini 2012. A trophic cascade triggers collapse of a salt-marsh ecosystem with intensive recreational fishing. *Ecology* 93(6): 1402–1410.

Amos, C. L., A. Bergamasco, G. Umgiesser, S. Cappucci, D. Cloutier, L. DeNat, M. Flindt, M. Bonardi, and S. Cristante 2004. The stability of tidal flats in Venice Lagoon: The results of in-situ measurements using two benthic, annular flumes, *Journal of Marine Systems*, 51: 211–241.

Barbier, E. B., S. D. Hacker, C. Kennedy, E. W. Koch, A. C. Stier, and B. R. Silliman 2011. The value of estuarine and coastal ecosystem services. *Ecological Monographs* 81(2): 169–193.

Barbier, E. B., I. Y., Georgiou, I. Y., Enchelmeyer, and D. J. Reed 2013. The value of wetlands in protecting Southeast Louisiana from hurricane storm surges. *PLoS ONE* 8: e58715. https://doi.org/10.1371/journal.pone.0058715.

Beland, M., T. W. Biggs, D. A. Roberts, S. H. Peterson, R. F. Kokaly, and S. Piazza 2017. Oiling accelerates loss of salt marshes, southeastern Louisiana, *PLoS ONE* 12(8): e0181197.

Bell, F. W. 1997. The economic valuation of saltwater marsh supporting marine recreational fishing in the southeastern United States. *Ecological Economics* 21: 243–254.

Bendoni, M. 2015. *Salt marsh edge erosion due to wind-induced waves*. PhD Thesis. University of Florence-TU Braunschweig.

Bendoni, M., S. Francalanci, L. Cappietti, and L. Solari 2014. On salt marshes retreat: Experiments and modeling toppling failures induced by wind waves. *Journal of Geophysical Research Earth Surface*, 119: 603–620.

Bendoni, M., R. Mel, L. Solari, S. Lanzoni, S. Francalanci, and H. Oumeraci 2016. Insights into lateral marsh retreat mechanism through localized field measurements. *Water Resources Research*, 52: 1446–1464.

Bilkovic, D., M. Mitchell, J. Davis, E. Andrews, A. King, P. Mason, J. Herman, N. Tahvildari, and J. Davis 2017. *Review of boat wake wave impacts on shoreline erosion and potential solutions for the Chesapeake Bay*. STAC Publication Number 17-002, Edgewater, MD.

Boesch, D. F., and R. E. Turner, 1984. Dependence of fishery species on salt marshes: The role of food and refuge. *Estuaries* 7: 460–468.

Booij, N., R. C. Ris, and L. H. Holthuijsen 1999. A third-generation wave model for coastal regions: 1. Model description and validation. *Journal of Geophysical Research: Oceans* 104 (C4): 7649–7666.

Breugem, W. A., and Holthuijsen, L., 2007. Generalized shallow water wave growth from Lake George. *Journal of Waterway Port Coastal and Ocean Engineering* 133: 23–37.

Bromberg, K. D., and M. D. Bertness 2005. Reconstructing New England salt marsh losses using historical maps. *Estuaries* 28: 823–832.

Callaghan, D. P., T. J. Bouma, P. Klaassen, D. Van der Wal, M. J. F. Stive, and P. M. J. Herman 2010. Hydrodynamic forcing on salt-marsh development: Distinguishing the relative importance of waves and tidal flows. *Estuarine, Coastal and Shelf Science* 89 (1): 73–88.

Chauhan, P. P. S. 2009. Autocyclic erosion in tidal marshes. *Geomorphology* 110(3): 45–57.

Chen, Y., M. B. Collins, and C. E. L. Thompson 2011. Creek enlargement in a low-energy degrading saltmarsh in southern England. *Earth Surface Processes and Landforms* 36: 767–778.

Chen, Y., C. E. L. Thompson, and M. B. Collins 2012. Saltmarsh creek bank stability: Biostabilisation and consolidation with depth. *Continental Shelf Research* 35: 64–74.

Chmura, G. L., S. C. Anisfeld, D. R. Cahoon, and J. C. Lynch 2003. Global carbon sequestration in tidal saline wetland soils. *Global Biogeochemical Cycles* 17:1111.

Cola S., L. Sanavia, P. Simonini, and B. A. Schrefler 2008. Coupled thermohydromechanical analysis of Venice lagoon salt marshes. *Water Resources Research* 44(5): W00C05. https://doi.org/10.1029/2007WR006570

Coops, H., F.W.B. Van der Brink, and G. Van der Velde. 1996. Growth and morphological responses of four helophyte species in an experimental water-depth gradient. *Aquatic Botany* 54: 11–24.

Costanza, R., R. d'Arge, R. de Groot, S. Farber, M. Grasso, B. Hannon, K. Limburg, et al. 1997. The value of the world's ecosystem services and natural capital. *Nature* 387: 253–260.

Costanza, R., O. Pérez-Maqueo, M. L. Martinez, P. Sutton, S. J. Anderson, and K. Mulder 2008. The value of coastal wetlands for hurricane protection. *Ambio* 37: 241–248.

Coverdale, C. T., M. D. Bertness, and A. H. Altieri 2013. Regional ontogeny of New England salt marsh die-off. *Conservation Biology* 27(5): 1041–1048.

Davidson, T. M., and C. E. de Rivera 2010. Accelerated erosion of saltmarshes infested by a non-native burrowing crustacean *Sphaeroma quoianum*. *Marine Ecology Progress Series* 419: 129–136.

D'Alpaos, A., S. Lanzoni, M. Marani, and A. Rinaldo 2007. Landscape evolution in tidal embayments: Modeling the interplay of erosion, sedimentation, and vegetation dynamics. *Journal of Geophysical Research: Earth Surface* 112 (F1): https://doi.org/10.1029/2006JF000537.

Day, J. W., F. Scarton, A. Rismondo, and D. Are 1998. Rapid deterioration of a salt marsh in Venice Lagoon, Italy. *Journal of Coastal Research* 14: 583–590.

Deegan L.A., D. S. Johnson, R. S. Warren, B. J. Peterson, J. W. Fleeger, S. Fagherazzi and W. M. Wollheim 2012. Coastal eutrophication as a driver of salt marsh loss. *Nature* 490: 388–392.

Emerton, L., and L. Kekulandala 2003. *Assessment of the economic value of Muthurajawela wetland. Occasional Papers of IUCN Sri Lanka, IUCN-World Conservation Union*, Sri Lanka Country Office, Colombo (Sri Lanka), Volume 4.

Fagherazzi, S. and P. L. Wiberg 2009. Importance of wind conditions, fetch, and water levels on wave-generated shear stresses in shallow intertidal basins. *Journal of Geophysical Research*, Vol. 114, F03022: DOI: 10.1029/2008JF001139.

Fagherazzi, S., M. L. Kirwan, S. M. Mudd, G. R. Guntenspergen, S. Temmerman, A. D'Alpaos, J. Koppel, et al. 2012. Numerical models of salt marsh evolution: Ecological, geomorphic, and climatic factors. *Reviews of Geophysics* 50 (1): https://doi.org/10.1029/2011RG000359.

Feagin, R. A., J. L. Irish, I. Möller, A. M. Williams, R. J. Colón-Rivera, and M. E. Mousavi 2009. Does vegetation prevent wave erosion of salt marsh edges? *Proceedings of the National Academy of Sciences of the USA*, 106(25): 10109–10113.

FitzGerald, D., Z. Hughes, and P. Rosen 2011. *Boat wake impact and their role in shore erosion processes, Boston Harbor Islands National Recreation Area*, Natural Resource Report NPS/NERO/NRR-2011/403, National Park Service, Fort Collins, Colorado.

Francalanci, S., M. Bendoni, M. Rinaldi, and L. Solari 2013. Ecomorphodynamic evolution of salt marshes: Experimental observations of bank retreat processes. *Geomorphology* 195: 53–65.

Fredsoe, J., and R. Deigaard 1993. *Mechanics of Coastal Sediment Transport. Advanced Series on Ocean Engineering*, vol. 3, World Science, Singapore.

French, G. T. 1990. *Historical shoreline changes in response to environmental conditions in west Delaware Bay*. MA thesis. University of Maryland College Park.

Gabet, E. J. 1998. Lateral migration and bank erosion in a saltmarsh tidal channel in San Francisco Bay, California. *Estuaries* 21 (4): 745–753.

Gazetas, G. 1991. Foundation vibrations. Ed. by H. I. Fang *Foundation Engineering Handbook*. . Kluwer Academic Publishers, Massachusetts, USA, pp. 553–593.

Gedan, K. B., B. R. Silliman, and M. D. Bertness 2009. Centuries of humandriven change in salt marsh ecosystems. *Annual Review of Marine Science* 1: 117–141.

Gedan, K. B., M. L. Kirwan, E. Wolanski, E. B. Barbier, and B. R. Silliman 2011. The present and future role of coastal wetland vegetation in protecting shorelines: Answering recent challenges to the paradigm. *Climatic Change* 106 (1): 7–29.

Gosselink, J. G. and R. M. Pope 1974. The Value of The Tidal Marsh. (LSUSG-74-03, Center for Wetland Resources, Baton Rouge: Louisiana StateUniversity.

Gray, D. H., and A. T. Leiser. 1982. *Biotechnical slope protection and erosion control.* Van Nostrand Reinhold Company Inc, New York.

Houser, C. 2010. Relative importance of vessel-generated and wind waves to salt marsh erosion in a restricted fetch environment. *Journal of Coastal Research*: 230–240.

Howes, N. C., D. M. FitzGerald, Z. J. Hughes, I. Y. Georgiou, M. A. Kulp, M. D. Miner, J. M. Smith, and J. A. Barras 2010. Hurricane-induced failure of low salinity wetlands. *Proceedings of the National Academy of Sciences of the USA* 107 (32): 14014–14019.

Huges, Z. J., D. M. FitzGerald, N. C. Howes, and P. S. Rosen 2007. The impact of natural waves and ferry wakes on bluff erosion and beach morphology, Boston Harbor, USA. *Journal of Coastal Research*, SI 50 (Proceedings of the 9th International Coastal Symposium), 497–501. Gold Coast, Australia, ISSN 0749.0208.

Hughes, Z. J., D. M. FitzGerald, C. A. Wilson, S. C. Pennings, K. Wieski, and A. Mahadevan 2009. Rapid headward erosion of marsh creeks in response to relative sea level rise. *Geophysical Research Letters* 36, L03602, doi:10.1029/ 2008GL036000.

Jadhav, R. S., and Q. Chen 2013. Probability distribution of wave heights attenuated by salt marsh vegetation during tropical cyclone. *Coastal Engineering* 82: 47–55.

Jadhav, R. S., Q. Chen, and J. M. Smith 2013. Spectral distribution of wave energy dissipation by salt marsh vegetation. *Coastal Engineering* 77: 99–107.

Karimpour, A., Q. Chen, and R. R. Twilley 2016. A field study of how wind waves and currents may contribute to the deterioration of saltmarsh fringe. *Estuaries and Coasts* 39: 935–950.

Kennish, M. J. 2001. Coastal salt marsh system in the U.S.: A review of anthropogenic impacts. *Journal of Coastal Research* 17(3): 731–748.

Keulegan, G. H. and L. H. Carpenter 1958. Forces on cylinders and plates in an oscillating fluid. *Journal of Research of the National Bureau of Standards* 60 (5): 423–440.

King, S. E., and J. N. Lester 1995. The value of salt marsh as a sea defence. *Marine Pollution Bulletin* 30 (3):180–189.

Kirwan, M. L., G. R. Guntenspergen, A. D'Alpaos, J. T. Morris, S. M. Mudd, and S. Temmerman 2010. Limits on the adaptability of coastal marshes to rising sea level. *Geophysical Research Letters* 37(23): https://doi.org/10.1029/2010GL045489

Kirwan, M. L., and A. B. Murray 2007. A coupled geomorphic and ecological model of tidal marsh evolution, *Proceedings of the National Academy of Science of the USA*, 104, 6118–6122.

Kirwan, M. L., and A. B. Murray 2008. Tidal marshes as disequilibrium landscapes? Lags between morphology and Holocene sea level change. *Geophysical Research Letters*, 35, L24401, doi:10.1029/ 2008GL036050.

Kirwan, M. and S. Temmerman 2009. Coastal marsh response to historical and future sea-level acceleration. *Quaternary Science Reviews* 28(17): 1801–1808.

Kobayashi, N., A. W. Raichle and T. Asano 1993. Wave attenuation by vegetation. *Journal of Waterway Port Coastal and Ocean Engineering* 119(1): 30–48.

Knutson, T. R., J. L. McBride, J. Chan, K. Emanuel, G. Holland, C. Landsea, I. Held, J. P. Kossin, A. K. Srivastava, and M. Sugi 2010. Tropical cyclones and climate change, *Nature Geoscience* 3: 157–163.

Le Hir, P., Monbet, Y., and Orvain, F., 2007. Sediment erodability in sediment transport modeling: Can we account for biota effects? *Continental Shelf Research* 27: 1116–1142.

Leonardi, N. and S. Fagherazzi 2014. How waves shape salt marshes. *Geology*, 42 (10): 887–890.

Leonardi, N. and S. Fagherazzi 2015. Effect of local variability in erosional resistance on large-scale morphodynamic response of salt marshes to wind waves and extreme events. *Geophysical Research Letters* 42: 5872–5879, doi:10.1002/2015GL064730.

Leonardi, N., N. K. Ganju and S. Fagherazzi 2015. A linear relationship between wave power and erosion determines salt-marsh resilience to violent storms and hurricanes. *Proceedings of the National Academy of Sciences* 113 (1): 564–568.

Leonardi, N., Z. Defne, N. K. Ganju, and S. Fagherazzi 2016. Salt marsh erosion rates and boundary features in a shallow Bay. *Journal of Geophysical Research Earth Surface*, 121: 1861–1875.

Madsen, P. A., and O. R. Sørensen 1992. A new form of the Boussinesq equations with improved linear dispersion characteristics. Part 2: A slowly varying bathymetry. *Coastal Engineering*, 18, 183–204.

Malkin, A. Y. and A. Y. Isayev 2006. *Rheological Concepts, Methods and Applications*. ChemTech, Toronto.

Marani, M., A. D'Alpaos, S. Lanzoni, and M. Santalucia 2011. Understanding and predicting wave erosion of marsh edges. *Geophysical Research Letters* 38 (21): https://doi.org/10.1029/2011GL048995.

Mariotti, G. and S. Fagherazzi 2010. A numerical model for the coupled longterm evolution of salt marshes and tidal flats. *Journal of Geophysical Research: Earth Surface* 115 (F1): https://doi.org/10.1029/2009JF001326.

Mariotti, G. and S. Fagherazzi 2013. Critical width of tidal flats triggers marsh collapse in the absence of sea-level rise. *Proceedings of the National Academy of Sciences of the USA*, 110 (14): 5353–5356.

Mariotti, G. and S. Fagherazzi 2013. Wind waves on a mudflat: The influence of fetch and depth on bed shear stresses. *Continental Shelf Research*, 60, S99–S110.

Mariotti, G., S. Fagherazzi, P. L. Wiberg, K. J. McGlathery, L. Carinello, and A. Defina 2010. Influence of storm surges and sea level on shallow tidal basin erosive processes. *Journal of Geophysical Research*, 115, C11012. https://doi.org/10.1029/2009JC005892.

Maurmeyer, E. M. 1978. *Geomorphology and evolution of transgressive estuarine washover barrier along the western shore of Delaware Bay*. PhD thesis. University of Delaware, Newark.

Maynord, S. 2001. Boat waves on Johnson Lake and Kenai River, Alaska. Technical Report U.S. Army Corps of Engineers. (No. ERDC/CHL-TR-01-31.

McLoughlin, S. M., P. L. Wiberg, I. Safak, and K. J. McGlathery 2014. Rates and forcing of marsh edge erosion in a shallow coastal bay. *Estuaries and Coasts*, 38: 620–638.

Mendez, F. J., and I. J. Losada 2004. An empirical model to estimate the propagation of random breaking and nonbreaking waves over vegetation fields. *Coastal Engineering*, 51 (2): 103–118.

Minkoff, D. R., M. Escapa, F. E. Ferramola, S. D. Maraschin, J. O. Pierini, G. M. E. Perillo, and C. Delrieux 2006. Effects of crabe-halophytic plant interactions on creek growth in a S. W. Atlantic salt marsh: A Cellular Automata model. *Estuarine Coastal Shelf Science*, 69, 403–413.

Mirtskhoulava, T. E. 1991. Scouring by flowing water of cohesive and non cohesive beds. *Journal of Hydraulic Research*, 29 (3):341–354.

Möller, I., M. Kudella, F. Rupprecht, T. Spencer, M. Paul, B. K. van Wesenbeeck, G. Wolters, et al. 2014. Wave attenuation over coastal salt marshes under storm surge conditions. *Nature Geoscience*, 7 (10):727–731.

Möller, I. and T. Spencer 2002. Wave dissipation over macro-tidal salt marshes: Effects of marsh edge typology and vegetation change. *Journal of Coastal Research*, 36(1):506–521.

Morris, P. H., J. Graham, and D. J. Williams 1992. Cracking in drying soils. *Canadian Geotechnical Journal*, 29 (2): 263–277.

Morris, J. T., P. V. Sundareshwar, C. T. Nietch, B. Kjerfve, and D. R. Cahoon 2002. Responses of coastal wetlands to rising sea level. *Ecology*, 83 (10): 2869–2877.

Neumeier, U. and C. L. Amos 2006. The influence of vegetation on turbulence and flow velocities in European salt marshes. *Sedimentology*, 53: 259–277.

Ojea, E., J. Martin-Ortega, and A. Chiabai 2012. Defining and classifying ecosystem services for economic valuation: the case of forest water services. *Environmental Science and Policy*, 19–20: 1–15.

Ozeren, Y., D. G. Wren, and W. Wu 2013. Experimental investigation of wave attenuation through model and live vegetation. *Journal of Waterway, Port, Coastal, and Ocean Engineering* 140 (5): https://doi.org/10.1061/(ASCE)WW.1943-5460.0000251

Phillips, J. D. 1985. *Aspat Bay analysis of the shoreline erosion, Delaware Bay, New Jersey*. PhD thesis. Rutgers University, New Brunswick.

Price, F. 2006. Quantification, Analysis, and Management of Intracoastal Waterway Channel Margin Erosion in the Guana Tolomato Matanzas National Estuarine Research Reserve, Florida. National Estuarine Research Reserve Technical Report Series 2006:1.

Priestas, A. M. and S. Fagherazzi 2011. Morphology and hydrodynamics of wave-cut gullies. *Geomorphology* 131 (1): 1–13.

Riffe, K. C., S. M. Henderson, and J. C. Mullarney 2011. Wave dissipation by flexible vegetation. *Geophysical Research Letters* 38 (18): https://doi.org/10.1029/2011GL048773.

Schwimmer, R. A. 2001. Rates and processes of marsh shoreline erosion in Rehoboth Bay, Delaware, USA. *Journal of Coastal Research* 17 (3): 672–683.

Schwimmer, R. A. and J. E. Pizzuto 2000. A model for the evolution of marsh shorelines. *Journal of Sedimentary Research* 70 (5): 1026–1035.

Selby, M. J. 1993. *Hillslope Materials and Processes*. Oxford University Press.

Silliman, B. R., J. Van de Koppel, M. W. McCoy, J. Diller, G. N. Kasozi, K. Earl, P. N. Adams, and A. R. Zimmerman 2012. Degradation and resilience in Louisiana salt marshes after the BP–Deepwater Horizon oil spill. *Proceedings of the National Academy of Sciences of the USA*, 109 (28): 11234–11239.

Silinski, A., M. Heuner, J. Schoelynck, S. Puijalon, U. Schroder, E. Fuchs, P. Troch, T. J. Bouma, P. Meire, and S. Temmerman 2015. Effects of wind waves versus ship waves on tidal marsh plants: A flume study on different life stages of Scirpus maritimus. *PLoS ONE* 10(3): e0118687.

Smith, S. M. 2009. Multi-decadal changes in salt marshes of Cape Cod, MA: Photographic analyses of vegetation loss, species shifts, and geomorphic change. *Northeastern Naturalist* 16(2): 183–208.

Smith, S. M. 2015. Vegetation change in salt marshes of Cape Cod national seashore (Massachusetts, USA) between 1984 and 2013. *Wetlands*, 35(1): 127–136.

Smith, J. A. M. 2013. The role of *Phragmites australis* in mediating inland salt marsh migration in a mid-Atlantic estuary. *PLoS ONE* 8(5): e65091.doi:10.1371/journal.pone.0065091.

Sorensen, R. M. 1973. Water waves produced by ships. *Journal of the Waterways, Harbors and Coastal Engineering Division*, 99(2), 245–256.

Spencer, T., I. Moller, F. Rupprecht, T. J. Bouma, B. K. van Wesenbeeck, M. Kudella, M. Paul, et al. 2015. Salt marsh surface survives true-to-scale simulated storm surges. *Earth Surface Processes and Landforms* 41(4): 543–552.

Suzuki, T. 2011. *Wave dissipation over vegetation fields.* PhD thesis. Delft University of Technology, Netherlands.

Swisher, M. 1982. *The rates and causes of coastal erosion around a transgressive coastal lagoon*, Rehoboth Bay, Delaware. MA thesis. University of Delaware, Newark.

Temmerman, S., P. Meire, T. Bouma, P. M. J. Herman, T. Ysebaert, and H. J. De Vriend. 2013. Ecosystem-based coastal defence in the face of global change. *Nature* 504: 79–83

Thorne, C. R. and N. K. Tovey 1981. Stability of composite river banks. *Earth Surface Processes and Landforms* 6 (5): 469–484.

Tonelli, M., S. Fagherazzi, and M. Petti 2010. Modeling wave impact on salt marsh boundaries. *Journal of Geophysical Research: Oceans* 115 (C9). https://doi.org/10.1029/2009JC006026.

Trosclair, K. J. 2013. *Wave transformation at a saltmarsh edge and resulting edge erosion: observation and modeling.* PhD thesis, University of New Orleans Theses and Dissertations, Paper 1777.

Tuan, Q. T. and H. Oumeraci 2012. Numerical modelling of wave overtopping-induced erosion of grassed inner sea-dike slopes. *Natural Hazards* 63 (2): 417–447.

Valiela, I. and J. M. Teal 1979. The nitrogen budget of a salt marsh ecosystem. *Nature* 280 (5724): 652–656.

Van de Koppel, J., D. Van der Wal, J. P. Bakker, and P. M. J. Herman 2005. Self-organization and vegetation collapse in salt marsh ecosytems. *The American Naturalist* 165 (1): E1–12.

Van der Meer, J. W., H. J. Verheij, J. Lindenberg, A. Van Hoven, and G. J. C. M. Hoffmans 2007. *Wave overtopping and strenght of inner slopes of dikes. Tech. rep. 05i028.* in Dutch. WL|Delft Hydraulics, Geodelft.

Van Der Wal, Z., De Graaf, G., and Lasthuizen, K. 2008. What's valued most similarities and differences between the organizational values of the public and private sector? *Public Administration* 86: 465–482.

Van Eerdt, M. M. 1985. Salt marsh cliff stability in the Oosterschelde. *Earth Surface Processes and Landforms* 10 (2): 95–106.

Van Eerdt, M. M. 1985. The influence of vegetation on erosion and accretion in salt marshes of the Oosterschelde, The Netherlands. *Vegetation* 62 (1–3): 367–373.

Watson, E. B., A. J. Oczkowski, C. Wigand, A. R. Hanson, E. W. Dawey, S. C. Crosby, R. L. Johnson, and H. M. Andrews 2014. Nutrient enrichment and precipitation changes do not enhance resiliency of salt marshs to sea level rise in the Northeastern U.S. *Climatic Change* 125: 501–509.

Webster, P. J., G. J. Holland, J. A. Curry, and H. -R Chang 2005. Changes in tropical cyclone number, duration and intensity in a warming environment, *Science*, 309 (5742): 1844–1846.

Weston, N. B. 2014. Declining sediments and rising seas: an unfortunate convergence for tidal wetlands. *Estuaries and Coasts* 37: 1–23.

Wilson, C. A. and M. A. Allison 2008. An equilibrium profile model for retreating marsh shorelines in southeast Louisiana. *Estuarine, Coastal and Shelf Science* 80 (4): 483–494.

Winterwerp, J. C. and W. G. M. Van Kesteren 2004. Introduction to the physics of cohesive sediment dynamics in the marine environment. Ed. by T. Van Loon. Vol. 56. *Developments in Sedimentology*. Elsevier, Amsterdam, the Netherlands.

Winterwerp, J. C., W. G. M. Kesteren, B. C. Prooijen, and W. Jacobs 2012. A conceptual framework for shear flow–induced erosion of soft cohesive sediment beds. *Journal of Geophysical Research: Oceans* 117 (C10): https://doi.org/10.1029/2012JC008072.

Young, I. R. and L. A. Verhagen 1996. The growth of fetch limited waves in water of finite depth. Part 1. Total energy and peak frequency. *Coastal Engineering* 29 (1):47–78.

16

Upland Migration of North American Salt Marshes

DANTE D. TORIO AND GAIL L. CHMURA

16.1 Introduction

Tidal wetlands including salt marshes and mangroves tend to move inland with sea level rise as they have done in the past. In North America geological records from salt marsh deposits showed early evidence of marsh inland migration as far back as 3,000 years ago when rates of eustatic sea levels decreased (Lambeck and Bard 2000). Where this decrease was accompanied by a slow in isostatic adjustments, marsh vegetation was able to establish and enhance sediment deposition as described by FitzGerald et al. (2008). As the elevation of marshes increased, they migrated seaward and inland (Redfield 1972). At these historic-ally low rates of sea-level rise and stable sediment supply, the rate of marsh migration inland is controlled by the slope of the hinterland. Brinson et al. (1995) noted that, if bordered by steeply sloping hinterlands, marsh migration would be "stalled." In the Gulf Coast of Florida, Raabe and Stumpf (2016) used historical maps and satellite imagery to show that, despite retreat of marsh seaward edge with increased rates of sea level rise, the low slopes of the hinterland allowed extensive inland migration that compensated for that loss. In Chesapeake Bay, Hussein (2009) found that in low-relief submerging areas, coastal marshes accrete vertically and migrate laterally over adjacent forest soils to keep pace with sea-level rise. We are unaware of any studies that have determined the maximum slope required to prevent stalling of marsh migration under any scenario of sediment supply or sea level rise.

Global sea levels are expected to rise higher than historical levels (up to 2.5 meters [Church et al. 2013, Sweet et al. 2017]), with profound impacts to the redistribution of coastal habitats (Kelley et al. 1995, Lambeck and Bard 2000, FitzGerald et al. 2008). With sea level rise, many coastal habitats would be shifting in location as their optimum site characteristics like the slope of the adjacent upland changes (Hussein 2009). Some habitat could increase in area but others could be degraded through erosion (Hughes et al. 2009) or completely lost if those requirements are not present.

Inland migration is an important process of tidal wetlands redistribution that could offset losses with accelerated sea level rise (Kelley et al. 1995, Hussein 2009, Kirwan et al. 2016). However, not all inland areas might be available to accommodate wetlands. While most of the worlds coastlines today are highly populated and developed, impervious surfaces and heavily built-up shorelines pose physical barriers to wetland migration and lateral expansion. In addition, gentle sloping coastal zones are highly in demand for development and human settlement. As a result, the remaining undeveloped shorelines are in higher elevations and steeper slopes which are less suitable for wetland establishment. With climate change, tidal wetlands caught between rising sea level and physical barriers are subjected to increasing risk of coastal squeeze (Doody 2004, Schleupner 2008). Coastal squeeze is a phenomenon where intertidal wetland habitats are obstructed or stalled by physical barriers at high water mark, thus, preventing the landward edges (i.e., low water mark) to migrate (Doody 2004, Pontee 2013) and causing the seaward edges to drown. With the negative impacts of accelerated sea level rise on marsh sustainability (FitzGerald et al. 2008, FitzGerald and Hughes 2019), it is crucial to identify landward migration corridors and the risk coastal squeeze to tidal wetlands (Borchert et al. 2018) for restoration and management purposes.

With projected sea level rise, low and flat areas of the coastal zones will be at increasing risk from flooding. In particular, the low elevation coastal zones (LECZ), a contiguous land area located at or below 10 m elevation above mean sea level, is likely to be flooded if sea level rises to 2 m (McGranahan et al. 2007). LECZ consist some of the most productive tidal ecosystems and offers a huge potential as a migration corridor for tidal wetlands. However, it is also one of the most developed, populated, and protected zones in the world (Ewing 2015). From hard coastal defences to built impervious surfaces, intensive developments on the LECZ could become potential barriers to tidal wetland migration and contribute to coastal squeeze. This has been demonstrated in several studies in the US Pacific coast (Thorne et al. 2018), Gulf of Mexico (Enwright et al. 2016, Borchert et al. 2018), Florida Coast (Raabe and Stumpf 2016), and the Gulf of Maine (Torio and Chmura 2013). Other studies in Mexico (Martínez et al. 2014, Salgado and Martinez 2017) and Canada (Jolicoeur and O'Carroll, 2007, Bernatchez and Fraser, 2012) suggested similar impacts of physical barriers along LCEZ.

A compressive assessment of wetland migration opportunities and coastal squeeze in North America is useful to determine the sensitivity of existing marshes to coastal squeeze, thus their future functions and ecosystem services. Tidal wetlands capture and store carbon (also referred to as Blue Carbon) more efficiently than their terrestrial counterparts but at the same time can release huge amount of greenhouse gases when degraded (McLeod et al. 2011, Pendleton et al. 2010). Protecting potential wetland migration corridors (Borchert et al. (2018) therefore could offset losses of tidal wetland area and potential greenhouse gas emissions from degraded wetlands in places where losses are inevitable. In addition, tidal wetlands provide breeding, feeding, and staging area for migratory birds and fish that follow as great distances as the scale of North America (Zacheis et al. 2001, Nordlie 2003, Abraham et al. 2005). Other important ecosystem services that have continental significance (e.g., Costanza et al. 1997, Arkema and Samhouri 2012, Barbier 2012, Chmura et al. 2012, Abson et al. 2014) might be lost when the habitat is degraded. Furthermore, the USA, Mexico, and Canada are among the countries with largest amount and share of land in the low elevation coastal zone providing huge potential for wetland migration. Thus, a continental-scale study could initiate discussions about policies for protecting wetland beyond their current locations within and outside a country's jurisdiction and beyond static protected area boundaries, and to identify sites that could provide the greatest return in ecosystem services (White and Kaplan 2017) with minimal investment cost.

While there are policies protecting wetlands against direct human damage, threats to wetland migration corridors, such as coastal squeeze, has not been fully integrated in wetland restoration and conservation in North America. In the USA, for example, the Gulf of Maine Council has listed potential salt marsh restoration projects but does not consider whether some marshes may be threatened by coastal squeeze or have adequate migration corridor with respect to sea level rise. In Mexico, CONABIO (ComisiÓn National Para El Conocimiento Y Uso de la Biodiversidad) prioritizes conservation sites based on their present contribution to biodiversity. Although such criteria are essential to maintain and protect existing salt marshes, they do not guarantee resilience. To promote resilience of wetlands in a rapidly changing world, Erwin (2009) recommends policy goals that would significantly reduce non-climate stressors, protect habitat migration corridors, and plan for medium- and long-term climate variability in addition to set conditions in many existing treaties (e.g., UNFCCC 2015, RAMSAR, Aichi Biodiversity Target 11 [CBD 2015], National Framework for Canada's Marine Protected Areas, North American Marine Protected Areas Network [CEC 2015a]). Hyrenbach et al. (2006) also notes that protected areas with fixed boundaries are no longer adequate to ensure resilient and functional ecosystems. In some parts of the world, dynamic protected areas have been proposed to protect coral reefs and fisheries (Game et al. 2009, Hobday et al. 2010) because of the increasing threat from climate change, human pressure, and the possibility of range shifts (Lewison et al. 2015). Regional-scale studies (Borchert et al. 2018) would therefore inform policies to protect wetlands at a bigger scale such as North America.

At present, coastal squeeze and wetland migration prospects of North American salt marshes are unknown because a continental study has proven to be difficult. Some reasons for this difficulty in scaling include the lack of consistent elevation data that covers the entire continent, highly variable accretion rates, and sea level rise rates. Global elevation and impervious datasets that are free and publicly available are potential source of input data, but their performance has not been evaluated. In addition, performing spatial analysis over the whole continent demands high performance computers. In this study we applied a model that ranks the threat of coastal squeeze on topography and imperviousness along potential salt marsh migration corridors using global datasets.

16.2 Methods

16.2.1 Data Sources

We analyzed coastal squeeze of salt marshes and their migration corridor in the USA, Canada and Mexico (Fig. 16.1). We define migration corridor as the extent of potential habitat expansion or equal to the extent of LECZ according to McGranahan et al. (2007), that is ≤ 10 m above sea level. Spatial data on tidal wetland extent, protected areas, and priority restoration sites were obtained from varied sources (Table 16.1). In the USA, most of the salt marsh areas have been mapped as part of the National Wetlands Inventory. Maps of Canadian salt marsh areas were obtained from the provinces of New Brunswick, Nova

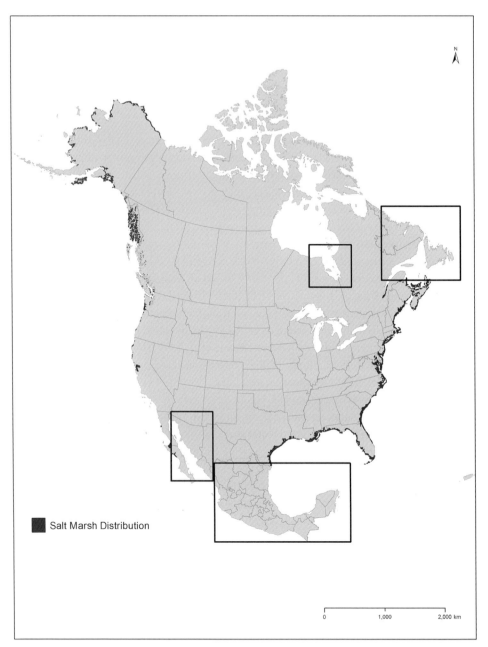

Figure 16.1 Current distribution of salt marshes in North America. Areas in black boxes have no saltmarsh coverage data.

Scotia, and British Columbia data repositories. Maps of Quebec salt marshes were obtained from Environment and Climate Change Canada. No data was available for the province of Newfoundland or the coasts of Hudson and James Bay. In Mexico, only the salt marshes on some parts of the Baja coast have been mapped.

Table 16.1 *Sources of spatial datasets on habitat and protected areas used in the analysis of coastal squeeze. *All states except Louisiana*

Habitat Area	Source Database	Creation Method	Temporal Range	Minimum Mapping Unit (m)	Active Link
Canada					
British Columbia	British Columbia ShoreZone 2014	Aerial photo interpretation and satellite image classification	2014	ND	www .princeedwardisland .ca/en
New Brunswick	GEONB Wetlands	Aerial photo interpretation and satellite image classification	1981–2009	ND	www.snb.ca/ geonb1/e/DC/ catalogue-E.asp
Prince Edward Island	2000 Wetland Inventory	Aerial photo interpretation and satellite image classification	to 2000	175	www.gov.pe.ca/gis/ index.php3? number=77581& lang=E
Nova Scotia	Nova Scotia Wetland Inventory	Aerial photo interpretation and satellite image classification	1980; 1990	100–	http://novascotia.ca/ natr/wildlife/ habitats/wetlands .asp
Quebec	St. Lawrence Wetlands	Aerial photo interpretation and satellite image classification	1990–2002	4	www.mcgill.ca/ library/find/maps/ stlwetlands
Mexico					
Pacific coast	US Geological Survey c/o Ward, David <dward@usgs .gov>	Aerial photo interpretation and satellite image classification			none
USA					
USA*	National Wetlands Inventory	Aerial photo interpretation	1983–2009	< 250	www.fws.gov/ wetlands/Data/ Data-Download .html
Louisiana	United States Geological Survey	Aerial photo interpretation	2013	No Data	http://pubs.usgs .gov/sim/3290/

Table 16.1 (*cont.*)

Habitat Area	Source Database	Creation Method	Temporal Range	Minimum Mapping Unit (m)	Active Link
Protected areas and restoration sites					
Marine protected areas of North America	Commission for Environmental Cooperation North American Environmental Atlas	Delineated boundaries	unknown	unknown	www.cec.org/Page .asp?PageID= 924andContentID= 2867
Bay of Fundy and Gulf of Maine potential restoration sites (New Brunswick and Nova Scotia)	Danika van Proosdij of Saint Mary's University (Halifax, Nova Scotia).	Field survey (represent diked and drained salt marshes)	NA	NA	NA
Priority Conservation Areas of Mexico	CONABIO (Arriaga et al. 1998)	unknown	unknown	unknown	www.conabio.gob .mx/conocimiento/ regionalizacion/ doctos/Mmapa.html

Maps of protected areas with salt marshes and potential salt marsh restoration sites were obtained for the coast of the Gulf of Maine (USA and Canada) and Mexico (Table 16.1). Potential restoration sites on the New Brunswick and Nova Scotia coasts of the Bay of Fundy (the upper part of the Gulf of Maine) represent diked and drained salt marshes mapped by Danika van Proosdij of Saint Mary's University (Halifax, Nova Scotia). For Mexico, the map of the priority sites with salt marsh was obtained from CONABIO. The maps for the North American marine protected areas with marshes were obtained from the Commission for Environmental Cooperation (CEC) North American Environmental Atlas (Commission for the Environment 2020).

16.2.2 Coastal Squeeze Calculation

Coastal squeeze was calculated in a series of steps and was based upon slope and percent imperviousness. First, a digital elevation model (DEM) and imperviousness data for North America was obtained from the CGIAR Consortium of Spatial Information (CGIAR 2015) and National Oceanic and Atmospheric Administration (NOAA 2010). The DEM was derived from the Shuttle Radar Topographic Mission (Jarvis et al. 2008, JPL 2015). It has a pixel size of 90 m and about 1–10 m vertical accuracy. The percent imperviousness data was obtained from NASA Socioeconomic Data and Application Centre. This dataset derived from VIIRS sensor night time lights and land cover derived from Landsat imagery and has a pixel size of 30 m (Elvidge et al. 2007, Brown de Colstoun et al. 2017). Because of the differences in resolution, and limited computing power, the DEM and imperviousness data were resampled at 250 m spatial resolution for faster processing. Second, subsets of elevation and imperviousness data that falls within LECZ were extracted from the original datasets. Third, the slope (in degrees) of the elevation subset was calculated using the slope function in ArcGIS 10.2.2 (Burrough and McDonnel 1999, ESRI 2015).

A model developed by Torio and Chmura (2013) was used to simulate marsh migration within the LECZ with simple assumptions that rate of accretion is constant, and that 100-year projected sea level rise of up to 2 m is linear although a nonlinear sea level rise is highly likely (Brinson et al. 1995, Raabe and Stumpf 2016). The following two equations were then applied to rank the subset of slope and imperviousness data.

$$\mu(coastal squeeze slope) = \frac{1}{1 + \left(\frac{slope}{11.5}\right)^{-3.95}} \tag{16.1}$$

$$\mu(coastal squeeze imperviousness) = \frac{1}{1 + \left(\frac{imperviousness}{15.4}\right)^{-5.0}} \tag{16.2}$$

After computing the two coastal squeeze components, they were combined to produce a cumulative coastal squeeze index in which both slope and imperviousness were assumed to have equal contributions. The resulting coastal squeeze indices (i.e., coastal squeeze slope, coastal squeeze imperviousness, and coastal squeeze cumulative) were reclassified into three classes representing three levels of threat; 1 being the lowest, and 3, the highest threat of coastal squeeze. The classes were based on thresholds values called tercile bins in which a squeeze index of <0.33 is assigned to bin 1 or lowest threat level, and a coastal squeeze index of >0.66 is assigned to Class 3, or the highest threat level. Intermediate values were assigned to Class 2. These intervals were based upon the Nature Conservancy's threat categories (Crain et al. 2008, The Natural Capital Project 2015) which represent the threshold of habitats under risk of multiple stressors. After creating the coastal squeeze classes, the resulting raster layer was converted to polygons using the raster to feature tool in ArcGIS 10.2.2. Then, using a spatial analysis (i.e., intersect operation), the salt marsh polygons were overlain to the threat polygons and the percent area under different threat categories was calculated. The same association procedure was implemented using the salt marsh restoration sites and protected areas (terrestrial and marine). This was implemented for all the threat classes of slope, imperviousness, and cumulative coastal squeeze.

16.3 Results

16.3.1 Cumulative Coastal Squeeze Threat

Our analysis indicates that ≈11% of the salt marsh area we analyzed is threatened by coastal squeeze (Fig. 16.2A). Of the total salt marsh area (13,474 km²) about 9% (or 1,213 km²) is under medium threat, and 2% (or 270 km²) marshes is under high threat.

The level of threat of coastal squeeze varies by political jurisdiction. In Mexico, the salt marshes in the states of Baja California Sur (40%) and Baja California Norte (70%) are

A)

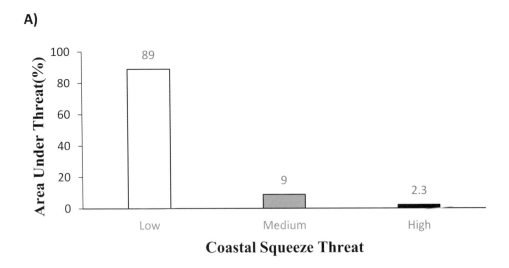

B)

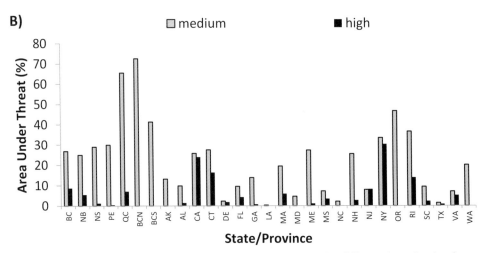

Figure 16.2 Percentage of all salt marshes area examined under different threat levels of cumulative coastal squeeze (A) and the proportion of salt marsh area of each state or province under medium and high threat levels (B).

QC: Quebec; NB: New Brunswick; NS: Nova Scotia; PE: Prince Edward Island; ME: Maine; NH: New Hampshire; RI: Rhode Island; CT: Connecticut; NY: New York; NJ: New Jersey; DE: Delaware; MD: Maryland; VA: Virginia; NC: North Carolina; SC: South Carolina; GA: Georgia; FL: Florida; AL: Alabama; MS: Mississippi; LA: Louisiana; TX: Texas; BC: British Columbia; AK: Alaska; WA: Washington; OR: Oregon; CA: California; BCN: Baja California Norte; BCS: Baja California Sur

under medium threat (Fig. 16.2B). No Mexican marshes are under high cumulative threat, but much of the coast has not been mapped. In Canada, more than 10% of the salt marsh area is under medium threat and at least 5% of the salt marshes in British Columbia, Quebec, and New Brunswick are under high threat. In the USA, states with large salt marsh

Table 16.2 *Number of sites (i.e., potential restoration and marine protected areas) with salt marshes under high threat of coastal squeeze*

| | Total | Coastal Squeeze Threat Variables | | Cumulative |
Sites	Sites	Slope	Imperviousness	Threat
Gulf of Maine restoration sites (Maine, USA)	**1084**	4	29	33
Bay of Fundy restoration sites (Nova Scotia and New Brunswick)	**322**	1	28	29
Marine Protected Areas (North America)	**840**		40	40

areas (i.e., >1,000 km^2, such as Louisiana, Texas, South Carolina, Florida, and Georgia) are least threatened. In contrast, the states with the least salt marsh area like New York, California, Connecticut, Rhode Island, and New Jersey stand out by having the greatest proportion of area (>10%) under medium and high threat.

Table 16.2 shows the percentage of potential salt marsh restoration sites and marine protected areas with salt marsh across North America under high threat of coastal squeeze. About 3% of the potential restoration sites on USA coast of the Gulf of Maine and 9% on the Canadian coast of the Gulf of Maine (Bay of Fundy) are under high threat. Of the marine protected areas 40 of 840 sites are under high threat.

16.3.2 Coastal Squeeze by Slope

Steep slopes, that is, ≥11.5° degrees could act as natural barriers to inland migration and stall lateral expansion of salt marshes. For the group of marshes analyzed in this study, slope posed a minor threat. In total, about 7 km^2 of salt marshes are highly threatened by topographic constraints. These represent only about 0.05% of the total North American salt marsh areas (Fig. 16.3).

Coastal squeeze by steep slope is high in Maine, California, and Washington although the area impacted is relatively small (≤1%). All the other US states have lower threat levels. The Mexican salt marshes that we analyzed are under medium threat. The state of Baja California Norte has the largest area under this threat category. In Canada, only New Brunswick and Nova Scotia have salt marshes limited by steep slopes. Figure 16.4 shows example of marshes threatened by steep slopes from surrounding hilly lands.

Few potential restoration sites or and marine protected areas are threatened by steep slopes (Table 16.2). In Maine, of the 33 sites under high cumulative coastal squeeze, four are due to steep slopes. Likewise, only one of 28 Bay of Fundy potential restoration sites faces a steep slope at its inland border.

16.3.3 Coastal Squeeze by Anthropogenic Barriers

Coastal squeeze by urban development (as determined by imperviousness) threatens more salt marshes than slope (Fig. 16.5). In North America, 262 km^2, or 2% of the total area of

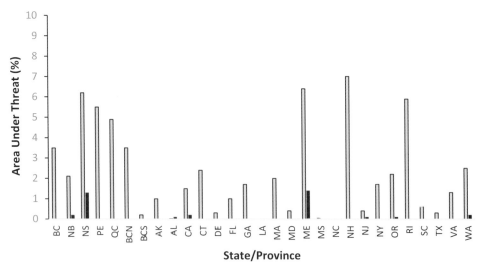

Figure 16.3 Percent of salt marsh area of each state or province under different threat levels of coastal squeeze by slope. (See Fig. 16.2 for explanation of state and province abbreviations.)

salt marsh is under high threat due to adjacent developed areas. The threat due to anthropogenic barriers is greater than that due to slope both in proportion of marsh area under threat and with respect to the number of states and provinces affected. In Canada, 3–5% of the marshes in New Brunswick, British Columbia, and Quebec face barriers due to inland development. In the USA, inland development threatens the future of 4%–30% of the marshes in New York, California, Connecticut, New Jersey, Rhode Island, Maryland, Virginia, and some part of Florida. Examples are Jamaica Bay, New York (Fig. 16.6A) and San Francisco Bay, California (Fig. 16.6B).

Contrary to the threat posed by steep slopes, the threat of anthropogenic barriers affects most of the potential restoration sites and protected areas under high cumulative coastal squeeze threat (Table 16.2). For example, 29 of the 33 highly threatened Gulf of Maine potential restoration sites and 28 of 29 highly threatened Bay of Fundy restoration sites are in highly developed areas. In Mexico, the one site under high cumulative coastal squeeze threat is mostly due to imperviousness. Overall, most marshes in marine protected areas are under high threat of coastal squeeze from development as they are surrounded by highly impervious surfaces. This implies that marshes inside protected areas are not protected from indirect threats like coastal squeeze. The extent to which marshes are threatened by coastal squeeze in these protected areas would benefit from more study at a finer spatial scale.

16.4 Discussion and Conclusions

Kirwan et al. (2016) and Feagin et al. (2010) showed that, with accelerated rates of sea level rise, there can be a net gain in marsh area if topographic and anthropogenic barriers do not

Figure 16.4 Location of salt marshes with the highest threat of coastal squeeze from steep slope (red, A). Boxes on the map show geographic location of an example at Quoddy Narrows, Lubec, Maine, USA (B) and Yarmouth, Nova Scotia, Canada (C). (A black and white version of this figure will appear in some formats. For the color version, please refer to the plate section.)

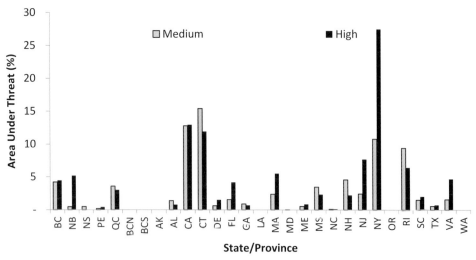

Figure 16.5 Percent of salt marsh area of each state or province under coastal squeeze by imperviousness. (See Fig. 16.2 for explanation of state and province abbreviations.)

prevent inland migration. Our results for North America indicate that anthropogenic barriers exert more influence than topography. About 10% of the total North American salt marshes are highly threatened by coastal squeeze. Of the two coastal squeeze variables, imperviousness threatens more salt marshes than slope. Less than 1% of the total salt marsh area (~7 km^2) is bordered by steep slopes while 2% (~262 km^2) is under high threat of imperviousness. In the USA, the proportion of salt marshes under high threat from inland development can be as high as 30%.

Steep slopes could prevent lateral migration of salt marshes (Hussein 2009). In North America, slopes do threaten the migration of a small percentage of marshes particularly in states or provinces that have relatively steep coastal areas such as New Brunswick, Nova Scotia, Maine, and some marshes in Washington and California. Our results for the California region agree with recent findings from Thorne et al. (2018) in which the state will lose wetlands in all projected sea level rise scenarios. Our results further suggest that these losses would be mostly driven by steep slope.

Marshes in less developed and relatively flat coastal areas (e.g., Florida, Louisiana, Texas, North and South Carolina, Alaska, and Virginia) have more potential for migration, expansion and thus permanence. Those salt marsh habitats under medium threat require careful strategic management to prevent further development to ensure connected migration corridors. In a recent study, Enwright et al. (2016) noted that much of the landward migration will concentrate along Florida, Louisiana, and Texas coastlines; however, future urbanization is expected to intensify in those same areas. Their findings agree with our results.

In Mexico, we have data only for the northern and southern Baja California States. As such, Mexican marshes are underrepresented in this study. However, a study of other intertidal habitats (Martínez et al. 2014) in the country implies coastal squeeze from topography and urbanization could threaten other important plant communities that may

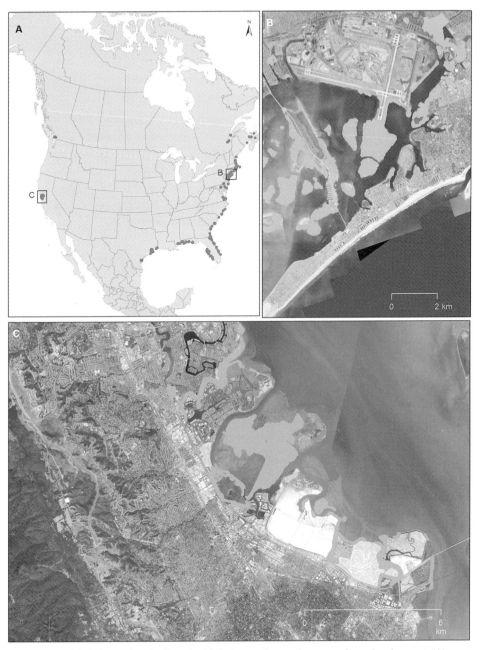

Figure 16.6 Salt marshes (red) under high threat of coastal squeeze from development (A), Jamaica Bay, New York (B) and San Francisco Bay, California (C). (A black and white version of this figure will appear in some formats. For the color version, please refer to the plate section.)

need to migrate inland with accelerated sea level rise. More research is needed to assess the threat of coastal squeeze to intertidal habitats in Mexico.

16.4.1 Implication for Management Decisions

Much of the salt marsh area of North America is found on the Gulf of Mexico and Atlantic coasts: regions that are most vulnerable to sea level rise (Pendleton et al. 2010) and where coastal development is intensive. As extensive areas are under medium threat, coastal squeeze will become an increasingly important factor affecting the permanence of salt marshes. In areas where threat of anthropogenic barriers is moderate, it should be possible to develop incentives to prevent further development of areas next to salt marshes. In other areas where protecting human settlement is inevitable, a combination of strategies to protect both human communities and maintain ecosystem services from wetlands should be developed (Salgado and Martínez 2017).

Within the USA, Florida, Louisiana, and Texas have the largest potential migration areas adjacent to tidal wetlands, but populations and the development within these areas are also high (Fig. 16.7). In Louisiana, large, less developed low-lying areas would be ideal for wetland migration but loss of marsh there is extensive due to other factors. In Canada, New Brunswick has considerable potential migration areas in addition to some low-lying areas in Quebec. Potential migration area in Alaska is underestimated because the elevation data does not cover the northern most part of the region but using a different dataset, we calculated an area of about 184,000 km^2 in low elevation coastal zone; the largest among the USA states.

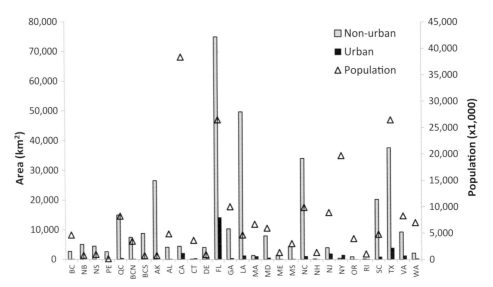

Figure 16.7 Distribution of low lying coastal areas associated with salt marshes (<30 m elevation) of North America and corresponding population. (See Fig. 16.2 for explanation of state and province abbreviations.)

Source: Population estimates from Mexico: Instituto Nacional de Estadistica Y Geografia, 2010: from Canada: Statistics Canada, 2014: from USA: US Census Bureau, 2014

Expansive low-lying hinterlands will not always guarantee future wetland migration inland because institutional and socio-economic barriers to migration may exist. As such it would be prudent to reconsider longstanding policies of shoreline hardening and shift to more hybrid approaches (Sutton-Grier et al. 2015). Titus and Neumann (2009) assert that policies to ensure that wetlands can migrate are likely to be less expensive if the planning takes place before development.

Salt marshes located in protected areas are not necessarily protected from rising sea levels and coastal squeeze. As such, static and geographically fixed protected areas will not be an effective management tool with climate change. Simple protection from direct disturbance will not be enough to ensure their future. Although, we may protect them from direct threats with fixed reserves, long term management of protected areas must consider future migration space and the rank of level of coastal squeeze can help prioritize use of limited conservation funds. Coastal squeeze analyses can guide the design of new protected areas and redesign of existing protected areas to include buffers, easements, or migration space to allow inland migration and maintain hydrological connections.

16.4.2 Implication for Restoration

The level of coastal squeeze of a salt marsh is an informative indicator of its restorability and permanence. As it has both present and future implications, the threat of coastal squeeze should be considered in prioritizing sites for restoration to prevent investments in sites that have a limited future. Marsh areas under low and medium threat with high potential for migration are ideal for generalized and long-term restoration. These may include prevention of further development or investment and keeping the present land undeveloped. On the other hand, marshes under high threat require specific and near-term management and prioritization. In these marshes, prioritization could be based on their potential to provide ecosystem services (e.g., important to migrating waterfowl, recreation, or as a storm buffer).

16.4.3 Implication for Ecosystem Services Provision

Preventing inland migration with sea level rise will lead to degradation of coastal habitats and their functions (Crooks 2004). For instance, the degradation and loss of salt marsh along migration corridors might starve populations of migratory waterfowl and diadromous fish. Similarly, if large areas of salt marshes are lost, the remaining small areas may not effectively protect human communities from storms. Maintaining coastal water quality is another important function of tidal wetlands which can diminish with decreasing marsh area. Most importantly, carbon capture and storage might decrease with marsh degradation.

Most of the salt marshes under high threat consist of small patches so it is possible that coastal squeeze may result in habitat fragmentation. With coastal squeeze, the size, shape, and spatial distribution of salt marsh patches can be modified in such a way that they may no longer be suitable as habitats to many endemic and migratory birds and fish (Hitch et al.

2011, Tomaselli et al. 2012). In addition, fragmented marshes may affect the predator-prey dynamics, especially between fishes that use marshes (Cooper et al. 2012).

16.4.4 Limitations, Caveats, and Prospects

The robustness of the coastal squeeze index at the continental scale is limited principally by the spatial resolution and vertical accuracy of the global elevation. In upscaling the index to North America, we found that the resampled elevation data underestimate or truncate the slope in narrow estuaries. To illustrate, an area with a slope of 11.5° in the LiDAR-based elevation has a slope of 1.5° in resampled DEM. For this reason, the midpoint parameter (f_2) or threshold slope for the coastal squeeze slope model requires adjustment to compensate for data loss (e.g., from 11.5 to 1.5 in Equation 16.1) especially in narrow estuaries. Using a limited area as a test site, we found that there is no need to adjust the midpoint parameter if the original resolution of 90 meter is used. For imperviousness (Equation 16.2), no modification was needed as the results of the validation tests matched the extent of urban areas in the LiDAR-based model. In this study, we found that imperviousness alone can be a good indicator of coastal squeeze potential at a North American scale.

Estimations of coastal squeeze by slope in narrow estuaries and coasts might be underestimated if elevation data is not spatially and vertically accurate. In British Columbia, for instance, the marsh area under high threat might be less certain. An elevation dataset with finer vertical and horizontal resolution, such as that obtained by radar interferometer, or by combining regional datasets with global datasets (e.g., US National Elevation Dataset, Canadian DEM, and refined Global DEM) would improve the analysis. In addition, coarse resolution elevation data does not cover areas below mean sea level or close to the water edge where most of the habitat polygons are situated. Absence of data on the near-shore area is a persistent problem that can be resolved by having seamless topographic and bathymetric data. Incomplete mapping of wetlands also contributes to the limitation of this study. For example, Mexican and Canadian marshes are underrepresented as the mapping was incomplete. In both countries, the threat classes remain underestimated and area estimates remains uncertain

Acknowledgements

Robert Houston of the US Fish and Wildlife Services provided maps of sites on the Gulf of Maine. Partial funding came from the Geomatics for Informed Decision (GEIODE) Project PIV-41, "The Participatory Geoweb for Engaging the Public on Global Environmental Change" (to R. Sieber), an NSERC Discovery grant, Commission for Environmental Cooperation grant (to G. Chmura), Natural Resources Canada grant (to S. Jolicoeur), Graduate Excellence Fellowship, Graduate Research Enhancement and Travel Awards from McGill University, and Global Environment and Climate Change Centre Award. McGill's Geographic Information Centre provided access to computing resources.

We also thank the two anonymous reviewers who provided constructive feedbacks and suggested important literatures that improved the clarity of the paper.

References

Abraham, K. F., Jefferies, R. L., and Alisauskas, R. T. 2005. The dynamics of landscape change and snow geese in mid-continent North America. *Global Change Biology*, 11 (6): 841–855.

Abson, D. J., von Wehrden, H., Baumgärtner, S., Fischer, J., Hanspach, J., Härdtle, W., Heinrichs, H. et al. 2014. Ecosystem services as a boundary object for sustainability. *Ecological Economics*, 103: 29–37.

Arriaga Cabrera, L., Vázquez Domínguez, E., González Cano, J., Jiménez Rosenberg, R., Muñoz López, E., and Aguilar Sierra, V. (coordinators). 1998. Regiones marinas prioritarias de México. Comisión Nacional para el Conocimiento y uso de la Biodiversidad. México.

Arkema, K. K., and Samhouri, J. F. 2012. Linking ecosystem health and services to inform marine ecosystem-based management. *American Fisheries Society Symposium*, 79: 9–25.

Barbier, E. B. 2012. Progress and challenges in valuing coastal and marine ecosystem services. *Review of Environmental Economics and Policy*, 6: 1–19.

Bernatchez, P., and Fraser, C. 2012. Evolution of coastal defence structures and consequences for beach width trends, Québec, Canada. *Journal of Coastal Research*, 28 (6): 1550–1566.

Borchert, S. M., Osland, M. J., Enwright, N. M., and Griffith, K. T. 2018. Coastal wetland adaptation to sea level rise: Quantifying potential for landward migration and coastal squeeze. *Journal of Applied Ecology*, 55: 2876–2887.

Brinson, M. M., Christian, R. R., and Blum, L. K. 1995. Multiple states in the sea-level induced transition from terrestrial forest to estuary. *Estuaries*, 18(4): 648–659.

Brown de Colstoun, E. C., Huang, C., Wang, P., Tilton, J. C., Tan, B., Phillips, J., Niemczura, S., Ling, P.-Y., and Wolfe, R. E. 2017. Global Man-made Impervious Surface (GMIS) Dataset from Landsat. Palisades, NY: NASA Socioeconomic Data and Applications Center (SEDAC). https://doi.org/10.7927/H4P55KKF. Accessed 01 01 2017.

Burrough, P. A., and McDonnel, R. A. 1999. *Principles of Geographical Information Systems*. Oxford University Press.

CBD. 2015. Convention on Biological Diversity. TARGET 11 - Technical Rationale extended (provided in document COP/10/INF/12/Rev.1). www.cbd.int/sp/targets/rationale/target-11/.

Commission for the Environment, 2020. North American Environmental Atlas. www.cec.org/tools-and-resources/north-american-environmental-atlas/map-files

CEC. 2015a. Commission for Environmental Cooperation. North American Protected Areas Network. www2.cec.org/nampan/.

CEC. 2015b. Commission for Environmental Cooperation. North America's Blue Carbon: Assessing the role of coastal habitats in the Continent's carbon budget. www.cec.org/bluecarbon.

CGIAR. 2015. Consultative Group for International Agricultural Research-Consortium of Spatial Information. Digital Elevation Model. www.cgiar-csi.org/data.

Chmura, G. L., Burdick, D., and Moore, G. 2012. Recovering salt marsh ecosystem services through tidal restoration, In: *Tidal Marsh Restoration*, ed. C.T. Roman and D.M. Burdick, 233–251. Washington, DC: Island Press.

Church, J. A., Clark, P. U., Cazenave, A., Gregory, J. M., Jevrejeva, S., A. Levermann, A., Merrifield M. A., et al. 2013. Sea level change. In: Climate Change 2013: The Physical Science Basis. Contribution of Working Group I to the Fifth Assessment Report of the Intergovernmental Panel on Climate Change [Stocker, T.F., D. Qin,

G.-K. Plattner, M. Tignor, S.K. Allen, J. Boschung, A. Nauels, Y. Xia, V. Bex and P.M. Midgley (eds.)]. Cambridge University Press, Cambridge, United Kingdom and New York, NY, USA.

Cooper, J. K., Li, J., and Montagnes, D. J. 2012. Intermediate fragmentation per se provides stable predator-prey metapopulation dynamics. *Ecology Letters*, 15: 856–863.

Costanza, R., dArge, R., deGroot, R., Farber, S., Grasso, M., Hannon, B., Limburg, K., et al. 1997. The value of the world's ecosystem services and natural capital. *Nature*, 387: 253–260.

Crain, C. M., Kroeker, K., and Halpern B. S. 2008. Interactive and cumulative effects of multiple human stressors in marine systems. *Ecology Letters*, 11: 1304–1315.

Crooks, S. 2004. The effect of sea-level rise on coastal geomorphology. *Ibis*, 146: 18–20.

Doody, P. J. 2004. "Coastal Squeeze": An historical perspective. *Journal of Coastal Conservation*, 10: 129–138.

Elvidge, C. D., Tuttle, B. T., Sutton, P. C., Baugh, K. E., Howard, A. T., Milesi, C., Bhaduri, B. and Nemani, R. 2007. Global distribution and density of constructed impervious surfaces. *Sensors*, 7: 1962–1979.

Enwright, N. M., Griffith, K. T., and Osland, M. J. 2016. Barriers to and opportunities for landward migration of coastal wetlands with sea-level rise. *Frontiers in Ecology and the Environment*, 14(6): 307–316.

ESRI. 2015. Environmental Systems Research Institute. How slope works. http://resources.arcgis.com/en/help/main/10.2/index.html#//009z000000vz000000.

Erwin, K. L. 2009. Wetlands and global climate change: The role of wetland restoration in a changing world. *Wetlands Ecology and Management*, 17: 71–84.

Ewing, L. C. 2015. Resilience from coastal protection. *Philosophical Transactions of the Royal Society A: Mathematical, Physical and Engineering Sciences*, 373(2053): 20140383.

Feagin, R. A., Martínez, M. L., Mendoza-Gonzalez, G., and Costanza, R. 2010. Salt marsh zonal migration and ecosystem service change in response to global sea level rise: a case study from an urban region. *Ecology and Society*, 15(4): 14.

FitzGerald, D. M., Fenster, M. S., Argow, B. A., and Buynevich, I. V. 2008. Coastal impacts due to sea-level rise. *Annual Review of Earth and Planetary Sciences*, 36: 601–647.

FitzGerald, D. M., and Hughes, Z. 2019. Marsh processes and their response to climate change and sea-level rise. *Annual Review of Earth and Planetary Sciences*, 47: 481–517.

Game, E. T., Grantham, H. S., Hobday, A. J., Pressey, R. L., Lombard, A. T., Beckley, L. E., Gjerde, K., Bustamante, R., Possingham, H. P., and Richardson, A. J. 2009. Pelagic protected areas: The missing dimension in ocean conservation. *Trends in Ecology and Evolution*, 24: 360–369.

Guerry, A. D., Polasky, S., Lubchenco, J., Chaplin-Kramer, R., Daily, G. C., Griffin, R., and Ruckelshaus, M., et al. 2015. Natural capital and ecosystem services informing decisions: From promise to practice. *Proceedings of the National Academy of Sciences* 112:7348–7355

Hitch, A. T., Purcell, K. M., Martin, S. B., Klerks, P. L., and Leberg, P. L. 2011. Interactions of salinity, marsh fragmentation and submerged aquatic vegetation on resident nekton assemblages of coastal marsh ponds. *Estuaries and Coasts*, 34: 653–662.

Hobday, A. J., Hartog, J. R., Timmiss, T. and Fielding, J. 2010. Dynamic spatial zoning to manage southern bluefin tuna (*Thunnus maccoyii*) capture in a multi-species longline fishery. *Fisheries Oceanography*, 19: 243–253.

Hughes, Z. J., FitzGerald, D. M., Wilson, C. A., Pennings, S. C., Więski, K., and Mahadevan, A. 2009. Rapid headward erosion of marsh creeks in response to relative sea level rise. *Geophysical Research Letters*, 36(3): https://doi.org/10.1029/2008GL036000.

Hussein, A. H. 2009. Modeling of sea-level rise and deforestation in submerging coastal ultisols of Chesapeake Bay. *Soil Science Society of America Journal* 73(1): 185–196.

Hyrenbach, K., Keiper, C., Allen, S., Ainley, D., and Anderson, D. 2006. Use of marine sanctuaries by far-ranging predators: Commuting flights to the California Current System by breeding Hawaiian albatrosses. *Fisheries Oceanography*, 15: 95–103.

Jarvis, A., Reuter, H. I., Nelson, A., and Guevara, E. 2008. Hole-filled SRTM for the globe version 4. CGIAR-CSI SRTM 90m database: http://srtm.csi.cgiar.org.

Jolicoeur, S., and O'Carroll, S. 2007. Sandy barriers, climate change and long-term planning of strategic coastal infrastructures, Îles-de-la-Madeleine, Gulf of St. Lawrence (Québec, Canada). *Landscape and Urban Planning*, 81(4): 287–298.

Jones, S., Bosch, A. C., and Strange, E. 2009. Vulnerable species: the effects of sea-level rise on coastal habitats. In Coastal Sensitivity to Sea-Level Rise: A Focus on the Mid-Atlantic Region. A report by the U.S. Climate Change Science Program and the Subcommittee on Global Change Research., ed. J. G. Titus, K. E. Anderson, D. R. Cahoon, D. B. Gesch, S. K. Gill, B. T. Gutierrez, E. R. Thieler and S. J. Williams, pp. 73–84: US Environmental Protection Agency, Washington, DC.

JPL. 2015. Jet Propulsion Laboratory. Shuttle Radar Topography Mission. www2.jpl.nasa .gov/srtm/statistics.html: National Aeronautics and Space Administration.

Kelley, J. T., Gehrels, W. R., and Belknap, D. F. 1995. Late Holocene relative sea-level rise and the geological development of tidal marshes at Wells, Maine, USA. *Journal of Coastal Research*, 11: 136–153.

Kirwan, M. L., Walters, D. C., Reay, W. G., and Carr, J. A. 2016. Sea level driven marsh expansion in a coupled model of marsh erosion and migration. *Geophysical Research Letters*, 43: 4366–4373.

Lambeck, K., and Bard, E. 2000. Sea-level change along the French Mediterranean coast for the past 30 000 years. *Earth and Planetary Science Letters*, 175(3–4): 203–222.

Lewison, R., Hobday, A. J., Maxwell, S., Hazen, E., Hartog, J. R., Dunn, D. C., Briscoe, D., et al. 2015. Dynamic ocean management: Identifying the critical ingredients of dynamic approaches to ocean resource management. *BioScience*, 65: 486–498.

Martínez, M. L., Mendoza-González, G., Silva-Casarín, R., and Mendoza-Baldwin, E. 2014. Land use changes and sea level rise may induce a "coastal squeeze" on the coasts of Veracruz, Mexico. *Global Environmental Change*, 29: 180–188.

McGranahan, G., Balk, D., and Anderson, B. 2007. The rising tide: Assessing the risks of climate change and human settlements in low elevation coastal zones. *Environment and Urbanization*, 19: 17–37.

McLeod, E., Chmura, G. L., Bouillon, S., Salm, R., Björk, M., Duarte, C. M., and Silliman, B.R. 2011. A blueprint for blue carbon: Toward an improved understanding of the role of vegetated coastal habitats in sequestering CO_2. *Frontiers in Ecology and the Environment*, 9(10): 552–560.

MEA. 2005. *Millennium Ecosystem Assessment. Ecosystems and Human Well-being*: Synthesis: Island Press Washington, DC.

NOAA. 2010. National Oceanic and Atmospheric Administration-National Center for Environmental Information. Global Distribution and Density of Constructed Impervious Surfaces. http://ngdc.noaa.gov/eog/dmsp/download_global_isa.html: NOAA.

Nordlie, F. G. 2003. Fish communities of estuarine salt marshes of eastern North America, and comparisons with temperate estuaries of other continents. *Reviews in Fish Biology and Fisheries*, 13(3): 281–325.

Pendleton, E. A., Thieler, E. R., and Williams, S. J. 2010. Importance of coastal change variables in determining vulnerability to sea-and lake-level change. *Journal of Coastal Research*, 26: 176–183.

Pontee, N. 2013. Defining coastal squeeze: A discussion. *Ocean and Coastal Management*, 84: 204–207

Raabe, E. A., and Stumpf, R. P. 2016. Expansion of tidal marsh in response to sea-level rise: Gulf Coast of Florida, USA. *Estuaries and Coasts*, 39(1): 145–157.

Redfield, A. C. 1972. Development of a New England salt marsh. *Ecological Monographs*, 42(2): 201–237.

Salgado, K., and Martínez, M. L. 2017. Is ecosystem-based coastal defense a realistic alternative? Exploring the evidence. *Journal of Coastal Conservation*, 21(6): 837–848.

Schleupner, C. 2008. Evaluation of coastal squeeze and its consequences for the Caribbean island Martinique. *Ocean and Coastal Management*, 51: 383–390.

Sutton-Grier, A. E., Wowk, K., and Bamford, H. 2015. Future of our coasts: The potential for natural and hybrid infrastructure to enhance the resilience of our coastal communities, economies and ecosystems. *Environmental Science and Policy*, 51: 137–148.

Sweet, W. V., Kopp, R. E., Weaver, C. P., Obeysekera, J., Horton, R. M., Thieler, E. R., and C. Zervas. 2017. Global and regional sea level rise scenarios for the United States. NOAA Technical Report NOS CO-OPS 083. NOAA/NOS Center for Operational Oceanographic Products and Services

Thorne, K., MacDonald, G., Guntenspergen, G., Ambrose, R., Buffington, K., Dugger, B., Freeman, C., et al. 2018. US Pacific coastal wetland resilience and vulnerability to sea-level rise. *Science Advances*, 4(2): 3270.

Titus, J. G., and Neumann, J. E.. 2009. Implications for decisions. In coastal sensitivity to sea-level rise: A focus on the mid-Atlantic region. A report by the U.S. Climate Change Science Program and the Subcommittee on Global Change Research., ed. J. G. Titus, K. E. Anderson, D. R. Cahoon, D. B. Gesch, S. K. Gill, B. T. Gutierrez, E. R. Thieler and S. J. Williams, pp. 141–156: US Environmental Protection Agency, Washington DC, USA.

Tomaselli, V., Tenerelli, P., and Sciandrello, S. 2012. Mapping and quantifying habitat fragmentation in small coastal areas: A case study of three protected wetlands in Apulia (Italy). *Environmental Monitoring and Assessment*, 184: 693–713.

Torio, D. D., and Chmura, G. L. 2013. Assessing coastal squeeze of tidal wetlands. *Journal of Coastal Research*, 29: 233–243.

Torio, D. D., and Chmura, G. L. 2015. Impacts of sea level rise on marsh as fish habitat. *Estuaries and Coasts*, 38: 1288–1303.

UNFCCC. 2015. United Nation Framework Convention on Climate Change. *NAMAs, Nationally Appropriate Mitigation Actions*. http://unfccc.int/focus/mitigation/items/ 7172.php.

White, E. P., and Brown, J. H. 2005. The template: patterns and processes of spatial variation. In *Ecosystem Function in Heterogeneous Landscapes*, ed. G. M. Lovett, C. G. Jones, M. G. Turner, and K. C. Weathers, pp. 31–47: Springer, New York, USA.

White, E. and Kaplan, D. 2017. Restore or retreat? Saltwater intrusion and water management in coastal wetlands. *Ecosystem Health and Sustainability*, 3(1): e01258.

Zacheis, A., Hupp, J. W., and Ruess, R. W. 2001. Effects of migratory geese on plant communities of an Alaskan salt marsh. *Journal of Ecology*, 89(1): 57–71.

17

Restoration of Tidal Marshes

JOHN DAY, DAVID M. BURDICK, CARLES IBÁÑEZ, WILLIAM J. MITSCH,
TRACY ELSEY-QUIRK, AND SOFIA RIVAES

17.1 Introduction

Salt marshes have been lost or degraded as the intensity of human impacts to coastal landscapes has increased due to agriculture, transportation, urban and industrial development, and climate change. Because salt marshes have limited distribution and embody a variety of ecological functions that are important to humans (see ecosystem services, Chapter 15), many societies have recognized the need to preserve remaining marshes, restore those that have been degraded, and create new marshes in areas where they have been lost. An emerging and critical threat to tidal marshes across the globe is increasing rates of sea level rise and other aspects of climate change, which complicates but also heightens the urgency for restoration. By restoration we mean re-establishing natural conditions and the processes needed to support their functions, especially self-maintenance (see Box 17.1). Typically, salt marshes are self-maintaining, with salt tolerant plants, mineral sediments, and tidal flooding interacting to maintain elevation and ecological functions under dynamic conditions (Chapters 4, 7, 8).

Box 17.1 The Society for Ecological Restoration (2004) defines restoration as: an intentional activity that initiates or accelerates the recovery of an ecosystem with respect to its health, integrity, and sustainability. Frequently, the ecosystem that requires restoration has been degraded, damaged, transformed, or entirely destroyed as the direct or indirect result of human activities.

While some equate restoration with creating habitats or ecosystems that existed prior to the loss or damage (typically arising from human actions) at a certain location, practitioners today understand that the historical ideal is most often unobtainable, not always desirable, and may be difficult to sustain under future conditions (Clewell and Aronson 2013). Instead, restoration ecologists recognize the need to establish conditions and processes (both physical and biological) that interact to develop and sustain ecosystems that are resilient in the face of current and future global change and inescapable human stressors. For salt marshes, this means choosing an appropriate location – one that once supported or now supports a degraded tidal marsh – and establishing processes that balance flooding and drainage, sedimentation and erosion, so that the system can self-organize and become self-sustaining. In some cases though, it is necessary to take an active role in ecosystem development such as promoting the appropriate vegetation community (e.g., planting or removing invasive species), or adding or removing sediment to achieve intertidal elevations.

There have been several books published on tidal marsh restoration. Roman and Burdick (2012), *Tidal Marsh Restoration: A Synthesis of Science and Management*, focuses on restoring tidal flow in marshes that had been tidally restricted, particularly in New England and Atlantic Canada, where tidal restrictions have been in place for decades. In *Approaches to Coastal Wetland Restoration: Northern Gulf of Mexico*, Turner and Streever (2002) detail construction techniques and costs of marsh restoration in an area undergoing the most rapid rate of coastal wetland loss in the contiguous USA. Craft (2015) takes a broader approach by linking ecological theory to restoration practice in both freshwater and estuarine wetlands in *Creating and Restoring Wetlands: From Theory to Practice*. From these works and others, a list of general principles has been developed for restoration practitioners to follow that applies to salt marsh restoration (Restore America's Estuaries 1999, Mitsch and Gosselink 2015, Society for Ecological Restoration 2004, Needelman et al. 2012) see Box 17.2. Our objective in this chapter is to highlight various restoration techniques and discuss issues surrounding large-scale coastal wetland restoration using three case studies. Two of the case studies are focused on river deltas where constraints imposed by human modification and development limit the systems' ability to adapt to sea-level rise. The third case study focuses on restoration in a large coastal plain estuary, Delaware Bay, where historic changes to tidal hydrology has been restored and invasive plant species have been controlled. Together, these case studies illustrate challenges and successes with attempting to restore appropriate hydrology, sedimentation, and vegetation in tidal marshes.

Box 17.2 Principles Taken and Adapted from Restore America's Estuaries 1999

Estuarine restoration plans should be developed at the estuary and watershed levels where possible and through open regional processes that incorporate all key stakeholders and the best scientific thinking to set vision, articulate goals, and incorporate an ecosystem perspective.

Individual project goals should be clearly stated, site specific, measurable, and long term – in many cases greater than 20 years, especially to incorporate potential climate change.

Site plans need to address off-site considerations, such as potential flooding and saltwater intrusion into wells, to be sure projects do not have negative impacts on nearby people and property.

Ecological engineering practices should be applied in implementing restoration projects by maximizing the use of natural processes and dynamics to achieve self-sustaining habitats and landscapes.

Performance criteria for projects need to include both functional and structural elements and be linked to suitable, local reference habitats. Scientifically based monitoring is essential to evaluate performance, conduct adaptive management and improve restoration science and techniques.

Public access to restoration sites should be encouraged wherever appropriate, but designed to minimize impacts on the ecological functioning of the site.

17.2 Human Impacts and Restoration Approaches

Human impacts on tidal marshes have been discussed throughout this volume (e.g., Chapters 5 and 16). Occupation of coastal areas by humans with agrarian lifestyles has led to diking of salt marshes and conversion to agriculture over many centuries. Some impounded sites were used as freshwater non-tidal wetlands (e.g., rice fields), whereas sediment was used as fill in others to reduce flooding. Increasing development of urbanizing areas led to filling of marshlands (e.g., Boston, Teal and Teal 1969, Bromberg and Bertness 2005), and building earthen levees for flood protection (e.g., Mississippi River distributaries, Boesch et al. 1994, Day et al. 2007, 2019) and dikes as coastal defenses (e.g., western Europe, Pethick 2002, Wolters et al. 2005, 2008). Transportation needs led to corridors built across marshes (fill for land transport; dredge for vessel passage and port development), further interfering with natural hydrology, sediment supply, and salinity regimes. Mineral (oil and gas) extraction has led to widespread subsidence along the US Gulf Coast (Day et al. 2000, 2007, Ko and Day 2004). Mineral extraction, industrialization, and dumping have led to pollution and chronic impacts damaging marshes almost everywhere (Culbertson et al. 2008, Deegan et al. 2012). Intentional and unintentional introductions of exotic species have led to ecological alteration of marshes and their interactions with coastal landscapes (e.g., Ford and Grace 1998, Sasser et al. 2018).

Both filling and dredging directly destroy tidal marshes, but remaining marshes continue to be degraded by a variety of indirect impacts that interfere with marsh processes. For example, hydrological impacts include the widespread practice of reducing or eliminating tides with undersized culverts, or physical alterations that attempt to control mosquito production (ditching, Open Marsh Water Management) or improve waterfowl habitat (impoundments, pool excavation, and ditch plugging; Elsey-Quirk and Adamowicz 2016). Dredging canals and creating spoil banks for oil extraction in Louisiana has disrupted hydrology and sediment supply, leading to massive marsh losses from subsequent drowning (Swensen and Turner 1987, Day et al. 2000, 2019). All of these impacts have caused damage to marshes that can be addressed at a local level, but since most interfere with ecological processes at a large scale, regional landscape approaches to restoration should be considered. With thoughtful monitoring, designed to test appropriate hypotheses, both large- and small-scale restoration projects can be natural experiments, ideally informed by the best science, and potentially refined and improved upon (i.e., adaptive management) as they are critically evaluated using science-based monitoring (NAS 2017).

17.2.1 Restoration of Hydrology to Supply Sediments and Regulate Salinity

17.2.1.1 Tidal Exchange

Semi-diurnal to monthly flooding brings salt, nutrients, and sediments to tidal marshes. The temporal and spatial pattern of salinity of the soils strongly limits the plants that can grow in the marsh and largely reflects tidal influence, but is also a product of evapotranspiration and the flow of the water table seaward (Mitsch and Gosselink 2015, e.g., Hackney et al.

1996). Sediments bring nutrients to the marsh and are captured by vegetation as flood waters slow, helping to build the marsh in elevation along coasts experiencing rising sea levels (see Chapters 7 and 16). Tidal restrictions interfere with marsh building processes, disadvantage native perennials, encourage invasive species, alter essential habitat and reduce biodiversity, interfere with fish passage and coastal food webs, and lead to loss of stored carbon resulting in subsidence. This has been especially extensive in the Mississippi delta (e.g., Couvillion et al. 2017). Human development along coasts often limits tidal exchange to marshes for agriculture (berms and dikes – see Ebro Delta and Delaware Bay case studies [17.3.2 and 17.3.3]) and transportation (causeways, etc.). Tidal flow through creeks is either cut off through tide gates or only partially accommodated by engineered structures such as bridges and culverts.

Although protected from direct impacts in the USA, the tidal flooding needs of tidal marshes were rarely been considered until the latter part of the twentieth century. Now, western Europe, eastern Canada, most regions of the USA, and many developing countries have federal, provincial, or state programs (e.g., "return the tides") to restore tides to diked marshlands or those with tide gates or undersized culverts. The general approach is to remove the tidal obstruction with the goal of allowing the full tidal potential upstream of the crossing: excavate the entire dike or berm, or remove sections at the blocked tidal channels; replace broken or undersized pipes or culverts; or replace tide gates with large culverts. Some sites may require excavation of the main tidal channels to initiate channel flow and the development of secondary creeks (see 17.3.3 Delaware Bay case study). Measurement of the potential tidal prism will support the use of hydrological models to aid in the selection of proper sizing of the structure (MacBroom and Schiff 2012), which becomes important when development upstream or landward of the restriction to be restored leads to concerns about tidal flooding of private property or infrastructure. In some cases, upstream development is threatened by full tidal extent, and therefore, the crossing is designed to increase tidal flow, but restrict the natural tidal range. Limiting the highest tides can also be accomplished with a self-regulating tide gate, but these solutions have mixed reviews in practice (Rozsa 1995, Simenstad et al. 2006, Reiner 2012, Adamowicz and O'Brien 2012). Examining 20 available projects in the Gulf of Maine, Konisky et al. (2006) found that only six tidal restoration projects had pre- and post-restoration hydrological data, measuring an average increase in tidal range from 38% to 74% of the potential tide, suggesting many crossings may still be inadequate to fully support natural tidal dynamics.

Once tides are restored, fish access is immediate, but other responses require varying amounts of time to achieve full functionality of tidally restored marshes (Rogers et al. 1992, Simenstad and Thom 1996, Burdick et al. 1997). Most projects allow natural revegetation with native perennial graminoids, but those used for agriculture (Wolters et al. 2005) or overrun with invasive species may need to include vegetation management as part of the project (see 17.3.3 Delaware Bay case study). Some marshes may have subsided too low to support native vegetation and partial restoration to establish habitat and rebuild elevation (Smith et al. 2009), or sediment additions have been used (Carnu and Sadro 2002). Where sediments are plentiful, as in the San Francisco Bay, natural sedimentation and channel development can 'self-design' marsh topography and hydrology, though natural colonization by vegetation and plant community development can take decades (Williams and Orr 2002).

17.2.1.2 Alluvial Flooding

Natural river flooding at seasonal or annual time scales brings sediments and freshwater that support the function and maintenance of deltaic coastal wetlands as well as other estuarine marshes. Although tidal restoration is apt for deltaic systems where human agricultural and economic activities have altered the hydrology of specific marshes, restoration of alluvial flooding may be more critical for the health and continued existence of deltaic marshes at a landscape scale (Syvitski et al. 2009, Day et al. 2011, 2016a,b; see Mississippi River Delta case study [17.3.1]). As riverbanks and floodplains have been developed, channels and distributaries have been walled off from their floodplains and deltas in order to reduce seasonal and annual flooding of human communities. The result is devastating losses of wetlands, both in area and ecological functions and services (Tockner et al. 2000, Syvitski et al. 2009). The management challenge for reintroducing alluvial flooding for marsh restoration is the protection of private property and infrastructure (roads, electricity) while directing floodwaters through marshes.

17.2.2 Fill Removal or Addition to Achieve Regular Flooding

17.2.2.1 Fill Removal

Many salt marshes and shallow intertidal shorelines have been filled with unconsolidated sediment for both marine-related activities (marinas, parking lots for water access, port terminals) and development (agriculture, residential, business, industrial and urban uses). If adjacent areas are still subject to tidal influence, fill can be removed to appropriate elevations and major creeks can be excavated (preferably to original locations and dimensions if records exist) to re-establish tidal flooding and provide a template upon which perennial graminoids can be planted (or allowed to return on their own) to restore the area to salt marsh. For natural revegetation, propagules or clonal fragments of desirable plants must be able to disperse to the site, germinate, and develop to maturity (Friess et al. 2012, Jones et al. 2016). Adjacent marshes have been found to be critical for natural revegetation, but if desired species do not exist in the adjacent marsh, seeding or planting is necessary (Wolters et al. 2008). Fill areas considered for restoration are often limited to parking lots and undeveloped fill. With sea level expected to increase rapidly over the next several decades (Nichols and Cazenave 2010, USGCRP 2017), it is important to consider the target elevation for the excavation (near future) as well as that of the recovered salt marsh (distant future), as well as provide areas for the marsh to migrate landward.

17.2.2.2 Pollution Remediation

Some fill may support built infrastructure that is not likely to be restored. Other fill areas may be contaminated with a variety of pollutants that would make restoration problematic.

If the polluted sediments are removed and replaced with clean fill to appropriate elevations or if pollutants can be capped with clean fill to substantially eliminate leaching of contaminants, restoration is possible. Since marshes will continue to grow in elevation as sea levels rise through surface accretion of fine-grained sediments, contaminant release becomes less likely over time. Documentation of such projects is not easily found in the literature (e.g., lead: Housatonic River CT: https://darrp.noaa.gov/hazardous-waste/lordship-point; mercury: South Bay Salt Ponds, San Francisco Bay CA: Valoppi 2018; heavy metals: Jamaica Island, Piscataqua River NH: www.epa.gov/superfund).

17.2.2.3 Sediment Addition Using Thin Layer Deposition

Some marshes are more vulnerable to sea level rise than others for a variety of factors, but particularly with respect to current elevation, local subsidence, loss of sediment supply, hydrological impairment (discussed above), erosion, and even low tidal range (< 1 m; see Mississippi River Delta case study [17.3.1]). One solution is to elevate the surface of the sediment to allow vigorous and sustainable plant growth by diverting sediment-rich river water (alluvial flooding is discussed previously [17.2.1.2]) or by bringing in sediment through local dredging and piping or spraying the sediments on the marsh surface. Termed thin layer deposition, the latter technique has been tried throughout the USA with mixed results (Wilber 1993, Slocum et al. 2005, Frame et al. 2006, Ray 2007, Graham and Mendelssohn 2013). Existing live plants can grow through sediment additions of less than 10 cm, but sediment additions much greater than 10–15 cm require reseeding from adjacent marshes (Wilber 1993). In all projects where it has been examined, sediment compaction has been reported, with thicker deposits leading to greater compaction of added sediment as well as underlying marsh sediments (Wilber 1993, Edwards and Mills 2005, Slocum et al. 2005) to the point of no added elevation in one study within a brackish marsh (Graham and Mendelssohn 2013). Even in sites with good to excellent plant regeneration, it is not clear how much this new sediment adds to the elevation of the marsh and its long-term sustainability (Ford et al. 1999, La Peyre et al. 2009, Graham and Mendelssohn 2013) and extended monitoring has been recommended (Edwards and Proffitt 2003).

17.2.2.4 Thick Sediment Addition

In contrast to thin layer deposition, some marshes lost to subsidence or erosion or saltwater intrusion have been restored through the placement of thick deposits of sediment in shallow water that creates intertidal habitat at appropriate elevations (Streever 2000, Edwards and Proffitt 2003, Staver 2015). These projects often use dredge spoil and may be large and remote, requiring planting (Streever 2000). The loss of the original marsh may be due to erosive forces and as such, these areas are often armored where erosion remains an issue (Poplar Island, MD: Staver 2015; and Jamaica Bay, NY: Frame et al. 2006).

17.2.2.5 River Diversions

In areas where salt marshes have been isolated from riverine input by levees and dikes, diversions of river water back into coastal marshes is being proposed and implemented.

River diversions are an important component of the large-scale restoration of the Mississippi delta (see Mississippi River Delta case study [17.3.1]). Diversions are also being planned for the Sacramento-San Juaquin delta, the Ebro delta, and other areas. Pont et al. (2017) reported on two large floods in the Rhone delta where levees failed leading to widespread flooding of the delta with deposition of up to 10 cm. The authors proposed that these floods served as an example of how diversions could be used to sustain the Rhone delta.

17.2.3 Living Shorelines

Early knowledge from marsh creation through use of dredge spoil laid the foundations for a new effort to restore or create new marshes between partially hardened (using a variety of sill materials) lower shorelines and uplands susceptible to erosion termed "living shore-lines" (NOAA 2015) – a shoreline managed to control erosion that has one or more living components (e.g., seagrass, oysters, marsh, dune grass). Armoring shorelines with riprap and seawalls that extend vertically from or just above the high tide line prevents marshes from being able to migrate landward as sea levels rise and eliminates the high diversity plant community at the upper marsh edge (Bozek and Burdick 2005). Instead, the living shoreline concept as applied to salt marshes moves the erosion-resistant material seaward to the base of the marsh and creates a gradual vegetated slope to the upland edge, using the vegetation to protect higher elevations from erosion (Morgan et al. 2009). As landowners become more concerned with erosion and conversion of their upland property to intertidal, the living shoreline approach could become a significant mechanism to promote marsh habitat that can effectively protect shorelines (Gittmann et al. 2014) and provide other benefits (e.g., habitat: Gittman et al. 2014; blue carbon: Davis et al. 2015).

With increasing rates of sea-level rise and significant losses of salt marshes reported across the USA (Day et al. 2007, Couvillion et al. 2017), there are medium and large-scale efforts to build and restore marshes using both thin and thick sediment addition in several regions: California, (e.g., Elkhorn Slough); Chesapeake Bay (e.g., Poplar Island); Delaware Bay (Prime Hook National Wildlife Refuge); Rhode Island (three national wildlife refuges). In Louisiana the Coastal Wetland Planning, Protection and Restoration Act has funded several such projects (LCWCRTF 2006, 2012) with similar projects as part of a $50 billion plan (CPRA 2012).

17.2.4 Invasive Species and Trophic Cascades

Removal of invasive species tends to be most effective when coupled with restoration of physical process that may limit the likelihood of reinvasion, such as increasing tidal hydrology for control of the Eurasian form of *Phragmites australis* in US marshes (Saltonstall 2002, Hazelton et al. 2014; see 17.3.3 Delaware Bay case study). Where exotic species have become established, such as the invasion of *Lepidium latifolium* and *Spartina alterniflora* along the US Pacific coast (Powers and Boyer 2013), treatment options may be limited to physical removal and herbicide use. Researchers are calling for coordinated regional efforts to control invasives to improve results when funding is limited (Hedges et al. 2003, Hazelton et al. 2014). In Washington, where the State Department of Agriculture has a coordinated program to control *S. alterniflora*, large stands have been eliminated (over 3,700 ha), yet over 20,000 occurrences of single or several plants are spread across 3,000 miles of coastline (2014–2016) (https://cms.agr.wa.gov/ WSDAKentico/Imported/355-SpartinaReport2011.pdf   http://agr.wa.gov/PlantsInsects/ Weeds/Spartina/default.aspx).

Coastal managers and scientists have very few tested approaches to restore marsh impacts associated with trophic cascades. Trophic cascades that impact vegetation are typically associated with runaway populations of herbivores when their predators are reduced in association with human actions (*Sesarma reticulatum* crabs in New England – Coverdale et al. 2012; lesser snow geese, *Anser caerulescens caerulescens*, in Hudson Bay, CA – Jefferies et al. 2006) or when they are introduced (*Littorina littorea* in New England – Tyrrell et al. 2008; nutria in Gulf coast marshes – Ford and Grace 1998, Sasser et al. 2018). One example of successful management of an invasive herbivore population is the Chesapeake Bay Nutria Eradication Project. Nutria are a semiaquatic invasive rodent from South America that feed on roots and leaves of marsh plants, and in their non-native range in the southern and mid-Atlantic USA and in Washington and Oregon, contribute to marsh deterioration, along with sea-level rise and erosion. Despite large population numbers, intensive removal of nutria has resulted in eradication in most of the coastal marshes of the Delaware–Maryland peninsula. Similar nutria eradication projects have been successful in England. Where the invader can be harvested for marginal gain, management actions to encourage taking (e.g., establishing a bounty program) has been found to reduce extent and severity of impacts (Ford and Grace 1998, Sasser et al. 2018).

17.3 Examples of restoration approaches

Along many coastlines, wetlands have been altered or degraded by human activities. Where rivers flow into the coastal zone, large-scale changes to river flow and connectivity such as in the Mississippi River Delta and the Ebro Delta have dramatically hindered wetland accretion processes. This increases the vulnerability of these systems to sea-level rise. Restoration of these deltaic systems, therefore, requires landscape-level management, including the diversion of river water and sediment to subsiding wetlands. Restoration is being considered for a number of deltas worldwide including in the northwestern Mediterranean (Day et al. 2016c, 2019), Grijalva–Usumacinta rivers and adjacent areas and the Karst coastal systems in the Yucatan Peninsula (Herrera-Silveira 2019), the Sacramento–San Juaquin delta system in California (Nagarkar and Raulund-Rasmussen 2016), and the Ganges–Bramhaputra–Meghna delta in Bangladesh (Nicholls et al. 2016). The Mississippi River Delta case study (17.3.1) highlights some of the overarching challenges with restoring a large deltaic system. The Ebro Delta case study (17.3.2) clearly shows how reducing freshwater and sediment input from the basin and isolating deltaic wetlands from river input has reduced the ability of these wetlands to survive sea-level rise. Conversely, enhancing river connectivity to wetlands allows positive plant–sediment–elevation feedbacks that increase resiliency to sea-level rise. In Delaware Bay, former diking and hydrological alteration of marshes for salt hay farming has required a comprehensive approach of dike removal, enhancement of tidal flow, and removal of invasive species. The Delaware Bay case study (17.3.3) illustrates the importance of tidal circulation, local source populations of native species, and management of invasive species. It also shows how self-design can work with tidal channel network development and marsh species establishment occurring naturally. These three case studies consider multiple scales and goals of wetland restoration that illustrate the challenges and benefits of redesigning our coast.

17.3.1 Mississippi River Delta Restoration

One of the largest coastal wetland restoration efforts in the world is underway in the Mississippi Delta in Louisiana, where the Mississippi River created a succession of delta lobes with vast areas of wetlands grading from freshwater forested swamps to brackish and salt marshes before reaching the Gulf of Mexico (Roberts 1997, Day et al. 2007, CPRA 2017). Remnants of the past river delta with its bays, bayous, marshes, barrier islands, and low relief uplands make up cover 25,000 km^2 (Peyronnin et al. 2017). Abandoned delta lobes and associated wetlands naturally subside over time with their existence dependent on sediments supplied by the river and by *in situ* organic soil formation (Hatton et al. 1983, Callaway et al. 1997). Human alterations of this system have contributed to large losses of wetlands and associated habitat value, ecological functions and services, and cultural and economic resources (Day et al. 2014, Day and Erdman 2018). The Mississippi Delta faces a variety of natural and human-made forcings that complicate delta restoration and that make sustainable restoration very difficult. At a global scale, Giosan et al. (2014) predicted that all mega deltas will shrink considerably in size due to a combination of accelerating sea-level rise and human impacts (see also Day et al. 2016c, 2019). Deltas developed over the past several thousand years, a period in which sea level was relatively stable and input from drainage basins was regular and predictable, and as open systems with a high degree of interaction among river, delta, and the coastal ocean. Human activity has changed all of this, endangering and greatly complicating sustainable management and restoration.

17.3.1.1 Need for Restoration

Today, approximately 30% of total Mississippi River flow flows down the Atchafalaya River, while the remaining 70% flows in the main channel of the river past New Orleans (Fig. 17.1).

The flow is contained within the main channel of the Mississippi River by levees to below New Orleans. For most of the river's length, levees have isolated the river from its floodplain and deltaic plain. Below New Orleans, however, free from levee restriction, a considerable amount of discharge of both water and sediments leaves the river above the Birdsfoot Delta (Allison et al. 2012). Aside from the relatively small Birdsfoot Delta and the outfall of the Atchafalaya Wax Lake Delta complex, net subsidence, erosion, and

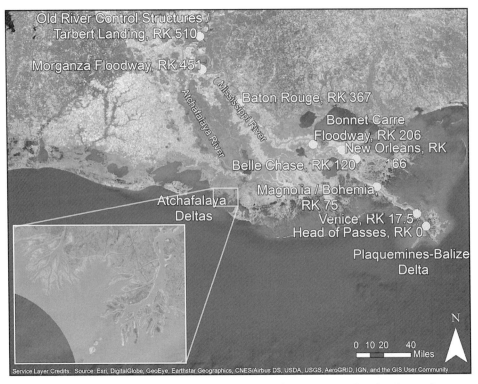

Figure 17.1 The lower Mississippi River and Atchafalaya courses. Landsat image from 2008 showing key features including river kilometers (RK) from the mouth of the Mississippi River and gauging stations. Source: Kemp et al. 2014

submergence are the processes dominating the Mississippi Rivers Deltaic wetlands. Canals and associated levee banks for oil and gas exploration and extraction, navigation, drainage, and logging occupy approximately 15,000 km in the Mississippi River Delta. These activities and hydrological impoundments alter wetland hydrology, salinity, and sediment transport and deposition (Swenson and Turner 1987, Day et al. 2007, Keddy et al. 2007). Spoil banks along canals reduce surface flooding and drainage and associated sediment movement. Canals themselves act as conduits for saltwater into brackish and freshwater wetlands. Since European settlement, approximately one-third of the original extent of the Mississippi River Delta wetlands have been lost (Craig et al. 1979, Gagliano et al. 1981, Day et al. 2000), with 4,900 km^2 of coastal wetlands, mostly marshes, lost between 1932 and 2010 (Couvillion et al. 2017). It has been suggested that without restoration, an additional 5,800 to over 10,000 km^2 of land area is predicted to be converted to open water by the end of the century (Fig. 17.1; CPRA 2017). Valuable ecological functions and services provided by these wetlands and barrier islands include habitat for fish and wildlife, high productivity, pollutant and carbon sequestration, nutrient transformation and removal, and protection from storms (Batker et al. 2014). Inhabitants and industries along the coast benefit from large expanses of coastal wetlands and barrier islands during storms (Day et al.

2007). Therefore, limiting coastal land loss and creating more land is of both economic and ecological importance.

17.3.1.2 Restoration Process

Restoring the Mississippi River Delta to its natural dynamic state is not a realistic goal given the extensive and intensive human alterations and development of the river and coast (Day et al. 2014, Day and Erdman 2018). The goal for restoration, therefore, is to best integrate natural processes and engineering to facilitate the creation of coastal land from freshwater and saline wetlands to barrier islands. Scientists, policy makers, land management agencies, landowners, oyster fishers, and other stakeholders involved in this process have different perspectives on how best to build land while sustaining current human activities and industries that have developed within a deteriorating delta (e.g., Day et al. 2016a). This poses a major challenge for planning and implementation of a restoration strategy focused on building land using natural processes inherent to a deltaic system. As a result of an intense planning processes guided by scientific expertise, four restoration approaches are being implemented in the Mississippi River Delta including (1) sediment diversions; (2) marsh creation and restoration using dredged sediments; (3) barrier island restoration; and (4) hydrological restoration such as backfilling canals and spoil bank removal (Day et al. 2007, 2014). The scientific consensus is that the most efficient and sustainable restoration approach is to reconnect the river to the delta (Day et al. 2007, 2014, 2016, Twilley and Rivera-Monroy 2009, Allison and Meselhe 2010, Paola et al. 2011, Nittrouer et al. 2012, DeLaune et al. 2013, Lopez et al. 2014). In practice, this involves diverting freshwater and sediments from the Mississippi River into receiving basins currently undergoing subsidence to create sub-deltas, islands, and crevasse splays, and a gradient of freshwater to saline wetlands (Kim et al. 2009, Rutherford et al. 2018). Because the extent of potential diversion influence is spatially limited relative to the extent of land loss, proactive wetland restoration and creation methods will also be necessary including the use dredged sediments from shipping channels and pumping of sand for the creation and enhancement of barrier islands.

The restoration process along with plans for flood protection is being guided by an evolving institutional framework that began in 1990 with the federal Coastal Wetlands, Planning and Protection Act, which provided $50 million per year for restoration. While additional funding and support was needed for large-scale restoration, it was not until after Hurricanes Katrina and Rita that the Coastal Protection and Restoration Authority (CPRA) was formed in 2005, which became the central authority to oversee and administer funding for coastal restoration and flood protection. Since 2012, CPRA has developed a Coastal Master Plan (CMP) every five years to guide restoration planning and implementation using scientific and engineering expertise.

The most recent 2017 CMP includes 79 restoration projects designed to build or maintain approximately 2,000 km^2 of land over the next 50 years through a combination of marsh creation, sediment diversions, and barrier island restoration (Fig. 17.2). The estimated cost of these restoration projects and additional efforts to protect coastal communities from flooding and storm surge risks is $50 billion dollars. Half of the $50 billion

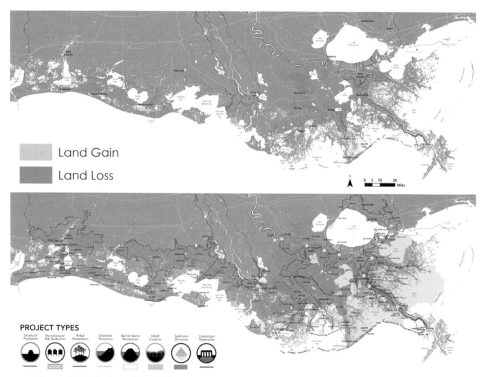

Figure 17.2 Projected land change along coastal Louisiana over the next 50 years if no action is taken (upper) and restoration projects implemented and planned under the 2017 Coastal Master Plan (lower). Projects include marsh creation (green), sediment diversion (white circles and brown area of influence), hydrological restoration (blue circles), barrier island restoration (red), and structural (pink) and non-structural risk reduction. (Modified from CPRA 2017) (A black and white version of this figure will appear in some formats. For the color version, please refer to the plate section.)

dollar price tag would go to restoration, of which 71% of restoration funding would be spent on marsh creation using dredge sediments (CPRA 2017). Under the 2017 CMP, 41 marsh creation projects and 7 sediment diversion projects are to be implemented. With both coastal restoration and protection activities, it is anticipated that $150 billion would be saved in the form of damage reduction over the next 50 years (CPRA 2017).

17.3.1.3 Sediment Diversions

While marsh creation using dredged sediments is a common restoration practice in many areas, river diversions as a restoration tool are unique to deltas. Two types of river diversions, freshwater and sediment, can be implemented with restoration goals of lowering salinity and building land, respectively. Sediment diversions have been proposed as a restoration technique since the 1970s (Day et al. 2000, Peyronnin et al. 2013, 2016a,b, Twilley et al. 2014, CPRA 2017) intended to reconnect the Mississippi River to its deltaic floodplain supplying subsiding wetlands and open water bodies with sediments from the river, to promote wetland growth and maintenance at an elevation above sea level (Kim et al. 2009, Paola et al. 2011). This would thereby restore a process similar to delta lobe building following shifts in the river channel. A diversion consists of a series of gates in the levee, which can be controlled to regulate the flow of river water, and sediment into receiving basins. Aside from control of the structures at the levee, the sediment diversion restoration process is passive, relaying on natural delta building processes (Paola et al. 2011), which makes this restoration technique different from those relaying on mechanical filling or grading. In 2012, the CMP proposed eight sediment diversions along the Mississippi River, with four considered high priority. Modelling of each potential project was implemented to predict the effects on land building, sediment transport, salinity, nutrients, and water levels. Based on the model results, two sediment diversions were put forth for implementation by CRPA in 2015, the Mid-Breton and Mid-Barataria diversions. Although large sediment diversions as a restoration technique have yet to be implemented in coastal LA, construction of the first, the Mid-Barataria Sediment Diversion is planned for 2021.

17.3.1.3.1 Operational Strategies

Recommendations for sediment diversion operation so as to limit negative consequences of increased flooding and high sediment loads to surrounding marshes, estuaries, and communities include a gradual increase in diversion operation, diversion opening during winter when sediment concentrations in the river are high and vegetation is dormant, and over short periods in spring and summer when sediment concentrations are high during rising floodwaters (Peyronnin et al. 2017). It is estimated that it could take 5–10 years for the distributary channel network to develop that would then allow the diversion to be operated at full capacity (Peyronnin et al. 2017).

If the Mississippi River Delta Plains historic functioning is used as a blueprint for restoration, much bolder action is required (Condrey et al. 2014, Day et al. 2016a, b). Saucier (1963) and Davis (1993, 2000) documented numerous crevasses along the lower Mississippi river prior to major anthropogenic alteration. For example, the Bonnet Carré crevasse functioned intermittently in the second half of the nineteenth century with discharge ranging from 2,000 to 6,500 m^3/s and built a crevasse splay of about 70 km^2 as well as filling in parts of western Lake Pontchartrain with up to 2 m of sediment (Saucier 1963; Davis 1993). Also, the 1927 artificial crevasse at Caernarvon resulted in a crevasse splay of about 130 km^2 with sediment deposition as high as 40 cm in only three months (Day et al. 2016b). Day et al. (2016a) suggested that diversions should be larger than proposed diversions (>5,000 m^3/s) and infrequent (active < once a year) to mimic historical functioning of crevasses and sub-delta lobe formation (see also Rutherford et al. 2018). Not only would this build more land but would have substantially lower impacts on water levels, salinity, nutrient load, and fisheries – controversial effects that have impeded implementation of diversions (Caffey and Schexnayder 2002, Day et al. 2016a). For maximum sediment capture and to mimic the geologic processes that build deltas from upstream to the coast, it has been recommended that diversions be placed in upstream reaches of the Mississippi Delta rather than far-downstream locations (Blum and Roberts 2009).

17.3.1.4 Challenges Associated with a Mississippi River Delta Restoration – Long-term Sustainability of the Deltaic Coast

The Mississippi River Delta, like many of the world's deltas, is experiencing a convergence of processes that heighten its vulnerability to loss and make restoration particularly challenging. A reduced sediment load in the Mississippi River, high rates of subsidence, accelerated sea-level rise, and more extreme weather events make continued land loss along the deltaic coast inevitable. Even for the use of sediment diversions, the Mississippi River now carries approximately 50% less sediment than prior to dam installation along its major tributaries (Keown et al. 1986; Kesel 1989; Kesel et al. 1992; Blum and Roberts 2009). And it is estimated that the maximum sediment available for the Mississippi Delta, as well as for many other deltas, is less than the sediment needed for the delta plain to keep with relative sea level (Blum and Roberts 2009, Giosan et al. 2014, Day et al. 2019). Kemp et al. (2016) reported that enhancing sediment supply from the lower Missouri River could potentially increase fine sediment input to the lower river by 100–200 Mt/yr. The Gulf coast is frequently battered with large tropical cyclones, which can provide sediment locally to wetlands along the coast (Cahoon et al. 1995, Nyman et al. 1995, Naquin et al. 2014), but can also cause saltwater intrusion and extensive damage to the vegetation and soil (Chabreck and Palmisano 1973, Guntenspergen et al. 1995, Couvillion et al. 2017).

Even with the $25 billon dollar restoration plan outlined under the 2017 CMP, more land will be lost than gained or maintained (Fig. 17.3). This may be considered better than a do-nothing strategy, but ultimately many coastal communities and infrastructure along the coast will need to be moved inland on stable continental foundation.

Another concern is whether there are enough economic resources necessary to accomplish the CMP. The CMP is envisioned at a $50 billion cost for a 50-year effort. But recent studies indicate that the actual cost would be closer to $100 billion (Barnes et al. 2015). In addition, increasing energy costs could increase the total cost even more for the highly energy intensive CMP (Wiegman et al. 2017). Growing climate impacts in terms of sea-level rise, more frequent category 4 and 5 hurricanes, and increasing peak Mississippi River discharge, and increasing resource scarcity will likely make restoration even more costly. Over time, energy costs and accelerating sea-level rise will lead to significant increases in the cost of marsh creation using dredged sediments, perhaps by an order of magnitude (Wiegman et al. 2017).

17.3.2 Restoration and Management of Salt Marshes in the Ebro Delta

The Ebro River is 910 km long with a drainage area of 85,362 km^2 and terminates in the Ebro Delta, one of the largest deltas (320 km^2) in the northwestern Mediterranean. About 65% of the delta area is rice fields and natural areas cover about 35% or 80 km^2. These natural areas include coastal lagoons and coastal wetlands (Fig. 17.4).

Coastal marsh restoration in the Ebro Delta involves working at two scales: that of individual salt marshes and at the level of the Ebro River Basin and Delta. At the basin scale, human activities have had a dramatic impact. Currently the Ebro Delta is undergoing coastal retreat at the river mouth because wave erosion and elevation loss in the delta are no longer offset by new sediments coming from the Ebro River. Sediment has been reduced by

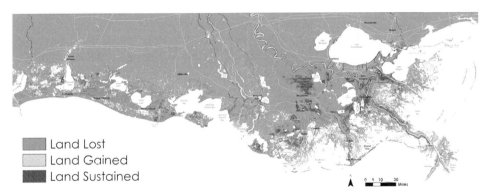

Figure 17.3 Projected 50 year outcome of fully enacted CMP under moderate environmental scenario (modified from CPRA 2017). (A black and white version of this figure will appear in some formats. For the color version, please refer to the plate section.)

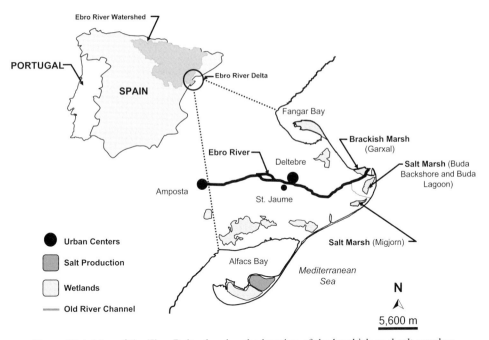

Figure 17.4 Map of the Ebro Delta showing the location of the brackish an,d salt marshes discussed and the Ebro River watershed. (From Ibáñez et al. 2010)

about 99% due to the construction of 170 dams upriver of the delta (Ibáñez et al. 1996a, 1997). The sediment deficit in the delta created by the dams, coupled with land subsidence from compaction, organic soil decomposition, accelerated sea level rise, and the already low elevation of the delta plain, puts the delta and its wetlands at major risk for submergence, saltwater intrusion damaging rice production, and coastal erosion (Ibáñez and

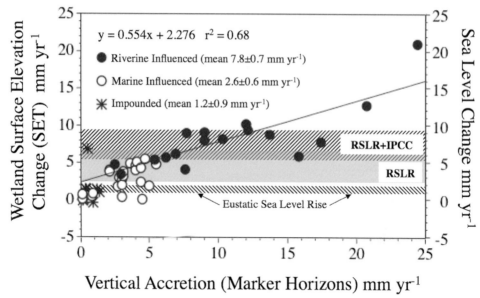

Wetland Surface Elevation Change (SET) mm yr^{-1}

Sea Level Change mm yr^{-1}

$y = 0.554x + 2.276$ $r^2 = 0.68$

● Riverine Influenced (mean 7.8±0.7 mm yr^{-1})

○ Marine Influenced (mean 2.6±0.6 mm yr^{-1})

✳ Impounded (mean 1.2±0.9 mm yr^{-1})

RSLR+IPCC

RSLR

Eustatic Sea Level Rise

Vertical Accretion (Marker Horizons) mm yr^{-1}

Figure 17.5 Wetland vertical accretion vs. surface elevation change for coastal Mediterranean riverine (black circle), marine (white circle), and impounded (black star) sites in the Ebro, Rhone, and Po deltas, and Venice Lagoon. Mean surface elevation changes for the three marsh types were compared to forcings causing water level rise: twentieth century global eustatic sea-level rise (ESLR); relative sea-level rise (RSLR) for the different sites due to subsidence plus ESLR; and RSLR plus twenty-first century predicted ESLR from the IPCC (2007) (RSLR + IPCC; see text for references). To survive rising sea level, coastal wetlands must grow at a rate ≥water level increase, implying that only sites with high sediment input will survive predicted sea-level rise. (From Day et al. 2011)

Prat 2003, Genua-Olmedo et al. 2016). From the perspective of the drainage basin, the most important management action for sustainable management of the delta is the restoration of the sediment flux in the river, which implies the remobilization of sediments trapped in reservoirs and increasing freshwater discharge to the delta via controlled floods. It has been shown that coastal marshes in the northwestern Mediterranean that receive river input are likely to survive accelerated sea-level rise while coastal marshes isolated from river input will not (Day et al. 2011, Fig. 17.5).

The major salt marshes in the delta occur along the backshore of the outer coast, especially around the Ebro River main mouth (Garxal) and secondary mouth (Migjorn), which together delineate Buda Island. The Buda Backshore marsh is located in a marine-influenced backshore area of Buda Island (Fig. 17.4). Salt marsh communities are dominated by *Arthrocnemum glaucum* and have low cover (10–20%) and mean vegetation height. The marsh is situated at an intermediate elevation. The Buda Lagoon marsh is dominated by *Sarcocornia fruticosa* with nearly 100% cover. This marsh is at the lowest elevation of the three salt marsh sites. The Migjorn marsh is located just behind the beach/dune system, close to the mouth of the abandoned

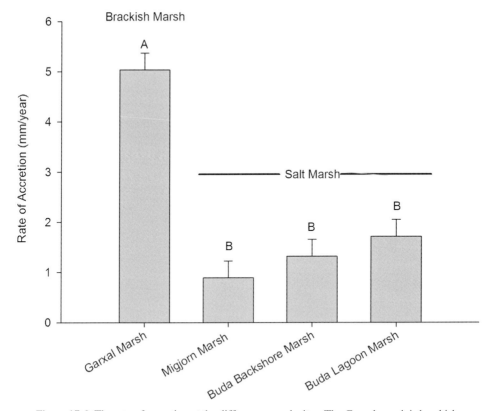

Figure 17.6 The rate of accretion at the difference marsh sites. The Garxal marsh is brackish while Migjorn, Buda Backshore, and Buda Lagoon are salt marsh. The higher accretion at the brackish site is strongly impacted by river input. (Ibañez et al. 2010)

Migjorn River channel (Fig. 17.4) and is at the highest relative elevation. The marsh is dominated by *S. fruticosa*.

An extensive brackish marsh occurs at Garxal and is strongly influenced by the Ebro River and receives periodic influx of fresh water, nutrients, and sediments. Formed over the last six decades as a result of the most recent change of the river mouth, the marsh is dominated by *P. australis* but *S. maritimus* also occurs. The rate of accretion at the Garxal site is higher than the three salt marshes due to river input (Fig. 17.6). Because of variations in sediment input, subsidence, and organic soil formation, some of the marshes have the potential to survive sea-level rise while others do not (Fig. 17.7).

Restoration of salt and brackish marshes and associated coastal lagoons in the Ebro Delta has taken several approaches. As early as the 1990s the goal in the Encanyissada Lagoon was to improve water quality and restore submerged vegetation by changing the hydrology. Wetlands were created in abandoned rice fields to reduce nutrient levels in water draining from rice fields and to improve habitat for protected bird species. Recently, new coastal lagoons, reedbeds, and salt marshes have been restored within

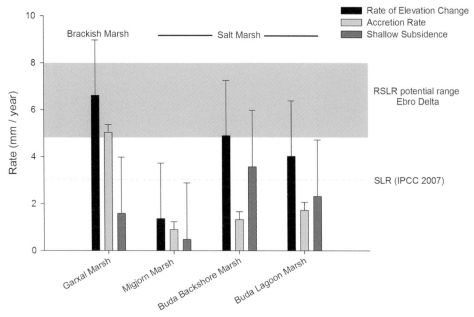

Figure 17.7 Comparison of mean elevation change, vertical accretion, and shallow subsidence for the different marsh sites. The dashed line is the IPCC 2007 projection for global sea level rise (SLR) rates, and the shaded area represents relative sea level rise rate (RSLR) projections estimated by Ibáñez et al. (1996b) for the Ebro Delta. (From Ibañez et al. 2010)

the framework of a European Union funded Life Project (Delta Lagoon). Functional restoration of Encanyissada Lagoon began with a bypass to divert drainage water from rice fields (with high nutrient and pesticide load) directly to Alfacs Bay and to the sea. Also, freshwater from the Ebro River was introduced using an existing irrigation canal (Forès et al. 2002). This allowed the control of freshwater inputs while the connections with the sea were maintained. At the same time, a small lagoon (0.56 km^2) that is part of the Encanyissada lagoon system but is separated by a floodgate was drained, drying out for several months to oxygenate the sediments. It was then refilled with irrigation water from the Ebro River, which allowed a quick recovery of submerged aquatic vegetation and bird populations (Forès at al. 2002).

Another project consisted of the abandonment of rice fields along the shore of the Tancada Lagoon (2.6 km²) that were allowed to colonize with brackish marsh species for nitrogen and phosphorus removal from agricultural runoff used to flood the restored wetlands (Comín et al. 2001). Nutrient concentrations were reduced by the restored wetlands with removal rates of 50%–95% for total nitrogen but less than 50% for total phosphorus resulting in improvement of water quality in this coastal lagoon. Vegetation in the restored rice fields was mostly fresh to brackish vegetation (*P. australis*, *S. maritimus*, *Typha latifolia*, and *Scirpus lacustris*). This demonstrates that wetland restoration can achieve multiple benefits including water quality improvement and wetland habitat recovery. A large restoration in Buda island

(1,280 ha) also involved the conversion of rice fields to brackish and salt marsh for habitat restoration favoring the conservation of priority bird species, but also for the reduction of nutrients and improvement of water quality in coastal lagoons.

More recently, a number of projects were implemented to construct green filters. In 2010, two large constructed wetlands (ca. 200 ha in total) were created to improve the quality of the water draining from surrounding cultivated fields and to enhance carbon sequestration and increase soil elevation. This was part of an ambitious plan of environmental restoration called the Plan for the Integral Protection of the Ebro Delta (PIPDE), which included the restoration of wetlands, riparian forests and river islands, among other actions.

Finally, the Life Delta Lagoon Project involved the restoration of the Alfacada and Tancada lagoons and marshes. The habitat restoration in the Alfacada Lagoon involved the re-naturalization of the hydrology by connecting the lagoon to the Ebro River and reconnecting the salt marshes to the lagoon. Then, nesting areas for gulls, terns, and waders were created by building two new islands in the salt marsh zone. Salt marsh was allowed to revegetate naturally. The project converted some 60 ha of rice fields back to coastal lagoon and reedbed habitat.

In the Tancada Lagoon area, an old aquaculture facility of 16 ha was allowed to convert back to salt marsh and shallow coastal ponds with a goal to provide habitat for the endangered Spanish toothcarp (*Aphanius iberus*) (Prado et al. 2017) and several protected bird species. Artificial islets were constructed in the lagoon for endangered bird species to breed, including slender-billed gull (*Croicocephalus genei*), Audouins's gull (*Larus audouinii*), little tern (*Sterna albifrons*), and gull-billed tern (*Gelochelidon nilotica*).

17.3.3 Delaware Bay Salt Marsh Restoration

A large restoration project in Delaware Bay on the Atlantic coast of USA involves the restoration, enhancement, and preservation of 5,000 ha of salt marshes (Fig. 17.8; Peterson et al. 2005, Mitsch and Gosselink 2015).[1] The reasoning was that the impact of once-through cooling for a power plant on fin fish, through entrainment and impingement, could be offset by increased fisheries production from restored salt marshes. Because of uncertainties involved in this kind of ecological trading, the area of salt marsh restoration necessary to compensate for the impacts of the power plant on fin fish production was set at four times what was estimated to be necessary to offset the fisheries loss.

Three distinct approaches are being utilized to restore the Delaware Bay coastline.

(1) Reintroduction of tidal flooding. The most important type of restoration involves the reintroduction of tidal inundation to about 1,800 ha of former diked salt-hay farms. Many marshes along Delaware Bay have been isolated by dikes from the bay, sometimes for centuries, and put into the commercial production of "salt hay" (*Spartina patens*). Hydrological restoration was accomplished by excavating breaches

---

[1] This restoration being carried out by New Jersey's electric utility (Public Service Enterprise Group [PSEG]), with advice from a team of scientists and consultants, was undertaken as mitigation for the potential impacts of once-through cooling from a nuclear power plant operated by PSEG on the Bay.

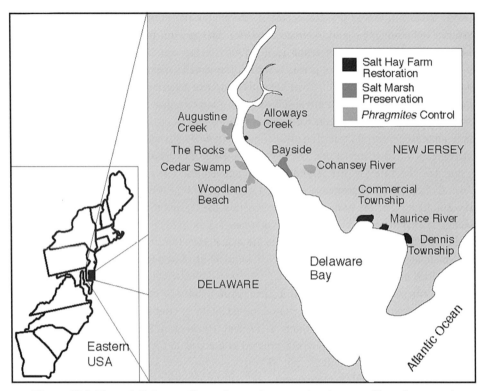

Figure 17.8 Map of Delaware Bay salt marsh restoration between New Jersey and Delaware from 1995 through present, showing locations of 5,800-ha salt marsh restoration and preservation that is being carried out to mitigate the loss of fin fish due to entrainment and impingement caused by once-through cooling at a Delaware Bay nuclear power plant. Wetlands are being preserved, restored from salt-hay farms by reintroducing flooding, and enhanced by removal of *P. australis* (reprinted from Mitsch and Gosselink 2015, with permission from John Wiley & Sons, Inc. Hoboken NJ.) (A black and white version of this figure will appear in some formats. For the color version, please refer to the plate section.)

in the dikes and, in most cases, connecting these new inlets to a system of recreated tidal creeks and existing canal systems.

(2) Re-excavation of tidal marsh creeks. Additional restoration involves enhancing drainage by re-excavating higher-order tidal creeks in these newly flooded salt marshes, thereby increasing tidal circulation. This is particularly important in marshes that were formerly diked, because the isolation from the sea has led to the filling of former tidal creeks. After initial tidal creeks were established, it was expected that the system would self-design more tidal channels and increase the channel density.

(3) Reduction of invasive *Phragmites* domination. In another set of restoration sites in Delaware and New Jersey, restoration involves the reduction in cover of the aggressive and invasive reed grass (*P. australis*) in 2,100 ha of non-impounded coastal wetlands. Alternatives that were investigated include hydrological modifications such as channel excavation, breaching remnant dikes, microtopographic changes, mowing, planting, and herbicide application.

Results of this study were reported in several presentations and reports, journal articles including Teal and Weinstein (2002), and several papers in a special issue of *Ecological Engineering* (Peterson et al. 2005). From a hydrodynamic perspective, in marshes where tidal exchange was restored, the development of an intricate tidal creek density from the originally constructed tidal creeks has been impressive. Figure 17.9 illustrates the development of a stream network at two of the newly restored marsh sites – Dennis Township and Commercial Township. The "order" of the stream channels increased from 5 or less to well over 20 over 9 years (1996 through 2004) in the Dennis Township site and over 30 over 17 years (1997 through 2013) at the larger Commercial Township site. The number of small tributaries increased from "dozens" to "hundreds" at all three salt-hay farm sites that were reopened to tidal flushing. Hydrological design occurred in a "self-design" fashion after only initial cuts by construction of the largest (class 1) channels.

For the salt-hay farms that were flooded, typical goals include a high percentage cover of desirable vegetation such as *S. alterniflora*, a relatively low percentage of open water, and the absence of the invasive reed grass *P. australis*. The success of this coastal restoration project, subject to a combination of legal, hydrological, and ecological constraints, is also being estimated through comparison of restored sites to natural reference marshes. Results of this part of the project after almost two decades are encouraging. At the formerly diked salt-hay farms, re-establishment of *S. alterniflora* and other favorable vegetation has been rapid and extensive.

In Dennis Township, approximately 70% of the site was dominated by *S. alterniflora* after only two growing seasons and almost 80% by the fifth year after construction (Fig. 17.10). Tidal restoration was completed at the Maurice River site, which is twice the size of the Dennis Township site in early 1998. Major revegetation by *S. alterniflora* and some *Salicornia* has occurred, with 71% of the site showing desirable vegetation after four growing seasons. At the third and the largest salt-hay farm restoration site at Commercial Township, which is five times larger than Dennis Township site, revegetation is occurring rapidly from the bayside. This study has shown that the speed with which salt marsh restoration takes place is dependent on three main factors:

(1) The degree to which the tidal "circulatory system" works its way through the marsh
(2) The size of the site being restored
(3) The initial presence of *Spartina* and other desirable species.

No planting was necessary on these sites, as *Spartina* seeds arrive by tidal fluxes from nearby salt marshes, but the design of the sites to allow that tidal connectivity (and hence the importance of appropriate site elevations relative to tides) was critical. Self-design works when the proper conditions for propagule disbursement are provided. Extensive ponding in some areas of the marshes, especially at Commercial Township, which has the highest ratio of area to edge, initially impeded the reestablishment of *Spartina* in some locations (Teal and Weinstein 2002). Creating additional streams or waiting for the tidal forces to cause the same effect eventually allows these areas to develop tidal cycles and *Spartina* to establish itself.

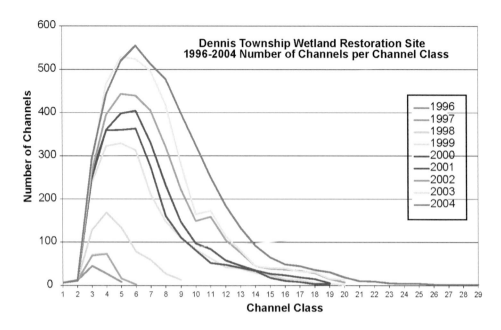

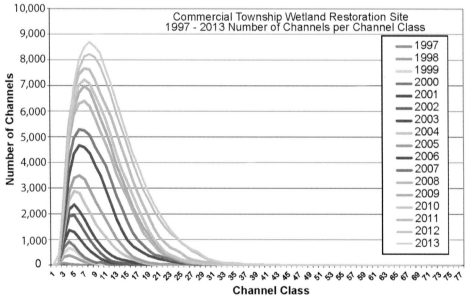

Figure 17.9 Total number of stream channels by channel class at Dennis Township (1996–2004) and Commercial Township (1997–2013) restored salt marshes on Delaware Bay (Provided with permission, Kenneth A. Strait, PSEG Service Corporation, Salem, NJ, and reprinted from Mitsch and Gosselink 2015, with permission from John Wiley & Sons, Inc. Hoboken NJ.) (A black and white version of this figure will appear in some formats. For the color version, please refer to the plate section.)

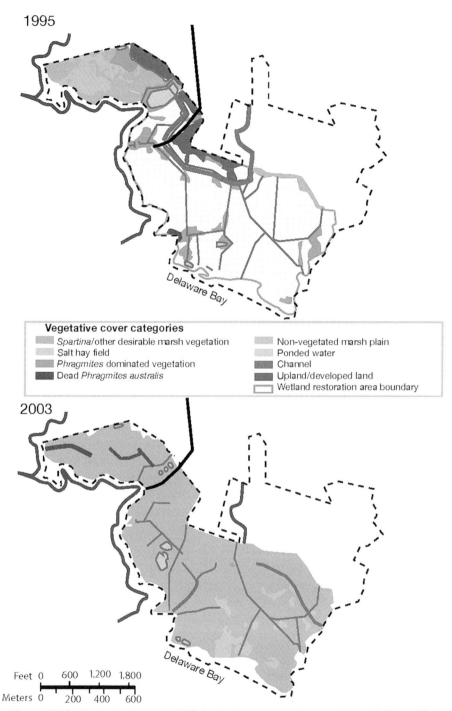

Figure 17.10 Vegetative cover in 1995 (top) prior to restoration at the Dennis Township restored salt marshes and in 2003 (bottom), seven years after low-channel-class tidal creeks were restored and dikes were breached. Light colored area indicates unvegetated region. Predominant dark gray areas are restored salt marshes of *Spartina alterniflora* and other desirable marsh vegetation. (Revised from Hinkle and Mitsch 2005 and reprinted from Mitsch and Gosselink 2015, with permission from John Wiley & Sons, Inc. Hoboken NJ)

Reducing Phragmites domination in another set of brackish marsh restoration sites in Delaware and New Jersey has required more years of effort, mainly through the use of herbicides, but there are now substantial areas of significant recovery of the marsh to desirable vegetation, including S. alterniflora. The *Phragmites* cover in the marsh at the Alloway Creek watershed in New Jersey decreased from 60 percent cover in 1996 to less than 5 percent cover in 2013.

17.4 Summary and Vision for the Future

Salt marshes are under threat from a variety of direct (e.g., dredging, filling) and indirect (e.g., hydrological impairments, sediment deficits, and subsidence) human activities, but none more widespread and devastating than impacts from climate change, especially rapidly rising sea levels. Climate impacts interact with other human impacts to make restoration more challenging (Day et al. 2016c, 2019). Salt marshes have been impacted by diking, impoundment, filling, isolation from riverine inputs, coastal protection defenses, dredging, pollution with high levels of nutrients and toxins, introduced species, and shoreline hardening with seawalls, riprap, and filling. Restoration approaches include dike removal, reestablishment of tidal creeks, fill removal, levee breaches, living shorelines, bridge and culvert additions, restored hydrology, river diversions, use of dredge material to restore and create marshes, and control of introduced species. Case studies have been presented for the Ebro and Mississippi deltas, and large-scale salt marsh restoration in Delaware Bay. Restoration of basin inputs are critical for delta restoration as are river diversions and creation of marshes with dredged sediments. These large-scale projects are challenged by scale, political will, and funding. In the Mississippi River Delta, coastal wetland loss will continue even with active restoration. The Delaware study showed that when tidal exchange was restored, a tidal network quickly developed. In the Ebro delta restoration of hydrology promoted salt marsh enhancement and development, improved water quality, and provided habitat for threatened and important species. Salt marsh protection and restoration are needed along with regional efforts to allow marsh expansion landward (migration). We see salt marsh restoration as a critical component in the overall conservation effort and recognize the utility and benefits of a variety of techniques adapted for specific geographical and geomorphic regions.

References

Adamowicz, S. C. and K. M. O'Brien. 2012. Drakes Island tidal restoration, science, community and compromise. In Roman, C. T. and D. M. Burdick (eds.) *Tidal Marsh Restoration: A Synthesis of Science and Practice.* pp. 315–332. Washington, DC: Island Press.

Allison, M. A., and Meselhe, E. A. 2010. The use of large water and sediment diversions in the lower Mississippi River (Louisiana) for coastal restoration. *Journal of Hydrology,* 387: 346e360.

Allison, M. A., Demas, C. R., Ebersole, B. A., Kleiss, B. A., Little, C. D., Meselhe, E. A., Powell, N. J., Pratt, T. C., and Vosburg, B. M. 2012. A water and sediment budget for the lower Mississippi–Atchafalaya River in flood years 2008–2010: Implications for sediment discharge to the oceans and coastal restoration in Louisiana. *Journal of Hydrology*, 432: 84–97.

Barnes, S., Bond, C., Burger, N., Anania, K., Strong, A., Weilant, S., and Virgets, S. 2015. Economic evaluation of coastal land loss in Louisiana. Lousiana State University and the Rand Corporation. Published online. http://coastal.la.gov/economic-evaluation-of-land-loss-in-louisiana/

Batker, D., de la Torre, I., Costanza, R., Day, J., Swedeen, P., Boumans, R., and K. Bagstad. 2014. The threats to the values of ecosystem goods and services of the Mississippi Delta. In Day, J., Kemp, P., Freeman, A., and Muth, D. (eds.) *Perspectives on the Restoration of the Mississippi Delta*. pp. 155–174. Springer, New York.

Baumann, R., Day, J. W., and Miller, C. 1984. Mississippi deltaic wetland survival: Sedimentation versus coastal submergence. *Science*, 224: 1093–1095.

Blum, M. D., and Roberts, H. H. 2009. Drowning of the Mississippi Delta due to insufficient sediment supply and global sea-level rise. *Nature Geoscience*, 2: 488–449.

Boesch, D. F., Josselyn, M. N., Mehta, A. J., Morris, J. T., Nuttle, W. K., Simenstad, C. A., and Swift, D. J. Scientific assessment of coastal wetland loss, restoration and management in Louisiana. *J. Coast. Res.* 1994, i–v, 1–103.

Bozek, C. and Burdick, D. M. 2005. Impacts of seawalls on saltmarsh plant communities in the Great Bay Estuary, New Hampshire USA. *Wetlands Ecology and Management*, 13: 553–568.

Bromberg, K. D., and Bertness, M. D. 2005. Reconstructing New England salt marsh losses using historical maps. *Estuaries*, 28: 823–832.

Burdick, D. M., Dionne, M., Boumans, R. M., and Short, F. T. 1997. Ecological responses to tidal restorations of two northern New England salt marshes. *Wetlands Ecology and Management*, 4: 129–144.

Caffey, R. H. and Schexnayder, M. 2002. Fisheries implications of freshwater diversions., in: *An Interpretive Topic Series on Louisiana Coastal Wetland Restoration, Coastal Wetland Planning, Preservation, and Restoration Act (eds.), National Sea Grant Library* No. LSU-G-02-003.

Cahoon, D. R., Reed, D. J., Day, Jr., J. W., Steyer, G. D., Boumans, R. M., Lynch, J. C., McNally, D., and Latif, N. 1995. The influence of Hurricane Andrew on Sediment Distribution in Louisiana Coastal Marshes. *Journal of Coastal Research Special Issue*, 21: 280–294.

Callaway, J. C., DeLaune, R. D. and Patrick, Jr., W. H. 1997. Sediment accretion rates from four coastal wetlands along the Gulf of Mexico. *Journal of Coastal Research*, 13: 181–191.

Chabreck, R. H, and Palmisano, A. W. 1973. The effects of Hurricane Camille on the marshes of the Mississippi River Delta. *Ecology*, 54: 1118–1123.

Clewell, A. F., and Aronson, J. 2013. *Ecological Restoration: Principles, Values, and Structures of an Emerging Profession*. 2nd ed. Island Press, Washington, DC, USA.

Coastal Protection and Restoration Authority of Louisiana (CPRA). *Integrated Ecosystem Restoration and Hurricane Protection: Louisiana's Comprehensive Master Plan for a Sustainable Coast*. Baton Rouge, LA, USA, 2007.

CPRA (Coastal Protection and Restoration Authority of Louisiana). *Louisiana's Coastal Master Plan for a Sustainable Coast*. Baton Rouge, LA, USA, 2012.

CPRA (Coastal Protection and Restoration Authority of Louisiana. *Louisiana's Coastal Master Plan for a Sustainable Coast; Coastal Protection and Restoration Authority of Louisiana*: Baton Rouge, LA, USA, 2017.

Comín, F. A., Romero, J. A., Hernández, O., and Menéndez, M. 2001. Restoration of wetlands from abandoned rice fields for nutrient removal, and biological community and landscape diversity. *Restoration Ecology*, 9 (2): 201–208.

Condrey, R., Hoffman, P., and Evers, D. 2014. The last naturally active delta complexes of the Mississippi River (LNDM): Discovery and implications. In Day, J., Kemp, P., Freeman, A., and Muth, D. (eds.) *Perspectives on the Restoration of the Mississippi Delta. Estuaries of the World.* pp. 33–50. Springer, New York.

Cornu, C. E., and Sadro, S. 2002. Physical and functional responses to experimental marsh surface elevation manipulation in Coos Bay's South Slough. *Restoration Ecology*, 10: 474–486.

Couvillion, B. R., Barras, J. A., Steyer, G. D., Sleavin, W. , Fischer, M., Beck, H., Trahan, N., Griffin, B., Heckman, D. 2017. *Land Area Change in Coastal Louisiana from 1932 to 2010, U.S. Geological Survey*: Reston, VA, USA.

Coverdale, T. C., Altieri, A. H., and Bertness, M. D. 2012. Belowground herbivory increases vulnerability of New England salt marshes to die-off. *Ecology*, 93: 2085–2094.

Craig N. J., Turner R. E., and Day J. W. 1979. Land loss in coastal Louisiana (U.S.A.). *Environmental Management*, 3: 133–144.

Craft, C. 2015. *Creating and Restoring Wetlands: From Theory to Practice.* 1st edn. Elsevier, Inc. The Netherlands.

Culbertson, J. B., Valiela, I., Pickart, M., Peacock, E. E., and Reddym, C. M. 2008. Long-term consequences of residual petroleum on salt marsh grass. *Journal of Applied Ecology*, 45: 1284–1292.

Davis, D. W. 1993. Crevasses on the Lower Course of the Mississippi River. *Coastal Zone'93*, vol. 1, pp. 360–378, July 19–23, New Orleans, Louisiana.

Davis, D. W. 2000. Historical perspective on crevasses, levees, and the Mississippi River. In Colten, C. E. (ed.) *Transforming New Orleans and its Environs.* pp. 84–106. Pittsburgh: University of Pittsburgh Press.

Davis, J. L., Currin, C. A., O'Brien, C. Raffenburg, and Davis, C. A. 2015. Living shorelines: Coastal resilience with a blue carbon benefit. *PLoS ONE* 10:e0142595. doi:10.1371/journal.pone.0142595

Day, J., Agboola, J., Chen, Z., D'Elia, C., Forbes, D., Giosan, L., Kemp, P., et al. 2016c. Approaches to defining deltaic sustainability in the 21st century. *Estuarine, Coastal and Shelf Science*, 183: 275–291.

Day, J. W., Boesch, D. F., Clairain, E. J., Kemp, G. P., Laska, S. B., Mitsch,W. J., Orth, K., et al. 2007. Restoration of the Mississippi Delta: Lessons from Hurricanes Katrina and Rita. *Science*, 315: 1679–1684.

Day J. W., Britsch L., Hawes S., Shaffer G., Reed D., and Cahoon D. 2000. Pattern and process of land loss in the Mississippi Delta: A spatial and temporal analysis of wetland habitat change. *Estuaries and Coasts*, 23:425–438.

Day, J., Cable, J., Lane, R., and Kemp, G. 2016b. Sediment deposition at the Caernarvon crevasse during the great Mississippi flood of 1927: Implications for coastal restoration. *Water*, 8 (38): doi:10.3390/w8020038.

Day, J., Colten, C., and Kemp, G. P. 2019. Mississippi Delta restoration and protection: shifting baselines, diminishing resilience, and growing non-sustainability. In: Wolanski, E. . Day, J, Elliott, M., and Ramnachandran, R. (eds.) *Coasts and Estuaries: The Future.* pp. 167–186. Elsevier, New York.

Day, J. W. and J. Erdman. 2018. *Sustainable pathways for Mississippi delta restoration – Pathways to a Sustainable Future.* Springer, Cham, Switzerland.

Day, J., Kemp, G. P., Freeman, A. M., and Muth, D. (Eds.). 2014. *Perspectives on the Restoration of the Mississippi Delta – The Once and Future Delta.* Springer, New York.

Day, J. W., Lane, R. R., D'Elia, C. F., Wiegman, A. R. H., Rutherford, J. S., Shaffer, G. P., Brantley, C. G., and Kemp, G. P. 2016a. Large infrequently operated river diversions for Mississippi delta restoration. *Estuarine, Coastal and Shelf Science*, 183: 292–303.

Deegan, L. A., Johnson, D. S., Warren, R. S., Peterson, B. J., Fleeger, J. W., Fagherazzi, S. and Wollheim, W. 2012. Coastal eutrophication as a driver of salt marsh loss. *Nature* 490: 388–392.

DeLaune, R. D., Kongchum, M., White, J. R., and Jugsujinda, A. 2013. Freshwater diversions as an ecosystem management tool for maintaining soil organic matter accretion in coastal marshes. *Catena*, 107: 139e144.

Edwards, K. R., and Proffitt, C. E. 2003. Comparison of wetland structural characteristics between created and natural salt marshes in southwest Louisiana, USA. *Wetlands*, 23 (2): 344–356.

Edwards, K. R., and Mills, K. P. 2005. Aboveground and belowground productivity of *Spartina alterniflora* (Smooth Cordgrass) in natural and created Louisiana salt marshes. *Estuaries*, 28: 252–265.

Elsey-Quirk, T., and Adamowicz, S. 2016. Influence of physical manipulations on short-term salt marsh morphodynamics: Examples from the north and mid-Atlantic coast, USA. *Estuaries and Coasts*, 39:423–439.

Elsey-Quirk, T., Middleton, B. A. and Proffitt, C. E. 2009. Seed dispersal and seedling emergence in a created and a natural salt marsh on the Gulf of Mexico coast in Southwest Louisiana, USA. *Restoration Ecology*, 17(3): 422–432.

Ford, M. A., Cahoon, D. R., and Lynch, J. C. 1999. Restoring marsh elevation in a rapidly subsiding salt marsh by thin-layer deposition of dredged material. *Ecological Engineering*, 12: 189–205.

Ford, M. A., and Grace, J. B. 1998. Effects of vertebrate herbivores on soil processes, plant biomass, litter accumulation and soil elevation changes in a coastal marsh. *Journal of Ecology*, 86: 974–982.

Forès, E. 1992. Desecación de la laguna de la Encanyissada: un procedimiento para disminuir los niveles de eutrofia. *Butlletí del Parc Nat Delta de l'Ebre*, 7: 26–31.

Forès, E., Espanya, A., and Morales, F. 2002. Regeneración de la laguna costera de La Encanyissada (Delta del Ebro). Una experiencia de biomanipulación. *Escosistemas* 2002/2.

Frame, G. W., M. K. Mellander, D. A. Adamo. 2006. Big Egg Marsh experimental restoration in Jamaica Bay, New York. In: Harmon, D. (ed.) *People, Places, and Parks: Proceedings of the 2005 George Wright Society Conference on Parks, Protected Areas, and Cultural Sites*. pp. 123–130. Hancock, Michigan. The George Wright Society.

Friess, D. A., Krauss, K. W., Horstman, E. M., Balke,T., Bouma, T. J., Galli, D. and Webb, E. L. 2012. Are all intertidal wetlands naturally created equal? Bottlenecks, thresholds and knowledge gaps to mangrove and saltmarsh ecosystems. *Biological Reviews*, 87: 346–366.

Gagliano S. M., Meyer-Arendt K. J., and Wicker K. M. 1981. Land loss in the Mississippi River deltaic plain. *Gulf Coast Association of Geological Societies Transactions*, 31:295–300

Genua-Olmedo, A., Alcaraz, C., Caiola, N., and Ibáñez, C. 2016. Sea level rise impacts on rice production: The Ebro Delta as an example. *Science of The Total Environment*, 571: 1200–1210.

Giosan L., Syvitski J., Constantinescu S., and Day J. 2014. Protect the world's deltas. *Nature*, 516: 31–33.

Gittman, R. K., Popowich, A. M., Bruno, J. F., and Peterson, C. H. 2014. Marshes with and without sills protect estuarine shorelines from erosion better than bulkheads during a category 1 hurricane. *Ocean & Coastal Management*, 102: 94–102.

Gittman, R. K., Peterson, C. H., Currin, C. A., Fodrie, F. J., Piehler, M. F. and Bruno, J. F. 2015. Living shorelines can enhance the nursery role of threatened estuarine habitats. *Ecological Applications*, 26: 249–263.

Graham, S. A., and Mendelssohn, I. A. 2013. Functional assessment of differential sediment slurry applications in a deteriorating brackish marsh. *Ecological Engineering*, 51: 264–274.

Guntenspergen, G. R., Cahoon, D. R., Grace, J., Stayer, G. D., Fournet, S., Townson, M. A., and Foote, A. L. 1995. Disturbance and recovery of Louisiana coastal marsh landscape from the impacts of Hurricane Andrew. *Journal of Coastal Research Special Issue*, 21: 324–339.

Hackney, C. T., Brady, S., Stemmy, L., Boris, M., Dennis, C., Hancock, T., O'Bryon, M., Tilton, C. and Barbee, E. 1996. Does intertidal vegetation indicate specific soil and hydrologic conditions. *Wetlands*, 16: 89–94.

Hatton, R. S., DeLaune, R. D., and Patrick, Jr, W. H. 1983. Sedimentation, accretion, and subsidence in marshes of Barataria Basin, Louisiana. *Limnology and Oceanography*, 28: 494–502.

Herrera-Silveira, J., Lara-Domíinguez, A., Day, J., Yáñez-Arancibia, A., Ojeda, S. M., Hernández, C. T., and Kemp, G. P. 2019. Ecosystem functioning and sustainable management in coastal systems with high reshwater input in the Southern Gulf of Mexico and Yucatan Peninsula. In: Wolanski, E., Day, J., Elliott, M., and Ramnachandran, R. (eds.) *Coasts and Estuaries: The Future*. pp. 377–397. Elsevier, New York.

Hazelton, E. L. G., Mozdzer, T. J., Burdick, D. M., Kettenring, K. M., and Whigham, D. F. 2014. *Phragmites australis* management in the United States: 40 years of methods and outcomes. *AoB Plants*, 6: plu001.

Hedges, P., Kriwoken, L. K. and Patten, K. 2003. A review of *Spartina* management in Washington State, US. *Journal of Aquatic Plant Management*, 41: 82–90.

Hinkle, R., and Mitsch, W. J. 2005. Salt marsh vegetation recovery at salt hay farm wetland restoration sites on Delaware Bay. *Ecological Engineering*, 25: 240–251.

Ibáñez, C., Canicio, A., Curcó, A., Day, J. W., and Prat, N. 1996b. *Evaluation of vertical accretion and subsidence rates*. MEDDELT Final Report, Ebre Delta Plain Working Group. University of Barcelona, Barcelona, Spain.

Ibáñez, C., Canicio, A., and Day, J. W. 1997. Morphologic development, relative sea level rise and sustainable management of water and sediment in the Ebre Delta, Spain. *Journal of Coastal Conservation* 3:191–202.

Ibanez, C., James, P. Day, J. W., Day, J. N., and Prat, N. 2010. Vertical accretion and relative sea level rise in the Ebro delta wetlands (Catalonia, Spain). *Wetlands*, 30: 979–988.

Ibáñez, C., and Prat, N. 2003. The environmental impact of the Spanish Hydrological Plan on the lower Ebro river and delta. *Water Resources Development*, 19(3): 485–500.

Ibáñez, C., Prat, N., and Canicio, A. 1996a. Changes in the hydrology and sediment transport produced by large dams on the lower Ebro river and its estuary. *Regulated Rivers*, 12(1): 51–62.

Jefferies, R. L., Jano, A. P. and Abraham, K. F. 2006. A biotic agent promotes large-scale catastrophic change in the coastal marshes of Hudson Bay. *Journal of Ecology*, 94: 234–242.

Jones, S. F., Stagg, C. A. Krauss, K. W. and Hester, M. W. 2016. Tidal saline wetland regeneration of sentinel vegetation types in the northern Gulf of Mexico: An overview. *Estuarine, Coastal and Shelf Science*, 174: A1–A10.

Keddy, P. A., Campbell, D., McFalls, T., Shaffer, G. P., Moreau, R., Dranguet, C., and Heleniak, R. 2007. The wetlands of Lakes Pontchartrain and Maurepas: Past, present and future. *Environmental Reviews*, 15: 43–77.

Kemp, G. P., Day, J. W. and Freeman, A. M. 2014. Restoring the sustainability of the Mississippi River Delta. *Ecological Engineering*, 65: 131–146

Kemp, P., Day, J., Rogers, D., Giosan, L., and Peyronnin, N. 2016. Enhancing mud supply to the Mississippi River delta: Dam bypassing and coastal restoration. *Estuarine, Coastal and Shelf Science*, 183: 304–313.

Keown, M. P., Dardeau, E. A., Jr., and Causey, E. M. 1986. Historic trends in the sediment flow regime of the Mississippi River. *Water Resources Research*, 22: 1555–1564.

Kesel R. H. 1989. The role of the Mississippi River in wetland loss in southeastern Louisiana, U.S.A. *Environmental Geology*, 13:183–193.

Kesel, R. H., Yodis, E. G., and McCraw, D. J. 1992. An approximation of the sediment budget of the lower Mississippi River prior to major human modification. *Earth Surface Processes and Landforms*, 17: 711–722.

Kim, W., Mohrig, D., Twilley, R., Paola, C., and Parker, G. 2009. Is it feasible to build new land in the Mississippi River Delta? *Eos, Transactions, American Geophysical Union*, 90: 373–374

Kirwan, M. L., and Blum, L. K. 2011. Enhanced decomposition offsets enhancedIbáñez C, Prat N (2003) The environmental impact of the Spanish Hydrological Plan on the lower Ebro river and delta. *Water Resources Development*, 19(3): 485–500.

Ko, J., and Day, J. 2004. A review of ecological impacts of oil and gas development on coastal ecosystems in the Mississippi delta. *Ocean and Coastal Management*, 47: 671–691.

Konisky, R. A., Burdick, D. M., Dionne, and Neckles, M. H. A. 2006. A regional assessment of salt marsh restoration and monitoring in the Gulf of Maine. *Restoration Ecology*, 14: 516–525.

La Peyre, M. K., Gossman, B., and Piazza, B. P. 2009. Short- and long-term response of deteriorating brackish marshes and open-water ponds to sediment enhancement by thin-layer dredge disposal. *Estuaries and Coasts*, 32: 390–402.

LCWCRTF (Louisiana Coastal Wetlands Conservation and Restoration Task Force). 2006. The 2006 Evaluation Report to the U.S. Congress on the Effectiveness of Coastal Wetlands Planning, Protection and Restoration Act Projects. U.S. Army Corps of Engineers, New Orleans.

LCWCRTF (Louisiana Coastal Wetlands Conservation and Restoration Task Force). 2012. The 2012 Evaluation Report to the U.S. Congress on the Effectiveness of Coastal Wetlands Planning, Protection and Restoration Act Projects. U.S. Army Corps of Engineers, New Orleans.

Lopez, J., Henkel, T., Moshogianis, A., Baker, A., Boyd, E., Hillmann, E., Connor, P., and Baker, D. B., 2014. Examination of deltaic processes of Mississippi River outlets Caernarvon Delta and Bohemia Spillway in southeast Louisiana. *Gulf Coast Association of Geological Societies Transactions*, 64: 707e708.

MacBroom, J. G., and Schiff, R. 2012. Predicting the hydrologic response of salt marshes to tidal restoration. In Roman, C.T. and D.M. Burdick (eds.) *Tidal Marsh Restoration: A Synthesis of Science and Practice*. pp. 315–332. Washington, DC: Island Press.

Mitsch, W. J. and Gosselink, J. G. 2015. *Wetlands*, 5th ed. John Wiley & Sons, Inc., Hoboken, NJ.

Morgan, P. A., Burdick, D. M., and Short, F. T. 2009. The functions and values of fringing salt marshes in northern New England, USA. *Estuaries and Coasts*, 32: 483–495.

Nagarkar M., and Raulund-Rasmussen, K. 2016. An appraisal of adaptive management planning and implementation in ecological restoration: Case studies from the San Francisco Bay Delta, USA. *Ecology and Society*, 21(2): 43.

Naquin, J. D., Lui, K. B, McCloskey, T. A., and Bianchette, T. A., 2014. Storm deposition induced by hurricanes in a rapidly subsiding coastal zone. *Journal of Coastal Research Special Issue*, 70: 308–313.

NAS (National Academies of Sciences, Engineering, and Medicine). 2017. *Effective Monitoring to Evaluate Ecological Restoration in the Gulf of Mexico*. Washington, DC: The National Academies Press.

Needelman, B. A., Crooks, S., Shumway, C. A. J., Titus, G. Takacs, R. and Hawkes, J. E. 2012. *Restore-Adapt-Mitigate: Responding to Climate Change through Coastal Habitat Restoration*. Washington, DC: Restore America's Estuaries.

Nicholls, R. J., and Cazenave, A. 2010. Sea-level rise and its impact on coastal zones. *Science*, 328: 1517–1520.

Nicholls, R. J., Hutton, C. W., Lazar, Allan, A., Adger, W., Adams, H., Wolff, J., Rahman, M. and Salehin, M. 2016. Integrated assessment of social and environmental sustainability dynamics in the Ganges-Brahmaputra-Meghna delta, Bangadesh. *Estuarine, Coastal and Shelf Science*, 183: 370–381.

Nittrouer, J. A., Best, J. L., Brantley, C., Cash, R. W., Czapiga, M., Kumar, P., and Parker, G., 2012. Mitigating land loss in coastal Louisiana by controlled diversion of Mississippi River sand. *Nature Geoscience*, 5: 534–537.

NOAA. 2015. *Guidance for Considering the Use of Living Shorelines*. Living Shorelines Workgroup. www.habitatblueprint.noaa.gov/living-shorelines/

Nyman, J. A., Crozier, C. R., and DeLaune, R. D. 1995. Roles and patterns of Hurricane sedimentation in an estuarine marsh landscape. *Estuarine, Coastal and Shelf Science*, 40, 665–679.

Paola, C., Twilley, R., Edmonds, D., Kim,W., Mohrig, D., and Parker, G. 2011. Natural processes in delta restoration: Application to the Mississippi delta. *Annual Reviews of Marine Science*, 3: 67–91.

Peterson, S. B., Teal, J. M., and Mitsch, W. J. eds. 2005. Delaware Bay Salt Marsh Restoration. *Ecological Engineering, Special issue*, 25: 199–314.

Pethick, J. 2002. Estuarine and tidal wetland restoration in the United Kingdom: policy versus practice. *Restoration Ecology*, 10: 431–437.

Peyronnin, N. S., Green, M., Richards, C. P., Owens, A., Reed, D., Chamberlain, J., Groves, D. G., Rhinehart, K., and Belhadjali, K. 2013. Louisiana's 2012 coastal master plan: Overview of a science-based and publicly informed decision-making process. *Journal of Coastal Research*, 67, 1–15.

Peyronnin, N. S., Caffey, R. H., Cowan Jr., J. H., Justic, D., Kolker, A. S., Laska, S. B., McCorquodale, A., et al. 2017. Optimizing sediment diversion operations: Working group recommendations for integrating complex ecological and social landscape interactions. *Water*, 9: 368. DOI: 10.3390/w9060368

Pont, D., Day, J. and Ibáñez, C. 2017. The impact of two large floods (1993–1994) on sediment deposition in the Rhone delta: Implications for sustainable management. *Science of the Total Environment*, 609: 251–262.

Powers, S. P., and Boyer, K. E. 2013. Marine restoration ecology. In Bertness, M. D., J. Bruno, B. Silliman, and J. Stachowicz et al. (eds.). *Marine Community Ecology and Conservation*. pp. 495–516. Sunderland, MA: Sinauer Associates.

Prado, P., Alcaraz, C., Jornet, L., Caiola, N. and Ibáñez, C. 2017. Effects of enhanced hydrological connectivity on Mediterranean salt marsh fish assemblages with emphasis on the endangered Spanish toothcarp (*Aphanius iberus*). *PeerJ* 5:e3009. DOI 10.7717/peerj.3009.

Proffitt, C. E., Chiasson, R. L., Owens, A. B., Edwards, K. R., and Travis, S. E. 2005. *Spartina alterniflora* genotype influences facilitation and suppression of high marsh species colonizing an early successional salt marsh. *Journal of Ecology*, 93(2): 404–416.

Proffitt, C. E., and Young, J. 1999. Salt marsh plant colonization, growth and dominance on large mudflats created using dredged sediments. In L. R Rozas et al. (eds.) *Recent Research in Coastal Louisiana: Natural System Function and Response to Human Influence*. pp. 218–228. Louisiana Sea Grant College Program, Baton Rouge, Louisiana.

Reiner, E. L. 2012. Restoration of tidally restricted salt marshes at Rumney Marsh, Massachusetts. In Roman, C. T. and D. M. Burdick (eds.) *Tidal Marsh Restoration a Synthesis of Science and Practice*. pp. 355–370. Washington, DC: Island Press.

Ray, G. L. 2007. *Thin layer disposal of dredged material on marshes: A review of the technical and scientific literature*. ERDC/EL Technical Notes Collection (ERDC/EL TN-07-1), Vicksburg, MS: U.S. Army Engineer Research and Development Center. 2007.

Restore America's Estuaries. 1999. *Principles of Estuarine Habitat Restoration*. Arlington Virginia. www.edc.uri.edu/restoration/html/resource/rae-erf.pdf

Roberts, H. H. 1997. Dynamic changes of the holocene Mississippi River delta plain: The delta cycle. *Journal of Coastal Research*, 13(3): 605–627.

Rogers, B. D., Herke, W. H., and Knudsen, E. E. 1992. Effects of three different water-control structures on the movement of coastal fishes and macrocrustaceans. *Wetlands*, 12: 106–120.

Roman, C. T. and Burdick, D. M. 2012. *Tidal Marsh Restoration: A Synthesis of Science and Management*. Island Press, Washington, DC.

Rozsa, R. 1995. Tidal restoration in Connecticut. In G. D. Dreyer and W. A. Niering (eds.) *Tidal Marshes of Long Island Sound: Ecology, History and Restoration*. Bulletin no. 34. pp. 51–65. The Connecticut College Arboretum, New London, Connecticut.

Rutherford, J., Day, J., D'Elia, C., Wiegman, A., Willson, C., Caffey, R., Shaffer, G., Lane, R., and Batker, D. 2018. Evaluating trade-offs of a large, infrequent sediment diversion for restoration of a forested wetland in the Mississippi delta. *Estuarine, Coastal and Shelf Science*, 203: 80–89.

Saltonstall, K. 2002. Cryptic invasion by a non-native genotype of the common reed, Phragmites australis, into North America. *Proceedings of the National Academy of Sciences of the USA* 99: 2445–2449.

Sasser, C., Holm,G., Evers, E., and Shaffer, G. 2018. The nutria in Louisiana: A current and historical perspective. In J. Day and J. Erdman (eds.) *Mississippi Delta Restoration – Pathways to a Sustainable Future*. pp. 39–60. Springer, Glam, Switzerland.

Saucier, R. R. 1963. *Recent Geomorphic History of the Pontchartrain basin*. Coastal Studies Series No. 9. Louisiana State University Press, Baton Rouge, LA.

Simenstad, C. A., Reed, D. and Ford, M. 2006. When is restoration not? Incorporating landscape-scale processes to restore self-sustaining ecosystems in coastal wetland restoration. *Ecological Engineering*, 26: 27–39.

Simenstad, C. A., and Thom, R. M. 1996. Functional equivalency trajectories of the restored Gog-Le-Hi-Te estuarine wetland. *Ecological Applications*, 6: 38–56.

Slocum, M. G., Mendelssohn, I. A., and Kuhn, N. L., 2005. Effects of sediment slurry enrichment on salt marsh rehabilitation: plant and soil responses over seven years. *Estuaries*, 28: 519–528.

Smith, S. M., Roman, C. T., James-Pirri, M., Chapman, K., Portnoy, and Gwilliam, J. E. 2009. Responses of plant communities to incremental hydrologic restoration of a tide-restricted salt marsh in southern New England (Massachusetts, USA). *Restoration Ecology*, 17: 606–618.

Society for Ecological Restoration International Science & Policy Working Group. 2004. *The SER International Primer on Ecological Restoration*. Tucson Az: Society for Ecological Restoration International.

Staver, L. W. 2015. *Ecosystem dynamics in tidal marshes constructed with fine grained, nutrient rich dredged material*. Dissertation, University of Maryland, College Park, MD.

Streever 2000. *Spartina alterniflora* marshes on dredge material: a critical review of the ongoing debate over success. *Wetlands Ecology and Management*, 8: 295–316.

Swenson, E. M., and Turner, R. E. 1987. Spoil banks: effects on a coastal marsh water-level regime. *Estuarine, Coastal and Shelf Science*, 24:599–609.

Syvitski, J., Kettner, A., Overeem, I., Hutton, E., Hannon, M. Brakenridge, G., Day, J., et al. 2009. Sinking deltas due to human activities. *Nature Geoscience*, 2: 681–686.

Teal, J., and Teal, M. 1969. *Life and Death of the Salt Marsh*. Ballantine Books, New York.

Teal, J. M, and Weinstein, M. P. 2002. Ecological engineering, design, and construction considerations for marsh restorations in Delaware Bay, USA. *Ecological Engineering*, 18: 607–618.

Tockner, K., Malard, F., and Ward, J. V. 2000. An extension of the flood pulse concept. *Hydrological Processes*, 14: 2861–2883.

Turner, R. E. and Streever, B. 2002. *Approaches to Coastal Wetland Restoration: Northern Gulf of Mexico*. Kugler Publications, The Netherlands.

Twilley, R. R., Couvillion, B. R., Hossain, I., Kaiser, C., Owens, A. B., and Steyer, G. D. 2014. Coastal Louisiana Ecosystem Assessment and Restoration Program: The role of ecosystem forecasting in evaluating restoration planning in the Mississippi River Deltaic Plain. American Fisheries Society Symposium. v. 64, p. 29–46.

Twilley, R. R., Couvillion, B. R., Hossain, I., Kaiser, C., Owens, A. B., Steyer, G. D., and Visser, J. M. 2018 Coastal Louisiana ecosystem assessment and restoration program: The role of ecosystem forecasting in evaluating restoration planning in the Mississippi river deltaic plain. *American Fisheries Society Symposium*, 64, 29–46.

Twilley, R. R., and Rivera-Monroy, V. 2009. Sediment and nutrient tradeoffs in restoring Mississippi River Delta: restoration vs. eutrophication. *Journal of Contemporary Water Research & Education*, 141: 39–44.

Tyrrell, M., Dionne, M. and Edgerly, J. 2008. Physical factors mediate effects of grazing by a non-indigenous snail species on saltmarsh cordgrass (Spartina alterniflora) in New England marshes. *ICES Journal of Marine Science*, 65: 746–752.

USGCRP. 2017. *Climate Science Special Report: Fourth National Climate Assessment*, Volume I [Wuebbles, D.J., D.W. Fahey, K.A. Hibbard, D.J. Dokken, B.C. Stewart, and T.K. Maycock (eds.)]. U.S. Global Change Research Program, Washington, DC, USA

Valoppi, L., 2018, Phase 1 studies summary of major findings of the South Bay Salt Pond Restoration Project, South San Francisco Bay, California: U.S. Geological Survey Open-File Report 2018–1039, 58 p., plus appendixes, https://doi.org/10.3133/ofr20181039.

Wiegman, A., Day, J., D'Elia, C., Rutherford, J., Morris, J., Roy, E., Lane, R., Dismukes, D., and Synder, B. 2017. *Modeling impact of sea-level rise, oil price, and management strategy on the costs of sustaining Mississippi delta marshes with hydraulic dredging*. STOTEN.

Wilber, P. 1993. Managing Dredged Material Via Thin-Layer Disposal in Coastal Marshes. USACE Technical Note EEDP-01-32.

Williams, P. B., and Orr, M. K. 2002. Physical evolution of restored breached levee salt marshes in the San Francisco Bay Estuary. *Restoration Ecology*, 10: 527–542.

Wolters, M., Garbutt, A. and Bakker, J. P. 2005. Salt-marsh restoration: evaluating the success of de-embankments in north-west Europe. *Biological Conservation*, 123: 249–268.

Wolters, M., Garbutt, A., Bekker, R. M., Bakker, J. P, and Carey, P. D. 2008. Restoration of saltmarsh vegetation in relation to site suitability, species pool and dispersal traits. *Journal of Applied Ecology*, 45: 904–912.

18

Impacts of Climate Change and Sea Level Rise

ZOE J. HUGHES, DUNCAN M. FITZGERALD, AND CAROL A. WILSON

Recent estimates of global salt marsh area sit at 5.5 million hectares (Mcowen et al. 2017). Conservatively, this translates to $1 trillion of ecosystem services per annum, potentially as much as $5 trillion (De Groot et al. 2012, Mehvar et al. 2018), equivalent to the entire US federal budget for 2019. There can be little debate as to the value of salt marshes, both in terms of the ecosystem services they provide and the key part they play in helping us understand past climate and sea level trends. This chapter summarizes the preceding work and draws together some key observations and notable knowledge gaps highlighted in the previous chapters. We provide a focus on the expected response of salt marshes to the stresses created by a changing climate.

Given their inherent lack of topography and, by their very nature, close proximity to high tide, marshes are clearly endangered by rising sea level. As water depth subtly increases relative to the marsh surface, the marsh takes longer to drain, increasing the period of time it is submerged (hydroperiod). This impacts rates of mineral sedimentation and the productivity of vegetation (Morris et al. 2002; Chapter 7; Chapter 8). As mentioned in Chapter 6, however, this is but one of many challenges they face. As global climate changes, we expect increases in temperature, changes in precipitation, and a shift in many weather systems, altering both storm pattern and frequency. This will have consequent impacts on freshwater and nutrient delivery to the coast, as well as destructive high-energy events and flooding. In addition to changes in climate, environmental alterations by humans will have long-term impacts on sediment supply and the potential inland migration of marshes (Chapter 16). A more subtle threat to the existing systems is that of changing ecosystem structure. This may occur purely as the ecotones which define saltmarsh environments shift, or as new, invasive species out compete native biota and, in doing so, change the very nature and ecosystem functions of the salt marsh.

Salt marshes are unique environments, which experience a combination of conditions that are not seen in any other landscape. Their evolution is driven by constantly fluctuating forces, which strive toward equilibrium. They undergo frequent flooding, in many cases twice daily. In stark contrast to fluvial and submarine systems, the flows which shape marshes are bidirectional, reversing entirely with each tide and reaching a maximum velocity at mid-stage. There are clear feedbacks amongst the tidal flows, geomorphology, and ecology within salt marshes, and although we have begun to explore many of these interactions, complexities increasingly become more apparent. The interactions of flora and fauna, amongst themselves and with the landscape are a key area still under investigation, and research that includes nutrients dynamics within these feedbacks is still in its infancy. However, it is clear that studying marsh landscapes and understanding how their function may change in the future, requires a holistic approach.

Certain factors stand out as key to understanding future marsh resiliency. These include sediment supply, net productivity, tidal range, storm frequency, shifts in dominant species, and proximal upland slopes and level of development. These factors dominant marsh accretionary processes and maintenance of their net areal extent. The balance between marsh sustainability and deterioration depends on rates and patterns of accretion. As the contribution of fluvial sediment to marshes decreases (Fig. 18.1; Weston, 2014) and the demand for suspended sediment to counter rising sea level increases, salt marsh must either rely increasingly on organic contributions, or cannibalize existing platform to support a smaller area of marsh.

It is well accepted that marsh biomass production, and therefore organic accretion, exhibit a parabolic response to sea level rise (Morris et al. 2002). However, the impacts of increasing atmospheric CO_2, freshwater inputs, and temperature will add complexity to the sea-level-driven accretion response. We learn in Chapter 5 that marshes are likely to vary spatially in productivity and that our understanding of behavior at one site will not necessarily translate to other areas, because of local differences in these drivers. While warmer temperatures will increase photosynthesis, respiration, and, thus, plant productivity, not all plants respond equally. It is likely that some marshes are already close to their optimum temperature and their productivity will actually begin to decrease as temperatures rises (Chapter 13). Parallel to responses in productivity, as temperatures increases decomposition of peat biomass will also increase. This process has the potential to overwhelm increases in productivity, dampening initial impacts on organic sediment accumulations. However, this is interaction itself is dampened by sea level rise which simultaneously boost productivity compared to decomposition, thus, the rate at which sea level rise accelerates is especially influential. Likewise, elevated CO_2 has a positive effect on productivity; however, the impact may be short-lived as plants become accustomed to the condition, and again the influence of small-scale local factors mean that the impact on the landscape is not yet predictable.

The capacity of a vegetated marsh to maintain its elevation within its optimum elevation range, also depends on the tidal amplitude. On this basis, Chapter 7 concludes that microtidal marshes are unable to withstand sea level rise of >2.5 mm yr^{-1}, meaning that microtidal marshes will drown before mesotidal marshes. The feedbacks between abiotic factors, vegetation, and elevation are subtle and complex, depending the local conditions and vegetation. Long term, however, sea level rise will detrimental to marsh accretion, and the faster sea level rise *accelerates*, the more of a threat it poses.

A key point raised in Chapter 12 is that, although we do not yet fully understand the sensitivity of coastal ecosystems to expected changes, many models and paradigms assume that sediment–hydrobiological interactions will remain unchanged or behave linearly. The combined effects of these factors are likely to reveal differences in the way sediment is

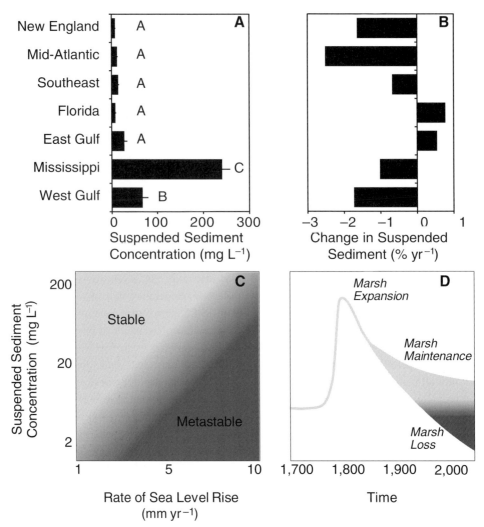

Figure 18.1 (A) Regionally averaged suspended sediment concentrations in rivers draining to the East and Gulf Coasts of the USA. Rivers that do not share the same letter have significantly different sediment concentrations and (B) proportional change in sediment concentration over time (in percent per year). Sediment supply is decreasing in nearly all regions of the USA This reduction will ultimately cross a threshold as marshes will not be able to create organic sediment or cannibalize from within at a sufficient rate to keep pace with sea level rise. (C) Conceptual model of tidal marsh stability as a function of suspended sediment concentration and rate of sea level rise (modified from Kirwan and Murray 2007) and D() conceptual diagram of changes in suspended sediment concentration in a U.S. river over time. (from Weston 2014). (A black and white version of this figure will appear in some formats. For the color version, please refer to the plate section.)

transported or how organisms interact with each other and their environment. For example, both salinity and temperature may increase decomposition rates, but their combined impact may be greater than the sum of the two alone (Wu et al. 2017).

The importance and impact of vegetation on saltmarsh landscape dynamics is well accepted, albeit not yet fully described. Part of this influence occurs by altering marsh hydrodynamics, both in terms of local flows over the marsh platform and, on a larger scale, creek evolution. Much of what we know about direct interaction between marsh vegetation and hydrodynamics is based on laboratory experiments (Chapter 12). Presently, we are unable to capture the morphological diversity of vegetation in numerical models and Chapter 11 concludes that more comparative studies of the dynamics of vegetation canopies geographically are needed to develop general relationships. That said, models and laboratory experiments are invaluable in understanding these systems for the very reason that they are difficult to measure. These two approaches both offer useful insights because of their ability to isolate parameters, control factors that are impractical to control in the field, and can be applied at scales, which are impossible to measure effectively (Chapters 11 and 12).

Improving our understanding of the range and extent of how vegetation impacts flow is particularly important, as marshes provide an effective means of wave attenuation during storms. However, again, the capacity of these systems to reduce storm surge varies significantly from site to site. Recent research also indicates that storms are valuable mechanism for marsh resilience to sea-level rise, by delivering large quantities of mineral sediment. As investments in ecosystem-defenses continue to grow, research on marsh–storm relationships is also needed.

Storm erosion, primarily at the platform edge, leads marshes fronted by large open-water areas to be in a state of continual flux. The relationship between wave power and erosion is confounded by the impact of water level and wave characteristics, and the controls on rates of retreat are not well understood. The influence of vegetation on erosion is a function of soil properties and plant density, and species. However, bank morphology and biological activity (micro- and macrofauna) also play a part in how fast the marsh edge slumps or erodes. Additional research is required to fully quantify these relationships.

The ultimate aim of exploring complex feedbacks in marsh systems is to provide efficient and accurate models or screening tools that can be applied quickly and with relative ease by coastal managers. This requires reducing the number of variables in models to only the significant parameters thereby simplifying the problem and creating a framework that is easy to use. But the key to doing this is identifying all the (non-negligible) feedbacks amongst variables, particularly those whose effects may not be evident.

As sea level rises and temperatures increase, changes are expected within salt marsh ecosystems. This includes the onshore translation of existing ecotones (boundaries between freshwater tidal marshes, tidal forests, high marsh, and low marsh) as hydroperiod and salinity conditions change. In some areas, shifts in foundation species are predicted. For example, in the northeastern USA, marshes dominated by *Spartina patens* are being invaded by *S. alterniflora* marshes (Donnelly and Bertness 2001), and in Florida and Texas, saltmarsh is being outcompeted as mangroves move northward with increases in temperature (Armitage et al. 2015). Changing vegetation will lead to alterations in accretion and offer different ecosystem services. While some invasive species may be benign, others may be less beneficial to marsh resilience that those they are replacing, or some may even be destructive. Plant or animal invasions may become a stressor that make a marsh more vulnerable to other stressors, such as sea-level rise. The impact of both native and invading fauna, even those operating on very small spatial scales, can impact, either positively or negatively, the entire landscape (Chapter 8).

This shifting of ecotones translates to the migration of marshes inland; this may allow the marsh some respite if the area created is equivalent to the area lost to edge erosion or submergence. However, the extent to which this process is possible is controlled by the slope of coastal plain and this can be limiting especially where marshes are backed by relatively steep coastal areas such as Nova Scotia, Maine, and Rhode Island (Chapter 16).

Presently, salt marshes face pressure from human activities at unprecedented scales and intensities (He et al. 2014; Chapter 5). Salt marshes have been impacted by diking, impoundment, filling, isolation from riverine inputs and by coastal protection defenses, dredging, pollution with high levels of nutrients and toxins, introduced species, and shoreline hardening (Chapter 17). Some of these human impacts are direct (e.g., dredging, filling) while others are indirect (e.g., hydrologic impairments, sediment deficits and subsidence). Moreover, most have a negative impact on marsh resilience. The presence of human development is often the critical impediment to marsh migration (Chapter 16).

Perhaps the most important factor impacting marsh survival and the most difficult to predict is the socioeconomic and political climate. Appreciation of the value of coastal wetlands and recognition of the threat to them from climate change varies country by county, state by state, and town by town. Channeling investment into these systems is not always seen as a priority, but some legislation has focused on the long-term protection of salt marsh systems. For example, efforts in Louisiana, which has a Coastal Master Plan totaling $50 billion over the next 50 years, includes $20 billion to fund to create and restore marsh systems (CPRA 2017). Chapter 17 discusses potential for marsh restoration, including dismantling dikes, reestablishment of tidal creeks, fill removal, levee breaches, living shorelines, bridge and culvert additions, restored hydrology, river diversions, beneficial use of dredge material to restore and create marshes, and control of introduced species. The authors recommend a landscape or regional scale approach: focusing on the preferred habitat of the invader and including treatments to recolonize with native plants quickly can help prevent future invasions. However, climate and human impacts interact to make restoration more challenging (Day et al. 2016, 2019).

It is not yet clear whether marshes will have the ability to persist into the twenty-second century, and if they do whether the landscapes will be recognizable in comparison to those that exist today. However, the consequences of the loss of salt marshes are undeniable and researchers will be important communicators and advocates to prolong their survival.

References

Armitage, A. R., Highfield, W. E., Brody, S. D. and Louchouarn, P. 2015. The contribution of mangrove expansion to salt marsh loss on the Texas Gulf coast. *PLoS ONE* 10(5): e0125404. doi:10.1371/journal.pone.0125404.

CPRA (Coastal Protection and Restoration Authority of Louisiana). 2017. Louisiana's Comprehensive Master Plan for a Sustainable Coast. Coastal Protection and Restoration Authority of Louisiana. Baton Rouge, LA.

Day, J., Agboola, J., Chen, Z., D'Elia, C., Forbes, D., Giosan, L., Kemp, P. et al. 2016. Approaches to defining deltaic sustainability in the 21st century. *Estuarine, Coastal and Shelf Science*. 183: 275–291.

Day, J., Colten, C. and Kemp, G. P. 2019. Mississippi Delta restoration and protection: Shifting baselines, diminishing resilience, and growing non-sustainability. In: E. Wolanski, J. Day, M. Elliott and R. Ramnachandran (eds). *Coasts and Estuaries: The Future*. Elsevier, Amsterdam, pp. 173–192.

De Groot, R., Brander, L., Van Der Ploeg, S., Costanza, R., Bernard, F., Braat, L., Christie, M., Crossman, N., Ghermandi, A., and Hein, L. 2012. Global estimates of the value of ecosystems and their services in monetary units. *Ecosystem Services.*, 1: 50–61.

Donnelly, J. P., and Bertness, M. D. 2001. Rapid shoreward encroachment of salt marsh cordgrass in response to accelerated sea-level rise. *Proceedings of the National Academy of Sciences of the USA*, 98:14218–14223.

He, Q., Bertness, M. D., Bruno, J. F., Li, B., Chen, G., Coverdale, T. C., Altieri, A. H., et al. 2014. Economic development and coastal ecosystem change in China. *Scientific Reports*, 4:5995.

Mcowen, C., Weatherdon, L., Bochove, J., Sullivan, E., Blyth, S., Zockler, C., Stanwell-Smith, D., Kingston, N. and Martin, C. 2017. A global map of saltmarshes. *Biodiversity Data Journal*, **5** (5): e11764. doi:10.3897/bdj.5.e11764

Mehvar S., Filatova, T., Dastgheib, A., de Ruyter van Steveninck E., and Ranasinghe R. 2018, Quantifying economic value of coastal ecosystem services: A review. *Journal of Marine Science and Engineering*, 6: 5; doi:10.3390/jmse6010005

Morris, J. T., Sundareshwar, P. V., Nietch, C. T., Kjerfve, B., and Cahoon, D. R. 2002. Responses of coastal wetlands to rising sea level. *Journal of Ecology*, 83 (10):2869–2877.

Weston, N.B. 2014. Declining sediments and rising seas: an unfortunate convergence for tidal wetlands *Estuaries and Coasts*, 37:1–23

Wu, W., Huang, H., Biber,P., and Bethel, M. 2017. Litter decomposition of *Spartina alterniflora* and *Juncus roemerianus*: implications of climate change in salt marshes. *Journal of Coastal Research*, 33(2): 372–384.

Index

Abiotic factors, 32, 82–83, 88, 98, 376, 477
Aboveground biomass, 185, 191, 194, 301, 339, 343, 354–355, 398
Accommodation space, 32, 35, 42, 226, 229, 245, 402
ADCIRC (Advanced CIRCulation finite element model), 292
Aggradation, 12–15
Algarve, Portugal, 2
Allochthonous organic matter, 171
    Accumulation, 14
    Refractory, 158
Amazon River Basin, Brazil, 133
Ammonia volatilization, 118
Anthropocene, 99
Anthropogenic barriers, 431–432
Argentina, 96
Arthrocnemum subterminale, 95
Atchafalaya River, Louisiana USA, 451
Atchafalaya Wax Lake Delta, Louisiana USA, 451
Atmospheric CO_2, 10, 22, 238, 353, 355, 477
Australia, 31, 185, 202, 239, 342, 371
    Moreton Bay, 165
Autocompaction, 12, 16, 189
Autocyclic marsh growth, 402

Baccharus halimifolia, 17–18
Bahía Blanca Estuary, Argentina, 186, 202, 220
Ballistic Momentum Flux, 319
Basal sea-level index points, 240
Basin hypsometry, 64
Bayhead delta, 41–42
Beaulieu River estuary, UK, 395
Bed friction, 70
Bed roughness, 311, 314
Bed/Bottom shear stress, 55, 63, 65, 73, 192, 206, 301, 312, 315, 333, 392, 401, 415
Benthic infauna, 192
Bidirectional flow, 53, 57
Biloxi, Louisiana USA, 57
Biodiversity-ecosystem functioning, 85
Biofilms, 191–192, 312–313
Biogeochemical cycles, 195, 319, 373

Biostabilisation, 73, 311–313
Biotic factors, 32
Bioturbation, 2, 55, 96, 115, 161, 195–205, 208, 312–318
    Crab, 179, 192, 196–201, 210
Blue Carbon, 424
Blue crabs, 90
Boat wakes, 391
Bogue Banks, North Carolina USA, 38
Boltzmann-like probability, 286
Boston Harbor, Massachusetts USA, 1, 399
Bottom-up controls, 86, 99
BP-Deepwater Horizon oil spill, 399
Breton Sound, Louisiana USA, 121, 191, 401
British Columbia, 96, 247, 427, 430, 438
British Isles, 18, 31, 35
Bulk density, 119, 133, 158, 174, 191, 198, 314, 320, 409
Burrowing species, 316, 320

Caernarvon, Louisiana USA, 121
Caernarvon Diversion, Louisiana USA, 355
Cantilever profiles, 408
Cape Romain, South Carolina USA, 2, 59, 201, 208–209
Capo Carvallo, Corsica, 132
Carbon accumulation, 31, 44, 95, 131, 337, 344, 389, 452, 461
Carbon dioxide, 351–355
Carbon sequestration, see Carbon accumulation
Carcinus maenas (European green crab), 16, 377
Carex spp., 19, 345
Cascadia subduction, 247–249
Cattails, 18, see *Typha angustifolia*
Cauchy number, 308
Changes in freshwater, nutrient, and sediment inputs, 355
Cheeseman Inlet, North Carolina USA, 38
Chesapeake Bay, USA, 13, 16, 188, 347, 352, 423, 449
Chézy friction parameter, 59
Chezzetcook, Nova Scotia, 3

Climate change, 2, 31, 179, 258, 323, 337, 357–358, 388, 424, 476
  Accretion, 343
  Carbon dioxide, 351
  Change in species composition, diversity, 342
  Decomposition, 339–342, 350
  Drought, 342
  Elevation change, 347
  Mangrove expansion, 342
  Plant adaptations to salinity, 348
  Plant morphology, 350
  Productivity/biomass, 339, 344, 349
  Root response, 349
  Salinity, 348–351
  Species composition, 346
  Temperature, 337
Coastal squeeze, 429
Coefficient of drag, 67, 70, 397–398
Cohesive Strength Meter, 192, 312
Colorado River Delta, Sonora Mexico, 59
Columbia River, USA, 2
Common reed, see *Phragmites australis*
Competition, 32, 91–93, 98, 180, 196, 284, 376
Cordgrass, see *Spartina alterniflora*
Coring, 11
Creek bank failure, 16, 191, 258, 400, 410, 479
  Cantilever profiles, 408
  Mass failure, 410
  Mechanisms, 401
  Toppling failure, 411

Decomposition, 2, 16–17, 95, 115, 158, 182, 191, 200–201, 269, 280, 337, 339–342, 346, 350–351, 353, 358, 399, 457, 477
Decreased stem elongation, 350
Degradation, 12, 15–17
Delaware Bay, 461–466
Delaware Bay, USA, 13, 444, 449, 451
Delft-3D, 290
Dendritic networks, 55
Denitrification, 119–120, 140
Diadromous fish, 437
Diatoms, 131, 138, 236, 312
Disease and parasitism, 90
Dissection of tidal channel, 55, 403, 410
Distichlis spicata, 18, 180, 234, 339, 342
Distributary mouth bars, 41
Disturbance, 84, 96–97, 137, 159, 195–205, 229, 239, 262, 268, 374
Drainage density, 63, 72, 208, 220
Drought, 342, 355

East China Sea, 132
East River Marsh, Guilford, Connecticut USA, 264
Eastern Scheldt, Netherlands, 410
Eastern Shore, Virginia USA, 3, 392
Ebro Delta, Spain, 446, 449–450
Ecogeomorphology, 178–179

Ecological destabilizers, 195–205
Ecological feedbacks, 32
Ecological stabilizers, 183–195
Ecosystem services, 157, 388–389, 424, 437, 476
Ecosystem/Ecological engineers, 83, 95–96, 178, 182–205
Ecotones, 33, 180, 476, 479
Ectothermic herbivores, 88, 91
Edge, see Marshes: Edge
Elbe River, Germany, 2
Elymus repens (Quackgrass), 367
Equilibrium elevation, 35, 157, 171
Estuarine restoration plans, 444
Etang de Toulvern, Bretagne, France, 2
European colonization, 13
Exploratory models, 279

Facilitation and mutualism, 93, 195
Feldspar marker layers, see Ground feldspar
Fill removal, 447
Flood-tidal delta, 33, 38–40, 45
Foraminifera, 234–236
Foraminiferal zonation, 18–19, 234–235, 242
Fort Pulaski National Monument, Georgia USA, 409
Frisian Islands, Germany, 2
Fundamental niche, 82
FVCOM (Finite-Volume Coastal Ocean Model), 292

Geographic patterns, 98, 117, 134, 271, 295
GEOMBEST, 285
Goldenrod, 17, see *Solidago sempirvirens*
Gouldsboro, Maine USA, 2
Great Marsh, Massachusetts USA, 137, 370, 374, 376, 379
Great Sippewissett Marsh, Massachusetts USA, 118, 139
Great South Bay, New York USA, 231
Green crab, see *Carcinus maenas* (European green crab)
Ground feldspar, 11, 171
Guana Tolomato Matanzas National Estuarine Research Reserve, Florida USA, 409

Habitat resilience, 347
Halophytes, 9–10, 55, 178, 227, 233, 388
Herbivory, 84, 86–88, 90, 179, 201, 347, 399
Heysham, UK, 54
Ho Bugt, Denmark, 3
Holocene sea-level change, 10, 246
Hortonian drainage density, 63, 72
Hudson Bay, Canada, 377, 450
Hudson River, New York USA, 3
Human impacts, 445
Hurricanes
  1938, 259
  1954, 259
  Andrew, 266
  Audrey, 266

Hurricanes (cont.)
　　Florence, 271
　　Hilda, 266
　　Hugo, 265
　　Katrina, 191, 260, 266, 268–269, 398, 453
　　Lili, 266
　　Michael, 271
　　Rita, 71, 266, 268–270, 453
　　Sandy, 230, 257, 265, 270
Hydrodynamics, 57
Hydro-MEM, 292

Ice rafting, 2, 15, 33, 97, 131, 265
Indicative meaning, 12, 19, 232–238, 242, 249
Infaunal species, common, 192
Intertidal-flat pioneer zone, 32
Invading organisms, 367
Invasive species, 367–381, 449–450
　　Fauna, 376–379
　　Plants, 369
Iva frutescens, 17–18, 84, 101

Jadammina macrescens, 19, 234, 248
Jamaica Bay, New York USA, 270, 432, 450
John Ripley Freeman, 233
Jumbile Cove, Galveston Bay, Texas USA, 409
Juncus gerardii, 18, 93
Juncus roemerianus, 18, 66–67, 180, 232, 234, 244

Labile organic matter, 16, 119, 158, 169
Laboratory flume experiments, 301–303
Lake Pontchartrain, Louisiana USA, 455
Lateral retreat, 389
Latitudinal variation, 88
Lepidium latifolium (Perennial pepperweed),
　　372–373
Lerez Estuary, Spain, 2
Limonium spp., 66
Little Manatee River estuary, Florida USA,
　　241, 244
Littorina littorea, 91, 202, 378, 380, 450
Living shorelines, 449
Long Island, New York USA, 9, 231, 257, 265, 399
Long-term ecological research site
　　Virginia Coast Reserve, 292
Louisiana Coastal Master Plan, 453–454, 480
Low elevation coastal zone, 424, 436
Lythrum salicaria (Purple loosestrife), 367

Mangrove expansion, 342
Marsh
　　Edge, 97
　　Inland migration, 423–425
Marsh Equilibrium Model, 157–159
Marsh organs, 161, 282, 344
Marsh platform, 1, 3, 33, 53–57, 62, 65–66, 68, 70, 73,
　　92, 97, 113, 120–121, 140, 178, 194, 201, 205,
　　227, 262, 279, 286, 292, 374, 401, 479

Marshes
　　Anthropogenic barriers, 431–432
　　Arctic, 10
　　Atlantic, 10
　　Boreal, 10
　　Broad, 12
　　Deltaic, 33, 41–43, 230, 291, 447, 450, 456
　　Diking, 445, 451, 466, 480
　　Edge, 2, 5, 53, 65, 73, 97, 190, 205, 266, 268, 272,
　　　　284, 388–415, 479
　　Fluvial-minor, 13
　　Fringing, 12, 17, 33–37, 140, 378
　　Patch, 33, 37–41, 437
　　Temperate, 10
　　Tropical, 10
Mass failure, 410
Mattapoisett, Massachusetts USA, 263
Meander cutoff, 55
Meandering channels, 16, 55
MEM (Marsh Evolution Model), 292
Microfossils, 3, 234–237
Migration inland
　　anthropogenic barriers, 431–432
Migration inland, 423–425
Miliammina fusca, 19, 234, 243, 247
Minas Basin, Bay of Fundy, Canada, 204
Mississippi River delta, Louisiana USA, 2, 15, 17, 42,
　　267, 355, 357, 383, 450–456
Mobile Bay, Alabama USA, 194
Models
　　ADCIRC (Advanced CIRCulation finite element
　　　　model), 292
　　Cellular automata, 288, 406
　　Delft-3D, 290
　　Empirical, 278, 282, 344
　　Exploratory, 279
　　FVCOM (Finite-Volume Coastal Ocean Model), 292
　　Hydro-MEM, 292
　　MEM (Marsh Evolution Model), 292
　　One-dimensional, 279
　　Simulation, 279
　　Three-dimensional (high resolution), 290
　　Two-dimensional (planar), 285–290
　　Two-dimensional (transect), 283–285
　　Vertical, 279–283
　　Zero-dimensional, 279
Modulus of elasticity, 308
Momentum equations, 59
Mont St. Michael Bay, France, 2
Morse River, Maine USA, 13
Mosquito control, 262, 445
Mudge model, 227
Mullet Pond, Apalachee Bay, Florida USA, 264
Multiple-stressor approach, 323
Muskrat, see *Ondatra zibethica*
Mutualism
　　mycorrhizal, 96
Myocastor corpus (Nutria), 179, 204, 368, 377, 450

Narragansett Bay, Rhode Island USA, 126, 133, 135, 137
Navier–Stokes equations, 278
Newport River, New Carolina USA, 42
Nitrogen
    Accumulation, 86, 120
    Atmospheric deposition, 117–118
    Burial (organic), 119
    Cycling, 116–122
    Fixation, 118
    Groundwater inputs, 116–117
Non-trophic processes, 91–97
Non-vegetation biogenic roughness, 311
Norfolk, UK, 185, 226
North Inlet, South Carolina USA, 162, 175, 198
Nutria, see *Myocaster corpus*

O'Brien-Jarrett-Marchi law, 63
Ondatra zibethica (Muskrat), 16, 204
Ontogeny, 33, 286
Optical backscatter sensor (OBS), 315
Outwelling, 113
Overwash fan, 259, 266
Oyster reef, 33, 38, 44, 181, 194

Pamlico River, North Carolina USA, 169
Paraglacial, 9
Parrett Estuary, UK, 57
Pathogens
    Bacterial, 90, 131, 312
    Fungal, 90, 131, 202, 378–379
Pattagansett River Marsh, Connecticut USA, 260
Pb isotopes, 246
Periwinkles, 91, 202, 378, 380, 450
Perennial pepperweed, *see Lepidium latifolium*
Phippsburg, Maine USA, 3
Phosphorus, 122–131, 141, 355, 388, 460
    Recycling, 128
Phragmites australis (Common reed), 17–18, 84, 95, 125, 368–370, 376, 449
Physiological controls, 82–83, 180, 341
Piracy of channel, 55
Plant adaptations to salinity, 348
Planthoppers, 89, 91–92
Plum Island, Massachusetts USA, 54, 122, 160, 162, 190, 292, 406
Poisson equation, 60
Pollution remediation, 447
Positron emission tomography (PET), 320
Pounawea, New Zealand, 3, 245–246
Predation, 85, 88–89, 92, 96, 193, 202, 367, 377, 399, 450
Predators
    Primary, 85
    Top, 85, 90
Primary production, 83–85, 96, 98, 113, 349
Productivity, 339
Pucinellia maritima, 18
Puerto Rosales, Argentina, 192

Purple loosestrife, see *Lythrum salicaria*, 367
Purple marsh crab, see *Sesarma reticulatum*

Q_{10} temperature coefficient, 341
Quackgrass, 367, see *Elymus repens*

Radio-isotopic dating, 3, 10–11, 225, 239, 244, 248, 263
Redfield model, 227, 265
Rehobooth Bay, Delaware USA, 36
Relative sea level, see Sea level: Relative
Restoration of hydrology, 445–447
Reynolds number, 67
Ribbed mussels, 93, 195
River Diversions, 448
Rockefeller Wildlife Refuge, Louisiana USA, 402
Romney Marsh, Massachusetts USA, 1
Root response, 349
Rosa rugosa, 17
Rowley, Massachusetts USA, 137
RTK (Real Time Kinematic Global Positioning System), 231
Ruppia maritima, 19
Rushes (*Juncaceae*), 17

Salicornia spp., 18–19, 66–67, 197, 261, 267, 346, 373, 375, 463
Salinity change, 348–351
Salt Pond, Falmouth, Massachusetts USA, 264
Saltmarsh macrophytes, 120, 178, 182, 201, 357
San Felice Island, Venice, Italy, 54
San Francisco Bay, California USA, 171, 370, 372–373, 432, 446
Sanborn Cove, Maine USA, 242–243
Scheldt estuary, Netherlands, 72
Schoenoplectus spp., 18, 345, 353
Scirpus maritimus, 68, 459
Scirpus tabernaemontani, 68
Scotland, 3
Sea Breeze Marsh, New Jersey USA, 261
Sea level
    Channel network response, 63
    Marsh elevation, 18, 53, 157–172, 344–348, 401, 448, 477
    Marsh equilibrium, see Sea level: Marsh position
    Reconstruction, 10, 229–240
    Relative, 32, 225–240
    Rise, 1, 38, 264, 344–348, 413, 423–424, 429, 460, 477
Secondary production, 85
Sedges (*Cyperaceae*), 17, 134, 338
Sediment
    Diversions, 456
    Erodibility, 315
    Resuspension, 35, 73, 160, 282, 315
    Suspended, 14, 33, 73, 158, 168, 170, 179, 185, 265, 282, 347, 477
    Trapping, 38, 72, 95, 159, 183–186, 265, 282, 291, 294, 343, 396

Sediment Elevation Table (SET), 3, 353
Sediment–fluid–biological interactions, 323
Self-organized criticality, 288
Sesarma reticulatum (Purple marsh crab), 16, 92, 201, 208, 316, 377, 399, 450
Severn Estuary, UK, 36, 58, 268, 400
Sewage input, 116–117, 122
Shaler model, 227
Shallotte River Estuary, North Carolina USA, 38
Shear strength, 189–191, 267–268, 311, 314, 316, 320, 356, 391, 398
Significant wave height, 72, 402
Silica, 131–138
    Amorphous, 131, 133
Silicon, 141, see Silica
Simulation models, 279
SLAMM (Sea Level Affecting Marsh Model), 290
Slowstand, 21
Slump, see Creek bank failure
Snell's law, 68
Snow goose, 88, 377
    Greater, 214
    Lesser, 450
Soil strength, see Shear strength
Solidago sempirvirens (Goldenrod), 17
Solway Firth, UK, 69, 268, 400
South Bay Salt Pond Restoration Project, San Francisco, California USA, 44, 171, 448
South Korea, 3, 185
Spartina alterniflora (Cordgrass), 10, 17, 44, 66, 68, 84, 89, 91, 93, 115, 118, 125, 134, 161, 185, 189, 200–202, 205, 208, 295, 303, 339, 343, 346, 368, 379, 397, 399, 449, 463
Spartina anglica, 67, 69, 185, 339, 370
Spartina patens, Saltmarsh Hay, 17
Spartina spp. (invasive), 370
Species composition, 346
Sprague River, Maine USA, 13
Square crab, see *Sesarma reticulatum*
St Helena Sound, South Carolina USA, 194
Storm surge, 259, 265, 267, 270, 309, 388, 394
    Attenuation, 71, 189, 258, 269, 388, 479
Stratigraphic evidence of storms in marshes, 259–264
Stratigraphic sequence, 227, 262
Stress tolerance, 92

Succotash Marsh, East Matunuck, Rhode Island USA, 260
Suwannee River delta, Florida USA, 57

Tagus Estuary, Portugal, 115
Temperature change, 337–344
Terrebonne Bay, Louisiana USA, 394
Testate amoebae, 237
The Wash, England, 2
Thick sediment addition, 448
Thin layer deposition, 448
Tidal prism, 55, 62, 64, 72, 446
Tijuana River estuary, California USA, 349
Tillingham, Essex, UK, 261, 267, 318
Top-down controls, 86–91, 377
Toppling failure, 411
Total organic carbon, 238
Transfer functions, 12, 237–238
Transgressive backbarrier marsh, 21, 232, 249
Trophic structure, 82
Typha angustifolia (Cattails), 18

U.S. National Oceanographic and Atmospheric Administration (NOAA), 41, 183, 428
Uca spp. (fiddler crabs), 16, 83, 93, 198, 200, 216, 316

Venice Lagoon, Italy, 63
Vertical accretion, 13, 33, 157–159, 163–171, 178, 185, 205, 350, 396, 458
Virgina Coast Reserve (LTER), 292

Wave attenuation, 71, 186, 258, 265, 272, 301, 312, 479
Wax Lake Delta, Louisiana USA, 451
Wells, Maine USA, 3
West Brittany, France, 3
Westerschelde estuary, Netherlands, 206, 261, 389, 392
Willapa Bay, Washington USA, 371
Wolf-spider, 91

Yangtze River Delta, China, 44, 185
Yellow Sea, 132
Yukon River, Alaska USA, 2

Zostera marina, 19, 304